RWTHedition

RWTH Aachen

Additive Manufacturing Center

Reinhart Poprawe
Editor

Tailored Light 2

Laser Application Technology

 Springer

Editor
Prof. Dr. rer. nat. Reinhart Poprawe, M.A.
Fraunhofer-Institut für
Lasertechnik (ILT)
Steinbachstr. 15
52074 Aachen
Germany
reinhart.poprawe@ilt.fraunhofer.de

ISSN 1865-0899 e-ISSN 1865-0902
ISBN 978-3-642-01236-5 e-ISBN 978-3-642-01237-2
DOI 10.1007/978-3-642-01237-2
Springer Heidelberg Dordrecht London New York

Library of Congress Control Number: 2009926871

© Springer-Verlag Berlin Heidelberg 2011

This work is subject to copyright. All rights are reserved, whether the whole or part of the material is concerned, specifically the rights of translation, reprinting, reuse of illustrations, recitation, broadcasting, reproduction on microfilm or in any other way, and storage in data banks. Duplication of this publication or parts thereof is permitted only under the provisions of the German Copyright Law of September 9, 1965, in its current version, and permission for use must always be obtained from Springer. Violations are liable to prosecution under the German Copyright Law.

The use of general descriptive names, registered names, trademarks, etc. in this publication does not imply, even in the absence of a specific statement, that such names are exempt from the relevant protective laws and regulations and therefore free for general use.

Cover design: deblik, Berlin

Printed on acid-free paper

Springer is part of Springer Science+Business Media (www.springer.com)

Preface

LASER – probably ever since this acronym has been invented for "Light amplification by stimulated emission of radiation" it generates fascination. The light of the special kind, monochromatic, extremely focusable, at first inspired science but increasingly is seen paving the road towards global relevant innovations in our modern world.

The paradigm that associates laser with a device is changing. Laser radiation is energy in one of its purest forms rather comparable with electricity. This also holds in terms of the quality of energy, i.e. the entropy of both forms of energy approaches zero in the limit, which means practically, it can be transformed into other forms of energy with maximum efficiency. Laser energy however is characterized by even higher degrees of freedom and thus can be adopted even more efficiently and more effectively to the demands of processes considered. Especially the extreme possibilities of modulation in time, space and frequency enable exact adoption to the demands of process technology (Fig. 1).

As a credo we came to the title of this book "Tailored Light". Consequently laser technology can be applied in a vast field of applications e.g. cutting, drilling, joining, ablation, soldering, hardening, alloying, cladding, polishing, generating and marking. We see applications in products of all relevant areas of our society, in mobility, energy, environment, health or production technology in general. The actual trend shows an increasing number of new applications every year. Thus the technology is seen to provide the momentum for innovations which are necessary to meet the global challenges.

The roots of this book can be found in a scriptum of a laser course at RWTH Aachen about 12 years ago. The content is based on part 2 "Laser Applications" and has been completed with actual comments and examples of relevance for the innovative engineer. The book splits up in two parts: Part 1: Fundamentals of laser materials processing, where the relevant physical phenomena are displayed, which form the basis of laser material processing by laser radiation. In part 2, Applications, a brief survey on the most important laser beam sources is given. Chapters 10–17 contain applications in manufacturing and production technology. Broadly diffused and practically integrated applications can be found as well as new perspectives like

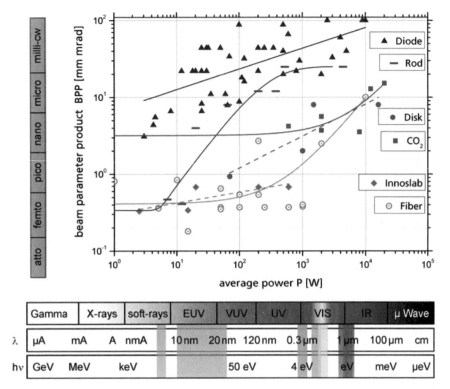

Fig. 1 Commercially obtainable lasers (R&D laser systems are not shown in the graph) are available from continuous operation (cw) down to pulse lengths below fs (10^{-15} s) and wavelengths from the far infrared (IR) down to a few nanometers. The spatial beam quality (focussability) is increasing continuously over the years in both, the temporal and the frequency spectrum

laser generating, laser polishing or laser cleaning. On top, Chap. 19 gives a survey on relevant processes in laser metrology.

The book has been compiled from contributions of many scientists in the Fraunhofer-Institute for Laser Technology (ILT) in Aachen and the University chair for Laser Technology at RWTH Aachen University (LLT). The generalized results from industrial and fundamental research projects have been included together with actual examples from real world industrial applications and thus shall close the loop from application oriented fundamentals to industrial and societal relevance.

Special thanks deserves Dr. rer. nat. Rolf Wester, who formulated in precise and comprehensive manner part 1 of the book "The Fundamentals". Also Dr. rer. nat. Torsten Mans deserves special mentioning, because his coordination and editing of the second part of this book was a tremendous contribution absolutely necessary in combining the knowledge of so many. The authors of the individual chapters deserve special acknowledgement. Here they are listed in alphabetical order:

Preface

Ü. Aydin	K. Klages	F. Schmitt
G. Backes	K. Kowalick	F. Schneider
A.L. Boglea	T. Kuhlen	T. Schwendt
L. Bosse	A. Lamott	B. Seme
K. Boucke	T. Mans	O. Steffens
M. Brajdic	W. Meiners	A. Temmler
C. Farkas	R. Noll	K. Walther
A. Gasser	A. Olowinsky	W. Wawers
A. Gillner	C. Over	A. Weisheit
J. Gottmann	D. Petring	R. Wester
A. Horn	B. Regaard	E. Willenborg
C. Janzen	A. Roesner	K. Wissenbach
C. Johnigk	U.A. Russek	N. Wolf
S. Kaierle	F. Sari	

Should the reader feel the wish for further, more intense consideration of individual parts of the content he is welcome to contact the individuals. This may concern a desire for clarification of fundamental questions and in depth explanation as well as adoption of that content to current questions and new applications. In this manner we want to contribute to global competitiveness and improve our role in applying laser technology as competent partner of the innovators.

Aachen
October 2010

Prof. Dr. rer. nat. Reinhart Poprawe M.A.

Contents

1. **Introduction** 1
   Rolf Wester

2. **The Behavior of Electromagnetic Radiation at Interfaces** 5
   Rolf Wester

3. **Absorption of Laser Radiation** 15
   Rolf Wester

4. **Energy Transport and Heat Conduction** 43
   Rolf Wester

5. **Thermomechanics** 63
   Rolf Wester

6. **Phase Transformations** 69
   Rolf Wester

7. **Melt Flow** 77
   Rolf Wester

8. **Laser-Induced Vaporization** 93
   Rolf Wester

9. **Plasma Physics** 113
   Rolf Wester

10. **Laser Beam Sources** 155
    Torsten Mans and Konstantin Boucke

11	**Surface Treatment** 173
	Konrad Wissenbach, Andreas Weisheit, Edgar Willenborg, Andre Temmler, Andreas Weisheit, Gerhard Backes, Andres Gasser, and Jens Gottmann

12	**Forming** 241
	Alexander Olowinsky

13	**Rapid Prototyping and Rapid Tooling** 253
	Christoph Over and Wilhelm Meiners

14	**Joining** 265
	Norbert Wolf, Dirk Petring, A.L. Boglea, Andreas Roesner, Ulrich Andreas Russek, Fahri Sari, Luedger Bosse, Felix Schmitt, Kilian Klages, and Alexander Olowinsky

15	**Ablation** 343
	Arnold Gillner, Alexander Horn, and Carsten Johnigk

16	**Drilling** 365
	Kurt Walther, Mihael Brajdic, and Welf Wawers

17	**Cutting** 395
	B. Seme, Frank Schneider, and Arnold Gillner

18	**System Technology** 427
	Kerstin Kowalick, Stefan Kaierle, Boris Regaard, and Oliver Steffens

19	**Laser Measurement Technology** 473
	André Lamott, Reinhard Noll, Csaba Farkas, Christoph Janzen, Tilman Schwendt, Tobias Kuhlen, and Ümit Aydin

A	**Optics** 537

B	**Continuum Mechanics** 547

C	**Laser-Induced Vaporization** 575

D	**Plasma Physics** 579

E Glossary of Symbols and Constants 585

F Übersetzung der Bildbeschriftungen 595

Index 603

Contributors

Ümit Aydin
Messtechnik, Fraunhofer-Institut für Lasertechnik, 5207 Aachen, Germany,
aydin.uemit@bcg.com

Gerhard Backes
Oberflächentechnik, Fraunhofer-Institut für Lasertechnik, 5207 Aachen, Germany,
Gerhard.backes@ilt.fraunhofer.de

Andrei Lucian Boglea
Mikrotechnik, Fraunhofer-Institut für Lasertechnik, 5207 Aachen, Germany,
andrei.boglea@ilt.fraunhofer.de

Luedger Bosse
Mikrotechnik, Fraunhofer-Institut für Lasertechnik, 5207 Aachen, Germany,
Luedger.bosse@de.TRUMPF-Laser.com

Konstantin Boucke
Laser und Laseroptik, Fraunhofer-Institut für Lasertechnik, 5207 Aachen,
Germany; Oclaro Photonics Inc. Tucson 85706, USA,
konstantin.boucke@spectraphysics.com; konstantin.boucke@oclaro.com

Mihael Brajdic
Fraunhofer-Institut für Lasertechnik, 5207 Aachen, Germany,
Mihael.brajdic@ilt.fraunhofer.de

Csaba Farkas
Messtechnik, Fraunhofer-Institut für Lasertechnik, 5207 Aachen, Germany,
csaba.farkas@ilt.fraunhofer.de

Andres Gasser
Oberflächentechnik, Fraunhofer-Institut für Lasertechnik, 52074 Aachen, Germany,
andres.gasser@ilt.fraunhofer.de

Arnold Gillner
Mikrotechnik, Fraunhofer-Institut for Laser Technology, 52074 Aachen, Germany,
arnold.gillner@ilt.fraunhofer.de

Jens Gottmann
Fraunhofer-Institut für Lasertechnik, 5207 Aachen, Germany,
jens.gottmann@ilt.fraunhofer.de

Alexander Horn
Mikrotechnik, Fraunhofer-Institut für Lasertechnik, 5207 Aachen, Germany,
alexander.horn@ilt.fraunhofer.de

Christoph Janzen
Messtechnik, Fraunhofer-Institut für Lasertechnik, 5207 Aachen, Germany,
christoph.janzen@ilt.fraunhofer.de

Carsten Johnigk
Oberflächentechnik, Fraunhofer-Institut für Lasertechnik, 5207 Aachen, Germany,
carsten.johnigk@ilt.fraunhofer.de

Stefan Kaierle
Systemtechnik, Fraunhofer-Institut für Lasertechnik, 5207 Aachen, Germany,
stefan.kaierle@ilt.fraunhofer.de

Kilian Klages
Mikrotechnik, Fraunhofer-Institut für Lasertechnik, 5207 Aachen, Germany,
kilian.klages@ilt.fraunhofer.de

Kerstin Kowalick
Systemtechnik, Fraunhofer-Institut für Lasertechnik, 5207 Aachen, Germany,
kerstin.kowalick@ilt.fraunhofer.de

Tobias Kuhlen
Messtechnik, Fraunhofer-Institut für Lasertechnik, 5207 Aachen, Germany,
tobias.kuhlen@ilt.fraunhofer.de

André Lamott
Messtechnik, Fraunhofer-Institut für Lasertechnik, 5207 Aachen, Germany,
a.lamott@aa-technologies.de

Torsten Mans
Laser und Laseroptik, Fraunhofer-Institut für Lasertechnik, 52074 Aachen,
Germany, torsten.mans@ilt.fraunhofer.de

Wilhelm Meiners
Oberflächentechnik, Fraunhofer-Institut für Lasertechnik, 5207 Aachen, Germany,
wilhelm.meiners@ilt.fraunhofer.de

Reinhard Noll
Lasermess- und Prüftechnik, Fraunhofer-Institut für Lasertechnik, 52074 Aachen,
Germany, reinhard.noll@ilt.fraunhofer.de

Alexander Olowinsky
Mikrotechnik, Fraunhofer-Institut für Lasertechnik, 5207 Aachen, Germany,
alexander.olowinsky@ilt.fraunhofer.de

Contributors

Christoph Over
Inno-shape c/o C.F.K. GmbH, 65830 Kriftel, Germany, over@inno-shape.de

Dirk Petring
Trenn- und Fügeverfahren, Fraunhofer-Institut für Lasertechnik, 5207 Aachen, Germany, dirk.petring@ilt.fraunhofer.de

Boris Regaard
Systemtechnik, Fraunhofer-Institut für Lasertechnik, 5207 Aachen, Germany, boris.regaard@clt.fraunhofer.de

Andreas Roesner
Oberflächentechnik, Fraunhofer-Institut für Lasertechnik, 5207 Aachen, Germany, Andreas.roesner@ilt.fraunhofer.de

Ulrich Andreas Russek
Mikrotechnik, Fraunhofer-Institut für Lasertechnik, 5207 Aachen, Germany, russek@rfh-koeln.de

Fahri Sari
Mikrotechnik, Fraunhofer-Institut für Lasertechnik, 5207 Aachen, Germany, fahri.sari@ilt.fraunhofer.de

Felix Schmitt
Mikrotechnik, Fraunhofer-Institut für Lasertechnik, 52074 Aachen, Germany, felix.schmitt@ilt.fraunhofer.de

Frank Schneider
Trenn- und Fügeverfahren, Fraunhofer-Institut für Lasertechnik, 5207 Aachen, Germany, frank.schneider@ilt.fraunhofer.de

Tilman Schwendt
Lasermess- und Prüftechnik, Fraunhofer-Institut für Lasertechnik, 52074 Aachen, Germany, tilman.schwendt@ilt.fraunhofer.de

B. Seme
Trenn- und Fügeverfahren, Fraunhofer-Institut für Lasertechnik, 5207 Aachen, Germany

Oliver Steffens
S&F Systemtechnik GmbH, 5207 Aachen, Germany, oliver.steffens@ilt-extern.fraunhofer.de

Andre Temmler
Oberflächentechnik, Fraunhofer-Institut für Lasertechnik, 5207 Aachen, Germany, andre.temmler@ilt.fraunhofer.de

Kurt Walther
Fraunhofer-Institut für Lasertechnik, 52072 Aachen, Germany, kurt.walther@ilt.fraunhofer.de

Welf Wawers
Mikrotechnik, Fraunhofer-Institut für Lasertechnik, 5207 Aachen, Germany,
welf.wawers@trw.com

Andreas Weisheit
Oberflächentechnik, Fraunhofer-Institut für Lasertechnik, 5207 Aachen, Germany,
andreas.weisheit@ilt.fraunhofer.de

Rolf Wester
Fraunhofer-Institut für Lasertechnik, 5207 Aachen, Germany,
rolf.weister@ilt.fraunhofer.de

Edgar Willenborg
Oberflächentechnik, Fraunhofer-Institut für Lasertechnik, 52074 Aachen, Germany,
edgar.willenborg@ilt.fraunhofer.de

Konrad Wissenbach
Oberflächentechnik, Fraunhofer-Institut für Lasertechnik, 5207 Aachen, Germany,
konrad.wissenbach@ilt.fraunhofer.de

Norbert Wolf
Trenn- und Fügeverfahren, Fraunhofer-Institut für Lasertechnik, 5207 Aachen, Germany, norbert.wolf@ilt.fraunhofer.de

Chapter 1
Introduction

Rolf Wester

The first laser was presented by THEODORE MAIMAN in the year 1960 [2]. Since 1963 the interaction of laser light with matter and its applications in the field of material processing are under ongoing investigation [3, 1]. Today a large array of laser material processing schemes are applied in industrial production. Although today material processing with laser light is used by engineers and technicians there are still many physicists involved in the further development of laser material processing. This is due to historical reasons and the fact that laser material processing only recently found its way into the engineering sciences. But the main reason for this is that the physical processes that are involved in the interaction of laser light with matter are quite complicated and are still not fully understood. So laser technology still remains a challenge for physicists and will also be so in the future.

In Fig. 1.1 the processes during material processing with laser light are depicted schematically. The parameters listed comprise only a small part of all the influencing variables that are involved. Besides these parameters dynamical processes in the workpieces like fluid dynamics and plasma formation have an important impact. Processes with such a large number of influencing parameters cannot be optimized just by trial and error but make mandatory a fundamental investigation of the underlying physical processes and their assured comprehension. The laser light hitting the workpiece is partly absorbed and partly reflected. The absorbed energy is given by

$$W_A = \int \int A(\lambda; I) \, I(\lambda, r, t) \, d^2r \, dt \qquad (1.1)$$

$A(\lambda, I)$ is the intensity and wavelength dependent absorption coefficient and $I(\lambda, r, t)$ the space and time dependent intensity at the wavelength λ. The absorption depends not only on the wavelength and polarization of the laser light but also on the material properties and the characteristics and geometry of the workpiece surface. The beam cross section depends on the wavelength, the beam quality, and the focusing optics. Because of heat conduction a heat front moves into the material. The temperature in

---

R. Wester (✉)
Fraunhofer-Institut für Lasertechnik, 5207 Aachen, Germany
e-mail: rolf.weister@ilt.fraunhofer.de

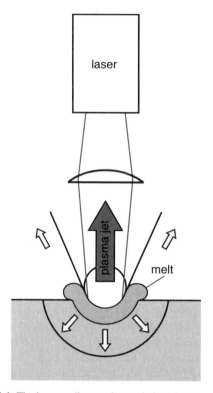

Fig. 1.1 The beam radius on the workpiece is determined by the wavelength, the beam quality, and the focusing optics. The absorption of the laser radiation in the workpiece depends on the wavelength-dependent material properties and the characteristics and geometry of the workpiece surface. The absorbed radiation causes heating, melting, and evaporation of the material. The melt is driven by shear stresses (gradient of surface tension or gas jet) and by pressure forces. The pressure forces can be caused by a gas jet or the off flowing vapor. The vapor can absorb a substantial part of the laser energy and by this shield the workpiece against the laser beam

the workpiece depends on the absorbed intensity $I_{\text{abs}} = A_\lambda(I) I_\lambda$, the duration of the interaction t_p, the beam radius on the surface $2 r_B$, the velocity v_p of the workpiece relative to the laser beam, and the thermo-physical parameters of the material like heat conductivity K, heat capacity c:

$$T = T(I_{\text{abs}}, t_L, r_B, v_P, K, c) \qquad (1.2)$$

After reaching the melting temperature the melting enthalpy has to be supplied by the laser beam. During laser alloying melt flow is forced by surface tension gradients. With processes like ablation or cutting with laser light the melt is driven out by a gas jet supplied externally. With deep penetration welding, the melt flows around the pin hole and solidifies. Melt flow is thus an important physical process during many laser material treatment processes. With further increase in the material temperature evaporation sets in. The evaporation enthalpy has to be supplied by the laser beam too. The off flowing vapor can reach pressures of up to 10^8 Pa. At these pressures the melt can be driven out partly or totally. Material removal is in that case not only accomplished by evaporation but also by self-induced driving out of the molten material. This can save part of the evaporation enthalpy in case of some material removal processes such as laser drilling.

The hot plasma can absorb a substantial part of the incident laser beam. The free electrons absorb energy from the laser field by the process of inverse bremsstrahlung and transfer part of this energy to the heavy particles like atoms and ions. By this the atoms and ions are excited and ionized and their kinetic energy increases. The initially small number of free electrons grows exponentially. This further increases the absorption of the laser light. A part of the absorbed energy is transferred to the workpiece surface by radiation or by transport of kinetic, excitation, or ionization energy. Plasma absorption can be so strong that no laser radiation reaches the surface any more. The plasma is further heated and expands rapidly. A laser-driven shock-front is created that moves toward the laser beam. This case of total plasma shielding has to be avoided. On the other hand the plasma can be utilized for the energy transfer in some cases; plasma supported energy transfer is used for all highly transparent media.

For a comprehension of the above-mentioned processes a fundamental understanding of the underlying physical processes is of great importance. In Chap. 2 the interaction of electromagnetic radiation at material surfaces is treated, the FRESNEL formulae and their applications are discussed. In Chap. 3 the absorption of non-electrically conducting materials, plasmas, and metals is investigated. The influence of the temperature and the surface characteristics on the absorptivity of metals is discussed. Chapter 4 deals with the problem of heat conduction. Solutions of the heat conduction problem are discussed for some special cases that allow at least a qualitative understanding of the processes during material processing. In Chap. 5 some results of thermo-mechanics are presented. Elastic and plastic deformations are treated. In Chap. 6 phase transformations of steel, a still much used material, is discussed. In Chap. 7 the basics of flow dynamics are presented. In Chap. 8 several models for describing nonthermal equilibrium evaporation are outlined that allow to evaluate the density and pressure of the off flowing vapor. In Chap. 9 some results of thermal equilibrium plasmas like the SAHA equation and transport processes in plasmas are compiled.

References

1. W. I. Linlor, Ion Energies Produced by Giant-Pulse, Appl. Phys. Lett., Vol. 3, p. 210, 1963
2. T. H. Maiman, Optical and Microwave-Optical Experiments in Ruby, Phys. Rev. Lett., Vol. 94, p. 564, 1960
3. J. F. Ready, Development of Plume of Material Vaporized by Giant Pulse Laser, Appl. Phys. Lett., Vol. 3, p. 11, 1963

Chapter 2
The Behavior of Electromagnetic Radiation at Interfaces

Rolf Wester

2.1 The FRESNEL Formula

Most of the processes that occur during the interaction of laser radiation with matter start at the surfaces. At small intensities the interaction of the electromagnetic field with the material at the surface is determined by the FRESNEL formulae. The FRESNEL formulae describe the reflection and transmission of a plane harmonic wave incident on an infinitely extended ideal plane surface. Reflection r and transmission t of the field amplitudes are defined by the following relations:

$$r = \frac{E_r}{E_i} \qquad (2.1)$$

$$t = \frac{E_t}{E_i} \qquad (2.2)$$

with

E_i – electric field of the incident wave;

$E_{r,t}$ – electric field of the reflected and transmitted waves, respectively.

It has to be emphasized that the FRESNEL formula only describes the behavior at the surface; there is nothing said about the transmission or absorption within the specimen.

The FRESNEL formula can be deduced from the Maxwell equations considering the boundary conditions for the fields at the surface. It has to be distinguished between perpendicular ($\perp$) and parallel ($\parallel$) polarization. In case of perpendicular polarization the electric field vector is perpendicular to the plane that is spanned by the incident and reflected wave vectors, whereas in case of parallel polarization the field vectors are parallel to this plane. The wave vectors $\vec{k}_i$, $\vec{k}_r$, and $\vec{k}_t$ all lie in one

---

R. Wester (✉)
Fraunhofer-Institut für Lasertechnik, 5207 Aachen, Germany
e-mail: rolf.weister@ilt.fraunhofer.de

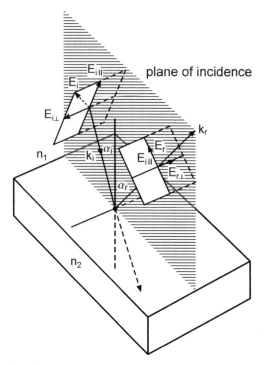

Fig. 2.1 Incident, reflected, and transmitted beam lie in one plane. The electric field vector can be split into two components, one component is parallel and the other component is perpendicular to this plane

plane. The vector of the electric field can have components perpendicular ($\perp$) or parallel ($\parallel$) to this plane. Reflection and transmission depend on the direction of the electric field vector relative to the plane of incidence (Fig. 2.1).

For practical applications the polarization of the laser radiation is often chosen so that the field strength is either perpendicular ($\perp$) or parallel ($\parallel$) to the plane of incidence (in special cases circular polarization is used). In these cases, the decomposition of the field vectors in their perpendicular and parallel components can be omitted and the mathematical treatment is simplified. Figure 2.2 shows the situation in the case of normal incidence.

The angle α that the wave vectors of the incident and reflected waves, respectively, make with the surface normal and the angle β of the refracted wave are connected by SNELL's law:

$$n_1 \sin \alpha = n_2 \sin \beta \tag{2.3}$$

n_1 and n_2 are the indices of refraction of the two media. The FRESNEL formulae for perpendicular polarization read [2][1]

---

[1] For a derivation see Appendix A.1.

2 The Behavior of Electromagnetic Radiation at Interfaces

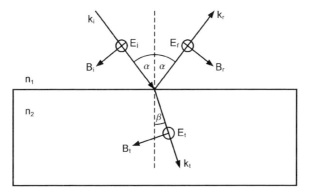

Fig. 2.2 Polarization perpendicular to the plane of incidence

$$\frac{E_r}{E_i} = r_s = \frac{n_1 \cos\alpha - \frac{\mu_1}{\mu_2}\sqrt{n_2^2 - n_1^2 \sin^2\alpha}}{n_1 \cos\alpha + \frac{\mu_1}{\mu_2}\sqrt{n_2^2 - n_1^2 \sin^2\alpha}} \quad (2.4)$$

$$\frac{E_t}{E_i} = t_s = \frac{2 n_1 \cos\alpha}{n_1 \cos\alpha + \frac{\mu_1}{\mu_2}\sqrt{n_2^2 - n_1^2 \sin^2\alpha}} \quad (2.5)$$

with

$\mu_{1/2}$ – magnetic permeability of media 1 and 2, respectively.

Figure 2.3 shows the situation for parallel polarization. The FRESNEL formulae for parallel polarization read [2][2]

$$\frac{E_r}{E_i} = r_p = \frac{\frac{\mu_1}{\mu_2} n_2^2 \cos\alpha - n_1 \sqrt{n_2^2 - n_1^2 \sin^2\alpha}}{\frac{\mu_1}{\mu_2} n_2^2 \cos\alpha + n_1 \sqrt{n_2^2 - n_1^2 \sin^2\alpha}} \quad (2.6)$$

$$\frac{E_t}{E_i} = t_p = \frac{2 n_1 n_2 \cos\alpha}{\frac{\mu_1}{\mu_2} n_2^2 \cos\alpha + n_1 \sqrt{n_2^2 - n_1^2 \sin^2\alpha}} \quad (2.7)$$

In the case of normal incidence, i.e., $\alpha = 0$, the plane of incidence cannot be defined uniquely any more and the difference between $\perp$ and $\parallel$ polarization vanishes. In that case Eqs. (2.4) and (2.6) and Eqs. (2.5) and (2.7) give the same results. The amplitudes of the reflected and transmitted waves in case of normal incidence are given by

---

[2] For a derivation see Appendix A.1.

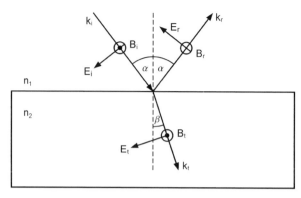

Fig. 2.3 The field vector of the reflected wave is often given in the opposite direction. In that case Eq. (2.7) has to be multiplied by -1

$$\frac{E_r}{E_i} = \frac{n_1 - n_2}{n_2 + n_1} \qquad (2.8)$$

$$\frac{E_t}{E_i} = \frac{2 n_1}{n_2 + n_1} \qquad (2.9)$$

When $n_2 > n_1$ the reflected wave undergoes a phase change of 180° (this holds for real indices of refraction).

2.1.1 FRESNEL *Formulae with Absorption*

The FRESNEL formulae are valid for real and for complex indices of refraction. The imaginary part describes the absorption of the wave in the medium. In what follows only the case that n_2 is complex but n_1 is real will be discussed. SNELL's law still applies but it has to be recognized that a complex value of n_2 implies that the angle β is complex valued too[3] and so does not have a simple physical interpretation any more. The determination of the angle that the phase front normal makes with the surface normal in the case of complex n_2 is more complicated as in the case of real n_2.[4] For the determination of the coefficients of reflection and transmission the knowledge of this angle is not necessary, so its determination will be omitted here. The complex index of refraction can be written as

$$n_c = n + i\,k \qquad (2.10)$$

with

n – real part of the index of refraction n_2;
k – imaginary part of the index of refraction n_2 or absorption coefficient.

---

[3] See Eq. (2.3).
[4] See BORN, WOLF [1].

2 The Behavior of Electromagnetic Radiation at Interfaces

When the phase differences between the fields are not of interest it is advisable not to use the amplitudes but the squares of the absolute values of the amplitudes:

$$R_s = |r_s|^2 \tag{2.11}$$
$$R_p = |r_p|^2 \tag{2.12}$$
$$T_s = |t_s|^2 \tag{2.13}$$
$$T_p = |t_p|^2 \tag{2.14}$$

The ratio of the squares of the absolute values of the amplitudes of two waves in the same medium equal the ratio of their intensities; R_s and R_p thus give the ratio of the intensities of the incident and the reflected waves, respectively. Contrary to this T_s and T_p do not equal the intensity ratio of the refracted and incident waves, respectively (a rigorous treatment of power conservation leads to the POYNTING vector which is given by the cross product of the $\vec{E}$ and $\vec{H}$ fields, respectively). For unpolarized radiation it follows that

$$R = \frac{R_\perp + R_\parallel}{2} \tag{2.15}$$

Here it is assumed that on an average the randomly distributed polarization directions are equally represented. The absorption coefficient A, defined as that part of the incident intensity that is not reflected, is given by

$$A = 1 - R \tag{2.16}$$

In the case of normal incidence and with $n_1 = 1$ and $n_2 = n + i\kappa$, R and T become

$$R = \frac{(n-1)^2 + \kappa^2}{(n+1)^2 + \kappa^2} \tag{2.17}$$

$$T = \frac{4}{(n+1)^2 + \kappa^2} \tag{2.18}$$

2.1.2 Analysis of the FRESNEL Formula and BREWSTER Effect

Because of SNELL's law R and T as well as r and t are only functions of the angle of incidence α (with $n_1, n_2 = n + i\kappa$ and the radiation wavelength λ kept constant). Figure 2.4 shows $R_\perp(\alpha)$, $R_\parallel(\alpha)$, and $1/2(R_\perp(\alpha) + R_\parallel(\alpha))$ with $n = 1.5$ and $\kappa = 0$. It can be seen that $R_\perp(\alpha)$ is a monotonically increasing function of α whereas $R_\parallel(\alpha)$ becomes zero at the BREWSTER angle α_B. At the BREWSTER angle α_B the refracted and reflected wave vectors are perpendicular to each other. The dipoles that oscillate in the direction of the incident wave vector cannot emit any power in that direction. This is the case if $\alpha + \beta = 90°$. With SNELL's law (Eq. (2.3)) it follows that

$$\alpha_B = \arctan\left(\frac{n_2}{n_1}\right) \tag{2.19}$$

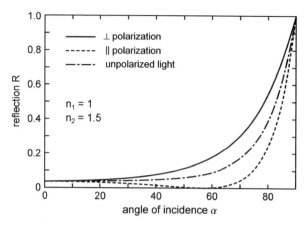

Fig. 2.4 Reflection in the case of perpendicular, parallel, and unpolarized light as a function of the angle of incidence. $n_1 = 1, n_2 = 1.5$

Figure 2.5 shows the intensity reflectivity of the reflection at a metal surface for $\perp$ and $\parallel$ polarization, respectively, as a function of the angle of incidence. The reflectivity is much higher compared to Fig. 2.4. In the case of $\perp$ polarization the reflectivity increases monotonically with increasing angle of incidence. Contrary to this behavior the reflectivity in case of $\parallel$ polarization decreases at first and only begins to increase above ca. 88°. The small reflectivity at this angle is caused by the BREWSTER effect although the reflectivity does not vanish as it does in the case of non-dissipating media. When the real part of the index of refraction is zero the reflectivity of $\parallel$ polarized light vanishes at the BREWSTER angle (see Fig. 2.4).

The BREWSTER effect can be utilized to polarize a light wave. Even when the angle of incidence does not equal the BREWSTER angle, the polarization plane is rotated in general. Figure 2.6 shows the electric field vectors before and after the

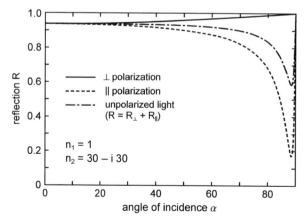

Fig. 2.5 Reflectivity in case of $\perp$ and $\parallel$ polarized light as a function of the angle of incidence. $n_1 = 1, n_2 = 30 + i\,30$. The value of the index of refraction n_2 corresponds quite well to Al

2 The Behavior of Electromagnetic Radiation at Interfaces

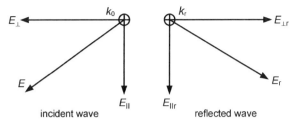

Fig. 2.6 Rotation of the plane of polarization during reflection at a denser medium

reflection. The wave vectors point into the plane of the figure (in case of normal incidence the two wave vectors are antiparallel). The ⊥ and ∥ electric field components undergo a phase shift of 180°. In the direction of the respective wave vectors this means that the vector of the parallel component keeps its direction whereas the vector of the perpendicular component reverses its direction. As a result the plane of polarization is rotated.

2.1.3 Total Reflection

When a wave is reflected at an interface with $n_2 < n_1$ the wave is totally reflected for angles of incidence greater than a limiting angle α_0 which is given by the condition $\beta = 90°$. With SNELL's law it follows that

$$\sin \alpha_0 = \frac{n_2}{n_1} \tag{2.20}$$

$$\alpha_0 = \arcsin\left(\frac{n_2}{n_1}\right) \tag{2.21}$$

With $n_1 = 1.5$ (glass) and $n_2 = 1$ (air) $\alpha_0 = 42°$.

2.2 Applications of the FRESNEL Formulae in the Field of Laser Technology

In the field of laser technology in particular the BREWSTER effect and total internal reflection are utilized.

2.2.1 BREWSTER Effect

The BREWSTER effect is exploited for

- polarization of the laser light;
- minimizing reflection losses by the use of BREWSTER windows;
- increasing the absorption during cutting, welding etc. by the use of angles of incidence near the BREWSTER angle.

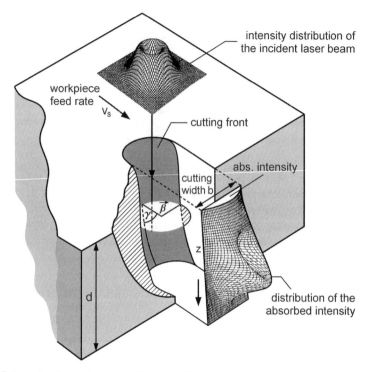

Fig. 2.7 Intensity distribution at a cutting front [3]

As an example for the utilization of the BREWSTER effect[5] Fig. 2.7 shows the distribution of the absorbed intensity at the cutting front during laser cutting of Al. The distribution is the result of convolving the intensity distribution of the incident laser beam with the incident angle-dependent absorption coefficient A (Eq. (2.16)).

2.2.2 Total Internal Reflection

Total internal reflection is predominantly used for beam guiding and shaping. Examples are

- low loss radiation transport through a fiber;
- cavity dumping.

---

[5] The BREWSTER effect occurs at the BREWSTER angle. The reduced reflectivity of ∥ polarized radiation at angles different from the BREWSTER angle is often called BREWSTER effect too.

References

1. M. Born, E. Wolf, Principles of Optics, Cambridge University Press, 1999
2. J. D. Jackson, Classical Electrodynamics, John Wiley & Sons, New York, 1975
3. D. Petring, Anwendungsorientierte Modellierung des Laserstrahlschneidens zur rechnergestützten Prozessoptimierung, Dissertation RWTH Aachen, Shaker 1993

Chapter 3
Absorption of Laser Radiation

Rolf Wester

During material processing with laser radiation the result of laser treatment is predominantly determined by the power that is absorbed within the workpiece. A measure of the power that is available for the material treatment process P_{abs} is the absorption A or absorptivity. The absorptivity is the ratio of power that is deposited within the workpiece and the power of the incident radiation:

$$A = \frac{P_{abs}}{P} \quad (3.1)$$

The absorptivity A can have any value between 0 and 1. The absorbed radiation energy is in general transformed to heat energy. This energy conversion can pass several stages that can possibly be utilized for material processing. The absorption of laser radiation can for example lead to the dissociation of molecules. Before this non equilibrium state relaxes to an equilibrium state, i.e., transformation of the absorbed energy to heat energy, the dissociated molecules can be removed. In that case material is ablated. The above-defined absorptivity is a global value that in general contains no information about where in the workpiece the radiation energy is deposited. In the case of metals the absorption always takes place in a thin surface layer, i.e., the absorption is localized. This information is not included in A.

The absorptivity can be determined directly by measuring the power of the incident laser radiation and the temperature increase of the workpiece (with known heat capacity) or indirectly by measuring the power of the reflected light P_r and the power that is transmitted through the workpiece P_t[1]:

$$P_{abs} = P - P_r - P_t \quad (3.2)$$

R. Wester (✉)
Fraunhofer-Institut für Lasertechnik, 5207 Aachen, Germany
e-mail: rolf.weister@ilt.fraunhofer.de

[1] Here the transmitted power is defined as the power that is transmitted through the whole workpiece contrary to the definition in Chap. 2 where P_t is the power that is transmitted through the surface of the workpiece.

If the radiation that is transmitted through the workpiece surface is totally absorbed within the workpiece the absorptivity is given by

$$A = 1 - R \tag{3.3}$$

In the case of non-conducting materials or very thin metal films part of the radiation that penetrates into the workpiece can leave the workpiece again. Describing this part by the transmittance T it holds that

$$A = 1 - R - T \tag{3.4}$$

The parameters

- index of refraction n
- index of absorption κ
- electric conductivity σ
- heat conductivity K
- specific heat c

are material-specific parameters. They solely depend on the properties of the material at hand and the radiation wavelength and can be calculated approximately for simple materials. The absorptivity not only depends on the material properties. Some further influencing factors are

- properties of the laser beam like wavelength, polarization;
- the ambient conditions (process gas, material that surrounds the workpiece, etc.);
- the surface properties (roughness, morphology, etc.);
- the geometry of the workpiece (thickness, boundaries of the workpiece, etc.);
- the changes of the workpiece and the environment that are induced by the absorbed laser power (local heating, phase changes, laser-induced plasma).

3.1 Description of the Phenomena

The description of the interaction of electromagnetic radiation with matter can be done at different model levels. Matter consists of electrons and atomic kernels. For spatial dimensions $r \gg 10^{-15}$ m the atomic kernels can be considered to be point charges, and for spatial dimensions greater than the classical electron radius $r \gg r_0 \sim 2.8 \times 10^{-15}$ m the electrons can be considered to be point charges too. These point charges interact with electromagnetic fields and excite spatially and temporally fast fluctuating fields on their part. The interaction between electromagnetic fields and electrically charged particles is treated rigorously within the framework of quantum electrodynamics. The quantum theoretical treatment of electromagnetic fields leads to the concept of photons, i.e., there can only be an integer number of photons being emitted or absorbed. If the particles that interact with the electromagnetic fields are bound these too can only absorb or emit certain energy quanta. In the framework of an atomistic model this is described as the absorption of a photon and the creation of a phonon (energy quantum of lattice oscillations).

In the framework of a classical description the spatio-temporal evolution of electromagnetic fields is described by the microscopic MAXWELL equations in vacuum interacting with point charges (positively charged atomic kernels and negatively charged electrons). The atoms can be described as dipoles that are excited by the external radiation and that emit radiation on their part that interferes with the primary radiation. This process can be interpreted as coherent scattering. If radiation hits a surface of a solid the radiation that is emitted by the dipoles at the solid surface consists of three parts. The first one corresponds to the reflected wave. The second one is emitted in the same direction as the incident wave. According to the EWALD–OSEEN extinction theorem [3] modulus and phase of this wave are such that the incident wave and this wave extinct each other exactly within region 2 (see Fig. 2.2). The third part corresponds to the refracted wave.

A further level of describing electromagnetic phenomena is to average over macroscopic small but microscopic large spatial regions. In this way one gets the macroscopic MAXWELL equations.[2] The macroscopic MAXWELL equations treat the matter as a continuum whose electromagnetic properties are described by material parameters. These parameters can either be measured or calculated using microscopic models. The macroscopic MAXWELL equations read as follows:

$$\vec{\nabla} \times \vec{E} = \frac{\partial \vec{B}}{\partial t} \tag{3.5}$$

$$\vec{\nabla} \times \vec{H} = \vec{j} + \frac{\partial \vec{D}}{\partial t} \tag{3.6}$$

$$\vec{\nabla} \cdot \vec{D} = \rho \tag{3.7}$$

$$\vec{\nabla} \cdot \vec{B} = 0 \tag{3.8}$$

For a solution of these equations one needs

- the relationship between $\vec{D}$ and $\vec{E}$ and $\vec{H}$ and $\vec{B}$ as well as the relationship between $\vec{j}$ and $\vec{E}$ (these relationships describe the material behavior);
- the geometry of the workpiece and its environment;
- the boundary conditions for $\vec{E}$, $\vec{D}$, $\vec{H}$, and $\vec{B}$.

In the following space charges are neglected, i.e., $\rho = 0$.

---

[2] See JACKSON [6]; LANDAU and LIFSCHITZ [8]

3.1.1 Field Relationships

The relationships between $\vec{E}$ and $\vec{D}$ and $\vec{B}$ and $\vec{H}$ are in general nonlinear. But in many cases of material processing with laser radiation the relationships can be very well approximated by linear models. The relationships depend in general on the frequency of the radiation field. In case of time harmonic fields the FOURIER-transformed field quantities are related according to

$$\vec{D}(\vec{r},\omega) = \varepsilon_0 \varepsilon(\omega) \vec{E}(\vec{r},\omega) \qquad (3.9)$$

$$\vec{B}(\vec{r},\omega) = \mu_0 \mu(\omega) \vec{H}(\vec{r},\omega) \qquad (3.10)$$

Many crystals have orientation-dependent material properties; these crystals are said to be anisotropic. In that case the dielectric constant ε and the magnetic permeability μ are no longer scalar quantities but tensors. This means that the electric field $\vec{E}$ and the electric displacement $\vec{D}$ in general are not collinear as is the case in vacuum and isotropic materials. This is associated with the phenomenon of birefringence. In the following anisotropic material behavior is neglected. The magnetic permeability μ only deviates significantly from the vacuum value 1 in case of ferromagnetic materials. In the following it will be assumed that

$$\mu(\omega) = 1 \qquad (3.11)$$

Inverse FOURIER transformation of Eq. (3.9) results in

$$\vec{D}(\vec{r},t) = \vec{E}(\vec{r},t) + \int_{-\infty}^{\infty} G(\tau) \vec{E}(\vec{r},t-\tau) d\tau \qquad (3.12)$$

$$G(\tau) = \frac{1}{2\pi} \int_{-\infty}^{\infty} [\varepsilon(\omega) - 1] e^{-i\omega\tau} d\omega \qquad (3.13)$$

In [6] it is shown that due to causality $G(\tau)$ vanishes for $\tau < 0$, i.e., the electric displacement $\vec{D}(t)$ at time t depends on the past history of the electric field $\vec{E}(t)$ but not on future values of $\vec{E}(t)$. Equation (3.12) is nonlocal in time but local in space. This local approximation only holds as long as the spatial scale of variations of the electric field is large compared to the scale of the atomic polarization. In case of dielectric materials with bound electrons this holds as long as the wavelength of the radiation is large compared to atomic dimensions but in conductors the free path length of electrons can exceed the wavelength or penetration depth of the radiation fields.[3] This will not be considered further.

---

[3] This is, e.g., the underlying mechanism of the anomalous skin effect.

3 Absorption of Laser Radiation

The material properties are accounted for in the MAXWELL equations by $\varepsilon(\omega)$. In case of small frequencies ε is constant. In this frequency region the polarization of the medium instantly follows the electric field. At higher frequencies the response of the material lags behind the temporal change of the electric field. At very high frequencies the material response vanishes. In frequency space it holds that

$$\lim_{\omega \to 0} \varepsilon(\omega) = \varepsilon_{\text{stat}} \tag{3.14}$$

$$\lim_{\omega \to \infty} \varepsilon(\omega) = \varepsilon_\infty \to 1 \tag{3.15}$$

3.1.2 Wave Equation

With Eqs. (3.5), (3.6), (3.9), and (3.10) the wave equation

$$\vec{\nabla}\left(\vec{\nabla} \cdot \vec{E}\right) - \Delta \vec{E} = -\mu_0 \varepsilon_0 \varepsilon \frac{\partial^2 \vec{E}}{\partial t^2} \tag{3.16}$$

results. In homogeneous media and with zero space charges $\vec{\nabla} \cdot \vec{E} = 0$. With $\mu_0 \epsilon_0 = 1/c^2$ Eq. (3.16) becomes

$$\Delta \vec{E} = \frac{\epsilon}{c^2} \frac{\partial^2 \vec{E}}{\partial t^2} \tag{3.17}$$

A solution of this equation is the plane wave

$$\vec{E} = \vec{E}_0 \exp\left[i\left(k\,z - \omega\,t\right)\right] \tag{3.18}$$

with

- k – complex wave number;
- ω – real angular frequency.

The complex wave number is

$$k = k_0 \sqrt{\epsilon} = k_0 n = k_r + i k_i \tag{3.19}$$

where n is the complex index of refraction. The plane wave solution can also be cast into the form

$$\vec{E} = \vec{E}_0 \exp\left[i\left(k_r\,z - \omega\,t\right)\right] \exp\left[-k_i z\right] \tag{3.20}$$

If the imaginary part of the complex wave number $k_i > 0$ the wave decays exponentially within the material.

3.1.3 Geometry of the Workpiece

In most cases in textbooks the temporal change of the geometry of the workpiece is neglected. In the following this approach will be adopted. But it has to be emphasized that during material processing the geometry of the workpiece can change considerably due to melt flow and evaporation.

3.2 Isolators

The main difference between isolators on the one side and plasmas and metals on the other side is that in isolators electrons are bound (at least as long as the incident intensities are below the threshold for ionization) whereas in plasmas and conductors the electrical properties are mostly determined by the free (or as in the case of metal quasi free) electrons. The dielectric constant of isolators is composed of three parts:

- electronic polarization: the electrons are pulled away from their rest position at the atomic sites and oscillate around these points;
- ionic polarization: differently charged ions are deflected with respect to each other, i.e., the effective ion charges of the crystal sub-lattices oscillate;
- orientation polarization of permanent dipoles: permanent dipoles are aligned by the electric field (e.g., H_2O).

Orientation polarization is negligible for frequencies in the infrared region and above. Electronic polarization is the main mechanism in the near-infrared down to the X-ray region. Due to their larger mass the oscillating frequencies of the ions are much smaller compared to the oscillating frequencies of the bound electrons. Because of this the ionic part of the dielectric constant dominates in the infrared region. The polarization can be described by harmonic oscillators. Harmonic oscillators are characterized by their eigenfrequency ω_0 and by the damping constant δ. The deflection x of an oscillator with effective charge e^* with respect to its rest position induces a dipole moment:

$$p = e^* x \tag{3.21}$$

The equation of motion in frequency space of an harmonic oscillator subject to an external field reads

$$-\omega^2 x(\omega) - i\,\delta\,\omega\, x(\omega) + \omega_0^2\, x(\omega) = \frac{e^*}{m} E(\omega) \tag{3.22}$$

The dipole moment is thus

$$p = \varepsilon_0\, \alpha(\omega)\, E(\omega) \tag{3.23}$$

$$\alpha = \frac{e^{*2}}{m\,\varepsilon_0}\, \frac{1}{\omega_0^2 - \omega^2 - i\,\delta\,\omega} \tag{3.24}$$

where α is the polarizability. The polarization P is given by

$$P = \frac{N}{V} p \qquad (3.25)$$

N/V is the density of harmonic oscillators; thus the polarization is given by

$$P = \frac{N}{V} \frac{e^{*2}}{m} \frac{1}{\omega_0^2 - \omega^2 - i\,\delta\,\omega} E(\omega) \qquad (3.26)$$

The field strength in the above equation is the field strength at the atomic location. In general this does not equal the average (averaged over microscopic large scales) field strength. With

$$P = \varepsilon_0 \left(\varepsilon(\omega) - 1 \right) E(\omega) \qquad (3.27)$$

it follows that

$$\varepsilon(\omega) = 1 + \frac{N}{V} \frac{e^{*2}}{m\,\varepsilon_0} \frac{1}{\omega_0^2 - \omega^2 - i\,\delta\,\omega} \qquad (3.28)$$

The damping can be caused by different physical processes. In case of atomic excitation the damping is due to radiation damping. In case of oscillations of sub-lattices of a crystal with respect to each other damping is caused by the coupling of the lattice vibration to other degrees of freedom. In both cases normally $\delta \ll \omega_0$ holds.

3.2.1 Electronic Polarization

Equation (3.28) applies to single oscillators. Atoms and ions in general have more than one electron and more than one eigenfrequency and damping constant. The eigenfrequencies correspond to the energy gaps between atomic levels. The transition probabilities determine the polarizability and can only be computed within a quantum mechanical framework. In doing this the form of Eq. (3.28) remains virtually unchanged:

$$\varepsilon(\omega) = 1 + \frac{N}{V} \alpha_{el}(\omega) \qquad (3.29)$$

$$\alpha_{el}(\omega) = \frac{e^2}{m\,\varepsilon_0} \sum \frac{f_{ij}}{\omega_{ij}^2 - \omega^2 - i\,\delta_{ij}\,\omega} \qquad (3.30)$$

The sum runs over all transitions ($i \to j$). The f_{ij} are the oscillator strengths of the transitions. The electronic polarizability α_{el} depends not only on the element at hand but also on the atomic or ionic environment in the solid or fluid. Equation (3.29) only holds for sufficiently rarefied gases. This is due to the fact that the field strength in Eq. (3.23) that induces the atomic dipole moment does not equal the averaged field strength at the atomic position but is affected by the polarization of the medium. The field strength E' that has to be inserted in Eq. (3.23) in the case of cubic symmetry is given by (with other symmetries the deviations are generally small) [14, 15]

$$E' = E + \frac{1}{3\,\varepsilon_0}\,P \tag{3.31}$$

With this it follows that

$$P = \frac{N}{V}\,p = \frac{N}{V}\,\alpha\,\varepsilon_0\,E' = \frac{N}{V}\,\alpha\,\varepsilon_0\left(E + \frac{1}{3\,\varepsilon_0}\,P\right) \tag{3.32}$$

Resolving with respect to P gives

$$P = \frac{\varepsilon_0\,\alpha\,\dfrac{N}{V}}{1 - \dfrac{\alpha\,\dfrac{N}{V}}{3}}\,E \tag{3.33}$$

With

$$D = \varepsilon_0\,E + P = \varepsilon_0\,\varepsilon\,E \tag{3.34}$$

the dielectric constant is

$$\epsilon = \frac{1 + \dfrac{2}{3}\,\alpha\,\dfrac{N}{V}}{1 - \dfrac{1}{3}\alpha\,\dfrac{N}{V}} \tag{3.35}$$

and the polarizability is

$$\alpha = \frac{3}{\dfrac{N}{V}}\,\frac{\varepsilon - 1}{\varepsilon + 2} \tag{3.36}$$

This is the equation of CLAUSIUS–MOSOTTI or the LORENTZ–LORENZ law. With $N/V = 10^{29}$ m^{-3} and $\varepsilon = 3$ the electronic polarizability is $\alpha = 1.5 \times 10^{-29}$ m^3.

3.2.2 Ionic Polarizability

In the following only the case of a single ion resonance frequency is discussed. The resonance frequencies of the ion sub-lattices are in the infrared region. The orientation polarization can be neglected in this region and the electronic polarization is almost frequency independent. With Eq. (3.28), $\omega_0 = \omega_T$, and the limiting cases $\varepsilon(0)$ and $\varepsilon(\infty)$ it follows that

$$\varepsilon(\omega) = \varepsilon(\infty) + \frac{\omega_T^2}{\omega_T^2 - \omega^2 - i\delta\omega}(\varepsilon(0) - \varepsilon(\infty)) \qquad (3.37)$$

$\varepsilon(\infty)$ contains the contribution of the hull electrons. $\varepsilon(\omega)$ is a complex quantity, $\varepsilon = \varepsilon_1 + i\varepsilon_2$. The frequency dependence of the real and imaginary parts of $\varepsilon(\omega)$ of a single harmonic oscillator is shown in Fig. 3.1. The frequency ω_T was chosen to be 1.9×10^{14} s ($\lambda = 10\,\mu$m) and the damping $\delta = 0.01\,\omega_T$. The dielectric constant at $\omega = 0$ is $\varepsilon = 3$; at large frequencies it is $\varepsilon = 2$. When approaching the resonance frequency ω_T coming from small frequencies the real and imaginary parts of ω_T increase, that is, the region of normal dispersion. After exceeding the resonance frequency Re $[\varepsilon(\omega)]$ decreases and becomes negative. This region of decreasing Re $[\varepsilon(\omega)]$ is the region of anomalous dispersion. In case of zero damping (Im $[\varepsilon] = 0$) and negative real part of ε (Re $[\varepsilon] < 0$) the real part of the complex index of refraction

$$n_c = n + i\kappa = \sqrt{\varepsilon} \qquad (3.38)$$

vanishes which implies that the wave cannot propagate in the medium but is totally reflected at the interface. With finite damping this behavior is attenuated. When exceeding the frequency ω_L Re $[\varepsilon(\omega)]$ becomes positive again. If the harmonic oscillator corresponds to a lattice vibration the frequency ω_L at which Re $[\varepsilon(\omega)] = 0$ is

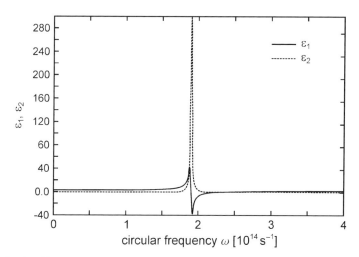

Fig. 3.1 Real and imaginary parts of the dielectric constant of a crystal with ion lattice. Computed according to Eq. (3.37) with $\omega_T = 1.9 \times 10^{14}$ s^{-1} which corresponds to $\lambda = 10\,\mu$m, $\delta = 0.01\,\omega_T$. The dielectric constant at $\omega = 0$ is $\varepsilon = 3$; at high frequencies it is $\varepsilon = 2$

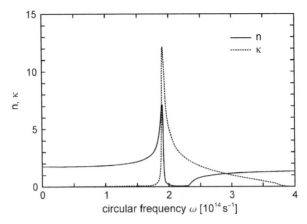

Fig. 3.2 Real and imaginary parts of the index of refraction computed using Eq. (3.38) with the dielectric constant given in Fig. 3.2

the resonance frequency of longitudinal lattice vibrations. In case of zero damping the LYDDANE–SACHS–TELLER relation follows from Eq. (3.37):

$$\left(\frac{\omega_T}{\omega_L}\right)^2 = \frac{\varepsilon(\infty)}{\varepsilon(0)} \quad (3.39)$$

The index of refraction is shown in Fig. 3.2. The reflectivity in case of normal incidence on a plane interface is

$$R = \frac{(n-1)^2 + \kappa^2}{(n+1)^2 + \kappa^2} \quad (3.40)$$

and is shown in Fig. 3.3. The reflectivity is quite large in the range of frequencies where the real part of $\varepsilon(\omega)$ is negative (in case of zero damping R approaches 1). With $\varepsilon(\omega) = 1$ and $\delta = 0$ $R = 0$.

Considering the restrictions in treating absorption mentioned above there is a further meaningful distinction:

- The optical penetration depth δ_{opt} is small compared to the workpiece thickness s. Then there is no transmission through the workpiece and the absorption is given by Eq. (3.3), i.e., absorption can be computed using the reflectivity only. The boundary conditions at the interface between workpiece and environment determine the absorptivity A.
- The optical penetration depth is large or comparable to the thickness of the material considered. The material is partially transparent. This case can be of importance in the field of surface treatment of metals when absorption-enhancing layers are used or when oxide layers exist.

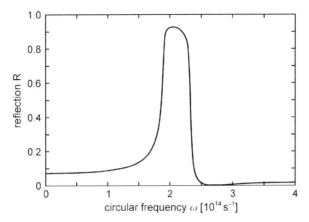

Fig. 3.3 Intensity reflectivity computed using Eq. (3.40) and the index of refraction depicted in Fig. 3.2

3.2.3 Supplementary Substances in Polymers

The principal optical properties of polymers are in many cases not determined by the polymer molecules but by supplementary substances such as color pigments, softeners. A polymer without supplementaries can be transparent at a certain wavelength but can be strongly absorbing with a supplementary added. Utilizing this effect makes it possible to weld polymer parts laying on top of each other without melting one of the outside surfaces.

3.3 Dielectric Properties of Plasmas

A plasma is a mixture of free electrons, positively charged ions (in some cases also negatively charged ions are present), and neutral particles (atoms and molecules). The charged particles can interact with electric and magnetic fields. Because of their much larger masses compared to electrons the interaction of the ions with electromagnetic fields is considerably less important, the energy exchange between plasma and field is almost entirely due to the electrons. The free electrons in metals can with some minor modifications be treated similar to plasma electrons so that the following is also applicable to metals, at least partially. With the assumption that the photon energy is small compared to the mean electron energy the plasma conductivity is given by[4]

$$\sigma = \frac{e^2 n_e}{m_e \nu_m} \frac{\nu_m}{\nu_m - i\omega} \qquad (3.41)$$

---

[4] See Appendix A.2.

with

- e – electron charge;
- n_e – electron density;
- m_e – electron mass;
- ν_m – momentum transfer frequency;
- ω – angular frequency of the electromagnetic field.

The momentum transfer frequency ν_m is defined as $1/\tau_m$ with τ_m being the mean time of transfer of electron momentum to other particles during collisions. ν_m consists of three parts, electrons collide with neutral particles, with ions, and with other electrons. The consideration of electron–electron collisions makes the computations a lot more complicated. But this contribution does not change the conductivity considerably so that in many cases it is sufficient to only consider the electron–neutral and electron–ion collisions.[5] Because of the large-range COULOMB interaction the electron–ion collision cross section is in general much larger than the electron–neutral cross section. The dielectric constant is related to the conductivity as

$$\varepsilon = 1 - \frac{\sigma}{i\omega\varepsilon_0} \tag{3.42}$$

With this it follows that

$$\varepsilon = 1 - \frac{\omega_p^2}{\omega^2 + \nu_m^2} + i\frac{\nu_m}{\omega}\frac{\omega_p^2}{\omega^2 + \nu_m^2} \tag{3.43}$$

The electron plasma frequency is

$$\omega_p = \sqrt{\frac{e^2 n_e}{\varepsilon_0 m_e}} \tag{3.44}$$

The index of refraction is given by the MAXWELL relation:

$$n_c = \sqrt{\varepsilon} = n + i\kappa \tag{3.45}$$

with

$$n^2 = \frac{1}{2}\sqrt{\left(1 - \frac{\omega_p^2}{\omega^2 + \nu_m^2}\right)^2 + \left(\frac{\nu_m}{\omega}\frac{\omega_p^2}{\omega^2 + \nu_m^2}\right)^2} + \frac{1}{2}\left(1 - \frac{\omega_p^2}{\omega^2 + \nu_m^2}\right) \tag{3.46}$$

---

[5] LORENTZ model of a plasma.

$$\kappa^2 = \frac{1}{2}\sqrt{\left(1 - \frac{\omega_p^2}{\omega^2 + \nu_m^2}\right)^2 + \left(\frac{\nu_m}{\omega}\frac{\omega_p^2}{\omega^2 + \nu_m^2}\right)^2} - \frac{1}{2}\left(1 - \frac{\omega_p^2}{\omega^2 + \nu_m^2}\right) \quad (3.47)$$

In Eq. (3.43) two limiting cases can be distinguished:

- $\nu_m = 0$: the dielectric constant is real;
- $\nu_m \gg \omega$: the real part of the dielectric constant is small compared to its imaginary part, which holds, for example, in case of metals.

3.3.1 Collision-Free Plasma

In the collision-free case the dielectric constant is given by

$$\varepsilon = 1 - \frac{\omega_p^2}{\omega^2} \quad (3.48)$$

The wave number of a plane electromagnetic wave is

$$|\vec{k}| = k_0 \, n_c \quad (3.49)$$

with

$$n_c = \sqrt{\varepsilon} = \sqrt{1 - \frac{\omega_p^2}{\omega^2}} \quad (3.50)$$

$$n_c = \sqrt{1 - \frac{\omega_p^2}{\omega^2}} \quad (3.51)$$

$$k_0 = \frac{\omega}{c} \quad (3.52)$$

with

k_0 – vacuum wave number;
c – vacuum velocity of light.

When ω_p/ω starting from small values approaches unity the wave number approaches zero which means that the wavelength

$$\lambda = \frac{2\pi}{k} \quad (3.53)$$

becomes infinite. If the plasma frequency exceeds the critical value ($\omega_p > \omega$) the index of refraction and the wave number become imaginary. This implies that the

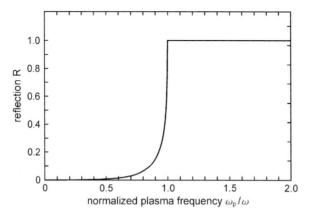

Fig. 3.4 Reflection of an electromagnetic wave at a vacuum collision-less plasma interface. If the plasma frequency ω_p exceeds the critical value, namely the frequency ω of the electromagnetic wave, the wave is totally reflected

wave cannot propagate within the plasma any more and because of this it is totally reflected at plasma interfaces.[6] The electrical field strength decays exponentially within the plasma.

$$\vec{E} = \vec{E}_0 \, e^{-i\omega t} \, e^{-k_i x} \quad (3.54)$$

k_i is the imaginary part of the wave number. The time-averaged absolute value of the POYNTING vector of a plane wave in a plasma is given by (Appendix A.3)

$$I = \frac{1}{2} \frac{k_r}{k_0} \frac{|E_0|^2}{Z_0} e^{-2 k_i x} \quad (3.55)$$

$$Z_0 = \sqrt{\frac{\mu_0}{\varepsilon_0}} \quad (3.56)$$

with

k_r – real part of the wave number;
Z_0 – wave impedance of free space.

If the wave number is purely imaginary, i.e., $k_r = 0$, then $I = 0$. This means that there is no time-averaged energy transport into the plasma and the decrease of the electric field is only due to reflection. Figure 3.4 shows the reflection at a collision-less plasma computed with Eq. (3.40) and the index of refraction according to Eq. (3.51). Above $\omega_p/\omega = 1$ total reflection occurs. In the case of finite collision frequency ν_m the decrease of the electric field within the plasma is due to reflection

---

[6] See Eq. (3.40), $R = 1$ for $n = 0$ and k finite.

and also absorption, i.e., the incident radiation is partly absorbed by the plasma. When the plasma frequency ω_p is small compared to the wave frequency ω, the imaginary part of the wave number becomes

$$k_i = \frac{1}{2} \frac{\nu_m \, \omega_p^2}{c \, (\omega^2 + \nu_m^2)} \tag{3.57}$$

3.3.2 Collision-Dominated Plasma

If the momentum transfer frequency ν_m is large compared to the wave frequency ω and the imaginary part of the plasma dielectric constant ε_i is large compared to its real part ε_r the dielectric constant is approximately given by

$$\varepsilon = i \, \frac{\omega_p^2}{\omega \, \nu_m} \tag{3.58}$$

With Eqs. (3.41) and (3.44) it follows that

$$\varepsilon = i \, \frac{\sigma}{\varepsilon_0 \, \omega} \tag{3.59}$$

The index of refraction then becomes

$$n_c = \sqrt{\frac{\sigma}{2 \, \omega \, \varepsilon_0}} \, (1 + i) \tag{3.60}$$

which means $n = \kappa$. With this the reflectivity Eq. (3.40) is

$$R = 1 - \frac{4 \, n}{(n+1)^2 + n^2} \tag{3.61}$$

With Eq. (3.61) and $n \gg 1$ the HAGEN–RUBENS relation follows:

$$R \simeq 1 - \frac{2}{n} = 1 - \sqrt{\frac{8 \, \omega \, \varepsilon_0}{\sigma}} \tag{3.62}$$

The requirements are in general fulfilled in the case of metals. As an example let $n_c = 30 + i \, 30$ (according to Eq. (3.60) this corresponds to an electric conductivity of $\sigma = 3 \times 10^6 \Omega^{-1} \mathrm{m}^{-1}$). With this $R = 93\%$. In spite of the high collision frequency and associated with this the strong damping an electromagnetic field is not entirely absorbed by the plasma but a great part of the incident wave energy is reflected. The complex wave number is given by

$$k = k_0 \, n_c \tag{3.63}$$

With the above-given example values the real and imaginary parts have equal values and the wave number exceeds the vacuum value by a factor of more than 30. The real part of the wave number is

$$k_r = \frac{2\pi}{\lambda} \tag{3.64}$$

λ is the wavelength within the plasma and is about 30 times smaller compared to the vacuum wavelength λ_0. The imaginary part of the wave number is

$$k_i = \frac{1}{\delta_s} \tag{3.65}$$

δ_s is the damping length of the wave in the plasma or the skin depth. With Eqs. (3.41) and (3.44) and the assumption of large collision frequency the plasma frequency can be expressed as

$$\omega_p = \sqrt{\frac{\sigma \, \nu_m}{\varepsilon_0}} \tag{3.66}$$

With this and with Eqs. (3.56), (3.58), and (3.60) the skin depth amounts to

$$\delta_s = \sqrt{\frac{2}{\omega \sigma \mu_0}} \tag{3.67}$$

Using the above-given example values of the index of refraction the skin depth is $\delta_s = \lambda_0/188$. With a vacuum wavelength λ_0 of 10.6 µm this gives $\delta_s = 53$ nm.

In addition to the above-described processes electromagnetic fields can excite different kinds of plasma waves. This especially holds in the presence of a static magnetic field. These processes normally are not important during material processing with laser radiation, so this will not be treated here.

3.4 Absorption of Metallic Materials

Most metals show large reflectivity and accordingly low absorption of electromagnetic radiation. Contrary to atoms and molecules the absorption spectra do not have discrete lines but are continuous from the far-infrared region (**FIR**) down to short wavelengths in the ultraviolet region (**UV**). In the **UV** there are also discrete bands of absorption. The explanation for this phenomenon is given by the electron theory of metals due to DRUDE [4]. Models that rely on classical physics like the DRUDE model can describe the absorption qualitatively quite well but in some cases their predictions are in conflict with experience. This holds, for example, for the contribution of the free electrons to the heat capacity. With quantum mechanical corrections, especially with the introduction of the FERMI statistic, these contradictions can be

3 Absorption of Laser Radiation

removed and the performance of the electron theory of metals is thus improved substantially. For example, the band structure of the absorption of metals in the **UV** can be attributed to transitions between different energy bands.

Even today modified versions of the DRUDE model are used to describe the absorption of electromagnetic radiation and the corresponding optical constants of metals. This is mainly due to its simplicity and its clarity. It can be shown that especially in the infrared region even quantum mechanical results are approximately reproduced if suitable parameters are introduced into the DRUDE theory [10, 16]. The absorption of metals at a given wavelength depends mainly on four parameters:

- the optical constants of the material;
- the physical condition of the surface (roughness, etc.);
- the chemical composition of the surface (oxide layers, etc.);
- the temperature that in turn influences the other three parameters.

The absorption of materials that are used in production environments can in general only be computed approximately because the physical and chemical surface properties cannot be determined with the necessary precision. Computational models are thus only reasonable for materials under ideal conditions, i.e., materials that comply with the assumptions of the theory and that have an ideal surface. Because of this in practical situations precise values of the absorption can only be determined experimentally. Despite this the knowledge of the physical mechanisms is crucial because the measurements can only be performed within small parameter fields and thus the functional dependence of the absorption on application-specific parameters have to be extrapolated from theory. For metal specimens with a large thickness compared to the skin depth it holds that $T \simeq 0$ and the absorption is given by

$$A = 1 - R \tag{3.68}$$

The index of refraction n and the absorption index κ depend on the wavelength and thus also A depends on the wavelength. The wavelength-dependent absorption $A(\lambda)$ as a function of the optical parameters at normal incidence is given by

$$A(\lambda) = 1 - \frac{(n(\lambda) - 1)^2 + \kappa(\lambda)^2}{(n(\lambda) + 1)^2 + \kappa^2(\lambda)} \tag{3.69}$$

The dependence of $A(\lambda)$ on the angle of incidence α and the polarization direction (parallel or perpendicular polarization) is described by the FRESNEL formulae.[7] For the computation of the optical parameters of metals their frequency-dependent electrical conductivity has to be determined. This is done in the following chapter in the frame of the DRUDE theory. The DRUDE theory largely corresponds to the plasma model of Sect. 3.3.

---

[7] See Chap. 2.

3.5 The DRUDE Model of Absorption

Within the framework of the DRUDE model the electrons in a metal are treated as a free electron gas moving under the action of an electrical potential. The electrical, thermal, combined thermo-electrical and magnetic characteristics are derived using gas kinetic models. Under the exposure of laser radiation the free electrons are accelerated and in return emit radiation corresponding to their acceleration. Free electrons cannot absorb energy on a time-averaged basis from an electric field oscillating periodically in time. The reason for this is that the electron velocity in steady state is $\pi/2$ phase shifted relative to the electric field. But this phase relation is distorted by collisions of the electrons with the periodic lattice potential and with phonons. The DRUDE model can particularly be applied to simple metals with isotropic crystal structure, like Na and Ka, and allows to compute at least approximately the optical constants n and κ based on the following three assumptions:

- the electromagnetic radiation only interacts with the free electrons in the conduction band; the polarization of atomic kernels and bound electrons and all oscillations of the crystal lattice are neglected;
- the free electrons in the metal obey OHM's law;
- all free electrons of a given metal can be characterized by a single effective mass m^* and a single collision frequency v_m (this hypothesis is based on the assumption that the FERMI statistic allows for unique values corresponding to m^* and v_m, an assumption that is only valid in some special cases).

The optical constants n and κ of a metal are related to macroscopically measurable electric and magnetic parameters:

$$(n + i\kappa)^2 = \left(\varepsilon - \frac{\sigma}{i\omega\varepsilon_0}\right)\mu \qquad (3.70)$$

with

σ – electric conductivity at frequency ω;
ε – dielectric constant, contribution of the bound electrons and ions;
μ – magnetic permeability.

If the interaction only takes place between radiation and free electrons it holds that $\varepsilon \simeq 1$ and $\mu \simeq 1$ in the **IR** region. It thus follows that

$$(n + i\kappa)^2 = \left(1 - \frac{\sigma}{i\omega\varepsilon_0}\right) \qquad (3.71)$$

The applicability of OHM's law

$$\vec{j} = \sigma \vec{E} \qquad (3.72)$$

implies that the movement of the free electrons in the metal can be described by a local relation between electric field and current density. The interaction with the crystal lattice and the degeneracy of the electrons have to be treated quantum mechanically. In the frame of the model of a quasi free electron gas the electrons can be treated as free particles if the free electron mass is substituted in the equation of motion by an effective mass that can be smaller or larger than the free electron mass. The equation of motion of the quasi free electrons in a metal conforms to the equation of motion of electrons in a plasma and reads[8]

$$m^* \frac{d\vec{v}}{dt} + m^* \nu_m \vec{v} = -e\vec{E} \quad (3.73)$$

with

m^* – effective mass of the electrons;
$\vec{v}$ – mean velocity of the electrons;
ν_m – momentum transfer frequency;
$\vec{E}$ – electric field of the laser radiation.

$(1/\nu_m)$ is the mean time for the transfer of electron momentum to phonons or lattice impurities during collisions. The velocity $\vec{v}$ of the electrons follows from Eq. (3.73):

$$\vec{v} = -\frac{e}{m^*} \frac{1}{\nu_m - i\omega} \vec{E} \quad (3.74)$$

The third assumption allows to use a unique electron density n_e in the expression for the current density:

$$\vec{j} = -e\, n_e \vec{v} \quad (3.75)$$

Combining Eqs. (3.72), (3.74), and (3.75) provides for the electric conductivity:

$$\sigma = \frac{e^2 n_e}{m^* (\nu_m - i\omega)} \quad (3.76)$$

Splitting into real and imaginary parts results in

$$\sigma = \sigma_1 + i\,\sigma_2 \quad (3.77)$$

$$\sigma_1 = \sigma_0 \frac{\nu_m^2}{\nu_m^2 + \omega^2} \quad (3.78)$$

---

[8] See Appendix A.2.

$$\sigma_2 = \sigma_0 \frac{\nu_m \, \omega}{\nu_m^2 + \omega^2} \tag{3.79}$$

with the **DC conductivity**

$$\sigma_0 = \frac{e^2 \, n_e}{m^* \, \nu_m} \tag{3.80}$$

With Eqs. (3.71) and (3.77), (3.78), (3.79), and (3.80) it follows for the index of refraction and the index of absorption of a metal within the frame of the DRUDE model that

$$n^2 = \frac{1}{2} \sqrt{\left(1 - \frac{\omega_p^2}{\omega^2 + \nu_m^2}\right)^2 + \left(\frac{\nu_m}{\omega} \frac{\omega_p^2}{\omega^2 + \nu_m^2}\right)^2} + \frac{1}{2}\left(1 - \frac{\omega_p^2}{\omega^2 + \nu_m^2}\right) \tag{3.81}$$

$$\kappa^2 = \frac{1}{2} \sqrt{\left(1 - \frac{\omega_p^2}{\omega^2 + \nu_m^2}\right)^2 + \left(\frac{\nu_m}{\omega} \frac{\omega_p^2}{\omega^2 + \nu_m^2}\right)^2} - \frac{1}{2}\left(1 - \frac{\omega_p^2}{\omega^2 + \nu_m^2}\right) \tag{3.82}$$

$$\omega_p^2 = \frac{e^2 \, n_e}{\epsilon_0 \, m_e}$$

Equations (3.81) and (3.82) correspond to Eqs. (3.46) and (3.47) of the index of refraction and the index of absorption of a plasma, respectively. In Eqs. (3.81) and (3.82) the effective electron mass m^*, the density n_e of the quasi free metal electrons, and the momentum transfer frequency ν_m of the electrons in the metal have to be known. The momentum transfer frequency ν_m is composed of two parts, the electrons transfer momentum

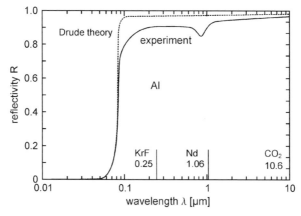

Fig. 3.5 Reflectivity of an Al surface. The theoretical curve is computed with the help of Eqs. (3.40), (3.81), (3.82), and (3.80) as well as the values $\sigma = 3.6 \times 10^7 \, \Omega^{-1} \text{m}^{-1}$, $n = 1.8 \times 10^{29} \, \text{m}^{-3}$, and $\nu_m = 1.3 \times 10^{14} \, \text{s}^{-1}$. The experimental values are taken from [5]

3 Absorption of Laser Radiation

- during collisions with phonons and
- during collisions with impurities.

3.6 Temperature Dependence of the Absorption of Metals

Experimentally usually an increase of the absorption with increasing temperature is observed (Fig. 3.6a). Figure 3.6b shows the specific electric resistance as a function of temperature of stainless steel and Fe. The similarity of the curves of Fig. 3.6a and Fig. 3.6b indicates similar causes of the temperature dependence. The DRUDE model explains both dependencies with the increase of the electron lattice collision frequency v_m. The temperature dependence of the electric conductivity can be extracted from the law of WIEDEMANN–FRANZ (Sect. 4.2):

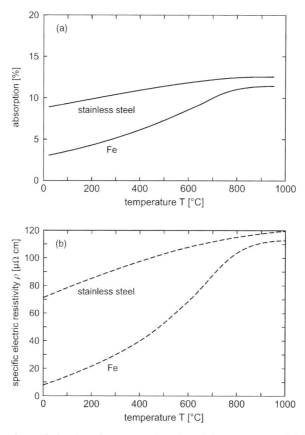

Fig. 3.6 Absorption and electric resistance as a function of the temperature. (**a**) Increase of the absorption of stainless steel and Fe with increasing temperature; (**b**) specific electric resistance of stainless steel and Fe

$$\frac{\sigma(T)}{K(T)} = \frac{1}{LT} \tag{3.83}$$

with

L – LORENZ number;
$L(\text{Al}) - 2.4 \times 10^{-8}$ V^2(mK^2)$^{-1}$;
$L(\text{Fe}) - 2.8 \times 10^{-8}$ V^2(mK^2)$^{-1}$.

The temperature dependence of the collision frequency follows immediately:

$$v_m = \frac{e^2 n_e}{m_e} \frac{LT}{K(T)} \tag{3.84}$$

Figure 3.7 shows the momentum transfer frequency $v_m(T)$ of Fe [17]. Only the contribution of the electron–phonon collisions has been included in Fig. 3.7 neglecting the contribution of the collisions between electrons and impurities. The application of Eq. (3.84) requires the law of WIEDEMANN–FRANZ to hold, i.e., collisions of the electrons must equally contribute to the thermal and electric resistance. This prerequisite is in general valid at high temperatures.

For a given material the index of absorption κ and the index of refraction n depend only on the electron collision frequency v_m and the wavelength of the radiation. Figure 3.8 shows n and κ as a function of the electron collision frequency in the plasma approximation.

Because the temperature dependence of the absorption and electric conductivity can be attributed to the same underlying physical mechanisms $A(T)$ can be determined from measurements of $\sigma(T)$ which can be measured easily and with higher precision. This means that if the electric conductivity $\sigma(T)$ and the absorptivity $A(\lambda; T)$ are known at $T = 20\,°$C, then $A(\lambda, T)$ can be calculated as a function of

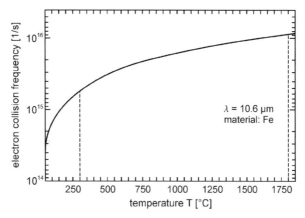

Fig. 3.7 Dependence of the electron collision frequency on the temperature [17]. Here only the contribution of the electron–phonon collisions has been considered while neglecting the collisions with atomic impurities

3 Absorption of Laser Radiation

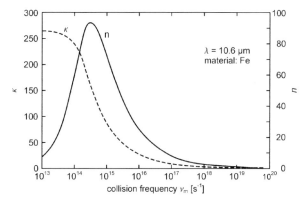

Fig. 3.8 Index of absorption and index of refraction as a function of the collision frequency [17]. κ is computed with the help of Eq. (3.82) and n with the help of Eq. (3.81) (plasma approximation). The density of the quasi free electrons n_e in Fe is estimated in [17] to be $\approx 6 \times 10^{29}$ m^{-3}. For the effective mass m^* the free mass m_e is used.

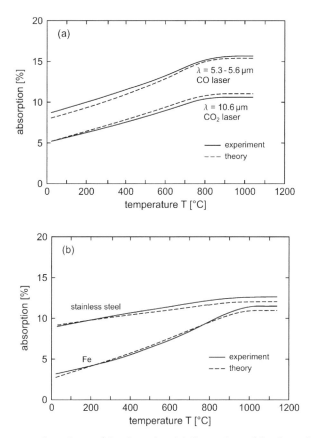

Fig. 3.9 Temperature dependence of the absorption. (**a**) Comparison of the absorption at different wavelengths of the laser radiation; (**b**) comparison of the absorption of different materials

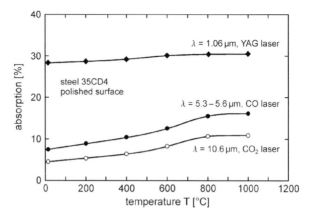

Fig. 3.10 Temperature dependence of the absorption at different wavelengths

temperature exploiting the temperature dependence of $\sigma(T)$. Figures 3.9 and 3.10 show the temperature dependence of the absorption for different materials and laser wavelengths.

3.7 Influence of the Surface Conditions

The roughness of surfaces has a significant influence on the spectral absorptivity of metals. In general the absorptivity increases with increasing roughness. Figure 3.11 shows the absorptivity of a polished and a grinded surface, respectively, of the same material. If the temperature approaches the melting point the higher absorptivity of the rough surface decreases to the value of the polished surface.

In treating the impact of the surface roughness quantitatively two limiting cases can be distinguished depending on whether the ratio of the mean quadratic height of the surface h and the wavelength λ is larger or smaller than 1. The mean quadratic height of the surface roughness is given by

$$h = \sqrt{\frac{1}{L} \int_0^L y^2 \, dx} \qquad (3.85)$$

with

λ – wavelength of the laser radiation.

In the limiting case $h/\lambda \ll 1$ the surface can be assumed to be ideal, then the well-known relations for reflection and refraction can be used. In the limit $h/\lambda \gg 1$ geometric optics can be applied [7, 1]. Geometric optics allows to estimate the

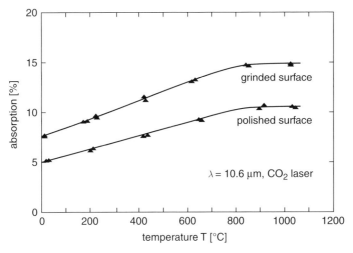

Fig. 3.11 Comparison of the absorptivity of a polished and grinded surface of the same material

absorptivity as a function of the angle of incidence of a rough surface by using the FRESNEL formulae for the computation of the absorption at the ideal material surfaces and by assuming a random distribution of surface inclinations. Scalar models have been developed by BECKMANN [2] and PORTEUS [9]. SACADURA [12] extended these models by considering polarization phenomena and the influence of a random distribution of the surface characteristics. SACADURA developed a model [11] in which the surface consists of a great number of V-formed microcavities with ideal flat walls and variable cone angles distributed according to a GAUIian distribution. With **Nd:YAG** laser radiation ($\lambda = 1.06\,\mu m$) the limiting case $h/\lambda \gg 1$ generally applies (except when using highly polished surfaces). In the case of **CO_2** laser radiation ($\lambda = 10.6\,\mu m$) this does not equally hold in general. SARI et al. [13] developed a model of plasma waves that is excited by surface irregularities.

In case of technical applications a measurement of the reflectivity is indispensable for capturing the influence of the surface structure. With intensities $I \ll I_p$ (I_p: process intensity) the absorption is determined by measuring the incident, diffusely, and directly reflected power (Fig. 3.12). Figure 3.13 shows the result of measurements for two different steel grades. With unpolarized radiation measured and computed values coincide quite well. The diffuse part amounts to about 10%, except for polished surfaces. With increasing roughness the direct reflection of sand-blasted targets decreases and the diffuse reflection increases.

In case of metals the reflectivity decreases with increasing temperature, which is attributed to the increasing electron–lattice collision frequency (Fig. 3.7). The increased reactivity at high temperatures can lead to irreversible modifications of the reflectivity by chemical reactions. Oxidation can take place when the hot surface is in contact with air. Oxide layers in general increase the absorptivity. This is of

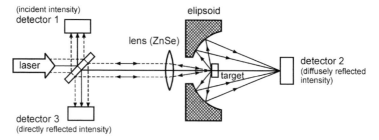

Fig. 3.12 Schematic of the experimental setup for measuring the direct and diffuse reflection

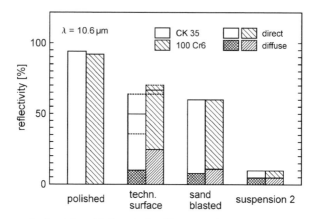

Fig. 3.13 Measured reflectivity with $I \ll I_p$ for different steel grades [17]

special importance when material is molten because the oxides are floating on the melt surface. On the other hand the original surface roughness does not influence the absorptivity any more because the laser beam hits the smooth melt surface.

References

1. A. Abdukadir, Spectral Directional Emittance of Roughned Metal Surfaces, 1973
2. P. Beckmann and A. Spizzichino, The Scattering of Electromagnetic Waves, Pergamon Press, New York, 1963
3. M. Born and E. Wolf, Principles of Optics, Cambridge University Press, 1999
4. P. Drude, Zur Elektronentheorie der Metalle, Annalen der Physik, Vol. 566, p. 1, 1900
5. R. E. Hummel, Optische Eigenschaften von Metallen und Legierungen, Vol. 22, Springer-Verlag, Berlin, 1971
6. J. D. Jackson, Classical Electrodynamics, John Wiley & Sons, New York, 1975
7. K. Kanayama and H. Baba, Directional Monochromatic Emittances of the Random Rough Surfaces of Metal and Nonmetals, Tr. J. S. M. E. Trans. 41, 1975
8. L. D. Landau and E. M. Lifschitz, Elektrodynamik der Kontinua, Vol. VIII, Akademie-Verlag, Berlin, 1984
9. J. O. Porteus, Relation Between Height Distribution of Rough Surface and the reflectance at normal, J. O. S. A., Vol. 53, p. 1394, 1963

10. S. Roberts, Optical Properties of Nickel and Tungsten and their Interpretation According to Drude's Formula, Phys. Rev., Vol. 114, p. 104, 1959
11. J. F. O. Sacadura, Influence de la rugosité sur le rayonnement thermique émis par les surfaces opaques, J. Heat Mass Transfer, Vol. 15, p. 1451, 1972
12. J. F. O. Sacadura, Modélisation et étude expérimentale du rayonnement thermique de sufaces métalliques microrugueuses, 1980
13. S. O. Sari, D. K. Cohen, and K. D. Scherkoske, Study of Surface Plasma-Wave Reflectance and Roughness Induced Scattering in Silver Foils, Phys. Rev. B, Vol. 21, p. 6, 1980
14. G. Stern and B. Gautier, Influence de la temperature et de l'etat de surface sur l'absorption spectrale $\lambda = 10.6\mu$ de differents alliages metalliques, Application au traitement thermique par laser CO_2 Rapport ISL R106/86, 1986
15. G. Stern and B. Gautier, Absorption d'un rayonnement laser CO_2 continu par differents alliages metalliques. Application au traitement thermique par laser, Clinqui journees europeennes optoelectroniques, 1985
16. T. J. Wieting and J. T. Schriempf, Free-electron theory and laser interactions with metals, Report of NRL Progress, 1972
17. K. Wissenbach, Härten mit CO_2-Laserstrahlung, Dissertation TH Darmstadt, 1985

Chapter 4
Energy Transport and Heat Conduction

Rolf Wester

The energy that is absorbed during laser material processing is mostly transformed to heat. The absorption within the specimen normally is not homogeneous. At metal surfaces the thickness of the layer in which the laser radiation is absorbed is only a fraction of the laser wavelength.[1] Because of the inhomogeneous absorption temperature gradients develop that cause heat fluxes. Heat conduction is thus a very important physical process during laser material processing.

4.1 Energy Transport Equation

In molten materials energy transport is not only due to heat conduction but also due to convection of melt. In many cases during laser material processing the workpiece and the laser beam are moved relative to each other.[2] Because of the relative movement of the laser beam and the workpiece the heat conduction problem formulated in the reference frame of the workpiece becomes time dependent, whereas in the reference frame of the laser beam it is stationary (assuming that the laser output is time independent) but with a conductive heat flow term added. The energy transport equation reads neglecting friction[3]

$$\frac{\partial \rho \, c_v \, T}{\partial t} + \vec{\nabla} \cdot (\rho \, c_v \, T \vec{v}) = -\vec{\nabla} \cdot \vec{q} + w(\vec{r}) \tag{4.1}$$

$$\vec{q} = -K \, \vec{\nabla} \, T \tag{4.2}$$

R. Wester (✉)
Fraunhofer-Institut für Lasertechnik, 5207 Aachen, Germany
e-mail: rolf.weister@ilt.fraunhofer.de

[1] See Sect. 3.3, p. 25.
[2] Generally the laser beam is fixed and the workpiece is moved, but with large or heavy pieces it is often more convenient to move the laser beam using a 'flying optic.'
[3] For a derivation see Appendix B or [3, 2].

with

T – temperature;
ρ – mass density;
c – specific heat;
w – absorbed energy per volume;
$\vec{v}$ – velocity;
$\vec{q}$ – heat flux density;
K – heat conductivity.

In case of virtually incompressible fluids the difference between c_v and c_p can be neglected in almost all cases of interest. The heat conductivity K is a macroscopic parameter that describes heat transport in a continuum. The microscopic mechanisms of heat conduction can be very different. Equation (4.1) is a parabolic differential equation that constitutes an initial boundary value problem, i.e., the problem is only well posed if correct initial and boundary conditions are prescribed (Fig. 4.1).

At a time instant t_0 the temperature has to be prescribed in the whole region of interest.[4] On the boundary of the region of interest for all times $t > t_0$ either the temperature, i.e., DIRICHLET boundary condition, or the normal derivative of the temperature, i.e., V. NEUMANN boundary condition, which is according to Eq. (4.2) equivalent to the heat flux, or a combination of both has to be prescribed.

The heat conduction equation is quasi-linear because ρ, c, and K can in general depend on the temperature but the highest derivative is linear.[5] The velocity $\vec{v}$ is in general space dependent. If the material is molten the heat conduction equation has in general to be solved together with the NAVIER–STOKES equation and the mass transport equation. The solution of the quasi-linear problem as well as the solution of the linear problem[6] in case of space-dependent flow velocity can in general only

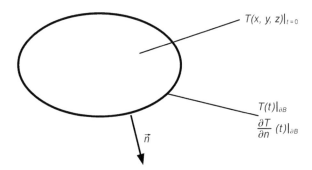

Fig. 4.1 Initial and boundary conditions for the solution of the heat conduction equation. B is the body and ∂B its boundary

---

[4] In the following $t_0 = 0$ will be assumed without loss of generality.
[5] A differential equation is only called nonlinear if the highest derivative is nonlinear.
[6] The heat conduction problem becomes linear when ρ, c, and K do not depend on the temperature.

be determined by means of numerical methods. With the simplifying assumption that ρ, c, and K and $\vec{v}$ have constant values, Eq. (4.1) together with Eq. (4.2) become a linear differential equation with constant coefficients. In this case general solutions can be found by means of the method of GREEN's functions, at least if simple boundary conditions can be applied.

4.2 Heat Conduction Mechanisms

Besides radiation heat transport, which is not going to be considered here, in solids two distinct mechanisms of heat transport occur. The first one is energy transport due to the heavy particles by lattice vibrations or phonons and the second one is due to free electrons. In non-conducting materials only phonon heat conduction occurs whereas in metals both phonon and electron heat conduction take place. In pure metals for all temperatures the electron heat conductivity is much larger than the phonon heat conductivity so that the latter can be neglected in metals. The highest value of heat conductivity can be observed in sapphire, a non-conductor. The strong binding forces between the atoms that are also responsible for the hardness of sapphire and, e.g., diamond cause the high phonon heat conductivity. The heat conductivity of sapphire is at its maximum,[7] $K = 2 \times 10^4$ WK^{-1} m^{-1} at 30 K [2]. In comparison the heat conductivity of Cu, which has the highest heat conductivity of all metals, is only $K = 5 \times 10^3$ WK^{-1} m^{-1} (at 20 K) [2]. The exact theoretical treatment of heat conductivity in solids is quite involved so that here only a simplified treatment based on elementary gas theory will be outlined.

From elementary gas theory the following expression for the heat conductivity of an ideal gas without excitation of inner degrees of freedom results:

$$K = \frac{1}{3} c v \lambda \tag{4.3}$$

with

c – heat capacity;
v – mean thermal velocity of the particles;
λ – mean free path between two consecutive collisions.

This expression can be used to approximate heat conduction in solids. In non-conductors $\vec{v}$ is the group velocity of the phonons and λ the mean free path of the phonons. The phonon mean free path of, e.g., quartz at $T = 0\,°C$ is 4 nm and that of NaCl is 2.3 nm [2]. In metals c is the heat capacity of the degenerate electron gas [2]:

---

[7] In general the heat conductivities have their maximal values at low temperatures; the heat capacity in Eq. (4.3) increases with increasing temperature (c vanishes at 0 K) whereas the mean free path decreases with the temperature.

$$C_{el} = \frac{\pi^2 n_e k_B T}{2 \varepsilon_F} \qquad (4.4)$$

$$\varepsilon_F = \frac{1}{2} m v_F^2 \qquad (4.5)$$

with

v_F – FERMI velocity;
ε_F – FERMI energy.

With the definition of the momentum transfer frequency

$$v_m = \frac{v_F}{\lambda} \qquad (4.6)$$

it follows that

$$K_{el} = \frac{\pi^2}{3} \frac{k_B^2}{e^2} \left[\frac{e^2 n_e}{m v_m} \right] T \qquad (4.7)$$

The expression in square brackets on the right-hand side is the electric conductivity, Eq. (3.80). With this the WIEDEMANN–FRANZ law follows (see Sect. 3.6):

$$\frac{K_{el}}{\sigma_{el}} = L T \qquad (4.8)$$

$$L = \frac{\pi^2 k_B^2}{3 e^2} \qquad (4.9)$$

with

L – LORENZ number.

L is not a real natural constant as Eq. (4.9) might suggest but varies slightly among different metals. The thermo-physical coefficients are functions of the temperature as well as the structure and phase of the material. The heat conductivity, e.g., is much smaller in the fluid phase than in the solid phase. When a phase change occurs the energy density is no longer a unique function of the temperature. In this case it is more convenient not to use the temperature but the enthalpy as dependent variable.

In the following some heat conduction problems with temperature-independent coefficients and constant convective velocity are presented. These problems are relevant for laser material processing but do not represent a complete treatment of the subject. A more complete treatment of heat conduction problems can, e.g., be found in CARSLAW and JAEGER [3].

4.3 Heat Conduction Equation with Constant Coefficients and the Method of GREEN's Functions

With the assumption of constant thermo-physical coefficients ρ, c, K and constant velocity $\vec{v}$ Eq. (4.1) becomes linear. Equations (4.1) and (4.2) then result in

$$\frac{\partial T}{\partial t} = \kappa \, \Delta T - \vec{v} \cdot \vec{\nabla} T + \frac{w}{\rho \, c} \tag{4.10}$$

$$\kappa = \frac{K}{\rho \, c} \tag{4.11}$$

with

κ – temperature conductivity.

If the source w in Eq. (4.10) is split into several sources then the resulting temperature distribution is given by the linear superposition of the temperature distributions that result from the individual sources. The method of GREEN's functions rests on this principle of superposition [6]. The GREEN's function is, despite a constant, the temperature distribution that results from a DIRAC delta source $\delta(t - t', \vec{r} - \vec{r}')$. The time t' and the location $\vec{r}'$ of the DIRAC delta source can have any value. The theory of GREEN's functions is not going to be outlined here, a comprehensive treatment of this topic can, e.g., be found in MORSE and FESHBACH [6] or in SOMMERFELD [8]. The GREEN's function of the heat conduction equation is a solution of Eq. (4.10) with a DIRAC delta source:

$$\frac{\partial G(\vec{r}, t | \vec{r}', t')}{\partial t} = \kappa \, \Delta \, G(\vec{r}, t | \vec{r}', t') - \vec{v} \cdot \vec{\nabla} \, G(\vec{r}, t | \vec{r}', t') + \delta(\vec{r} - \vec{r}', t - t') \tag{4.12}$$

The derivatives are taken with respect to t and $\vec{r}$. With the initial condition $T(t < t') = 0$ everywhere and the boundary condition $T(t, |\vec{r}| \to \infty) = 0$ the solution reads

$$G(\vec{r}, t | \vec{r}', t') = \frac{1}{[4 \, \pi \, \kappa \, (t - t')]^{3/2}} \, \exp\left(-\frac{[\vec{r} - (\vec{r}' + \vec{v} \, (t - t'))]^2}{4 \, \kappa \, (t - t')}\right) \tag{4.13}$$

The general solution for an arbitrary space- and time-dependent source that conforms to the same initial and boundary conditions as the GREEN's function Eq. (4.13) can be computed as a convolution of GREEN's function with the source distribution:

$$T(x, y, z, t) = \int_0^t \int_{-\infty}^{\infty} \int_{-\infty}^{\infty} \int_{-\infty}^{\infty} G(\vec{r}, t | \vec{r}', t') \, \frac{w(\vec{r}', t')}{\rho \, c} \, d^3 r' \, dt' \tag{4.14}$$

The GREEN's function in the time-independent case reads

$$G(\vec{r}, \vec{r}') = \frac{1}{4\pi\kappa} \frac{1}{|\vec{r}-\vec{r}'|} \exp\left(\frac{\vec{v}\cdot(\vec{r}-\vec{r}')}{2\kappa}\right) \exp\left(-\frac{|\vec{v}||\vec{r}-\vec{r}'|}{2\kappa}\right) \quad (4.15)$$

and the general solution is given by

$$T(x, y, z) = \int_{-\infty}^{\infty}\int_{-\infty}^{\infty}\int_{-\infty}^{\infty} G(\vec{r}, \vec{r}') \frac{w(\vec{r}')}{\rho c} d^3 r' \quad (4.16)$$

In Eqs. (4.14) and (4.16) the integration runs over the whole space. In most cases of practical importance the workpieces have finite dimensions. A first approximation to a finite workpiece is a workpiece with a plane surface that extends into half-space. The solution of the half-space problem can simply be derived from the solution of the whole space by mirroring the source at the plane that separates the two half-spaces. If the radiation is absorbed within a thin surface layer the absorption can be described by DIRAC delta functions (Fig. 4.2):

$$w(x, y, z) = I(x, y) \lim_{\epsilon \to 0} [\delta(z+\epsilon) + \delta(z-\epsilon)] \quad (4.17)$$
$$= I(x, y) \, 2\delta(z) \quad (4.18)$$

The factor of 2 must not be omitted when using Eqs. (4.14) or (4.16) for solving half-space problems, because these equations apply to the whole space problem.

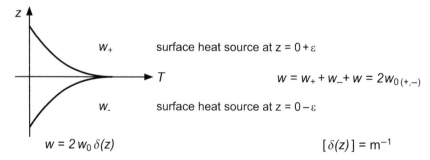

Fig. 4.2 Characterization of a surface source that applies to a half-space by a DIRAC delta volume source of the whole space

4.3.1 Point Source

In most cases of laser materials processing the laser radiation is focused onto the workpiece surface. The intensity that is absorbed at the surface is thus spatially concentrated. At distances that are large compared to the laser spot size on the surface the details of the laser intensity distribution is inessential. In that case the source can be assumed to be a point source. The point source is placed at the origin of the coordinate system and the power is switched from 0 to a constant value at $t = 0$:

$$w(x, y, z, t) = P_L(t)\, \delta(x)\, \delta(y)\, \delta(z) \tag{4.19}$$

P_L is the laser power that is absorbed in a half space. Without loss of generality the velocity is set to be

$$\vec{v} = \vec{e}_x\, v \tag{4.20}$$

With Eqs. (4.13) and (4.14) after integration over the space coordinates, which is trivial because of the DIRAC delta source distribution, the following temperature distribution results:

$$T(x, y, z, t) - T_\infty = \int_0^t \frac{2\, P_L(t')}{\rho\, c} \frac{1}{[4\, \pi\, \kappa\, (t - t')]^{3/2}} \exp\left(-\frac{[x - v\,(t - t')]^2 + y^2 + z^2}{4\, \kappa\, (t - t')}\right) dt' \tag{4.21}$$

The term T_∞ was added because GREEN's function Eq. (4.14) holds for zero temperature at infinity. The integrand is essentially the GREEN's function Eq. (4.14). A general analytical solution of the integral in Eq. (4.21) is not known. In the case of $v = 0$ one gets

$$T(r, t) - T_\infty = \int_0^t \frac{2\, P_L(t')}{\rho\, c} \frac{1}{[4\, \pi\, \kappa\, (t - t')]^{3/2}} \exp\left(-\frac{r^2}{4\, \kappa\, (t - t')}\right) dt' \tag{4.22}$$

$$r = \sqrt{x^2 + y^2 + z^2}$$

When the laser power at time $t = 0$ is switched from 0 to the constant value P_L, i.e.,

$$P_L(t) = P_L\, \Theta(t) \tag{4.23}$$

with

$\Theta(t)$ – HEAVISIDE step function,

the solution of the integral Eq. (4.22) reads[8]

$$T(r,t) - T_\infty = \frac{P_L}{2\pi \rho c \kappa} \frac{1}{r} \text{erfc}\left(\frac{r}{\sqrt{4\kappa t}}\right) \quad (4.24)$$

with

    erfc – error function.

In Fig. 4.3 isotherms are shown schematically.

In the time-independent case and with finite velocity the temperature distribution is given by [7]

$$T(x,y,z) - T_\infty = \frac{2 P_L}{\rho c} \frac{\exp\left(-\frac{|v|r - vx}{2\kappa}\right)}{4\pi \kappa r} \quad (4.25)$$

Figure 4.4 shows a schematic of the isotherms. The temperature decreases in positive x-direction because the velocity is assumed to be negative:

$$T \sim \frac{\exp\left(-\frac{|v|}{\kappa} x\right)}{r} \quad (4.26)$$

with

    $x > 0, \ y = 0, \ z = 0.$

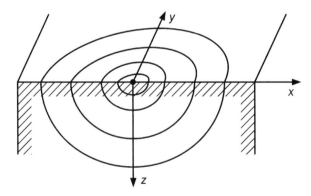

Fig. 4.3 Schematic of the isotherms for a non-moving point source

---

[8] See Appendix B.9.

4 Energy Transport and Heat Conduction

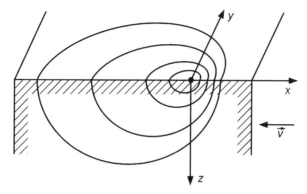

Fig. 4.4 Schematic of the isotherms of a moving source

In the negative x-direction one gets

$$T \sim \frac{1}{r} \quad (4.27)$$

with

$x < 0, \ y = 0, \ z = 0,$

and in the transverse direction at $x = 0$

$$T \sim \frac{\exp\left(-\frac{|v|\, r}{2\kappa}\right)}{r} \quad (4.28)$$

with

$x = 0, \ r = \sqrt{y^2 + z^2}.$

4.3.2 Line Source

The line source is a model for laser energy absorption and heat conduction in cases in which the coupling does not occur at the plane surface but within a keyhole that is initially not present but is formed by processes that are induced by the laser power, like evaporation. This is the case, e.g., during laser cutting and deep penetration welding. In case of cutting a cutting front is created that moves through the workpiece and in case of deep penetration welding a vapor-filled keyhole is formed that penetrates into the workpiece. In both cases the laser energy is absorbed at the walls of the cutting front and the keyhole, respectively. Assuming that there are no temperature gradients in the direction of the incident laser beam but only gradients normal to the keyhole or cutting front, respectively, the heat conduction can approximately be treated two-dimensionally. In the simplest case the keyhole is idealized to be a line. At least for distances sufficiently far away from the line source the temperature distribution can approximately be described by this model. The line source is given by

$$w = \frac{w'(t)}{\rho c} \delta(x)\, \delta(y) \tag{4.29}$$

$$w' = \frac{P_L}{s} \tag{4.30}$$

with

w' – absorbed power per length.

With Eqs. (4.15) and (4.16) the temperature distribution results:

$$T(x, y, t) - T_\infty = \int_0^t \frac{w'(t)}{\rho c} \frac{1}{4\pi\kappa(t-t')} \exp\left(-\frac{(x - v(t-t'))^2 + y^2}{4\kappa(t-t')}\right) dt' \tag{4.31}$$

In the case of zero velocity and constant power the temperature is given by[9]

$$T(r, t \geq 0) - T_\infty = \frac{w'}{\rho c} \frac{1}{4\pi\kappa} E_1\left(\frac{r^2}{4\kappa t}\right) \tag{4.32}$$

$$r^2 = x^2 + y^2$$

$$E_1(x) = \int_x^\infty \frac{e^{-\xi}}{\xi} d\xi \tag{4.33}$$

with

E_1 – exponential integral [1].

The isotherms are concentric circles centered at the origin ($x = 0, y = 0$). In the transient case and with finite velocity the temperature is given by [7]

$$T(x, y) - T_\infty = \frac{w'}{\rho c} \frac{1}{2\pi\kappa} K_0\left(\frac{|v|\, r}{2\kappa}\right) \exp\left(\frac{v x}{2\kappa}\right) \tag{4.34}$$

---

[9] See Appendix B.9.

with

K_0 – modified BESSEL function of the second kind.

Figure 4.5 shows isotherms computed using Eq. (4.34). The BESSEL function K_0 diverges at $r = 0$, i.e., the temperature becomes infinite at $r = 0$. For large arguments K_0 can be expanded [1]:

$$K_0(z) \simeq \sqrt{\frac{\pi}{2z}} \exp(-z) \left[1 - \frac{1}{8z} + \frac{3^2}{2!(8z)^2} - \frac{3^2 \cdot 5^2}{3!(8z)^3} - \cdots \right] \quad (4.35)$$

For large z the expression in parenthesis in Eq. (4.35) approximately equals 1. The asymptotic behavior for large positive values of x and with $v < 0$ is then given by

$$T \sim \frac{\exp\left(-\frac{|v|r}{2\kappa}\right) \exp\left(\frac{vx}{2\kappa}\right)}{\sqrt{r}} \quad (4.36)$$

with

$x > 0,\ y = 0$.

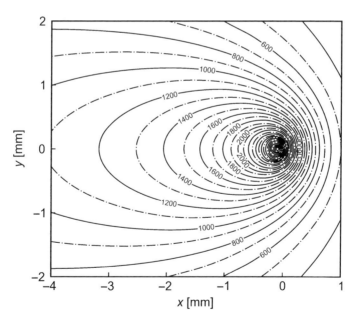

Fig. 4.5 Isotherms in case of a moving line source in Al computed with Eq. (4.34): $T_\infty = 300$ K, $v = 0.1$ m s^{-1}, $w' = 10^6$ W m^{-1}, $\kappa = 4.9 \times 10^{-5}$ m^{2}s^{-1}, $\rho = 2.7 \times 10^3$ kg m^{-3}, $c = 0.909$ kJ kg^{-1} K^{-1}

In negative x-direction it holds that

$$T \sim \frac{1}{\sqrt{r}} \tag{4.37}$$

with

$x < 0, \ y = 0,$

and in the transverse direction

$$T \sim \frac{\exp\left(-\frac{|v|}{2\kappa} y\right)}{\sqrt{r}} \tag{4.38}$$

with

$x = 0.$

4.3.3 Transversal Infinitely Extended Surface Source

In case of metal workpieces the laser radiation is in general absorbed within a thin surface layer, the thickness of which is much smaller than the lateral extent of the laser spot. There are cases in which also the heat penetration depth is small compared to the lateral extent of the laser beam at the surface, e.g., with large laser beam diameters and fast movement of the workpiece relative to the laser beam or with short laser pulses. In this case it can approximately be assumed that the source is transversal infinitely extended, i.e., the general three-dimensional problem reduces to a one-dimensional problem. The boundary region of the laser spot, however, cannot be described by this model (Fig. 4.6). The laser power density is given by

$$w = 2 I_L(t) \delta(z) \tag{4.39}$$

with

I_L – absorbed laser intensity.

The factor of 2 again has to be added because Eq. (4.14) holds for the whole space (see above). Inserting this into Eq. (4.14) and spatially integrating over the whole space and assuming zero velocity results in

$$T(z,t) - T_\infty = \int_0^t \frac{2 I_L(t')}{\rho c} \frac{1}{\sqrt{4\pi \kappa (t-t')}} \exp\left(-\frac{z^2}{4\kappa (t-t')}\right) dt' \tag{4.40}$$

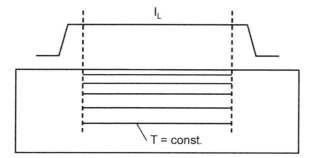

Fig. 4.6 One-dimensional heat conduction. When the diameter of the incident laser beam is large compared to the heat penetration depth approximately a one-dimensional heat conduction model can be adopted. The boundary regions of the laser beam, however, cannot be described by this simplified model

When the laser intensity at $t = 0$ is switched from 0 to the finite value I_L the solution of this integral reads[10]

$$T(z,t) - T_\infty = \frac{2 I_L}{\rho c} \sqrt{\frac{t}{\kappa}} \, \text{ierfc}\left(\frac{z}{\sqrt{4\kappa t}}\right) \tag{4.41}$$

$$\text{ierfc}(x) = \frac{1}{\sqrt{\pi}} e^{-x^2} - x \, \text{erfc}(x)$$

In the case of finite pulse duration t_L one gets

$$w = 2 I_L \, \delta(z) \, \Theta(t) \, \Theta(t_L - t) \tag{4.42}$$

and thus

$$T(z,t) - T_\infty = \frac{2 I_L(t)}{\rho c \sqrt{\kappa}} \times \left[\sqrt{t} \, \text{ierfc}\left(\frac{z}{\sqrt{4\kappa t}}\right) \right.$$
$$\left. - \Theta(t - t_L) \sqrt{t - t_L} \, \text{ierfc}\left(\frac{z}{\sqrt{4\kappa (t - t_L)}}\right) \right] \tag{4.43}$$

Figure 4.7 shows the temperature distribution at different instants of time.

In the stationary case with transversal infinitely extended surface source the differential equation can also be solved easily without using the method of GREEN's functions. The heat conduction equation without volume source reads

$$\frac{d^2 T}{dz^2} - \frac{v_z}{\kappa} \frac{dT}{dz} = 0 \tag{4.44}$$

---

[10] See Appendix B.9.

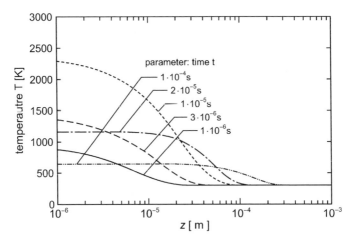

Fig. 4.7 Temperature as a function of the location at different time instances. Computed with Eq. (4.43). The material is Al, $I_L = 10^{10}\,\text{W m}^{-2}$, $t_L = 10^{-5}\,\text{s}$, $T_\infty = 300\,\text{K}$, $\kappa = 4.9 \times 10^{-5}\,\text{m}^2\,\text{s}^{-1}$, $\rho = 2.7 \times 10^3\,\text{kg m}^{-3}$, $c = 0.909\,\text{kJ kg}^{-1}\,\text{K}^{-1}$

At $z = 0$ the laser intensity I_L is coupled into the workpiece. In the above treated cases the absorption at the surface was treated using DIRAC delta volume sources. But this can also be modeled using appropriate boundary conditions: The heat flux at the surface has to equal the absorbed laser intensity (Figure 4.8a):

$$- K \left.\frac{dT}{dz}\right|_{z=0} = I_L \qquad (4.45)$$

with $v = -v_z$ and the boundary condition at infinity

$$T|_{z=\infty} = T_\infty \qquad (4.46)$$

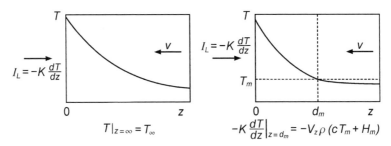

Fig. 4.8 Transversal infinitely extended surface source. Prescribed quantities: (**a**): heat flux at the surface, temperature at infinity. (**b**): Heat flux and temperature at d_m

the solution reads

$$T - T_\infty = \frac{I_L}{\rho c} \frac{1}{v_z} \exp\left(-\frac{v_z}{K} z\right) = \frac{I_L}{K} \frac{K}{v_z} \exp\left(-\frac{v_z}{K} z\right) \qquad (4.47)$$

If the material is assumed to be molten in the region $0 \le z \le d_m$ the heat flux at $z = d_m$ has to supply the energy for heating and melting the material:

$$-K \left.\frac{dT}{dz}\right|_{z=d_m} = -v_z \rho (c T_m + H_m) \qquad (4.48)$$

with

ρ – mass density
c – specific heat capacity
T_m – melting temperature
H_m – melting enthalpy.

The melting enthalpy is the energy that has to be supplied in order to transform a given amount of material at constant pressure and at the melting temperature T_m from the solid phase to the liquid phase. Because part of the heat energy is consumed by the phase transformation the heat flux at $z = d_m$ is discontinuous. With Eq. (4.48) and the further requirement

$$T|_{z=d_m} = T_m \qquad (4.49)$$

the temperature distribution is given by

$$T(z) = \left(T_m + \frac{H_m}{c}\right) \exp\left(\frac{v_z}{K} d_m\right) \exp\left(-\frac{v_z}{K} z\right) - \frac{H_m}{c} \qquad (4.50)$$

The intensity that has to be coupled into the workpiece surface at $z = 0$ is (Eq. (4.45))

$$I_L = v_z \rho (c T_m + H_m) \exp\left(\frac{v_z}{K} d_m\right) \qquad (4.51)$$

Figure 4.9 shows the temperature distribution for different velocity values and a melt film thickness of $d_m = 50\,\mu\text{m}$. This is a typical value, e.g., during laser cutting. The material constants are that of Al. The temperature at the surface increases with increasing velocity.

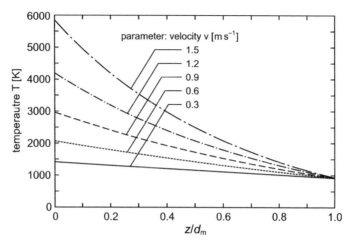

Fig. 4.9 Temperature as a function of the normalized coordinate z/d_m computed according to Eq. (4.50). The material is Al, $d_m = 50\,\mu\text{m}$, $v = 0.3, 0.6, 0.9, 1.2, 1.5\,\text{m s}^{-1}$, $\kappa = 4.9 \times 10^{-5}\,\text{m}^2\,\text{s}^{-1}$, $c = 0.909\,\text{kJ kg}^{-1}\,\text{K}^{-1}$, $T_m = 937\,\text{K}$, $H_m = 3.77 \times 10^2\,\text{kJ kg}^{-1}$

4.3.4 Transversal Infinitely Extended Volume Source

There are cases in which the penetration depth is finite and in which its impact on the energy absorption has to be taken into account. Especially plastics can have quite large absorption length. The absorbed power density is in this case

$$w(z,t) = \alpha\, I_L(t)\, \exp(-\alpha\, |z|) \tag{4.52}$$

with

α – absorption coefficient
α^{-1} – absorption length.

In the exponential function the absolute value of z is used because the absorbed energy density has to be symmetric with respect to the plane $z = 0$. With Eq. (4.14) and $v = 0$ the solution reads

$$T(z,t) - T_\infty = \frac{2}{\rho c} \int_0^t I_L(t') \frac{1}{\sqrt{4\pi \kappa (t-t')}} \exp\left(-\frac{z^2}{4\kappa(t-t')}\right) A(z,t,t',\alpha)\, dt' \tag{4.53}$$

$$A(z, t, t', \alpha) = \frac{\sqrt{\pi}}{2} \frac{\alpha a}{2} \left[\text{erfce} \left(\frac{\alpha a}{2} + \frac{z}{a} \right) + \text{erfce} \left(\frac{\alpha a}{2} - \frac{z}{a} \right) \right] \quad (4.54)$$

$$a = \sqrt{4\kappa(t-t')}$$

$$\text{erfce}(x) = \exp(x^2)\text{erfc}(x)$$

When the absorption coefficient tends to infinity then $A(z, t, t', \alpha) \to 1$ and the result of Eq. (4.40) is reproduced.

4.3.5 GAUSSian Intensity Distribution

With a GAUSSian intensity distribution at the surface the source is

$$w(x, y, z, t) = \frac{2 P_L}{\rho c} \frac{2}{\pi w_0^2} \exp\left(-\frac{2(x^2 + y^2)}{w_0^2} \right) \delta(z) \quad (4.55)$$

with

w_0 – beam waist.

Within the beam radius w 87% of the beam power is contained. The temperature distribution is given by

$$T(x, y, z, t) - T_\infty = \int_0^t \frac{2 P_L}{\rho c} \times \frac{1}{\sqrt{4\pi \kappa (t-t')}} \frac{1}{4\pi \kappa (t-t') + w_0^2/2}$$

$$\times \exp\left(-\frac{(x - v(t-t'))^2 + y^2}{4\kappa (t-t') + w_0^2/2} \right)$$

$$\exp\left(-\frac{z^2}{4\kappa (t-t')} \right) dt' \quad (4.56)$$

With $v = 0$ ($x = 0$, $y = 0$, $z = 0$), and constant laser power this simplifies to

$$T(0, 0, 0, t) = \frac{2 P_L}{\rho c} \frac{1}{\sqrt{2}\kappa \pi^{3/2} w_0} \arctan\left(\sqrt{\frac{8\kappa t}{w_0^2}} \right) \quad (4.57)$$

4.3.6 Finite Workpiece Thickness

In the case of finite workpiece thickness solutions of the heat conduction problem can be found using GREEN's function Eq. (4.15), which conforms to boundary conditions at infinity, by exploiting symmetries. This is achieved by choosing the source w distribution in such a way that the boundary conditions that have to be

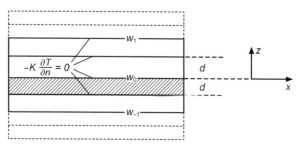

Fig. 4.10 Method of mirror sources for the calculation of temperature distributions in the case of finite workpiece thickness. The shaded region corresponds to the workpiece

imposed in the case of finite workpiece thickness are fulfilled automatically (method of mirror sources). Figure 4.10 shows this schematically. The heat fluxes normal to the workpiece surfaces vanish everywhere. This also holds where the laser beam hits the surface because w is modeled by two symmetric DIRAC delta functions, so that exactly at the interface at $z = 0$ the heat flux is zero. In order to get zero heat flux at the plane $z = -d$ another source is placed at the plane $z = -2d$ and so on. Similarly for $z + 2d$ and so on. For the exact solution an infinite number of delta sources need to be summed over:

$$w = \sum_{n=-\infty}^{\infty} w_n \qquad (4.58)$$

$$w_n = 2 P_L \delta(x - x_0) \delta(y - y_0) \delta(z - 2 n d) \qquad (4.59)$$

In the stationary case and with a point source (infinitely many point sources) the temperature is given by

$$T = \frac{2 P_L}{\rho c} \exp\left(-\frac{v x}{2 \kappa}\right) \sum_{n=-\infty}^{\infty} T_n \qquad (4.60)$$

$$T_n = \frac{\exp\left(-\frac{v}{2 \kappa} \sqrt{x^2 + y^2 + (z - 2 n d)^2}\right)}{4 \pi \kappa \sqrt{x^2 + y^2 + (z - 2 n d)^2}} \qquad (4.61)$$

For practical calculations it is sufficient to retain only a few terms in the sum.

4.4 Temperature-Dependent Thermo-physical Coefficients

In the preceding sections constant thermo-physical coefficients had been assumed. This is an idealization, the thermo-physical coefficients in general depend on the structure, phase, and temperature of the workpiece material. When phase transitions take place the energy density is no unique function of the temperature any more so that instead of using the temperature in the heat conduction equation the

4 Energy Transport and Heat Conduction

enthalpy is the relevant physical quantity. When the temperature dependence of the thermo-physical coefficients is taken into account the energy transport equation no longer is linear and analytical solutions are in general unavailable so that numerical algorithms have to be employed. The most frequently used methods are the **Finite Element Method**, the **Finite Difference**, and the **Finite Volume Method** [9].

Figure 4.11 shows the heat conductivity of Al, Cu and Fe. Cu shows the highest heat conductivity, Fe the smallest. The heat conductivities are highest at small temperatures. The sharp decrease of the Al heat conductivity occurs at the melting temperature. This is because the heat conductivity in the totally unordered liquid phase is much smaller than in the at least partially ordered solid phase.

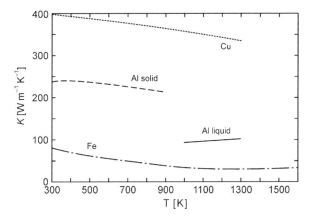

Fig. 4.11 Heat conductivity of Al, Cu, and Fe as a function of temperature [10]. The sharp decrease of the Al heat conductivity occurs at the melting temperature; the heat conductivity in the liquid phase is much smaller than in the solid phase

4.5 Heat Conduction in Case of Short Laser Pulse Durations

FOURIER's law Eq. (4.2) implies that the heat flux is proportional to the temperature gradient and that heat propagates with infinite velocity. But this is an approximation that can only be applied on large time scales. On shorter time scales heat conduction deviates from the results given by FOURIER's law. From non-equilibrium dynamics it follows that on short time scales FOURIER's law has to be augmented by a relaxation term [4]:

$$\tau_{\vec{q}} \frac{\partial \vec{q}}{\partial t} + \vec{q} = -K \vec{\nabla} T \tag{4.62}$$

This modification leads to a finite thermal propagation speed. The relaxation time constant τ_q and the heat conductivity K depend on the microscopic properties of the material like, e.g., collision frequencies of electrons and phonons. These quantities can be deduced from solutions of the BOLTZMANN equation [5]. The relaxation time constant τ_{qe} for the electron heat flux corresponds to the electron collision frequency and its value is in the (1–10) fs range. In case of very short pulses it also has to be considered that electron and phonon temperatures are not in equilibrium with each other. To treat this situation a two-temperature model has to be employed in which electron and phonon temperatures and their interaction have to be accounted for. If the laser beam diameter is large compared to the thermal penetration depth the problem can be treated one-dimensionally. With the assumption that the normal of the workpiece surface and the incident laser beam are parallel to the z-axis the heat conduction problem can be described by the following system of equations:

$$\frac{\partial U_e}{\partial t} + \frac{\partial q_e}{\partial z} = I(x, y, t) A \alpha \exp(-\alpha z) + h_{ex}(T_{ph} - T_e) \quad (4.63)$$

$$\frac{\partial U_{ph}}{\partial t} + \frac{\partial q_{ph}}{\partial z} = h_{ex}(T_e - T_{ph}) \quad (4.64)$$

$$\tau_{q_e} \frac{\partial q_e}{\partial t} + q_e = -K_e \frac{\partial T_e}{\partial z} \quad (4.65)$$

$$\tau_{q_{ph}} \frac{\partial q_{ph}}{\partial t} + q_{ph} = -K_{ph} \frac{\partial T_{ph}}{\partial z} \quad (4.66)$$

α is the volume absorption coefficient. The phonon heat conductivity can in general be neglected compared to the electron heat conductivity (for metals at least) so that these equations can be simplified. Because of the finite value of the relaxation time constants τ_q on short time scales wave-like phenomena can occur with the result that within short times heat energy can be transported from colder to warmer regions. To describe the interaction of fs laser pulses with solid materials the correction Eq. (4.62) and a two-temperature model has to be used whereas in the case of ps pulses the simpler FOURIER's law can be applied.

References

1. M. Abramowitz and I. A. Stegun, Pocketbook of Mathematical Functions, Harri Deutsch, Thun, 1984
2. E. Becker and W. Bürger, Kontinuumsmechanik, Teubner, Stuttgart, 1975
3. H. S. Carslaw and J. C. Jaeger, Conduction of Heat in Solids, Edition 2, Oxford University Press, 1959
4. D. Jou, J. Liebot, and J. Casas-Vazques, Irreversible-Thermodynamics Approach to Nonequilibrium Heat Fluctuations, Phys. Rev., Vol. A36, p. 508, 1982
5. V. Kostrykin, M. Nießen, J. Jandeleit, W. Schulz, and E. W. Kreutz, Picosecond Laser Pulses Induced Heat and Mass Transfer, SPIE Conference on High-Power Laser Ablation, Vol. SPIE 3343, 1998
6. P. M. Morse and H. Feshbach, Methods of Theoretical Physics, McGraw-Hill, New York, 1953
7. D. Rosenthal, The Theory of Moving Sources of Heat and its Application on Metall Treatments, Trans. ASME, Vol. 48, pp. 849–866, 1946
8. A. Sommerfeld, Theoretische Pysik, Vol. IV, Harri Deutsch, Thun, 1977
9. W. Törnig, Numerische Mathematik für Ingenieure und Physiker, Springer-Verlag, Berlin, 1979
10. R. C. Weast, Handbook of Chemistry and Physics, CRC Press, 1990

Chapter 5
Thermomechanics

Rolf Wester

5.1 Elastic Deformations

Within solid bodies there can be stresses due to, e.g., deformations or thermal loadings. The force per unit area $\vec{t}$ that acts within a solid body on an area element with normal $\vec{n}$ can be expressed using the stress tensor $\hat{T}$. In components the relation reads [1]

$$t_i = \tau_{ij} n_j \tag{5.1}$$

with

t_i – vector components of $\vec{t}$
n_j – vector components of $\vec{n}$
τ_{ij} – tensor components of $\hat{T}$

Here EINSTEIN's sum convention is applied, i.e., summation over double indices. The stress vector $\vec{t}$ does depend not only on the position but also on the normal $\vec{n}$, thus $\vec{t}$ is not a vector field, whereas $\hat{T}$ only depends on the position and because of this is a tensor field. The stress–strain relation for elastic materials is given by[1]

$$\tau_{ij} = \frac{E}{1+\nu}\left(\gamma_{ij} + \frac{\nu}{1-2\nu}\gamma_{kk}\delta_{ij}\right) \tag{5.2}$$

$$\gamma_{kk} = \gamma_{11} + \gamma_{22} + \gamma_{33}$$

with

γ_{ij} – components of GREEN's strain tensor
E – YOUNG's modulus
ν – POISSON's ratio.

R. Wester (✉)
Fraunhofer-Institut für Lasertechnik, 5207 Aachen, Germany
e-mail: rolf.weister@ilt.fraunhofer.de

[1] A derivation of this relation can be found in Appendix B.

The components of the strain tensor in geometric linear approximation, i.e., in case of sufficiently small deformations, are given by

$$\gamma_{ij} = \frac{1}{2}\left(\frac{\partial u_i}{\partial x_j} + \frac{\partial u_j}{\partial x_i}\right) \quad (5.3)$$

The u_i are the components of the displacement vectors of the material points of the solid with respect to a reference configuration. If the position of a material point in the reference configuration is given by $\vec{x}_0$ then the position of the material point after deformation is given by

$$\vec{x} = \vec{x}_0 + \vec{u} \quad (5.4)$$

The reversal of Eq. (5.2) reads

$$\gamma_{ij} = \frac{1+\nu}{E}\left(\tau_{ij} - \frac{\nu}{1+\nu}\tau_{kk}\delta_{ij}\right) \quad (5.5)$$
$$\tau_{kk} = \tau_{11} + \tau_{22} + \tau_{33}$$

5.1.1 Uniaxial Loading

It is assumed that a bar is loaded only longitudinally so that $\tau_{22} = \tau_{33} = 0$. Inserting into Eq. (5.5) yields

$$E = \frac{\tau_{11}}{\gamma_{11}} \quad (5.6)$$

$$\nu = -\frac{\gamma_{22}}{\gamma_{11}} = -\frac{\gamma_{33}}{\gamma_{11}} \quad (5.7)$$

This clarifies the meaning of YOUNG's modulus and of POISSON's ratio. YOUNG's modulus equals the slope of the stress–elongation curve in case of uniaxial loading. E has the dimension N/m^2, i.e., that of stress. POISSON's ratio ν is the ratio of the lateral to the longitudinal strain in case of uniaxial loading. ν is dimensionless and its value typically lies between 0.2 and 0.49. For most metals this value is about 0.3. POISSON's ratio is a measure for the compressibility of the material. With $\nu = 0.5$ the solid is incompressible, this means the solid volume does not change during arbitrary deformations. With $\nu = 0$ no lateral deformation occurs in case of only longitudinal loading.

5.1.2 Uniaxial Strain

Again it is assumed that a bar is loaded only longitudinally. Additionally the bar is assumed to be clamped laterally. Than $\gamma_{22} = 0$ and $\gamma_{33} = 0$. Inserting in Eq. (5.2) yields

5 Thermomechanics

$$\frac{\tau_{11}}{\gamma_{11}} = E \frac{(1-\nu)}{(1+\nu)(1-2\nu)} \tag{5.8}$$

$$\frac{\tau_{22}}{\tau_{11}} = \frac{\tau_{33}}{\tau_{11}} = \frac{\nu}{1-\nu} \tag{5.9}$$

5.2 Thermal Induced Stress

If a solid is heated it will expand in general. If the temperature is homogeneous and the expansion is not obstructed by external forces then there will be no stresses within the body (unless there already had been stresses prior to heating). An unobstructed, isotropic expansion due to thermal heating can be described by the spherical symmetric tensor:

$$\gamma_{ij}(\theta) = \alpha\,\theta\,\delta_{ij} \tag{5.10}$$

$$\theta = T - T_0$$

with

α – coefficient of thermal expansion ($\alpha_{steel} \approx 6 \times 10^{-6}/K$)
T – temperature
T_0 – temperature before heating.

Within the frame of a geometric and physical linear theory the total strain tensor is given by a superposition of this tensor with the tensor Eq. (5.3):

$$\gamma_{ij} = \frac{1+\nu}{E}\left(\tau_{ij} - \frac{\nu}{1+\nu}\tau_{kk}\delta_{ij}\right) + \alpha\,\theta\,\delta_{ij} \tag{5.11}$$

The inverse relation reads

$$\tau_{ij} = \frac{E}{1+\nu}\left(\gamma_{ij} + \frac{\nu}{1+\nu}\gamma_{kk}\delta_{ij}\right) - \frac{E}{1-2\nu}\alpha\,\theta\,\delta_{ij} \tag{5.12}$$

When the thermal induced expansion is totally obstructed, i.e., all components of the strain tensor are zero, the thermal induced stresses are given by

$$\tau_{ii} = -\frac{E}{1-2\nu}\alpha\,\theta \tag{5.13}$$

$$\tau_{ij,i\neq j} = 0$$

which is an isotropic stress state.

5.3 Plastic Deformation

In case of ideal elastic materials deformations are reversible when the prescribed loadings are released. Real solids show this kind of behavior only for sufficiently small deformations. When the elasticity limit is exceeded there remain deformations even when the loadings are released entirely. This kind of inelastic behavior can be time independent or time dependent. In the former case only the order of the loadings is of importance but not the rate of the loadings. This rate-independent behavior is called plastic deformation. In the time-dependent case the deformations not only depend on the loadings but also on their time rate. Examples are creeping and relaxation processes. Besides this cracks can occur. In the following plastic deformations are considered in more detail.

Elastic deformations are in general accompanied by volume changes whereas plastic deformations are generally volume preserving. A microscopic mechanism that underlies plastic deformation is the slipping of crystal layers along each other. But the shear stresses that are necessary for this to take place are about 100 times larger compared to those that have been observed experimentally during plastic deformation. This is due to lattice impurities that can translocate in the presence of shear stresses. Because only single atoms have to be displaced contrary to whole crystal layers the necessary forces are much lower. Impurities can be present before applying the loading or can be generated by the shear stresses, especially at the surfaces. Figure 5.1 schematically shows the strain–stress diagram of an ideal elastoplastic solid in the case of uniaxial loading. Up to the yield point the material is elastic, the slope of the diagram equals YOUNG's modulus E. From the yield point when the stress no longer increases, the solid is solely deformed plastically. When releasing the loading only the elastic contribution of the deformation is restored. This ideal behavior is not observed in general. The dislocation of impurities is partly obstructed by the plastic deformation, the material hardens. Because of this the stress increases even after reaching the yield point although with a smaller slope compared to the region of pure elastic deformation.

During elasto-plastic deformation the strain tensor can be separated into an elastic and a plastic contribution:

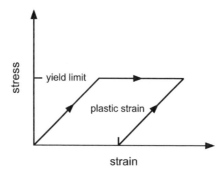

Fig. 5.1 Stress–strain diagram of an ideal elasto-plastic solid

$$\gamma_{ij} = \gamma_{ij}^e + \gamma_{ij}^p \qquad (5.14)$$

The elastic part of the strain is related to the stress according to the linear stress–strain relation Eq. (5.2). Plastic deformation takes place according to the VON MISES criterion if [2]

$$f(\tau_{ij}) = \sqrt{\frac{3}{2} S_{ij} S_{ij}} - Y(\lambda) = 0 \qquad (5.15)$$

$$S_{ij} = \tau_{ij} - \frac{1}{3} \tau_{kk} \delta_{ij} \qquad (5.16)$$

S_{ij} are the so-called stress deviators. They specify how much a stress state deviates from an isotropic stress state. Plastic deformations do not occur with isotropic stress states. When $f(\tau_{ij}) < 0$ the deformation is elastic, in case of $f(\tau_{ij}) = 0$ it is plastic. The plastic deformation takes place in such a way that the stress state always lies on the surface determined by $f(\tau_{ij}) = 0$ in the space of the stress components. λ is the integrated absolute value of the plastic deformation of the solid. The yield point depends on λ because according to the above-mentioned hardening of the material during plastic deformation the yield point can be shifted to higher values. The strain tensor is symmetric so that a principal axes transformation can always be accomplished. In the frame of principal axis all non-diagonal components of the stress tensor vanish. In that case one gets

$$f(\tau_{ij}) = \sqrt{\frac{1}{2}\left[(\tau_{11} - \tau_{22})^2 + (\tau_{11} - \tau_{33})^2 + (\tau_{22} - \tau_{33})^2\right]} \qquad (5.17)$$

5.3.1 Examples of Plastic Deformations

Plastic deformation of solid materials is used in many areas. All metal sheet bending procedures rely on irreversible plastic deformation. But plastic deformations can also occur undesirably. During welding part of the material is heated and even molten. During cooling and re-solidification the material suffers volume changes which induces stresses that are partly limited by plastic deformations and partly remain as residual stresses. During heat treatment volume changes can take place that in general are not restored entirely during cooling. In these situations residual stresses remain too.

References

1. E. Becker and W. Bürger, Kontinuumsmechanik, Teubner, Stuttgart, 1975
2. J. Betten, Kontinuumsmechanik, Springer, 1993

Chapter 6
Phase Transformations

Rolf Wester

There are many new metal alloys, ceramic materials, and plastics that have become important materials in many fields of industrial production. Despite this Fe-based materials still are widely used due to their versatile properties. Because of its low strength pure Fe is rarely used but Fe-C alloys allow to produce steels and cast Fe grades of a great variety of desired properties. Due to the steel making process there are besides C always other chemical elements present like Si, Mn, P, and S. Further chemical elements like Cr, Ni, Mo, V, W, etc., are often added to get distinct properties. Considering additionally heat treatment which is mostly used to change mechanical properties there is a great array of application areas for Fe-based materials.

There exist Fe-based materials of low up to the highest mechanical strengths (340–2000 N/mm^2), with excellent corrosion resistance, increased heat resistivity, good deformation properties even at low temperatures, high erosion resistance, good casting properties, weldability etc. Not all of these desired properties can be realized in a single steel grade, but in any grade there are a few of these properties that are accentuated. Principal distinctions are the following:

- steel or cast steel: these are Fe-C alloys that are ductile without any further treatment with less than 2.06% C
- cast Fe: these are Fe-C alloys with more than 2.06% C (mostly between 2.5 and 5% C) which cannot be forged but can only be casted

6.1 Fe-C Diagram

6.1.1 Pure Fe

Below the melting temperature Fe atoms compose crystals, though in the macroscopic realm no single crystals are formed but small crystallites. Fe exists in two

R. Wester (✉)
Fraunhofer-Institut für Lasertechnik, 5207 Aachen, Germany
e-mail: rolf.weister@ilt.fraunhofer.de

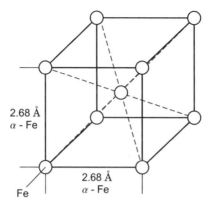

Fig. 6.1 Cubic body-centered α-Fe

distinct crystal structures. Below 911 °C the crystal structure is cubic body-centered (cbc) (Fig. 6.1). There are four Fe atoms at the four corners of the equilateral cube and a further one in the center of the cube. This form of Fe is called α-Fe or ferrite. Below the CURIE temperature of 769 °C Fe is ferromagnetic, above this temperature it is paramagnetic. The crystal structure does not change when crossing this temperature. Above 911 °C the crystal structure changes to be a cubic face-centered structure (cfc) (Fig. 6.2). In this case besides the four Fe atoms at the cube corners there are eight Fe atoms on the eight faces of the cube. This crystal structure possesses a smaller density compared to the cubic body-centered structure and is called γ-Fe or austenite. When getting at 1392 °C the cfc structure is transformed back to form again a cbc structure which is called δ-Fe or δ-ferrite. The reason for the transformations of the crystal structures at 911 and 1392 °C, respectively, is that the newly built crystal structures have lower total energy at the given temperatures. The transformations of the crystal structures are called allotropic transformations because the transformation does not take place simultaneously within the whole solid but in the transition region there exists a mixture of both crystal structures (Fig. 6.3). At 1536 °C the crystallites disintegrate and the material becomes liquid.

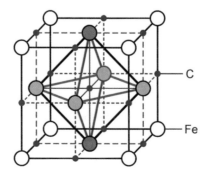

Fig. 6.2 Cubic face-centered γ-Fe. Included are locations where dissolved C can be placed

Commonly the transition temperatures are specially named. The names are of French origin:

Transition	During heating	During cooling
melt $\rightleftharpoons \delta$	A_c	A_r
$\delta \rightleftharpoons \gamma$	A_{c4}	A_{r4}
$\gamma \rightleftharpoons \alpha$	A_{c3}	A_{r3}
$\alpha_{param} \rightleftharpoons \alpha_{ferrom}$	A_{c2}	A_{r2}
austenite $\rightleftharpoons$ perlite	A_{c1}	A_{r1}

A : arrêt
    hold temperature
c : chaffage
    heating
r : refroidissement
    cooling

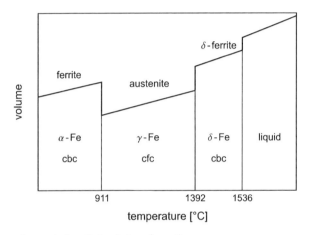

Fig. 6.3 Volume change during allotropic transformations

6.1.2 Fe-C Mixtures

When there is no pure Fe but a Fe-C mixture the situation becomes more involved. The C can exist in Fe alloys in different forms. First C can be dissolved in the Fe crystals (α, γ, δ) [2, 1]. Solutions with different C concentrations are called mixed crystals. The C solubility depends on the temperature and the crystal structure. The cfc lattice (γ-Fe) has a maximal C solubility of 2.01% whereas the cbc lattice (α-Fe and δ-Fe) has only a maximal solubility of 0.02%.

Besides being solved in the cfc or cbc Fe lattice C can exist in the form of Fe$_3$C, called Fe-carbide. In some cases C can also exist as an independent phase in the form of graphite. When the structure only consists of Fe-C mixed crystals and carbide the system is called metastable Fe-Fe$_3$C system. This structure is preferably formed during rapid cooling. In the Fe-C diagram (Fig. 6.4) the equilibrium lines are drawn through. The system consisting of Fe-C mixed crystals and C in the form of graphite is called stable Fe-C system. The equilibrium lines for this system are normally drawn dashed in the Fe-C diagram. In both cases the equilibrium lines are valid only for sufficiently slow temperature changes.

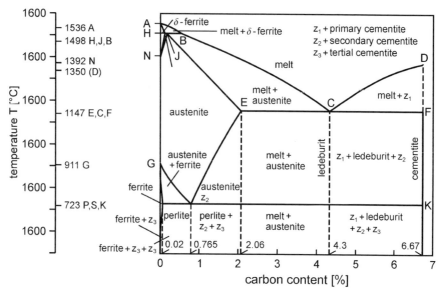

Fig. 6.4 Fe-C diagram

6.1.2.1 Metastable System

In the metastable Fe-Fe$_3$C system the C only exists in bound form up to a concentration-dependent temperature value. In the Fe-C diagram the separation line is given by the line QPSECD. Above this line C only exists in solved form.

Above the liquidus line ABCD the material is liquid. During cooling Fe-C mixed crystals begin to form along the line ABC and along the line CD primary crystallization (crystallization in the liquid phase) of Fe$_3$C sets in. After primary crystallization of γ mixed crystals or Fe$_3$C the melt solidifies when the line ECF (solidus line or eutectic line) is reached. Between liquidus and solidus lines a mixture of crystallites and melt coexists. Because the C concentrations of the primary melt and the crystallites differ the concentration of the melt is shifted toward the C concentration of the eutectic point C. The transition point C at which two phases are created out of a single one is called eutecticum. The word eutecticum has its origin in Greek and means the immediate and complete transition from liquid to solid state or in other words liquidus and solidus lines coincide. Fe-C mixed crystals and carbide crystallites form simultaneously which leads to a fine-grained and uniform structure that consists of two different kinds of crystals. An eutectic alloy always has the lowest possible melting temperature and the best casting properties. In the Fe-C diagram at a temperature of 1145 °C and a C concentration of 4.3% there is an eutecticum. The structure that forms at this point is called LEDEBURIT (after its discoverer A. LEDEBUR). Because of the transformation of the γ mixed crystals at 723 °C to perlite (see below) at normal temperature LEDEBURIT consists of perlite and carbide. At C concentrations between 2.06 and 4.3% and at normal temperatures there are besides LEDEBURIT perlite islands that originate from the primarily formed γ mixed crystals. At C concentrations above 4.3% there are extended carbide zones within the LEDEBURIT structure.

The Fe-C diagram shows a similar structure to the eutecticum at a C concentration of 0.83% C and a temperature of 723 °C. At this point the ferrite and carbide phases of the austenite structure segregate and form a structure called perlite. This "small eutecticum" is called eutectoid. Accordingly one speaks about eutectoidic, under eutectoidic (<0.83% C), and over eutectoidic (>0.83% C) steel.

During perlite formation the homogeneous γ mixed crystal (austenite, cfc lattice) with 0.8% C decays into a heterogeneous mixture of α-mixed crystals (0.02% C, ferrite, cbc lattice) and Fe_3C (6.67% C, carbide). This eutectoidic reaction is determined by diffusion which means that this process depends on the diffusion rate and thus on the temperature. The perlite formation itself happens by nucleation and crystal growth. Disk-like small carbide crystallites serve as nuclei that preferably are formed at austenite grain boundaries. In the immediate vicinity the C concentration in the austenite decreases and can easily be transformed to ferrite. Because of the reduced solubility of C in ferrite the C atoms are "squeezed out" of the ferrite regions and concentrate at the crystallite boundaries. This enhances the growth conditions for the C-rich Fe_3C that is subsequently formed at these sites. This is the reason for the disk-like structure of perlite. The disk width depends on the C diffusion rate which decreases with decreasing temperature. Thus the structure of perlite becomes much finer when the transformation takes place at low temperatures.

List of the most important Fe-C structures:

- austenite: cfc γ-Fe
- ferrite: cbc α-Fe
- carbide: Fe_3C
- perlite: ferrite + carbide in the form of striations created during eutectoid decay of the austenite at 723 °C and a C concentration of 0.8%
- LADEBURIT I: eutectic (γ-Fe + Fe_3C) is created from the melt at 1147 °C and a C concentration of 4.3%
- LADEBURIT II: γ-Fe decays to perlite at 723 °C

6.2 Hardening of Perlitic Structures

During hardening of perlite the steel has first to be transformed to the austenite structure. Because of the finite diffusion rates this make necessary a minimum holding time so that the C concentration in the carbide filaments can decrease from 6.67 to under 2.01%, the maximum C solubility of austenite (γ-Fe). The higher the temperature the higher the diffusion rate and thus the lower the necessary holding time. When the material is now cooled very rapidly no perlite can form any more because the C cannot diffuse out of the cubic face-centered ferrite (α-Fe). The C atoms that are captured in the Fe crystal deform the crystal which makes it very hard. This structure is called martensite and besides being very hard it is also difficult to deform plastically because the C obstructs the movement of impurities and dislocations.

6.2.1 C Diffusion

The atoms of a crystal lattice oscillate around their rest positions. Normally they do not leave their position within the crystal. At sufficiently high temperatures the energy of an atom can be large enough to be able to move within the crystal the atom is diffusing. The energy threshold for diffusion is called activation energy. Especially impurity atoms that are much smaller than the host atoms can diffuse, like, e.g., C in an Fe matrix. The diffusion flux density is according to FICK's first law given by

$$\vec{j}_D = -D\vec{\nabla}c \tag{6.1}$$

with

D – diffusion coefficient
c – concentration of the diffusing species.

The diffusion coefficient D depends on the temperature and the crystal structure:

$$D = D_0 \exp\left[-\frac{E_A}{k_B T}\right] \tag{6.2}$$

with

D_0 – frequency factor
E_A – activation energy.

The frequency factor D_0 describes the oscillation properties of the crystal lattice. The activation energy E_A is a measure of the energy threshold that atoms have to overcome in order to be able to move around. Figure 6.5 shows the diffusion coefficient of C in Fe as a function of the temperature. The temporal change of the concentration is according to FICK's second law:

$$\frac{\partial c}{\partial t} = \vec{\nabla} \cdot D\vec{\nabla}c \tag{6.3}$$

In the following it will be assumed that there exists a plane carbide disk in the region given by $-z_{\text{carbide}} < z < z_{\text{carbide}}$. With this assumption and the assumption that the diffusion coefficient does not depend on the C concentration, Eq. (6.3) can be simplified:

6 Phase Transformations

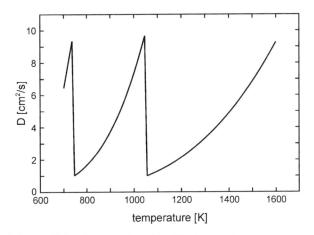

Fig. 6.5 C diffusion coefficient in Fe as a function of the temperature

$$\frac{\partial c}{\partial t} = D \frac{\partial^2 c}{\partial z^2} \tag{6.4}$$

The solution of Eq. (6.4) reads (see Appendix B.10)

$$c(z,t) - c_{\text{ferrite}} = \frac{1}{2}(c_{\text{carbide}} - c_{\text{ferrite}}) \left[\text{erf}\left(\frac{z + z_{\text{carbide}}}{\sqrt{4Dt}}\right) - \text{erf}\left(\frac{z - z_{\text{carbide}}}{\sqrt{4Dt}}\right) \right] \tag{6.5}$$

or in normalized form

$$c'(z',t') = \frac{c(z,t) - c_{\text{ferrite}}}{c_{\text{carbide}} - c_{\text{ferrite}}}$$
$$= \frac{1}{2}\left[\text{erf}\left(\frac{z'+1}{\sqrt{t'}}\right) - \text{erf}\left(\frac{z'-1}{\sqrt{t'}}\right)\right] \tag{6.6}$$
$$z' = \frac{z}{z_{\text{carbide}}}$$
$$t' = \sqrt{\frac{4Dt}{z_{\text{carbide}}^2}}$$

Figure 6.6 shows the normalized concentration c' as a function of the normalized coordinate z' at different normalized time instances t'.

The time constant τ for decay of the carbide disk follows from the requirement

$$c(0,\tau) = c_{\text{austenite}} = c_{\text{carbide}} \, \text{erf}\left(\frac{z_{\text{carbide}}}{\sqrt{4D\tau}}\right) \tag{6.7}$$

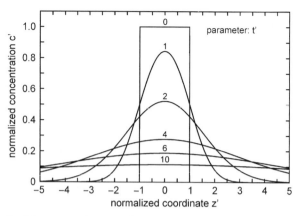

Fig. 6.6 Normalized C concentration according to Eq. (6.6) as a function of the normalized coordinate z' at different normalized time instances t'.

With $c_{\text{austenite}} = 2.1\%$ and $c_{\text{carbide}} = 6.67\%$ it follows that

$$\text{erf}\left(\frac{z_{\text{carbide}}}{\sqrt{4D\tau}}\right) \approx \frac{1}{3.2} \tag{6.8}$$

and further

$$\tau \approx 3\,\frac{z_{\text{carbide}}}{D} \tag{6.9}$$

This time constant determines the minimum holding time for complete transformation of the perlite structure to the austenite structure. On the other hand the material has to be cooled subsequently so rapidly that the austenite is totally transformed to martensite without forming unwanted crystal structures like perlite or beinite. This requires cooling rates of more than 10^3 K/s which implies short laser interaction times so that only that part of the workpiece is heated that has to be transformed whereas neighboring parts remain cold initially so that the heat can be transported rapidly from hot to cold regions. The necessary holding time can only be realized if the diffusion coefficient is sufficiently large which means that the transformation temperatures have to exceed by far the threshold of austenite formation. Equation (4.43) shows the time evolution of the temperature when applying a laser beam with a square pulse in time and in case that the assumption of one-dimensional heat conduction is justified.

References

1. Dieter Horstmann, Das Zustansdschaubild Eisen-Kohlenstoff, Stahleisen, 1985
2. Hans Stüdermann, Wärmebehandlung von Stahl, Carl-Hanser, 1967

Chapter 7
Melt Flow

Rolf Wester

During laser material processing the material is often heated so strongly that it melts. In the case of laser surface alloying the melt convection is forced by gradients of the surface tension which are utilized for the mixing of the basis material and the alloying material. During cutting and ablation the material is expelled partly or entirely as melt, during deep penetration welding the melt flows around the keyhole. A solution of the system of mass, momentum, and energy conservation equations is, as in the case of heat conduction, only possible when appropriate initial and boundary conditions are prescribed. Boundary and initial conditions strongly depend on the specific problem at hand. In the present presentation only simple models will be discussed that help to comprehend the main principles.

In the following the equations governing fluid flow are presented and the boundary conditions at the interfaces liquid–solid and liquid–ambient atmosphere are discussed. Then two simple solutions are treated in more detail. The first one is plane potential flow around a cylinder. This serves as a first approximate model of the melt flow around the keyhole during laser welding. The second one is boundary layer flow which occurs during cutting, drilling, and ablation.

7.1 Mass, Momentum, and Energy Conservation

Melt flow implies not only mass and momentum transport but energy transport as well. Thus in the mathematical treatment of melt flows an energy equation has to be included besides the equations for mass and momentum transport. A derivation of these three equations can be found in Appendix B. The melt is virtually incompressible, which means the density is almost constant. The mass conservation equation (Eq. B.50) reads in this case

$$\vec{\nabla} \vec{v} = 0 \qquad (7.1)$$

R. Wester (✉)
Fraunhofer-Institut für Lasertechnik, 5207 Aachen, Germany
e-mail: rolf.weister@ilt.fraunhofer.de

The momentum or NAVIER–STOKES equation in the case of incompressible fluids is given by

$$\frac{\partial \vec{v}}{\partial t} = -(\vec{v}\nabla)\vec{v} - \frac{\vec{\nabla} p}{\rho} + \nu \Delta \vec{v} + g \qquad (7.2)$$

with

p – pressure
η – kinematic viscosity
ρ – mass density
$\nu = \eta/\rho$ – dynamic viscosity
g – acceleration of gravity.

In the energy equation Eq. (B.107) the heat generation due to friction can be neglected. With Eq. (B.108) it follows that

$$\frac{\partial \rho c_v T}{\partial t} = \vec{\nabla} \cdot (K \vec{\nabla} T) - \vec{\nabla} \cdot (\vec{v}\ \rho c T) + w(\vec{r}, t) \qquad (7.3)$$

with

T – temperature
c – specific heat capacity
K – heat conductivity
w – absorbed power density.

7.2 Boundary Conditions

The flow problem is only well posed if appropriate boundary conditions are prescribed. Melts that exist during laser material interaction are bound by solid material as well as by the ambient atmosphere or vapor. At the melt–solid interface the boundary condition for the velocity field is

$$\vec{v}_L|_{\text{boundary}} = \vec{v}_S|_{\text{boundary}} \qquad (7.4)$$

The index L designates the liquid phase and the index S the solid material. This equation implies that the velocity is continuous at the boundary. The continuity of the normal component allows solid material to enter the molten region which is transformed to the liquid phase in doing so. The continuity of the tangential component ensures the no-slip condition that applies in case of a flow involving friction.

The boundary between melt and ambient atmosphere is a free surface. In general the contour of this surface is not fixed but adjusts itself according to the flow conditions and the ambient atmosphere. This has to be taken into account, the contour can, e.g., be determined iteratively. In the case of a fixed contour of the free surface the normal component of the velocity is

7 Melt Flow

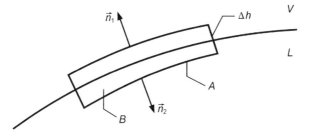

Fig. 7.1 Momentum conservation at a free boundary between liquid (L) and vapor (V)

$$(\vec{v} \cdot \vec{n})|_{\text{boundary}} = 0 \tag{7.5}$$

Equation 7.5 implies that there is no mass flow through the boundary (evaporation is neglected here). The momentum flux at the surface follows from Eq. (B.55) and Eq. (B.56). The integration volume B is shown in Fig. 7.1. If the height Δh becomes zero the volume integral in Eq. (B.56) vanishes. The surface integral in Eq. (B.56) together with Eqs. (B.61) and (B.62) results in

$$(\hat{T}\,\vec{n}_1)_V + (\hat{T}\,\vec{n}_2)_L = -(\hat{T}\,\vec{n}_2)_V + (\hat{T}\,\vec{n}_2)_L \tag{7.6}$$

$$\vec{n}_1 = -\vec{n}_2$$

The index V designates vapor or ambient atmosphere, respectively. $\hat{T}$ is the stress tensor. The two normal vectors point in opposite directions (Fig. 7.1). The melt behaves like a NEWTONian fluid but the NEWTONian fluid stress tensor $\hat{T}$ has to be augmented by contributions that act at the surface. The surface tension σ induces the capillary pressure and gradients of the surface tension induce tangential forces (shear forces) at the boundary. This is shown schematically in Fig. 7.2. The surface forces are given by

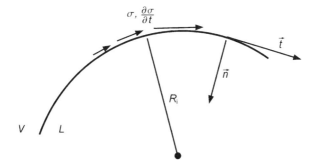

Fig. 7.2 Surface forces at boundaries between melt and gas. R_i is one of the two principal radii of curvature, σ is the surface tension, and $\frac{\partial \sigma}{\partial t}$ is the change of the surface tension in the tangential direction. $\vec{n}$ and $\vec{t}$ are unit vectors in normal and tangential directions, respectively

$$(\hat{T}\,\vec{n})_{\partial B} = p_K\,\vec{n} + \frac{d\sigma}{dt}\,\vec{t} \tag{7.7}$$

$$p_c = \sigma\left(\frac{1}{R_1} + \frac{1}{R_2}\right) \tag{7.8}$$

with

R_1, R_2 – principal radii of curvature of the boundary contour
p_c – capillary pressure.

The surface tension σ depends on the material at hand (small amounts of additives can have a significant impact) and the temperature. The interface shown in Fig. 7.2 is convex. In case of concave interfaces the radii of curvature have negative sign.

The contribution of the NEWTONian stress tensor follows from Eq. (B.82):

$$(\hat{T}\,\vec{n}) = -\bar{p}\,n_i + \eta\left(\frac{\partial v_i}{\partial x_j} + \frac{\partial v_j}{\partial x_i}\right) n_j \tag{7.9}$$

$$\bar{p} = p - \lambda\frac{\partial v_j}{\partial x_j} \tag{7.10}$$

with

η – dynamic viscosity
λ – volume viscosity.

EINSTEIN summation convention is adopted (summation over double indices). The volume viscosity acts like a change of the static pressure. This volume effect vanishes in case of incompressible fluids. The component of Eq. (7.9) that is parallel to the surface normal is given by

$$(\hat{T}\,\vec{n})\cdot\vec{n} = -\bar{p} + \eta\left(\frac{\partial v_i}{\partial x_j}\,n_i\,n_j + \frac{\partial v_j}{\partial x_i}\,n_i\,n_j\right) \tag{7.11}$$

$$= -\bar{p} + \eta\left(\frac{\partial}{\partial x_j}(v_i\,n_i)\,n_j + \frac{\partial}{\partial x_i}(v_j\,n_j)\,n_i\right)$$

With Eq. (7.5) the second term on the right side vanishes. With Eqs. (7.6), (7.7), and (7.11) it follows that

$$(\hat{T}\,\vec{n}_2)_V \cdot \vec{n}_2 = -\bar{p}_V \tag{7.12}$$
$$(\hat{T}\,\vec{n}_2)_L \cdot \vec{n}_2 = -p_L + p_C \tag{7.13}$$
$$\bar{p} = p_L + p_C \tag{7.14}$$

The component of Eq. (7.9) that is tangential to the boundary is given by

$$(\hat{T}\vec{n})\cdot\vec{t} = \eta\left(\frac{\partial v_i}{\partial x_j}n_j t_i + \frac{\partial v_j}{\partial x_i}n_j t_i\right) \tag{7.15}$$

$$= \eta\left(\frac{\partial}{\partial x_j}(v_i t_i)n_j + \frac{\partial}{\partial x_i}(v_j n_j)t_i\right)$$

The second term on the right side vanishes again because of Eq. (7.5). The first term is the derivative in normal direction of the tangential component of the velocity field. With this it follows that

$$(\hat{T}\vec{n})\cdot\vec{t} = \eta\frac{\partial v_t}{\partial n} \tag{7.16}$$

With Eq. (7.7)

$$(\hat{T}\vec{n}_2)_L\cdot\vec{t} = \eta_M\left(\frac{\partial v_t}{\partial n_2}\right)_L + \frac{d\sigma}{dt} \tag{7.17}$$

$$(\hat{T}\vec{n}_2)_V\cdot\vec{t} = \eta_G\left(\frac{\partial v_t}{\partial n_1}\right)_V \tag{7.18}$$

$$\eta_L\left(\frac{\partial v_t}{\partial n_2}\right)_L + \frac{d\sigma}{dt} = \eta_V\left(\frac{\partial v_t}{\partial n_1}\right)_V \tag{7.19}$$

The surface tension is a function of the surface temperature T and of the concentrations c of alloying additives or impurities:

$$\sigma = \sigma(T, c) \tag{7.20}$$

The gradient of the surface tension thus is given by

$$\frac{d\sigma}{dt} = \frac{\partial\sigma}{\partial T}\frac{\partial T}{\partial t} + \frac{\partial\sigma}{\partial c}\frac{\partial c}{\partial t} \tag{7.21}$$

The temperature dependence of the surface tension of molten metal is quite well known [2]. This does not equally hold for the concentration dependence. Additionally the concentrations of the substances that are present at the boundary are often unknown. Even in the case of additives the concentrations can differ from that in the bulk material because of selective vaporization. In the case of impurities like absorbed gases or oxides the uncertainty is even larger.

7.3 Plane Potential Flow

If in a frictionless flow the rotation of a velocity field is zero everywhere at a given time then according to HELMHOLTZ's vortex theorem it will be zero for all times [1]. With

$$\vec{\nabla}\times\vec{v} = 0 \tag{7.22}$$

the velocity field can be expressed as the gradient of a scalar potential field:

$$\vec{v} = \vec{\nabla}\phi \tag{7.23}$$

With

$$\vec{\nabla} \cdot \vec{v} = 0 \tag{7.24}$$

it follows that

$$\Delta \phi = 0 \tag{7.25}$$

The potential ϕ is thus a solution of LAPLACE's equation. For the solution of this kind of flow problem standard methods of electrostatics can be used (method of conformal mapping).

7.3.1 Source and Dipole Flow

The velocity field of a line source is shown in Fig. 7.3. With the definition of the source

$$Q := \iint \vec{v} \cdot \vec{n} \, dA \tag{7.26}$$

the velocity field is given by

$$\vec{v} = \frac{Q}{4\pi r^2} \vec{e}_r \tag{7.27}$$

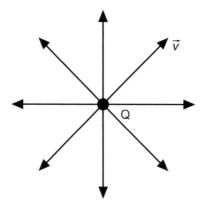

Fig. 7.3 Velocity field of a plane source flow

7 Melt Flow

With this the potential is

$$\phi = -\frac{Q}{4\pi r} \qquad (7.28)$$

Figure 7.4 schematically shows the velocity field of a plane source flow. If the sign of Q is negative Q is not a source but a drain. The superposition of the velocity field of a source and a drain results in a dipole flow whose potential is given by

$$\phi = -\frac{Q}{4\pi}\left(\frac{1}{r} - \frac{1}{r'}\right) \qquad (7.29)$$

With the law of cosines it follows that

$$r'^2 = r^2 - 2r\,\Delta x\,\cos\theta + \Delta x^2 \qquad (7.30)$$

The distance Δx between source and drain is assumed to be small compared to r and r', respectively. Then it follows that

$$r' = r\left(1 - \frac{\Delta x}{r}\cos\theta\right) \qquad (7.31)$$

and

$$\frac{1}{r'} = \frac{1}{r}\left(1 + \frac{\Delta x}{r}\cos\theta\right) + O\left(\frac{\Delta x^2}{r^2}\right) \qquad (7.32)$$

Thus with $\Delta x \to 0$ while leaving the dipole moment M constant the potential Eq. (7.29) approximately is given by

$$\phi = \frac{M}{4\pi r^2}\cos\theta \qquad (7.33)$$

With

$$\cos\theta = \frac{x}{r} \qquad (7.34)$$

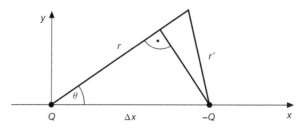

Fig. 7.4 Dipole flow

this results in

$$\phi = \frac{M}{4\pi r^3} x \qquad (7.35)$$

7.3.2 Flow Around a Cylinder

The superposition of the dipole flow Eq. (7.35) and a plane flow which has the potential

$$\phi = v_\infty x \qquad (7.36)$$

with the constant velocity v_∞ results in the potential

$$\phi = v_\infty x \left(1 + \frac{a^3}{2r^3}\right) \qquad (7.37)$$

$$\frac{M}{4\pi r^3 v_\infty} = \frac{a^3}{2} \qquad (7.38)$$

With Eq. (7.23) the velocity components are given by

$$v_x = v_\infty x \left(1 + \frac{a^3}{2r^3} - \frac{3}{2}\frac{a^3 x^2}{r^5}\right) \qquad (7.39)$$

$$v_y = -v_\infty \frac{3}{2}\frac{a^3 x y}{r^5} \qquad (7.40)$$

At $r = a$ it follows that

$$v_x = \frac{3}{2} v_\infty \left(1 - \frac{x^2}{a^2}\right) \qquad (7.41)$$

$$v_y = -\frac{3}{2} v_\infty \frac{x y}{a^2} \qquad (7.42)$$

The scalar product of the velocity and the position vector $\vec{r}$ at $|\vec{r}| = a$ vanishes:

$$(\vec{v} \cdot \vec{r})_{|\vec{r}|=a} = (v_x x + v_y y)_{|\vec{r}|=a} = 0 \qquad (7.43)$$

The position vector is parallel to the cylinder surface normal. Equation (7.43) thus implies that the component of the velocity normal to the cylinder surface at $|r| = a$ vanishes and thus the velocity field given by Eqs. (7.39) and (7.40) constitutes a flow field around a cylinder with radius a. Figure 7.5 shows streamlines of the flow field Eqs. (7.39) and (7.40). At the points $(x = -a, y = 0)$ and $(x = a, y = 0)$ the velocity is zero. These two points are the front and rear stagnation points,

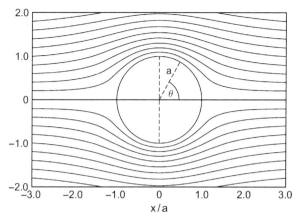

Fig. 7.5 Streamlines of a frictionless, incompressible flow around a cylinder

respectively. The velocity components as a function of the angle θ at the cylinder surface are given by

$$v_x = \frac{3}{2} v_\infty \sin^2 \theta \tag{7.44}$$

$$v_y = -\frac{3}{2} v_\infty \cos\theta \sin\theta \tag{7.45}$$

The absolute value of the velocity at the cylinder surface is thus

$$|v| = \frac{3}{2} v_\infty \sin\theta \tag{7.46}$$

According to BERNOULLI's law [1] the following relation holds along a streamline in case of a frictionless and incompressible fluid:

$$p + \frac{\rho}{2} |\vec{v}|^2 = \text{const.} \tag{7.47}$$

i.e., the sum of the static pressure and the dynamic pressure is constant along a streamline. Along the streamline that touches the cylinder surface it thus holds

$$p_\infty + \frac{\rho}{2} v_\infty^2 = p(\theta) + \frac{9}{4} \frac{\rho}{2} v_\infty^2 \sin^2\theta \tag{7.48}$$

At the stagnation points $\theta = 0$ and $\theta = \pi$, respectively, the velocity vanishes and the static pressure becomes

$$p = p_\infty + \frac{\rho}{2} v_\infty^2 \tag{7.49}$$

The static pressure is minimal at $\theta = \pi/2$:

$$p = p_\infty - \frac{5}{4} \frac{\rho}{2} v_\infty^2 \qquad (7.50)$$

while the velocity has its maximum value at this point which amounts to $3/2\, v_\infty$. The force that acts on a unit area of the cylinder surface is given by

$$d\vec{f} = -\vec{e}_r\, p(\theta)\, a\, d\theta\, dz \qquad (7.51)$$
$$\vec{e}_r = \cos\theta\, \vec{e}_x + \sin\theta\, \vec{e}_y \qquad (7.52)$$

Integration over the total cylinder surface, i.e., from $\theta = 0$ to $\theta = \pi$, yields the total force that is exerted on the cylinder by the flow. The result shows that the total force vanishes which is in contradiction to experience. The reason for this is the idealization of a frictionless flow that was presupposed in the above derivation. Real flows are not frictionless which especially shows at boundaries. At flow–solid boundaries the tangential flow component vanishes at the boundary due to friction (no-slip condition). The tangential flow component increases rapidly within a small layer away from the boundary up to (or nearly to) the value that a frictionless flow would have. Outside the boundary layer the flow can approximately be treated as being frictionless if the contour of the solid body is assumed to include the boundary layer. Then the methods for treating plane frictionless potential flows can be applied, while the boundary layers are treated within the frame of the boundary layer theory. But this approximate solution has its limitations. The boundary layers can detach from the surface and turbulences can emerge which cause forces on the body which is passed by the flow. In case of melt–gas (or melt–vapor) interfaces, like during laser welding, there is no no-slip condition of the melt. On the other hand during laser welding only a small part of the material around the keyhole is liquid whereas in the above-discussed case of flow around a cylinder the total region outside the cylinder was assumed to be liquid. The potential flow thus is only an approximation of the real melt flow around a keyhole.

7.4 Laminar Boundary Layers

When a solid is passed by a fluid then due to frictional forces boundary layers evolve. Within the boundary layer thickness the value of the tangential velocity component that is zero at the boundary increases up to (or nearly to) the value that the flow velocity would have without friction. Figure 7.6 schematically shows the boundary layer during flow around a plate.

The boundary layer theory rests on the assumption that the velocity component parallel to the surface is much larger than the component perpendicular to the surface and that the pressure gradient perpendicular to the surface can be neglected. The latter assumption makes it possible to use for the pressure gradient in x-direction the value outside the boundary. In the stationary case the x-component of the NAVIER–

7 Melt Flow

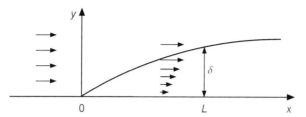

Fig. 7.6 Boundary layer at a plane plate. The plate begins at $x = 0$

STOKES equation reads

$$v_x \frac{\partial v_x}{\partial x} + v_y \frac{\partial v_x}{\partial y} = -\frac{1}{\rho}\frac{\partial p}{\partial x} + \nu \left(\frac{\partial^2 v_x}{\partial x^2} + \frac{\partial^2 v_x}{\partial y^2} \right) \quad (7.53)$$

with

ν – kinematic viscosity.

The boundary layer approximations are given by

$$|v_x| \gg |v_y| \quad (7.54)$$

$$\left.\frac{\partial p}{\partial x}\right|_{y=0} \simeq \left.\frac{\partial p}{\partial x}\right|_{y=\delta} \quad (7.55)$$

$$\frac{\partial v_x}{\partial x} = O\left(\frac{v_x}{L}\right) \quad (7.56)$$

$$\frac{\partial^2 v_x}{\partial x^2} = O\left(\frac{v_x}{L^2}\right) \quad (7.57)$$

$$\frac{\partial v_x}{\partial y} = O\left(\frac{v_x}{\delta}\right) \quad (7.58)$$

$$\frac{\partial^2 v_x}{\partial y^2} = O\left(\frac{v_x}{\delta^2}\right) \quad (7.59)$$

L is a characteristic length describing the extension of the boundary layer. The term

$$\frac{\partial^2 v_x}{\partial x^2} \quad (7.60)$$

can be neglected in the NAVIER–STOKES equation. On the other hand the term

$$v_x \frac{\partial v_x}{\partial x} \quad (7.61)$$

cannot be neglected because it can have, according to Eq. (7.59), the same order of magnitude as

$$v_y \frac{\partial v_x}{\partial y} \qquad (7.62)$$

With this the boundary layer approximation of the NAVIER–STOKES equation neglecting the lifting force is given by

$$v_x \frac{\partial v_x}{\partial x} + v_y \frac{\partial v_x}{\partial y} = -\frac{1}{\rho} \frac{\partial p}{\partial x} + \nu \frac{\partial^2 v_x}{\partial y^2} \qquad (7.63)$$

The continuity equation reads

$$\frac{\partial v_x}{\partial x} + \frac{\partial v_y}{\partial y} = 0 \qquad (7.64)$$

If the pressure gradient is known these two equations are sufficient to solve the problem. As mentioned above the pressure gradient can approximately be determined using the solution of the flow problem outside the boundary layer with friction neglected. The pressure gradients that occur during laser material processing are either determined by gas flows that are used to expel the melt or by the ablation pressure that emerges during evaporation of the material.

The boundary layer flows that exist during, e.g., ablation and cutting are slightly different from the one shown in Fig. 7.6. Figure 7.7 shows this schematically in case of laser cutting. The laser beam propagates from top to down and is partially absorbed at the melt film surface. The absorbed laser intensity is used for heating and melting of the material. The workpiece is moved to the left and the gas flow

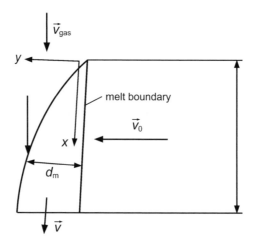

Fig. 7.7 Boundary layer flow during laser material processing

coming from the top exerts pressure and shear forces onto the melt film that lead to the expulsion of the melt. The material thus flows into the boundary layer at $y = 0$ (and melts in doing so) and leaves the boundary layer region at $x = d$. The mean velocity at $x = d$ can be determined using the workpiece thickness d, the melt film thickness d_m, and the feed rate:

$$\bar{v}_x(x = d) = \frac{d}{d_m} v_0 \qquad (7.65)$$

It is assumed that no melt is flowing through the melt surface. In this case the following kinematic condition has to be met by the melt thickness d_m [3]:

$$\frac{d\, d_m}{dx} = \left|\frac{v_y}{v_x}\right|_{y=d_m} \qquad (7.66)$$

$$d_m(0) = 0 \qquad (7.67)$$

The REYNOLDS number appropriate for the problem at hand is

$$\text{Re} = \frac{\rho\, v_0\, d_m}{\eta} \qquad (7.68)$$

The REYNOLDS number is a dimensionless number describing the ratio of the inertial forces to the friction forces within a flow. In the case of small REYNOLDS numbers ($\text{Re} \ll 1$) the inertial term can be neglected compared to the friction term in Eq. (7.63) whereas in the case of large REYNOLDS numbers ($\text{Re} \gg 1$) the friction term can be neglected.

7.4.1 Friction-Dominated Boundary Layer Flow

With $\text{Re} \ll 1$ and neglecting the inertial term in Eq. (7.63) it follows that

$$\frac{\partial p}{\partial x} = \eta \frac{\partial^2 v_x}{\partial y^2} \qquad (7.69)$$

The boundary condition at the melt surface is

$$-\eta \left.\frac{\partial v_x}{\partial y}\right|_{y=d} = \tau \qquad (7.70)$$

The shear stress τ that is exerted onto the melt surface by the gas jet and the pressure gradient is of the same order of magnitude [4]:

$$\tau = \sqrt{\frac{\eta_G \, \rho_G \, v_G^3}{d}} \tag{7.71}$$

$$\frac{\partial p}{\partial x} \simeq \frac{p_G}{d} \tag{7.72}$$

With the ansatz for the velocity v_x

$$v_x = v_{x_0} + c_1 \, y + c_2 \, y^2 \tag{7.73}$$

the following solution results:

$$v_x = v_{x_0} - \frac{1}{\eta} \left(\tau \, y + \frac{\partial p}{\partial x} \left(d \, y - \frac{1}{2} y^2 \right) \right) \tag{7.74}$$

Taking the derivative with respect to x yields

$$\frac{\partial v_y}{\partial x} = -\frac{1}{\eta} \frac{\partial v_y}{\partial x} \, y \, \frac{d \, d_m}{dx} \tag{7.75}$$

With the continuity equation

$$\frac{\partial v_y}{\partial y} = -\frac{\partial v_x}{\partial x} \tag{7.76}$$

and Eq. (7.75) it follows for the velocity component v_y that

$$v_y = v_{y_0} + \frac{1}{\eta} \frac{\partial p}{\partial x} \, y \, \frac{d \, d_m}{dx} \, \frac{y^2}{2} \tag{7.77}$$

Using the kinematic condition Eqs. (7.66) and (7.74) as well as (7.77) the following determining equation for the boundary layer thickness d_m results:

$$\frac{d_m^3}{d_p^3} + \frac{d_m^2}{d_\tau^2} = 1 \tag{7.78}$$

$$d_\tau^2 = \frac{2 \, \eta \, v_0 \, x}{\tau} \tag{7.79}$$

$$d_p^3 = \frac{3 \, \eta \, v_0 \, x}{-\dfrac{\partial p}{\partial x}} \tag{7.80}$$

With vanishing pressure gradient the boundary layer thickness is given by d_τ whereas with vanishing shear stress it is given by d_p.

7.4.2 Inertia-Dominated Boundary Layer Flow

If the REYNOLDS number is large compared to 1 (Re $\gg$ 1) the friction in Eq. (7.63) can be neglected. Additionally it will be assumed that

$$\frac{\partial v_y}{\partial y} = 0 \qquad (7.81)$$

i.e., the melt is not accelerated in the y-direction. Hence from the continuity equation it also follows that

$$\frac{\partial v_x}{\partial x} = 0 \qquad (7.82)$$

With this the momentum conservation Eq. (7.63) becomes

$$\frac{\partial v_x}{\partial y} = -\frac{1}{v_y \rho} \frac{\partial p}{\partial x} \qquad (7.83)$$

The solution of this equation with $v_y = v_0$ is given by

$$v_x = v_{x_0} - \frac{1}{v_0 \rho} \frac{\partial p}{\partial x} y \qquad (7.84)$$

Using the kinematic condition Eq. (7.66) the boundary layer thickness d_m follows:

$$d_m = \sqrt{\frac{2 \rho x}{-\frac{\partial p}{\partial x}} v_0} \qquad (7.85)$$

References

1. E. Becker and W. Bürger, Kontinuumsmechanik, Teubner, Stuttgart, 1975
2. A. E. Brandes, Smithells Metals Reference Book, Butterworths, London, 1983
3. M. Vicanek, G. Simon, H. M. Urbassek, and I. Decker, Hydrodynamical instability of melt flow in laser cutting, Journal of Physics D: Applied Physics, Vol. 20, pp. 140–145, 1987
4. M. Vicanek and G. Simon, Momentum and heat transfer of an inert gas jet to the melt in laser cutting, Journal of Physics D: Applied Physics, Vol. 20, pp. 1191–1196, 1987

Chapter 8
Laser-Induced Vaporization

Rolf Wester

There are laser material treatment applications that are associated with the vaporization of material. Examples are laser drilling, deep penetration laser welding, sublimation cutting and, in some cases, ablation.

The starting point of a physical description and mathematical treatment of evaporation processes is evaporation in the case of vapor-melt phase equilibrium. The equilibrium vapor pressure is given by the CLAUSIUS–CLAPEYRON equation. If thermodynamic equilibrium exists, there are as many particles leaving the melt surface, i.e., evaporate, as there are vapor particles that hit the melt surface and re-condensate. In a state of thermodynamic equilibrium, there is no net flow of particles in either direction. This means that during material ablation, when there is a net flow of particles, the system is necessarily in a non-equilibrium state.

In what follows, approximate expressions for the net particle fluxes as a function of temperature during non-equilibrium evaporation are derived, starting, with the flux of particles that leave the melt surface in the case of thermodynamic equilibrium vaporization. In doing so, it is assumed that the particles within the melt stay in a state of thermodynamic equilibrium. Using the derived vaporization rate as a function of temperature and the energy conservation equation, the evaporation rate as a function of the laser intensity is determined.

8.1 Vapor Pressure in Thermodynamic Equilibrium

A vessel partially filled with melt is assumed to be in a heat bath at the temperature T. The space above the melt is filled with vapor. In a state of thermodynamic equilibrium it holds [7] that

$$T_V = T_M = T \tag{8.1}$$

$$p_V = p_M = p \tag{8.2}$$

$$g_V = g_M \tag{8.3}$$

R. Wester (✉)
Fraunhofer-Institut für Lasertechnik, 5207 Aachen, Germany
e-mail: rolf.weister@ilt.fraunhofer.de

with

> T – temperature
> p – pressure
> g – free enthalpy per atom.

The saturation vapor pressure is a function of the temperature. The derivative of the saturation vapor pressure p_{SV} with respect to the temperature is given by the equation of CLAUSIUS–CLAPEYRON[1] [1, 4, 7]:

$$\frac{dp_{SV}}{dT} = \frac{H_V(T)}{(v_V - v_C)\,T} \tag{8.4}$$

with:

> v_V – specific volume in the vapor phase
> v_C – specific volume in the condensed phase
> H_V – evaporation enthalpy.

The evaporation enthalpy is the energy that has to be supplied in order to vaporize a given amount of material. The evaporation enthalpy is a function of the temperature.[2] The vapor is treated as ideal gas. With

$$v_V = \frac{1}{\rho} = \frac{R\,T}{p} \tag{8.5}$$

and Eq. (8.4), it follows:

$$\frac{dp_{SV}}{dT} = \frac{H_V(T)}{R\,T^2} p_{SV} \tag{8.6}$$

with

> ρ – mass density
> R – ideal gas constant.

Integrating Eq. (8.6) with constant H_V yields:

$$p_{SV} = p_0 \, \exp\left(\frac{H_V}{R\,T_0}\right) \exp\left(-\frac{H_V}{R\,T}\right) \tag{8.7}$$

with

> p_0 – vapor pressure at T_0.

---

[1] See Appendix C.1.
[2] See Appendix C.2.

8 Laser-Induced Vaporization

At $p_0 = 10^3$ hPa, the temperature T_0 corresponds to the evaporation temperature under normal conditions. With the abbreviation

$$p_{SV,\max} = p_0 \exp\left(\frac{H_V}{R T_0}\right) \tag{8.8}$$

it follows:

$$p_{SV} = p_{SV,\max} \exp\left(-\frac{H_V}{R T}\right) \tag{8.9}$$

The saturation vapor density is given by

$$\rho_{SV} = \frac{p_{SV,\max}}{R T} \exp\left(-\frac{H_V}{R T}\right) \tag{8.10}$$

and the particle density by

$$n_{SV} = \frac{p_{SV,\max}}{m R T} \exp\left(-\frac{H_V}{R T}\right) \tag{8.11}$$

8.2 Vaporization Rate

The velocity distribution of the vapor particles in thermodynamic equilibrium is given by the MAXWELLian distribution [7]:

$$f(v_x, v_y, v_z) = n_{SV} \left(\frac{m}{2 \pi k_B T}\right)^{3/2} \exp\left(-\frac{m}{2 k_B T} (v_x^2 + v_y^2 + v_z^2)\right) \tag{8.12}$$

with the saturation vapor density n_{SV} according to Eq. (8.11). The particle flux density of vapor particles hitting the melt surface and re-condensating there is given by (the x-axis is assumed to be normal to the surface):

$$j_{(V \to S)} = \int_{-\infty}^{0} v_x \, g(v_x) \, dv_x = -\frac{n_{SV}}{2} \frac{v_{TH}}{2} \tag{8.13}$$

$$g(v_x) = \int_{-\infty}^{\infty} f(v_x, v_y, v_z) \, dv_y \, dv_z = n_{SV} \sqrt{\frac{m}{2 \pi k_B T}} \exp\left(-\frac{m v_x^2}{2 k_B T}\right) \tag{8.14}$$

$$v_{TH} = \sqrt{\frac{8 k_B}{\pi m} T} = \sqrt{\frac{8}{\pi} R T} \tag{8.15}$$

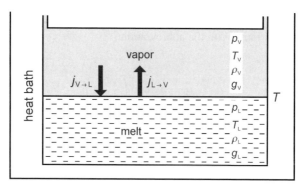

Fig. 8.1 Vapor pressure in thermodynamic equilibrium

V_{TH} is the mean thermal speed. In thermodynamic equilibrium, the flux of particles emerging from the melt surface equals the flux of vapor particles hitting the melt surface and re-condensating there (see Fig. 8.1). Considering the directions of the fluxes, it holds

$$j_{(M \to V)} = -j_{(V \leftarrow M)} = \frac{n_{SV}}{2} \frac{v_{TH}}{2} \qquad (8.16)$$

In thermodynamic equilibrium systems there is no net vaporization. In the case that the vapor state does not correspond to the equilibrium state, e.g., if the vapor is not enclosed but can freely expand, the particle flux density j of the particles emerging from the melt will still correspond to the equilibrium state at that temperature as long as the the melt surface stays in the thermodynamic equilibrium state that corresponds to the given temperature. That means that the velocity distribution of the particles in the melt surface equals the equilibrium distribution or deviates only slightly. The velocity distribution Eq. (8.14) with $v > 0$ is the velocity distribution of the particles at the melt-vapor interface just emerging from the melt.[3] This distribution function thus does not depend on the state of the vapor phase but only on the state of the melt surface. As long as the melt stays in this thermodynamic state of equilibrium, the emerging vapor flux can be described using this velocity distribution function, which is a half-MAXWELLian distribution. The distribution function is normalized so that the particle density is

$$n_V = \int_{-\infty}^{\infty} g(v_x)\, dv_x \qquad (8.17)$$

Because particles with velocity $v_x < 0$ do not exist in case of the half-MAXWELLian distribution, the density of the particles is only half of the saturation particle density:

---

[3] It is assumed here that all vapor particles that hit the melt re-condensate there.

$$n_V = \frac{n_{SV}}{2} \qquad (8.18)$$

The probability of finding a particles velocity within a given velocity interval with $v_x > 0$ doubles in the case of the half-MAXWELLian because of the lack of particles with $v_x < 0$. The distribution function of the off-flowing particles is thus given by

$$g(v_x) = \frac{n_{SV}}{2} \, 2 \sqrt{\frac{m}{2\pi k_B T}} \exp\left(-\frac{m v_x^2}{2 k_B T}\right) \Theta(x) \qquad (8.19)$$

with

$\Theta(x)$ – HEAVISIDE step function:

Figure 8.2 shows the velocity component $g(v_x)/n_V$ of the distribution function Eq. (8.19) together with the velocity component $g(v_x)/n_{SV}$ of the equilibrium distribution Eq. (8.14). The velocity of the off-flowing vapor equals the mean velocity in the x-direction. From the distribution function Eq. (8.19), it follows:

$$v_V = \frac{v_{TH}}{2} \qquad (8.20)$$

which is just half the thermal speed.

Describing the non-equilibrium vaporization using the saturation vapor density Eq. (8.11) that was derived under the assumption of thermodynamic equilibrium is only justified as long as the melt surface stays in a state of thermodynamic equilibrium. The equilibrium state of the melt surface therefore may only be disturbed slightly. This holds if sufficiently few particles leave the melt surface. Sufficiently few means that the time constant of evaporation of a particle is large compared to the

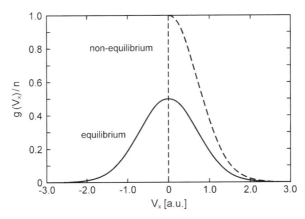

Fig. 8.2 Velocity distribution function, determined according to Eq. (8.14) (*drawn through line*) and Eq. (8.19) (*dashed line*), respectively. The distributions are normalized with respect to the densities

time constant of relaxation toward an equilibrium state in the melt. The relaxation time constant t_{relax} amounts to a few mean collision times of the particles in the melt. The mean collision time is in the ps range so that as an upper bound it can be estimated that $t_{\text{relax}} \approx 10$ ps. The vaporization time constant $t_{\text{evaporate}}$ follows from the ratio of the number of particles per area N_{LS} on the melt surface to the flux density of evaporating particles ($n_V \, v_V$):

$$t_{\text{evaporate}} = \frac{N_{LS}}{n_V \, v_V} \qquad (8.21)$$

Assuming the particle radius $r_a = 5 \cdot 10^{-10}$ m, one gets

$$N_{LS} \approx 10^{18} \, \text{m}^{-2} \qquad (8.22)$$

The equilibrium condition is given by

$$t_{\text{evaporate}} \gg t_{\text{relax}} \qquad (8.23)$$

The maximum evaporation rate should thus obey the inequality

$$n_V \, v_V \ll \frac{N_{LS}}{t_{\text{relax}}} = 10^{29} \frac{1}{m^2 s} \qquad (8.24)$$

for the assumption that the melt stays in a thermodynamic state of equilibrium to be justified. The value given in Eq. (8.24) corresponds to an ablation rate of about 1 m/s. Because of collisions among vapor particles, the velocity of some of the particles can become negative, so they move back to the melt surface. The particle flux density according to Eq. (8.16) is thus an upper bound of the net particle flux off the melt surface at the temperature T.

The above-described simple model is similar to the model proposed by AFANASEV and KROKHIN [3] that is based on the assumption that the evaporation can be described by a combustion wave. KROKHIN does not assume the velocity of the vapor particles at the melt surface as given by Eq. (8.20), but by the local velocity of sound.

The non-MAXWELLian velocity distribution of the vapor particles emerging from the melt surface is driven by collisions within a few free mean paths toward a MAXWELLian velocity distribution superimposed by a mean flow velocity. The transition layer is called KNUDSEN layer. The parameters density, temperature, and mean flow velocity that determine the MAXWELLIAN distribution have to be determined using kinetic models; this means that a solution of the BOLTZMANN equation for the off-flowing particles has to be determined. The velocity distribution function can, e.g., be computed by MONTE CARLO simulations [6] or by approximate solutions of the BOLTZMANN equation [8]. The parameters of the MAXWELLian distribution outside the KNUDSEN layer depend on the surface temperature of the melt and the ratio of the vapor pressure at this temperature to the pressure of the

8 Laser-Induced Vaporization

ambient atmosphere. If the ambient pressure equals the vapor pressure, no net evaporation takes place; this is the case of thermodynamic equilibrium. With increasing vapor pressure, the off-flow velocity increases and the temperature of the vapor particles decreases below the surface value. With increasing vapor pressure or decreasing ambient pressure, respectively, the velocity of the off-flowing vapor does only increase up to a certain value, which is given by the local value of the velocity of sound. The simple model described above mainly covers the case of strong evaporation. The predicted value of the velocity, however, is below the local velocity of sound and the predicted vapor temperature is too high. The deviations, however, are in many cases small enough to be tolerable; the vaporization rate ($n_V\, v_V$) anyway depends mainly on the absorbed laser intensity.

8.3 Particle and Energy Conservation During Laser-Induced Vaporization

In Sect. 8.2 an approximate expression for the evaporation rate as a function of the surface temperature was derived (Eq. (8.16) together with Eqs. (8.15) and 8.11). The surface temperature is determined by the laser intensity that is absorbed within a thin surface layer, by conduction of heat into the material, and by the off-flowing vapor (Fig. 8.3). In the stationary 1-dimensional case, the particle conservation can be written as

$$n_V\, v_V = n_0\, v_p \quad (8.25)$$

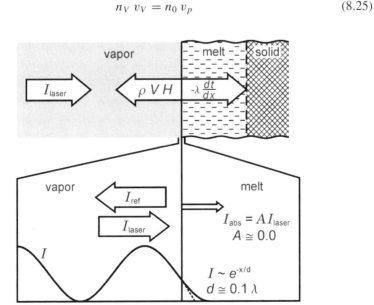

Fig. 8.3 Schematic drawing of the ablation of metals with laser radiation

with

n_V – vapor density
v_V – vapor velocity
n_0 – density of the melt
v_p – process or ablation rate.

Energy conservation is given by [2]

$$I_{abs} = m\, n_0\, v_p\, (H_V + H_m + c_p\, T) \tag{8.26}$$

with

I_{abs} – absorbed laser intensity
c_p – specific heat capacity for constant pressure
H_V – evaporation enthalpy
H_m – melting enthalpy
m – mass of the vapor particles.

In the energy conservation equation (8.26), the kinetic energy of the vapor has been omitted. With Eqs. (8.25), (8.26), (8.11), (8.20) and the assumption that the vapor temperature equals the surface temperature, it follows that

$$I_{abs} = I_0(T)\, \exp\left(-\frac{H_V}{R\,T}\right) \tag{8.27}$$

$$I_0(T) = \frac{p_{SV,\max}}{R\,T}\, (H_V + H_m + c\,T)\, \frac{1}{4}\, \sqrt{\frac{8}{\pi}\, R\,T} \tag{8.28}$$

The exponential factor in Eq. (8.27) varies much stronger as a function of the temperature than $I_0(T)$ does, so that $I_0(T) \approx I_0(T_V)$ can be assumed. Then Eq. (8.26) can be resolved with respect to the temperature:

$$T \simeq \frac{H_V}{R}\, \frac{1}{\ln\left(\dfrac{I_0(T_V)}{I_{abs}}\right)} \tag{8.29}$$

T_V is the temperature corresponding to the saturation vapor pressure of 1050 mbar. The temperature varies only slightly with the absorbed intensity. The particles density follows, using Eqs. (8.25), (8.26), and (8.20):

$$n_V = 2\, \frac{I_{abs}}{\sqrt{\dfrac{8}{\pi}\, R\,T}}\, \frac{1}{m\,(H_V + H_m + c\,T)} \tag{8.30}$$

In Eq. (8.30) the temperature according to Eq. (8.29) can be used. Because the temperature varies only slightly with the absorbed intensity the temperature T can approximately be assumed to be equal to the vaporization temperature T_v. Using the surface temperature that can be computed either by the implicit relation Eq. (8.27) or by the approximation, Eq. (8.29), all thermodynamic quantities of the vapor can be determined:

$$T_V = T \tag{8.31}$$

$$p_V = \frac{1}{2} p_{SV,\max} \exp\left(-\frac{H_V}{RT}\right) \tag{8.32}$$

$$n_V = \frac{1}{2} \frac{p_{SV,\max}}{mRT} \exp\left(-\frac{H_V}{RT}\right) \tag{8.33}$$

$$v_V = \frac{1}{2} \sqrt{\frac{8}{\pi} RT} \tag{8.34}$$

The process rate follows from the energy conservation Eq. (8.26):

$$v_p = \frac{I_{\text{abs}}}{m \, n_0 \, (H_V + H_m + c_p T)} \tag{8.35}$$

Figures 8.4, 8.5, 8.6, 8.7, 8.8 and 8.9 show T_V, p_V, n_V, v_V and v_p as a function of the absorbed intensity for Al and Fe.

The energy conservation equation (8.26) implies that in the 1-dimensional stationary case the absorbed laser energy is used to heat and subsequently melt and evaporate the material. In the general 3-dimensional case, part of the laser energy is lost by heat conduction to regions of the material that are not going to be vaporized. These losses reduce the surface temperature compared to the 1-dimensional case or make necessary a higher laser intensity in order to obtain the same surface temperature. Thus, in general, the effect of 3-dimensional heat conduction has to be taken

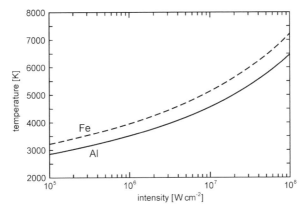

Fig. 8.4 Temperature as a function of the absorbed laser intensity determined using Eq. (8.27)

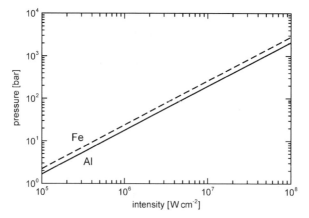

Fig. 8.5 Vapor pressure as a function of the absorbed laser intensity determined using Eq. (8.32)

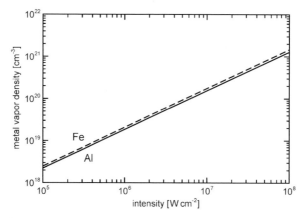

Fig. 8.6 Vapor density as a function of the absorbed laser intensity determined using Eq. (8.33)

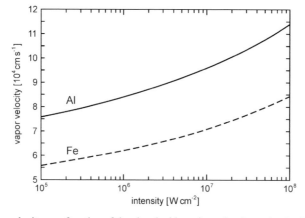

Fig. 8.7 Vapor velocity as a function of the absorbed laser intensity determined using Eq. (8.34)

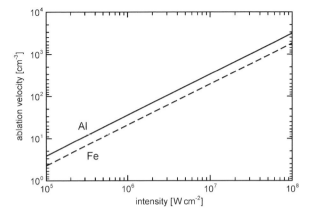

Fig. 8.8 Ablation rate as a function of the absorbed laser intensity determined using Eq. (8.35)

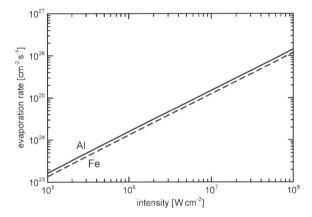

Fig. 8.9 Evaporation rate ($n_V\, v_V$) as a function of the absorbed laser intensity

into account in the energy conservation equation. This requires solving the time-dependent 3-dimensional heat conduction equation together with the appropriate application-dependent boundary conditions. Analytical solutions can only be found in special cases. Using numerical methods makes it possible to solve even very complicated problems, but at the expense of large implementation and computing efforts.

8.4 Description of the Evaporation Process as a Combustion Wave

The above-described evaporation model is to be understood as a first approximation. The particles flowing back to the melt surface were neglected and for the temperature and the vapor velocity, simple assumptions were made. If the saturation evaporation pressure only deviates slightly from the ambient pressure, the assumption that the vapor temperature equals the surface temperature is surely justified. On the other hand, in this case of weak vaporization, the number of particles flowing back to the melt surface cannot be neglected and the above presupposed value of the vapor velocity derived from the half-MAXWELLian distribution is too high. In the case of strong evaporation on the other hand, the temperature drops below the melt temperature due to the vapor expansion and acceleration within the KNUDSEN layer and the vapor velocity can reach the local velocity of sound [8]. The proposal of AFANASEV and KROKHIN to describe evaporation in analogy to combustion waves marks an improvement in the modeling of evaporation processes. Originally this model was applied to the case of strong evaporation. In the following, this idea is described in a slightly modified form.

In the theory of combustion waves, the combustion front is treated as a discontinuity (Fig. 8.10). The not-yet combusted material flows into the discontinuity where the combustion takes place. The chemical energy that is released during combustion is transferred into heating and accelerating the material. In the case of laser-induced evaporation, the chemical energy is replaced by the absorbed laser energy. The states of the material on the left- and right-hand sides of the discontinuity, respectively, are related by the jump conditions of RANKINE–HUGONIOT. The absorbed laser energy has to be included in the energy conservation equation. The RANKINE–HUGONIOT jump conditions read

$$\rho_1 v_1 = \rho_2 v_2 \tag{8.36}$$

$$\rho_1 v_1^2 + p_1 = \rho_2 v_2^2 + p_2 \tag{8.37}$$

$$I_V + \rho_1 \left(e_1 + \frac{p_1}{\rho_1} + \frac{v_1^2}{2} \right) = \rho_2 \left(e_2 + \frac{p_2}{\rho_2} + \frac{v_2^2}{2} \right) \tag{8.38}$$

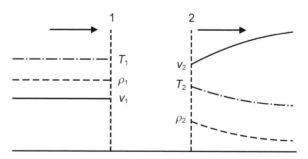

Fig. 8.10 Schematic drawing of a combustion front. The layer between the regions 1 and 2 is treated as an infinitely thin discontinuity. In reality, this layer has a finite thickness within which the local equilibrium state of region 1 relaxes to the local equilibrium state of region 2. Within this layer there exists no MAXWELLian distribution of particle velocities and thus no local thermodynamic equilibrium. On the right side of region 2 it is indicated that the state of the off-flowing gas is not constant in general, but where the states of the gas are always states of local thermodynamic equilibrium. This region is not included in the present treatment because this would make necessary solving the flow problem with the appropriate boundary conditions imposed by the ambient atmosphere

Index 1 designates the quantities within the melt and Index 2 the quantities within the vapor, respectively. The intensity I_V is that part of the laser intensity that is transferred into evaporation. The total absorbed laser intensity is composed of this part and the part that is used to heat and melt the material. With this it follows:

$$I_V = I_{abs} - \rho_1 \, v_1 \, (H_v + H_m + c \, T) \tag{8.39}$$

The internal energy e_1 is

$$e_1 = c_1 \, T \tag{8.40}$$

In e_2 the evaporation energy has to be taken into account

$$e_2 = c_2 \, T + H_V \tag{8.41}$$

It is assumed that within the vapor, the ideal gas law holds as

$$p_2 = \rho_2 \, R \, T_2 \tag{8.42}$$

AFANASEV and KROKHIN [3] assumed equal temperatures for melt and vapor and that the vapor velocity equals the local velocity of sound. ADEN [2] proposed a different assumption. It is well known from the theory of combustion waves that the local velocity of sound is the limiting velocity that cannot be exceeded [5]. This is expressed by the CHAPMAN–JOUGUET condition. During weak vaporization, the vapor velocity can be below this value. If the vapor velocity is below the local velocity of sound, the ambient vapor (or gas) acts back on the vapor flow in the vicinity of the surface and by this on the evaporation process. In that case, the vapor flow has to be calculated self-consistently. This will not be done here, but useful results can be gained even without a complete solution. In doing so, the vapor pressure is prescribed and all the other quantities, e.g., the vapor velocity, are computed using this quantity. The value of the ambient pressure at which the vapor velocity reaches local velocity of sound is the minimal pressure that can exist in front of the melt surface. Even when there is vacuum downstream the evaporated vapor itself causes this minimal pressure. In that case, the evaporation is not influenced by the ambient atmosphere. During expansion of the vapor within the KNUDSEN layer, the temperature decreases. In order to take this into account, the following additional condition for the single particle energy is required to hold [2]:

$$\frac{3}{2} k_B \, T_1 + \frac{m}{2} v_1^2 = \frac{3}{2} k_B \, T_2 + \frac{m}{2} v_2^2 \tag{8.43}$$

The sum of the thermal and the kinetic energy of a particle is conserved during evaporation. In thermodynamic equilibrium the pressure in the melt equals the saturation vapor pressure. This is composed of the momentum that the off-flowing vapor exerts on the melt surface and the momentum transfer of the particles that flow back from the vapor region onto the melt surface. If a part of the back-flowing particles are missing, the pressure in the melt decreases. In [2] this is taken into account by the following assumption:

$$p_1 = p_{SV}(T_1) - \rho_2 v_2^2 \tag{8.44}$$

The pressure within the melt is reduced, compared to the saturation vapor pressure by the momentum flux density of the net number of evaporated particles. The heat capacity of an ideal gas is given by

$$c = \frac{3}{2}\frac{k_B}{m} = \frac{3}{2} R \tag{8.45}$$

In crystal solids well above the DEBYE temperature, the heat capacity is twice as large, i.e., $c_1 = 6/2\ R$.[4] In fluids, the value of the heat capacity lies between these two values. Because the internal energy is much smaller than the evaporation enthalpy, the error introduced by Eq. (8.45) in the energy conservation equation (8.38) is negligible. Likewise, the pressure work p/ρ can be neglected on both sides of Eq. (8.38). With this and with Eq. 8.36 again the energy conservation equation (8.26) follows:

$$I_{abs} = \rho_2 v_2 (H_V + H_m + c\,T) \tag{8.46}$$

Neglecting the first term on the left side of Eq. (8.37) together with Eq. (8.44) results in

$$p_{SV} = 2\,\rho_2\,v_2^2 + p_2 \tag{8.47}$$

The velocity v_2 is normalized with respect to the local velocity of sound c_2:

$$s = \frac{v_2}{c_2} \tag{8.48}$$

$$c_2 = \sqrt{\kappa\,R\,T}$$

with:

κ – adiabatic coefficient

The vapor density is normalized with respect to the saturation vapor density

$$x = \frac{\rho_2}{\rho_{SV}} \tag{8.49}$$

$$\rho_{SV} = \frac{p_{SV}}{R\,T_1} \tag{8.50}$$

---

[4] Three degrees of freedom of the kinetic energy and three degrees of freedom of the potential energy.

and the pressure with respect to the saturation vapor pressure:

$$z_p = \frac{p_2}{p_{sv}} \tag{8.51}$$

Equation (8.47) becomes

$$2\kappa s^2 + z_p = 1 \tag{8.52}$$

or:

$$s^2 = \frac{1 - z_p}{2\kappa z_p} \tag{8.53}$$

If the vapor pressure equals the saturation vapor pressure, i.e., $z_p = 1$, then $s = 0$ and thus also $v_2 = 0$. In that case, there is no net evaporation. This is again the case of thermodynamic equilibrium. The vapor velocity cannot exceed local velocity of sound, which means $s_{max} = 1$. The minimal normalized vapor is then using $\kappa = 5/3$:

$$z_{p,min} = \frac{1}{1 + 2\kappa} \simeq 0.23 \tag{8.54}$$

The value of the pressure within the vapor thus cannot be smaller than $(0.23\, p_{SV})$. Neglecting the kinetic energy of the melt in Eq. (8.43) and applying the normalizations Eq. (8.48), Eq. (8.49), and Eq. (8.51) results in

$$\frac{3}{2} R T_1 = \frac{3}{2} R T_2 + \frac{1}{2} v_2^2 \tag{8.55}$$

$$s^2 = \frac{3}{\kappa}\left(\frac{x}{z_p} - 1\right) \tag{8.56}$$

With Eq. (8.52) this becomes

$$x = \frac{1}{6}(1 + 5 z_p) \tag{8.57}$$

With the minimal value of z_p (Eq. (8.54)), the minimal value of the normalized density is given by

$$x_{min} = 0.358 \tag{8.58}$$

The ratio of the temperatures T_2 and T_1 is

$$\frac{T_2}{T_1} = \frac{z_p}{x} = 6\,\frac{z_p}{1 + 5 z_p} \tag{8.59}$$

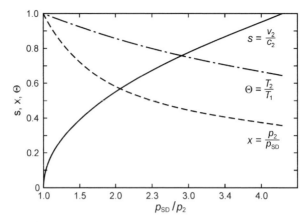

Fig. 8.11 Normalized velocity, normalized density, and normalized temperature as a function of the normalized vapor pressure, computed using Eq. (8.53), Eq. (8.57), and Eq. (8.59), respectively

The minimal value of the normalized temperature is thus

$$\left(\frac{T_2}{T_1}\right)_{min} \simeq 0.64 \tag{8.60}$$

Figure 8.11 shows the normalized velocity v_2/c_2, the normalized density n_2/n_{SV}, and the normalized temperature T_2/T_1 as a function of the normalized vapor pressure $p_{SV}/p_2 = 1/z_p$, with p_2 being the pressure at the KNUDSEN layer-vapor interface. The pressure p_2 does not equal the pressure of the ambient atmosphere but rather depends on the evaporation rate (and itself influences it) and on the density and pressure of the ambient atmosphere. For a more exact computation of the vaporization rate, the vapor density, the vapor temperature, the energy conservation within the solid, the evaporation process itself, and the vapor flow downstream of the KNUDSEN layer considering the boundary conditions imposed by the ambient atmosphere have to be solved self-consistently.[5]

8.5 Kinetic Model of the Evaporation and The KNUDSEN Layer

In the following, a kinetic model of the evaporation is outlined [8]. Again the assumption is made that in the case of non-equilibrium evaporation, the particles that leave the melt surface have the same velocity distribution as in the case of total thermodynamic equilibrium. Figure 8.12 schematically shows the evaporation process. The melt surface is located at $x = 0$. The distribution function of the off-flowing particles at the melt surface corresponds to a half-MAXWELLian distribution. The particle density equals half the saturation density at the given surface temperature:

---

[5] See also [2].

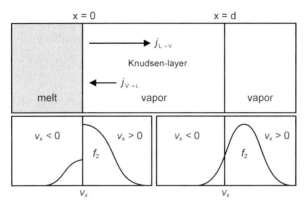

Fig. 8.12 Schematic of the velocity distribution function at the melt surface and the edge of the KNUDSEN layer. Within the KNUDSEN layer, the non-MAXWELLian distribution function of the vapor particles at $x = 0$ relaxes toward a MAXWELLian distribution with superimposed mean flow velocity. The thickness d of the KNUDSEN layer amounts to a few free-path lengths. The edge of the KNUDSEN layer at $x = d$ cannot be defined as sharply as it is depicted here

$$f_{SV}(0, \vec{v}) = n_{SV} \left(\frac{m}{2\pi k_B T_1(0)} \right)^{\frac{3}{2}} \exp\left(-\frac{m\left(v_x^2 + v_y^2 + v_z^2\right)}{2 k_B T_1(0)} \right) \quad : \quad v_x \geq 0 \tag{8.61}$$

The distribution function of the back-flowing particles ($v_x < 0$) is not known, so that a reasonable assumption has to be made for this. The total distribution function of off-flowing and back-flowing particles in general is not a MAXWELLian-distribution.[6] The vapor particles relax on their way through the KNUDSEN layer to a MAXWELLian distribution with superimposed mean flow velocity caused by collisions among the vapor particles:

$$f_2(d, \vec{v}) = n_2(d) \left(\frac{m}{2\pi k_B T_2(d)} \right)^{\frac{3}{2}} \exp\left(-\frac{m\left([v_x - v_2(d)]^2 + v_y^2 + v_z^2\right)}{2 k_B T_2(d)} \right) \tag{8.62}$$

The vapor density n_2, the temperature T_2, and the vapor velocity v_2 follow from the requirement of conservation of the particle, momentum, and energy flux densities. This signifies that the particle, momentum and energy flux densities respectively have to be equal if either computed using the velocity distribution function at the

---

[6] This only holds in case of thermodynamic equilibrium.

melt surface or at the point $x = d$.[7] For this, an assumption has to be made for the distribution function of the back-flowing particles at the melt surface. YTREHUS [8] proposed the following distribution function at the melt surface:

$$v_x > 0 : f_1(0, \vec{v}) = f_{SV}(0, \vec{v}) \tag{8.63}$$
$$v_x < 0 : f_1(0, \vec{v}) = \beta \cdot (f_2(0, \vec{v}) + f_{SV}(0, \vec{v})) \tag{8.64}$$

The distribution function of the back-flowing particles is the sum of the distribution function at the edge of the KNUDSEN layer Eq. (8.62) and the distribution function that corresponds to the equilibrium state (Eq. (8.61)) multiplied by a factor β. Given are the surface temperature as well as another quantity at the edge of the KNUDSEN layer at $x = d$. This can be either the pressure or, in the case of strong evaporation, the velocity, which in that case equals the local velocity of sound. The three conservation equations for particle density, momentum, and energy are three equations for the three unknowns n_2, v_2, and β, respectively, in the case that the pressure is prescribed or n_2, T_2, and β in the case that the velocity is prescribed (and equals the velocity of sound). Figure 8.13 shows the normalized velocity, the normalized density, and the normalized temperature as a function of the normalized pressure. Comparison with Fig. 8.11 shows that there are only minor differences between of the results gained from the combustion wave model and the kinetic model, respectively.

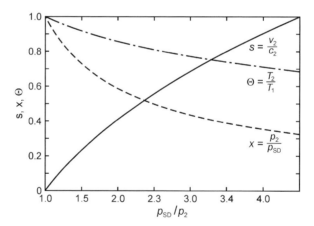

Fig. 8.13 Normalized velocity, normalized density, and normalized temperature as a function of the normalized pressure [8]

---

[7] For the computation of moments of the velocity distribution, see Appendix C.3.

References

1. G. Adam and O. Hittmair, Wärmetheorie, Vieweg, 1978
2. M. Aden, E. Beyer, and G. Herziger, Laser Induced Vaporization of a Metal Surface, J. Appl. Phys. D., Vol. 23, 1990
3. Y. Afanasev and O. N. Krokhin, Vaporisation of Matter Exposed to Laser Emission, Sov. Phys. JETP, Vol. 35, p. 639, 1967
4. R. Becker, Theorie der Wärme, Springer-Verlag, 1975
5. A. J. Chorin and J. E. Marsden, A Mathematical Introduction to Fluid Mechanics, Springer-Verlag, New York, 1979
6. M. Murukami and K. Oshima, Rarefied Gas Dynamics, DFVLR Press, 1974
7. F. Reif, Fundamentals of Statistical and Thermal Physics, McGraw Hill, 1965
8. T. Ytrehus, J. L. Potter Ed., Theory and Experiments on Gas Kinetics in Evaporisation, Int. Symp. Rarefied Gas Dynamics 1976, 10 Snowmass-at-Aspen Col., 1977

Chapter 9
Plasma Physics

Rolf Wester

The energy of the laser radiation that is applied to the workpiece during laser material processing has to be transformed effectively into usable process energy. The normal light absorptivity at those wavelengths for which high-power lasers are available today is quite low in case of dielectrics and in case of metals its value lies in the 10–40% region.[1] When exceeding a critical intensity in general an increase of the absorptivity can be observed [12, 14]. Figure 9.1 shows this behavior in case of processing Cu using a pulsed Nd:YAG laser. At small intensities the reflectivity corresponds to the normal reflectivity, after exceeding the critical value of the intensity the reflectivity decreases to a value of almost zero. The decrease of the reflectivity is accompanied by an increase of the energy coupling into the workpiece. The increased coupling can be ascribed to a plasma that develops at the workpiece surface. This plasma is created by the incident laser beam. The plasma absorbs a great part of the laser energy and partly transfers the energy to the workpiece [15]. All low absorbing materials that have a high heat conductivity like metals or diamond and sapphire are best laser processed in this plasma-supported regime [12].

For the optimal coupling of the plasma energy into the workpiece surface the plasma has to be located near the surface. This can be realized by short laser pulses or in vacuum. With longer laser pulses or with normal ambient atmosphere LSD (**L**aser **S**upported **D**etonation) waves are created that absorb the laser energy at the side remote from the specimen surface. The threshold intensity for LSD wave formation increases with the wavelength of the laser radiation. Figure 9.2 shows pictures made with a framing camera during exposing a steel specimen to CO_2 laser radiation. The individual frames are each separated in time by 50 ns. If the intensity is carefully adapted the glowing plasma stays in contact with the workpiece surface (lower row of Fig. 9.2). The laser energy is absorbed in the plasma near the surface and is transformed into heat and radiation in the UV and visible regions and is subsequently partly transferred to the specimen surface [15]. At higher intensities the vapor density and accordingly the absorption increase (upper row of Fig. 9.2).

R. Wester (✉)
Fraunhofer-Institut für Lasertechnik, 5207 Aachen, Germany
e-mail: rolf.weister@ilt.fraunhofer.de

[1] See Chap. 3, Fig. 3.5.

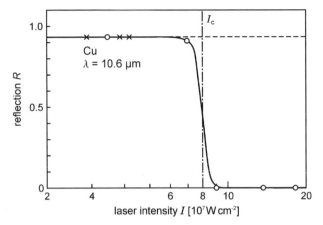

Fig. 9.1 Reflectivity R of a Cu target as a function of the laser intensity at normal incidence gained from drilling experiments with pulsed Nd:YAG-lasers with a pulse length of $t_L \simeq 10^{-7}$ s [14]

The laser energy is almost entirely absorbed in a thin plasma sheet at the side remote from the surface so that only a small fraction of the laser energy reaches the surface. The plasma is strongly heated and accelerated away from the workpiece. When it detaches laser processing is interrupted. Only after expansion and sufficient rarefaction of the plasma the laser beam can again reach the specimen surface and processing can go on.

On one hand the laser-induced plasma can be utilized to increase the energy coupling (this is called 'plasma-enhanced coupling'), but there can also be situations when the plasma shields the specimen so that the laser energy cannot reach the workpiece surface any more. The vapor plasma can not only absorb but also scatter the laser light, mostly due to material clusters that are either ejected from the melt surface or evolve by condensation within the vapor. A further effect of the hot plasma is the pressure force exerted onto the surface that can deform the surface so that a key hole forms that expands deeply into the workpiece.

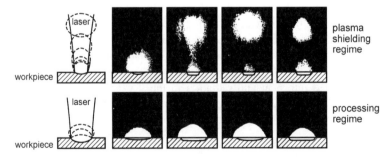

Fig. 9.2 Framing shoots of the laser-induced plasma above a steel specimen during processing with a pulsed CO_2 laser beam with 50 ns time gaps [2]. *Upper row*: plasma shielding, laser intensity $I > 2 \times 10^7$ W cm^{-2}, the plasma detaches from the workpiece surface. *Lower row*: laser processing, laser intensity 2×10^6 W cm$^{-2} < I < 10^7$ W cm^{-2}, the plasma stays in contact with the workpiece surface

9 Plasma Physics

The comprehension of all these phenomena makes necessary a knowledge of the plasma physics basics. Plasma physics is a large and complex field so that here only a short introduction can be presented. First the concept of DEBYE shielding is explained. In many cases of interest the laser-induced plasmas are in a state of (near) thermodynamic equilibrium. Equilibrium state variables are compiled. Of special importance is the SAHA equation that allows to calculate electron and ion densities. Furthermore, transport processes and the corresponding transport coefficients as well as the interaction of laser light with plasmas is discussed. Then some non-equilibrium processes are shortly presented. The last part covers the field of radiation that is emitted by plasmas.

9.1 Debye Radius and Definitions

A plasma is a many-particle system that in general consists of free electrons, neutral atoms, and positively as well as in some cases negatively charged ions. A principal characteristic of plasmas is their electric neutrality, i.e., in a plasma there exist as much positive as negative charges. However, this must not be the case for an arbitrarily chosen sub-volume of the plasma. When choosing the sub-volume to just include a single electron or ion, charge neutrality obviously is violated. Even if there are several particles within the chosen sub-volume charge neutrality need not necessarily exist. If there is a net charge within a sub-volume ambient charges of opposite sign are attracted while those of the same sign are repelled. By this the original net charge is compensated for. The length scale within which charge compensation takes place is called shielding length and depends on the charge density and the ratio of the COULOMB interaction energy and the mean thermal energy of the particles. In the following the screening length in the DEBYE approximation is derived. For this a single ion (the same holds for electrons) is considered which is treated as a point charge while the other charge carriers surrounding the ion are treated as a charged continuum. This corresponds to the self-consistent field approximation (or mean field approximation) in solid state physics and the HARTREE–FOCK approximation in atomic physics, respectively. The electrons are attracted by the ion while the other positive ions are repelled. Due to the COULOMB interaction the probability of finding a plasma electron closer to positive ions and farther away from electrons (and if present negative ions) is enhanced compared to the case of no COULOMB interaction. This holds correspondingly for ions too. The long distance COULOMB interaction leads to correlations among the spatial locations of the charged particles and thus enhances the order of the system compared to a system without COULOMB interaction which means that the entropy of a plasma is smaller than that of a system with no COULOMB interaction. The densities of electrons and ions surrounding the single ion are given by BOLTZMANN factors:

$$n_e = n_{e_0} \exp\left(\frac{e\phi}{k_B T}\right) \tag{9.1}$$

$$n_p = n_{p_0} \exp\left(-\frac{e\phi}{k_B T}\right) \tag{9.2}$$

with

  e – elementary charge
  k_B – BOLTZMANN constant
  T – plasma temperature.

n_{e_0} and n_{p_0} are the mean densities of electrons and ions, respectively. The distributions Eqs. (9.1) and (9.2) are equilibrium distributions. For the assumption of equilibrium to apply the time constants for relaxation toward the equilibrium state have to be sufficiently small. In the following it is assumed that the energy $e\phi$ is small compared to the mean thermal energy $k_B T$. Then the exponential factors in Eq. (9.1) and Eq. (9.2) can be linearized. The density of the single ion is described by a DIRAC delta function. The electric potential obeys POISSON's equation:

$$\Delta\phi = -\frac{e}{\varepsilon_0}\delta(r) + \frac{e}{\varepsilon_0}(n_e - n_p) \tag{9.3}$$

Inserting the linearized densities results in

$$\Delta\phi = -\frac{e}{\varepsilon_0}\delta(r) + \frac{e^2(n_{e_0} + n_{p_0})}{\varepsilon_0 k_B T}\phi \tag{9.4}$$

The solution of this equation reads

$$\phi = -\frac{e}{4\pi\varepsilon_0 r}\exp\left(-\frac{r}{r_D}\right) \tag{9.5}$$

with the DEBYE radius[2]

$$r_D = \sqrt{\frac{\varepsilon_0 k_B T}{e^2(N_{e_0} + n_{p_0})}} \tag{9.6}$$

The first part in Eq. (9.5) is the potential of the point charge of the single ion while the exponential describes the contribution of the surrounding plasma particles. The potential of a point charge in a plasma decays outside a distance given by the DEBYE radius much stronger than the potential of an isolated point charge. This means that a plasma is electrically neutral on scales that are large compared to the DEBYE radius. Figure 9.3 shows the DEBYE radius as a function of the electron density at a plasma temperature of 1 eV (11600 K). The electron densities in laser-induced

---

[2] The generalization in case of multiple charged ions can be found in Appendix D.2.

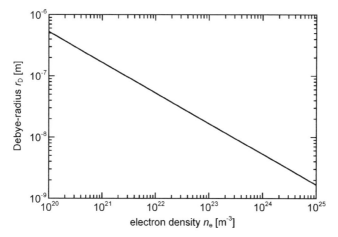

Fig. 9.3 DEBYE radius as a function of the electron density computed using Eq. (9.6) with a temperature of 1 eV (11600 K)

plasmas during laser material processing with CO_2 laser radiation with intensities in the range of several 10^{10} W m^{-2} are between 10^{22} and 10^{23} m^{-3} and the temperatures are about 0.5 to 2.0 eV [18]. With an electron density of 10^{23} m^{-3} and a temperature of 1 eV the DEBYE radius amounts to $r_D = 16$ nm. At higher intensities the values of electron density and temperature are in general higher too. The main assumption in deriving the above expression of the DEBYE radius was that the surrounding electrons and ions can be treated as a continuous charged cloud. This implies that there are many particles within a DEBYE sphere. The number of particles within a DEBYE sphere is given by

$$N_D = \frac{4}{3} \pi (n_e + n_i) r_D^3 \qquad (9.7)$$

This number has to be large compared to 1 for the DEBYE approximation to hold. When this requirement is not met the concept of shielding has to be modified. The shielding radius then becomes larger compared to the results of the DEBYE approximation [5]. With the above-mentioned values of the electron density and the temperature $N_D = 3.4$, which means that the prerequisites of the DEBYE approximation are only met approximately. This will be further discussed in Sect. 9.3. Expanding the exponential in Eq. (9.5) in case of small values of r_D results in

$$\phi = -\frac{e}{4\pi\varepsilon_0}\frac{1}{r} + \frac{e}{4\pi\varepsilon_0}\frac{1}{r_D} \qquad (9.8)$$

The first term on the right-hand side is the potential of the single ion and the second term is the potential that is caused by the other plasma particles at the single ion location.

A further important quantity in plasma physics is the LANDAU length l which is defined as the distance from a point charge at which the potential energy equals the mean thermal energy of the plasma particles:

$$l = \frac{e^2}{4\pi\varepsilon_0 k_B T} \tag{9.9}$$

A plasma approximately behaves like an ideal gas if the mean thermal energy of the particles by far exceeds the mean COULOMB interaction energy. This requirement is met in case that the mean distance between the plasma particles is large compared to the LANDAU length. The plasma parameter

$$\Gamma = l\,(n_e + n_i)^{1/3} \tag{9.10}$$

thus has to be small compared to 1 for a plasma to be an ideal plasma. Another important parameter is the ratio of the LANDAU length and the DEBYE radius:

$$\mu = \frac{l}{r_D} \tag{9.11}$$

This parameter too has to be small compared to 1. With Eq. (9.7) it follows that

$$\mu = \frac{1}{3N_D} \tag{9.12}$$

The last quantity mentioned here is the COULOMB logarithm:

$$\ln \Lambda_c = \ln\left[\frac{12\pi\varepsilon_0 r_D k_B T}{e^2}\right] \tag{9.13}$$

$$\Lambda_c = 9\,N_D = \frac{3}{\mu} \tag{9.14}$$

This quantity will be explained in more detail in one of the next sections.

9.2 Some Results from Thermodynamics and Statistics of a Plasma

In the framework of thermodynamics many-particle systems in thermodynamic equilibrium are treated macroscopically,[3] the microscopic nature of the systems is excluded. The equations of state either have to be determined empirically or have to be derived theoretically from 'first principles' within the framework of equilibrium statistics. The principal quantity that links the thermodynamic and the statistical treatment is the partition function. The partition function of a canonical ensemble is given by [17]

---

[3] Some useful thermodynamic relations are compiled in Appendix D.1.

9 Plasma Physics

$$Z = \sum_r \exp\left(-\frac{E_r}{k_B T}\right) \tag{9.15}$$

with

E_r – total energy of the system in state r.

The summation is over all states including degenerated states. The probability to find a system in state r with energy E_r is given by

$$P_r = \frac{\exp\left(-\frac{E_r}{k_B T}\right)}{Z} \tag{9.16}$$

Multiplying by E_r and summing over all states gives the mean or internal energy of the system:

$$E = \sum_r E_r \exp\left(-\frac{E_r}{k_B T}\right) Z^{-1} \tag{9.17}$$

or with Eq. (9.15)

$$E = k_B T \frac{\partial \ln Z}{\partial T} \tag{9.18}$$

The free energy of a system is defined by [17]

$$F = E - TS = -k_B T \ln Z \tag{9.19}$$

S is the entropy:

$$S = -\left(\frac{\partial F}{\partial T}\right)_{V,N} \tag{9.20}$$

with

N – number of particles $N_1, N_2, \ldots$ of the different particle species
V – volume.

The derivative is taken with respect to T while leaving the volume and all particle densities constant. Together with Eq. (9.19) it follows that

$$\frac{E}{T^2} = -\left(\frac{\partial}{\partial T}\frac{F}{T}\right)_{V,N} \tag{9.21}$$

The chemical potential of particle species j is defined as

$$\mu_j = \left(\frac{\partial F}{\partial N_j}\right)_{T,V,N} \tag{9.22}$$

During differentiation of the right-hand side with respect to the number N_j all other particle numbers, the volume, and the temperature are kept constant. Derivation of the free energy with respect to the volume, while keeping the temperature and all particle numbers fixed, gives the pressure:

$$p = -\left(\frac{\partial F}{\partial V}\right)_{T,N} \tag{9.23}$$

9.2.1 Partition Function of an Ideal Plasma

In computing the partition function Eq. (9.15) the sum runs over all total system states with E_r being the energy of the total system state r. In case of a system of non-interacting particles the energy E_r is given by the sum of the energies e_{r_i} of all particles. In computing the partition function the sum runs over all states of all particles. In case of non-distinguishable particles during summation states that are only distinguished by permutations of particles have to be counted only once:

$$Z = \sum_{r_1, r_2, \ldots} \exp\left(-\frac{1}{k_B T}(e_{r_1} + e_{r_2} + \cdots)\right) = \frac{z^N}{N!} \tag{9.24}$$

with

N – number of particles.

$N!$ corresponds to the number of possible permutations of the particles. In case of several particle species one gets

$$Z = \prod_i^m \frac{z_i^{N_i}}{N_i!} \tag{9.25}$$

with

m – number of different particle species.

The total energy of a particle is given by the sum of its kinetic and its internal energy. With the assumption that kinetic and internal energies are independent of each other it is possible to sum over all states of the different degrees of freedom independently. The single particle partition function is thus given by the product of the translational degrees of freedom and of the internal degrees of freedom:

9 Plasma Physics

$$z_i = \xi_i \, U_i \tag{9.26}$$

The partition function of the translational degrees of freedom is given by

$$\xi_i = \frac{V}{\Lambda^3} \tag{9.27}$$

with the thermal DE'BROGLIE wavelength

$$\Lambda = \frac{h}{\sqrt{2\pi m k_B T}} \tag{9.28}$$

with

h – PLANCK's constant.

The partition function of the internal degrees of freedom of an atom is given by

$$U_a = \sum_{j}^{\infty} g_j^a \exp\left(-\frac{E_j^a}{k_B T}\right) \tag{9.29}$$

with

g_j^a – degeneracy.

The COULOMB potential has infinitely many bound states that all have negative energies. The continuum limit has energy 0. The ground state has the lowest energy; with increasing main quantum number the energies increase and in the limit $n \to \infty$ become 0, i.e., the exponential factor in Eq. (9.29) becomes 1. Because of this the partition function Eq. (9.29) diverges. But this only holds in case of an isolated atom. An atom in a gas or plasma is not isolated, however. Due to the interaction of the particles with each other the states close to the continuum cannot be attributed to the individual atoms any more, the electrons in these states are quasi-free. In computing the partition function the summation has to be restricted to states with energies below an energy limit that depend on the particle density and the temperature.

When computing the partition function of ions the energy zero point has to be set with care. In case that the energy of the ground state of the neutral atom is taken as the zero point the ionization energy has to be added to the energies of the ion states when these are taken with respect to the ion ground state:

$$U_i = \sum_{j} g_j^i \exp\left(-\frac{E_{ion}^a + E_j^i}{k_B T}\right) \tag{9.30}$$

Separation of the ionization energy gives

$$U_i = \exp\left(-\frac{E^a_{ion}}{k_B T}\right) U'_i \qquad (9.31)$$

with

$$U'_i = \sum_j g^i_j \exp\left(-\frac{E^i_j}{k_B T}\right) \qquad (9.32)$$

The energies and degeneracies of all states have to be known for the computation of the partition function. In case of hydrogen these are known and the partition function can be determined quite easily. The summation must, however, not be performed over all (infinitely many) states of the isolated hydrogen atom. Due to the interaction with the other plasma particles the continuum limit is shifted toward smaller energies. If the particle densities are large this can lead to very high ionization degrees even at relatively small temperatures. This phenomenon is also called pressure ionization. The sum in the partition function must not run over all infinitely many states but has to be truncated at an energy below the continuum limit. In case of other species with more complicated level schemes the determination of the partition function is much more difficult. For quite a lot of elements the partition functions are tabulated in [4]. Figure 9.4 shows the partition function of an Al plasma as a function of the temperature assuming a reduction of the ionization energy of 0.14 eV. The change of the ionization energy is due to the COULOMB interaction. This will be discussed in more detail below in connection with the SAHA equation. The reduction of the ionization energy gives the limit for the summation in calculating the partition function.

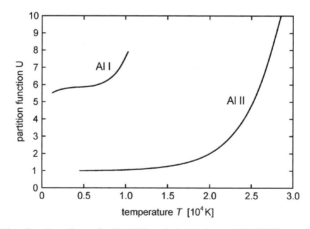

Fig. 9.4 Partition function of atomic Al (Al I) and singly charged Al (Al II) as a function of the temperature. The reduction of the ionization energy is $\Delta E_i = 0.1$ eV. The partition functions of Al and other metal plasmas are tabulated in [4]

9 Plasma Physics

The partition function of the internal degrees of freedom of electrons is due to the two possible spin states equal to 2:

$$U_e = 2 \qquad (9.33)$$

9.2.2 State Variables of an Ideal Plasma

The logarithm of the partition function Eq. (9.24) is

$$\ln Z = \sum_i^m \ln \frac{z_i^{N_i}}{N_i!} = \sum_i^m (N_i \ln z_i - \ln N_i!) \simeq \sum_i^m N_i (\ln z_i - \ln N_i + 1) \qquad (9.34)$$

In the last step the STIRLING formula was used:

$$\ln N! \simeq N \ln N - N \qquad (9.35)$$

For simplification in the following only a single particle species is considered:

$$\ln Z = N (\ln \xi + \ln U - \ln N + 1) = N \left(\ln \frac{V}{\Lambda^3} + \ln U - \ln N + 1 \right) \qquad (9.36)$$

In case of multiple particle species the contributions of all particle species are added. With Eq. (9.18) the internal energy is given by

$$E = k_B T N \left(\frac{T}{\xi} \frac{\partial \xi}{\partial T} + \frac{T}{U} \frac{\partial U}{\partial T} \right) \qquad (9.37)$$

The first term in parentheses is the contribution due to the translational degrees of freedom. With Eq. (9.27) and Eq. (9.28) this term becomes

$$\frac{T}{\xi} \frac{\partial \xi}{\partial T} = T \Lambda^3 \frac{\partial \Lambda^{-3}}{\partial T} = \frac{3}{2} \qquad (9.38)$$

The second term together with Eq. (9.29) becomes

$$\frac{T}{U} \frac{\partial U}{\partial T} = \frac{T}{U} \frac{1}{k_B T^2} \sum_j g_j E_j \exp\left(-\frac{E_j}{k_B T}\right) = \frac{\overline{E}}{k_B T} \qquad (9.39)$$

with

$\overline{E}$ – mean energy of the inner degrees of freedom.

In case of ions the ionization energy has to be added:

$$\overline{E}^i = E^a_{ion} + E^i_{ex} \tag{9.40}$$

with

E^i_{ex} – mean excitation energy of an ion
E^a_{ion} – ionization energy of the neutral atom.

With Eq. (9.37), (9.38), and (9.39) the internal energy is

$$E = \frac{3}{2} N k_B T + N \overline{E} \tag{9.41}$$

For the free energy it follows with Eq. (9.19) that

$$F = -k_B T N \left(\ln \frac{U}{\Lambda^3} - \ln \frac{N}{V} + 1 \right) \tag{9.42}$$

With this the chemical potential is given by (Eq. (9.22))

$$\mu = \left(\frac{\partial F}{\partial N} \right)_{T,V} = k_B T \ln \frac{N}{V} - k_B T \ln \frac{U}{\Lambda^3} \tag{9.43}$$

and the pressure by (Eq. (9.23))

$$p = -\left(\frac{\partial F}{\partial V} \right)_{T,N} = \frac{N k_B T}{V} \tag{9.44}$$

9.2.3 Coulomb Corrections

In the last two sections the COULOMB interaction among the plasma particles was neglected. The interaction leads to a modification of the state variables. As already mentioned above the interaction reduces the entropy. The coulomb interaction energy of a system consisting of free electrons and ions is given by the expression [11]

$$E_c = \frac{1}{2} \sum_{i(\text{Elec})} \sum_{j(\text{Elec})} \frac{e^2}{4\pi \varepsilon_0 r_{ij}} + \frac{1}{2} \sum_{i(\text{Ion})} \sum_{j(\text{Ion})} \frac{e^2}{4\pi \varepsilon_0 r_{ij}} - \sum_{i(\text{Ion})} \sum_{j(\text{Elec})} \frac{e^2}{4\pi \varepsilon_0 r_{ij}} \tag{9.45}$$

The sum in the first term on the right-hand side of Eq. (9.45) runs over all electrons whereas in the second term it runs over all ions. If the sums run over all indices

9 Plasma Physics

i and j the interaction energy between two particles is counted twice. Because of this the sum has to be multiplied by $1/2$. The third term comprises the electron–ion interaction. Because in this sum the indices belong to different particles the interaction is not counted twice. The r_{ij} are the distances between the respective particles. Performing the summations in Eq. (9.45) in a many-particle system is in general quite tedious and can only be done approximately. In the case that the distances r_{ij} are randomly distributed, i.e., the locations of the particles are not correlated, the contributions cancel and the interaction energy E_c becomes zero. But due to the interaction there are correlations so that E_c does in general not vanish. In the following E_c will be calculated in the DEBYE approximation.

The DEBYE approximation corresponds to the self-consistent field approximation, i.e., the summation over distinct particles is replaced by integration over particle densities or probability densities. Equation (9.45) then becomes

$$E_c = \frac{1}{2} n_{e_0}^2 \iint u_{ee} \, w_{ee}(r) \, dV \, dV + \frac{1}{2} n_{i_0}^2 \iint u_{ii} \, w_{ii}(r) \, dV \, dV + n_{e_0} n_{i_0} \iint u_{ei} \, w_{ei}(r) \, dV \, dV \qquad (9.46)$$

with the interaction energies

$$u_{ee} = u_{ii} = \frac{e^2}{4\pi \varepsilon_0 r} \qquad (9.47)$$

$$u_{ei} = -\frac{e^2}{4\pi \varepsilon_0 r} \qquad (9.48)$$

with

r – distance between particles
$w(r)$ – probability.

The $w(r)$ are the probabilities to find two particles that are separated by r. These probabilities follow from the densities Eq. (9.1) and (9.2), respectively, by dividing by the mean particle densities. With this and the potential Eq. (9.5) it follows that

$$w_{ee} = w_{ii} = 1 - \frac{e^2}{4\pi \varepsilon_0 k_B T} \frac{\exp(-\frac{r}{r_D})}{r} \qquad (9.49)$$

$$w_{ei} = 1 + \frac{e^2}{4\pi \varepsilon_0 k_B T} \frac{\exp(-\frac{r}{r_D})}{r} \qquad (9.50)$$

The probability to find a particle in the vicinity of another particle whose charge has the same sign is reduced compared to the non-interacting case and enhanced if the two particles have opposite sign. Because every particle interacts with every other particle the integration has to be done twice. The factor 1/2 in Eq. (9.46) again is introduced because otherwise the contribution of the interaction between two particles would be counted twice. With the condition of neutrality

$$n_{e_0} = n_{i_0} \tag{9.51}$$

it follows

$$E_c = -2 n_{e_0}^2 \int_V \int_V \frac{e^2}{4\pi\varepsilon_0} \frac{e^2}{4\pi\varepsilon_0 k_B T} \frac{\exp(-\frac{r}{r_D})}{r^2} dV\, dV \tag{9.52}$$

With the volume element

$$dV = 4\pi r^2\, dr \tag{9.53}$$

and the definition of the DEBYE radius Eq. (9.6) it results that

$$E_c = -n_{e_0} \int \frac{e^2}{4\pi\varepsilon_0} \frac{1}{r_D^2} \int_0^\infty \exp\left(-\frac{r}{r_D}\right) dr\, dV \tag{9.54}$$

The COULOMB interaction energy of a plasma in DEBYE approximation is thus given by

$$E_c = -\frac{1}{2} V \frac{e^2}{4\pi\varepsilon_0 r_D} (n_{e_0} + n_{i_0}) \tag{9.55}$$

The COULOMB interaction reduces the internal energy of a plasma compared to a system without interaction due to the fact that the particles are not totally free. If the free electrons would be bound to the ions the number of particles and because of this also the energy of the translational degrees of freedom would halve (in case of fully singly ionized plasma). Inserting Eq. (9.55) into Eq. (9.21) and integrating yields the COULOMB correction of the free energy:

$$F_c = -\frac{1}{3} V \frac{e^2}{4\pi\varepsilon_0 r_D} (n_{e_0} + n_{i_0}) \tag{9.56}$$

The free energy of a plasma is smaller too compared to non-interacting systems. With Eq. (9.20), (9.22), (9.23), and (9.56) the COULOMB corrections of the entropy, the pressure, and the chemical potential of the electrons and ions in the DEBYE approximation follows:

$$S_c = -\frac{\partial F_c}{\partial T}\bigg|_{V,N} = -\frac{1}{3} V \frac{e^2}{4\pi\varepsilon_0 r_D} (n_{e_0} + n_{i_0}) \tag{9.57}$$

$$p_c = -\frac{\partial F_c}{\partial V}\bigg|_{T,N} = -\frac{1}{3}\frac{e^2}{4\pi\varepsilon_0 r_D}(n_{e_0} + n_{i_0}) \tag{9.58}$$

$$\mu_{c_e} = -\frac{\partial F_c}{\partial N_e}\bigg|_{T,p,N_0,N_i} = -\frac{1}{2}\frac{e^2}{4\pi\varepsilon_0 r_D} \tag{9.59}$$

$$\mu_{c_i} = -\frac{\partial F_c}{\partial N_i}\bigg|_{T,p,N_0,N_e} = -\frac{1}{2}\frac{e^2}{4\pi\varepsilon_0 r_D} \tag{9.60}$$

As already mentioned above the correlations among the particles reduce the entropy as well as the pressure. In case of a fully ionized plasma with an electron density of 10^{24} m^{-3} and a temperature of 1 eV the pressure amounts to 1500 kPa when neglecting the COULOMB interaction and the reduction of the pressure due to the interaction amounts to 168 hPa. The chemical potential of the electrons and ions decreases too. The reduction of the chemical potential is equivalent to a reduction of the ionization energy which has significant consequences for the degree of ionization of a plasma. Generalizations of some of the formulas for multiple ionized ions can be found in Appendix D.2.

9.2.4 Law of Mass Action and SAHA Equation

The SAHA equation which describes the equilibrium ionization in a plasma is one of the most important equations in plasma physics and is a special case of the law of mass action. According to the second law of thermodynamics the entropy of an isolated system is maximal. From this it follows that in a system that is in contact with a heat bath and whose volume is fixed the free energy F acquires a minimum value. The free energy contains contributions from atoms (a) (in general there can also be molecules), ions (i), and electrons (e) and is a function of the temperature, the volume, and the particle densities:

$$F = F(T, V, N_a, N_e, N_i) \tag{9.61}$$

In an isolated system the sum of all atoms and ions is constant. Because the free energy acquires a minimum value the variation of the free energy with the particle densities while keeping temperature and volume fixed has to vanish in a state of thermodynamic equilibrium:

$$\frac{\partial F}{\partial N_a}dN_a + \frac{\partial F}{\partial N_e}dN_e + \frac{\partial F}{\partial N_i}dN_i = 0 \tag{9.62}$$

The differentials of the different particle species are not independent of each other. When a free electron is created an ion is created too and a neutral particle is annihilated:

$$dN_a = -dN_e = -dN_i \tag{9.63}$$

Inserting yields

$$\frac{\partial F}{\partial N_e} + \frac{\partial F}{\partial N_i} = \frac{\partial F}{\partial N_a} \tag{9.64}$$

The derivative of the free energy with respect to the density of a particle species while keeping constant all other quantities is the chemical potential of the particle species considered (Eq. (9.22)). Thus

$$\mu_e + \mu_i = \mu_a \tag{9.65}$$

Equation (9.65) expresses that the sum of the chemical potentials of an electron and an ion equals the chemical potential of a neutral atom. With the densities $n = N/V$ the chemical potentials are given by (Eq. (9.43), (9.59), (9.60))

$$\mu_a = k_B T \ln n_a - k_B T \ln \frac{U_a}{\Lambda_a^3} \tag{9.66}$$

$$\mu_e = k_B T \ln n_e - k_B T \ln \frac{U_e}{\Lambda_e^3} + \mu_{ce} \tag{9.67}$$

$$\mu_i = k_B T \ln n_i - k_B T \ln \frac{U_i}{\Lambda_i^3} + \mu_{ci} \tag{9.68}$$

$$\mu_{ce} = \mu_{ci} = -\frac{1}{2} \frac{e^2}{4\pi\varepsilon_0 r_D} \tag{9.69}$$

Inserting in Eq. (9.65) results in

$$\ln n_e + \ln n_i - \ln n_a = \ln \frac{U_e}{\Lambda_e^3} + \ln \frac{U_i}{\Lambda_i^3} - \ln \frac{U_a}{\Lambda_a^3} - (\mu_{ce} + \mu_{ci}) \tag{9.70}$$

With this the SAHA equation of ionization equilibrium in a plasma reads

$$\frac{n_e n_i}{n_a} = \frac{U_e}{\Lambda_e^3} \frac{\Lambda_a^3}{U_a} \frac{U_i}{\Lambda_i^3} \exp[-(\mu_{ce} + \mu_{ci})] = K(T)^{-1} \tag{9.71}$$

with

$K(T)$ – equilibrium constant.

The partition function of the internal degrees of freedom of an electron is 2 (two spin states). The DE'BROGLIE wavelength of ions and atoms, respectively, is almost identical:

$$\Lambda_i \simeq \Lambda_a \tag{9.72}$$

With Eq. (9.31) and (9.28) it follows that

$$\frac{n_e n_i}{n_a} = \frac{2}{\Lambda_e^3} \frac{U_i'}{U_a} \exp\left(-\frac{E_{\text{ion}}^a - \Delta E_{\text{ion}}^a}{k_B T}\right) \tag{9.73}$$

$$\Lambda_e = \frac{h}{\sqrt{2\pi m_e k_B T}} \tag{9.74}$$

$$\Delta E_{\text{ion}}^a = -(\mu_{ce} + \mu_{ci}) = \frac{e^2}{4\pi \varepsilon_0 r_D} \tag{9.75}$$

with

m_e – electron mass
E_{ion}^a – ionization energy
ΔE_{ion}^a – reduction of the ionization energy.

The reduction of the ionization energy is due to the COULOMB interaction. With an electron density of 10^{24} m^{-3} and a temperature of 1 eV the ionization energy is decreased by 0.29 eV. Because the DEBYE radius equation contains the electron density Eq. (9.73) is an implicit, nonlinear equation for determining the electron density, which can only be solved by iterative numerical methods. Even at low temperatures ($k_B T > 0.1 E_{\text{ion}}$) the degree of ionization is considerably high. This is due to the fact that the two particles, electron and ion, can occupy much more states than a single neutral atom. In Eq. (9.15) the sum runs over all states including the degenerate states. Even in case of small values of the exponential factors, the contributions of the higher energy levels can be large due to their great number. Equation (9.73) accordingly also applies in case of higher ionization stages. Instead of the atoms a the corresponding values of the ions with ionization stage ($i - 1$) have to be taken. With $i = 1$ Eq. (9.73) for the atom is reproduced.[4] Figure 9.5a–f shows solution of a system of three SAHA equations of the type Eq. (9.73) for three ionization stages in a Fe plasma.

9.3 Transport Characteristics of Plasmas

In the preceding sections the plasma was assumed to be in a state of thermodynamic equilibrium. Transport processes arise when there are deviations from thermodynamic equilibrium. If the relevant mean free paths that govern the transport processes are small compared to the spatial dimensions over which the plasma state varies locally states of thermodynamic equilibrium exist, at least approximately. The local quantities can be determined using the equilibrium equations. Particle diffusion and heat fluxes are functions of the local density and temperature gradients. The following treatment is restricted to those cases in which the electric current densities

---

[4] The generalization in case of multiple charged ions can be found in Appendix D.2.

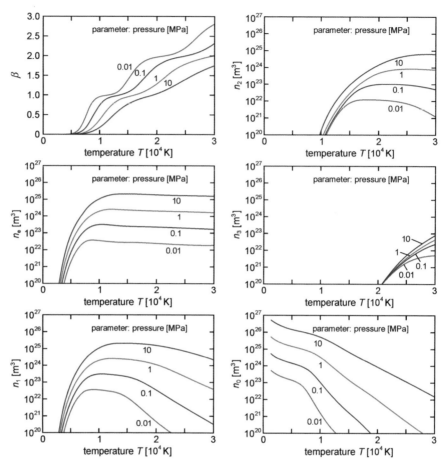

Fig. 9.5 Figures a–f show solutions of a system of three equations of the type Eq. (9.73) for three ionization stages of Fe. The generalization of the SAHA equation in case of multiple charged ions can be found in Appendix D.2. The ionization energies are FeI: 6.9 eV, FeII: 16.18 eV, and FeIII: 30.64 eV [4]. The plasma is neutral, i.e., $n_e = n_1 + 2 n_2 + 3 n_3$. n_1, n_2, and n_3 are the densities of the single, twofold, and threefold charged ions, respectively. Parameter is the pressure, i.e., $p(n_0 + n_e + n_1 + n_2 + n_3) k_B T$. The pressures are 0.01, 0.1, 1, and 10 MPa (0.1, 1, 10, 100 bar), respectively. Figure a shows the electron density normalized with respect to the sum of the densities of the heavy particles as a function of temperature. At the temperature values at which β increases more sharply the mean energy becomes high enough to ionize the next ionization stage. Figures b–e show the electron density, the density of the single, twofold, and threefold charged ions and figure f the density of the neutral particles as a function of the temperature

are only functions of the local electric field. The complex electric high-frequency conductivity is derived in Chap. 3 and Appendix A.2 (Eq. (3.41)):

$$\sigma = \frac{e^2 n_e}{m_e \nu_m} \frac{\nu_m}{\nu_m - i \omega} \qquad (9.76)$$

In Eq. (9.76) only the contribution of the electrons is included. The momentum transfer frequency ν_m is the inverse of the mean time for momentum transfer of the electrons to other particles during collisions. The momentum transfer frequency is given by[5]

$$\nu_m = n \,\overline{v\, \sigma_m(v)} \qquad (9.77)$$

with

n – density of collision partners
v – electron speed
$\sigma_m(v)$ – momentum transfer cross section.

The overline indicates time averaging. The momentum transfer cross section is given by

$$\sigma_m(v) = \int_0^{4\pi} \frac{d\sigma_c}{d\Omega}(v)\,(1-\cos\phi)\,d\Omega \qquad (9.78)$$

The product of differential cross section and the factor $(1-\cos\phi)$ is integrated over the solid angle from 0 to 4π. The factor $(1-\cos\phi)$ is the relative value of momentum transfer that occurs during an elastic collision with a collision partner whose mass by far exceeds the mass of the colliding particle when the colliding particle is scattered by an angle of ϕ. In case of forward scattering, i.e., $\phi = 0$, there is no momentum transfer; if the scattering angle equals $\pi/2$ the momentum transfer equals the momentum of the colliding particle and in case of backward scattering, i.e., $\phi = \pi$, twice the momentum of the colliding particle is transferred to the heavy collision partner.

The collision frequency in Eq. (9.76) in general consists of several contributions. As long as the mean electron energy is small compared to all excitation energies of the heavy particles there are much more elastic than inelastic collisions, so that it suffices to only consider the contributions of the elastic collisions to the momentum transfer frequency. In general a plasma is not fully ionized so that the collision frequency is composed of the contributions of the neutral particles and the ions:

$$\nu_m = \nu_{m,ea} + \nu_{m,ei} \qquad (9.79)$$

The differential cross section depends on the interaction potential between electron and collision partner. In case of neutral collision partners the interaction range is short and the interaction potential depends on the atomic structure. The cross sections are either measured or calculated theoretically within the frame of quantum

---

[5] Equation (9.76) is only strictly valid if the collision frequency $\nu_m(v)$ is constant; nevertheless Eq. (9.76) describes the conductivity also in case of non-constant $\nu_m(v)$ if an appropriate mean value of the collision frequency is used.

mechanics. In the following only simple models of electron-neutral collisions are presented. At small energies in many cases of interest the elastic cross section is approximately velocity independent. With a MAXWELLian distribution of electron velocities the collision frequency is thus given by

$$v_{m,ea} = n_a \sigma_0 \bar{v} = n_a \sigma_0 \sqrt{\frac{8\,k_B\,T}{\pi\,m_e}} \qquad (9.80)$$

The collision frequency Eq. (9.80) is a monotone function of the temperature. The collision cross section σ_0 is of the order of 10^{-19} m^2. With an electron temperature of about 1 eV it follows that

$$\frac{v_{m,ea}}{n_a} \simeq 10^{-13} \text{ m}^3 \text{ s}^{-1} \qquad (9.81)$$

At high energies the collision cross sections approximately decrease inversely proportional to the velocity so that the collision frequency is independent of the velocity and thus also independent of the temperature. In case of ions and at distances that are large compared to the ion radius the interaction potential can very well be described by a COULOMB potential. The details of the ion potential at small distances can be neglected as long as the COULOMB potential dominates the collisions. The differential cross section for a COULOMB collision between an electron and an ion is given by [10]

$$\frac{d\sigma_c}{d\Omega}(v) = \frac{Z^2\,e^4}{(4\,\pi\,\varepsilon_0)^2\,(m_e\,v^2)^2\,\sin^4(\phi/2)} \qquad (9.82)$$

with

Z – ion charge
v – velocity of the electron far away from the ion
ϕ – scattering angle.

The impact parameter b is connected with the scattering angle by the relation

$$\cot(\phi/2) = \frac{b\,m_e\,v^2\,4\,\pi\,\varepsilon_0}{Z\,e^2} \qquad (9.83)$$

Inserting the differential cross section Eq. (9.82) into Eq. (9.78) together with the solid angle element

$$d\Omega = 2\,\pi\,\sin\phi\,d\phi\,d\psi \qquad (9.84)$$

results in the momentum transfer cross section:

$$\sigma_{m,ei}(v) = \frac{4\pi \, Z \, e^4}{(4\pi\,\varepsilon_0)^2 \, (m_e \, v^2)^2} \int_{\phi_{min}}^{\pi} \frac{(1-\cos\phi)\sin\phi}{\sin^4(\phi/2)} \, d\phi \qquad (9.85)$$

The integral gives

$$\int_{\phi_{min}}^{\pi} \frac{(1-\cos\phi)\sin\phi}{\sin^4(\phi/2)} \, d\phi = \ln\left(\frac{1}{\sin(\phi_{min}/2)}\right) \qquad (9.86)$$

Equation (9.86) has a logarithmic divergence if $\phi_{min} \to 0$. Small scattering angles are associated with large impact parameters, i.e., the distance between the collision partners is large. The divergence is thus due to the infinite range of the COULOMB potential. As was pointed out when treating the DEBYE screening in Sect. 9.1 the potential of an ion in a plasma drops much faster than the pure COULOMB potential. In computing the cross section the screened potential Eq. (9.5) should have been used but there is no simple analytical solution for this. Another possibility to avoid the divergence is to cut off the COULOMB potential at the DEBYE radius. The maximal impact parameter then equals the DEBYE radius. With this the minimal scattering angle is using Eq. (9.83) given by

$$\phi_{min} = 2 \arctan\left(\frac{Z\,e^2}{4\pi\,\varepsilon_0\,r_D\,m_e\,v^2}\right) \qquad (9.87)$$

The argument of arctan is besides a factor of 2 the ratio of the potential energy in a COULOMB potential at the distance $r = r_D$ (r_D = DEBYE radius) and the kinetic energy of the impacting electron. When this ratio is large the scattering angle becomes large. An essential prerequisite of the DEBYE approximation is that the mean kinetic energy is large compared to the mean potential energy. Then the argument in the arctan function is small for most of the electrons if their velocities are MAXWELLian distributed and because of this the scattering angle is small too. Expanding the arctan function in case of small arguments, inserting in Eq. (9.86), and substituting v^2 with its mean value

$$\overline{v^2} = \frac{3\,k_B\,T}{m_e} \qquad (9.88)$$

results in

$$\ln\left(\frac{1}{\sin(\phi_{min}/2)}\right) = \ln \Lambda_c \qquad (9.89)$$

$$\Lambda_c = \frac{12\pi\,\varepsilon_0\,r_D\,k_B\,T}{Z\,e^2} \qquad (9.90)$$

with

$\ln \Lambda_c$ – COULOMB logarithm.

In deriving this expression the slow electrons were neglected and the cut off of the COULOMB potential at the DEBYE radius is rather somewhat arbitrary than justified exactly. But because this only enters logarithmically the error that is introduced by this approximation is relatively small. This holds at least as long as $\ln \Lambda_c$ is large compared to 1, which is the case if the number of particles within the DEBYE sphere is large. The ions and electrons within the DEBYE sphere can then be treated like a continuous background that shields the charge of an ion outside the region given by the DEBYE radius. This is the regime of ideal plasmas with the mean kinetic energy by far exceeding the mean interaction energy. If on the other hand the number of particles within the DEBYE sphere is close to 1, which implies that the interaction energy is not negligible any more, the concept of shielding has to be modified [18, 7]. This is the regime of non-ideal plasmas that can be described accurately only within the frame of quantum physics. In the following only some simple considerations are presented. In order to avoid the divergence of the COULOMB logarithm $\ln[\Lambda_c]$ can be substituted by $\ln[1 + \Lambda_c]$. The expression for the shielding length r_s has to be modified too. One possibility is to take the mean electron distance as the lower limit of the shielding length, i.e.,

$$r_s = r_D \quad : r_D > n_e^{-1/3} \tag{9.91}$$
$$r_s = n_e^{-1/3} \quad : r_D < n_e^{-1/3} \tag{9.92}$$

GÜNTHER et al. [7] derived the following expression for the shielding length from experiments with weakly non-ideal plasmas in the case of $N_D \geq 1$:

$$r_s = r_D \left(5 \exp\left(-\frac{2}{3} N_D\right) + 1\right) \tag{9.93}$$

with

N_D – number of electrons and ions within a DEBYE sphere (Eq. (9.8)).

The electron–ion momentum transfer cross section is using Eq. (9.85) and (9.89) given by

$$\sigma_{m,ei}(v) = 4\pi \frac{Z^2 e^4}{(4\pi \varepsilon_0)^2 (m_e v^2)^2} \ln \Lambda_c \tag{9.94}$$

The electron–ion momentum transfer frequency is with Eq. (9.77)

$$v_{m,ei}(v) = 4\pi \frac{Z^2 e^4}{(4\pi \varepsilon_0)^2 m_e^2 v^3} \ln \Lambda_c \, n_i \tag{9.95}$$

9 Plasma Physics

The COULOMB logarithm is only valid for not too small electron velocities. Because of this Eq. (9.95) is not averaged by integrating over all velocities but by substituting v^3 by its mean value. In case of a MAXWELLian distribution of the electron velocities this is given by

$$\overline{v^3} = 8\sqrt{\frac{2}{\pi}} \left(\frac{k_B T}{m_e}\right)^{3/2} \tag{9.96}$$

With this it follows that

$$\nu_{m,ei} = \left(\frac{\pi}{2}\right)^{3/2} \frac{Z^2 e^4}{(4\pi\varepsilon_0)^2 \sqrt{m_e} (k_B T)^{3/2}} \ln \Lambda_c \, n_i \tag{9.97}$$

$$\nu_{m,ei} = 1.7 \times 10^{-12} \, \text{s}^{-1} \, \text{m}^3 \left(\frac{k_B T}{e}\right)^{-3/2} \ln \Lambda_c \, n_i \, Z^2 \tag{9.98}$$

The DC ($\omega = 0$) electron conductivity of a fully ionized plasma without neutral particles and with only singly ionized ions is with Eq. (9.76) given by

$$\sigma_e = \left(\frac{2}{\pi}\right)^{3/2} \frac{(4\pi\varepsilon_0)^2 (k_B T)^{3/2}}{e^2 \sqrt{m_e}} \frac{1}{\ln \Lambda_c} \frac{n_e}{n_i} \tag{9.99}$$

$$\sigma_e \approx 1.646 \times 10^4 \frac{1}{\Omega \, \text{m}} \left(\frac{k_B T}{e} \frac{1}{V}\right)^{3/2} \frac{1}{\ln \Lambda_c} \frac{n_e}{n_i} \tag{9.100}$$

If the electron and ion densities are equal then the conductivity is independent of the density.[6] The conductivity increases proportional to $T^{3/2}$.

In Eq. (9.99) only the electron–ion collisions are accounted for. This plasma model is called LORENTZ plasma. If electron–electron collisions are taken into account too the electric conductivity is described by the SPITZER equation [8]:

$$\sigma_{sp} = 1.16 \left(\frac{2}{\pi}\right)^{3/2} \frac{(4\pi\varepsilon_0)^2 (k_B T)^{3/2}}{e^2 \sqrt{m_e}} \frac{1}{\ln \Lambda_c} \tag{9.101}$$

Equations (9.99) and (9.101) only differ by the prefactor of 1.16. For the sake of completeness the electron heat conductivity of a fully ionized plasma is given here. The heat conductivity of an ideal gas is [17]

$$K_e = \frac{5}{2} \frac{n \, k_B^2 \, T}{m \, \nu_m} \tag{9.102}$$

With Eq. (9.97) the electron heat conductivity of a fully ionized ideal plasma is then given by

---

[6] Neglecting the density dependence of the COULOMB logarithm.

$$K_e = \frac{5}{2} \left(\frac{2}{\pi}\right)^{3/2} \left(\frac{4\pi\varepsilon_0}{e^2}\right)^2 \frac{e^{5/2} k_B}{\sqrt{m_e}} \left(\frac{k_B T}{e}\right)^{5/2} \frac{n_e}{n_i} \frac{1}{\ln \Lambda_c} \quad (9.103)$$

$$K_e \approx 3.6 \, \frac{W}{m \, K} \left(\frac{k_B T}{e} \frac{1}{V}\right)^{5/2} \frac{n_e}{n_i} \frac{1}{\ln \Lambda_c} \quad (9.104)$$

9.4 Interaction Between Electromagnetic Waves and Plasmas

The absorption of electromagnetic radiation by plasma electrons is treated in Sect. 9.3 in the framework of the so-called microwave approximation. The absorption mechanism is also called inverse bremsstrahlung. The equation of motion of the mean electron momentum is

$$\frac{\partial \bar{v}}{\partial t} + \nu_m \bar{v} = -\frac{e}{m_e} E \quad (9.105)$$

This simple equation of motion for the mean electron momentum is only an approximation. A more exact calculation can only be done within the frame of kinetic models. But the basic relationships can also be drawn from the simple models so that kinetic models will not be discussed here. In case of time harmonic fields and using Eq. (9.105) it follows that

$$\bar{v} = -\frac{e}{m_e} E \frac{\nu_m + i\omega}{\nu_m^2 + \omega^2} \quad (9.106)$$

In the collision-free case ($\nu_m = 0$) and at steady state the electron velocity and the electric field are out of phase by 90°. This means that no net energy is absorbed during one field cycle, the energy that is gained by the electron during one half cycle is lost again by the electron during the next half cycle. The mean energy is

$$\varepsilon = \frac{m_e}{2} \bar{v}^2 = \frac{e^2}{2 m_e \omega^2} |E|^2 \quad (9.107)$$

With a field strength of $E = 2 \times 10^6$ V m^{-1} which corresponds to an intensity of 10^{10} W m^{-2} and the frequency of CO$_2$ lasers $\omega = 1.88 \times 10^{14}$ s^{-1} the mean electron energy is 10^{-5} eV, which is far below the mean thermal energy of a hot vapor of 0.3 eV ($T \approx 3000$ K). If there are collisions things change dramatically. The collisions disturb the phase relation between electron movement and electric field. When at the instant of field reversal the electron collides such that the direction of its velocity is reversed too the electron can gain energy in the next half cycle as well instead of losing energy. After the next half cycle the same holds again. The collision frequency in Eq. (9.106) does not imply that the collisions occur exactly periodically, the reciprocal of the collision frequency is just the time between two consecutive collisions (of the same electron) averaged over many collisions. Besides this not every collision results in a reversal of the direction of the velocity, rather every angle can be realized. Equation (9.106) describes the average over many collisions. The mean value of the power absorbed by an electron is

$$\frac{1}{2}\,\mathrm{Re}\left[\bar{v}(e\,E^*)\right] = \frac{1}{2}\frac{e^2}{m_e}\frac{v_m}{v_m^2+\omega^2}\,|E|^2 \qquad (9.108)$$

In the collision-free case the absorbed power vanishes as outlined above. In case of very large collision frequencies ($v_m \gg \omega$) the absorbed power decreases inversely proportional to the collision frequency. The maximum of the power absorption is reached at $v_m = \omega$. This corresponds to the above-sketched case of collisions that occur at the instant of field reversal. With a field strength of $E = 2\times 10^6$ V m^{-1} and the frequency of CO_2 lasers the mean absorbed power is 10^9 eV s^{-1}. The electron energies are in the 1 eV range so that the time constant for energy absorption is in the 1 ns range. The intensity absorptivity of an electromagnetic wave is twice the imaginary part of the complex wavenumber[7]:

$$\alpha = 2\,\mathrm{Im}\,[k] = 2\,\frac{\omega}{c}\,\mathrm{Im}(n_c) \qquad (9.109)$$

In the limit of small plasma frequencies (compared to the frequency of the radiation field) the absorption coefficient in the microwave approximation is given by (Eq. (3.56))

$$\alpha \simeq \frac{\omega_p^2\,v_m}{c\,\omega^2}\,\frac{1}{1+\frac{v_m^2}{\omega^2}} \qquad (9.110)$$

$$\omega_p^2 = \frac{e^2\,n_e}{m_e\,\varepsilon_0} \qquad (9.111)$$

The second factor on the right-hand side of Eq. (9.110) describes correlations between distinct collisions. These correlations cause a reduction of the absorption [16]. The microwave approximation is mainly based on the assumption that the photon energy $\hbar\omega$ is small compared to the mean thermal energy of the electrons. During interaction of electrons with electromagnetic fields both stimulated emission and absorption take place. Equation (9.109) represents the difference of both processes. All electrons can participate in absorption but only those electron can take part in emission whose energy exceeds the photon energy. If the photon energy is larger than the energy of a part of the electrons these electrons cannot participate in emission and thus the net absorption, which is the difference of absorption and stimulated emission, is larger than given by Eq. (9.110).

---

[7] See Sect. 3.3.

The impact of the collisions is described above classically as a disturbance of the phase relation between electromagnetic field and electron movement. Quantum mechanically the absorption is described as a collision process with the collision partners electron, photon, and another particle. A photon has the energy $\hbar\omega$ and the momentum $\hbar\omega/c$. If the electron takes the energy of the photon then the electron momentum increase corresponding to the energy $\hbar\omega$ is much larger than the photon momentum. The momentum excess has to be transferred to the third collision partner. During scattering of low-energy photons at free electrons the photon energy does not change. This scattering process is called THOMSON scattering (in case of high photon energies COMPTON scattering takes place, which in general is accompanied by energy transfer). When the photon energy is in the range of the mean electron energy or above not all electrons can emit photons because their energy is too low. The true absorption coefficient is given by the difference of absorption coefficient and coefficient of stimulated emission [1]:

$$\alpha_\omega = \alpha_{\omega A} - \alpha_{\omega S} \qquad (9.112)$$

The absorption coefficient follows by averaging the velocity-dependent absorption coefficient of an electron $\eta_{\omega A}(v)$ with respect to the velocity distribution of the electrons and multiplying by the electron density [1]:

$$\alpha_{\omega A} = n_e \int \eta_{\omega A}(v) f(v) 4\pi v^2 \, dv \qquad (9.113)$$

with

$\eta_{\omega A}(v)$ – coefficient of absorption of a photon of energy $\bar{h}\omega$ by an electron of velocity v.

The electron velocity after absorption is v'. Energy conservation demands

$$\frac{m_e}{2} v'^2 = \frac{m_e}{2} v^2 + \hbar\omega \qquad (9.114)$$

Correspondingly there is a similar relation for the coefficient of stimulated emission $\eta_{\omega S}(v')$. Here v' is the electron velocity before the emission and v the electron velocity after the emission:

$$\alpha_{\omega S} = n_e \int \eta_{\omega S}(v') f(v') 4\pi v'^2 \, dv' \qquad (9.115)$$

From the principle of detailed balancing a relationship follows between $\eta_{\omega A}$ and $\eta_{\omega S}$ [1]:

$$\eta_{\omega A}(v) v^2 \, dv = \eta_{\omega S}(v') v'^2 \, dv' \qquad (9.116)$$

9 Plasma Physics

$v^2 \, dv$ and $v'^2 \, dv'$, respectively, are, except a constant factor, the number of states within the velocity intervals $(v + dv)$ and $(v' + dv')$, respectively. In case of a MAXWELLian distribution of the electron velocities Eqs. (9.112), (9.113), (9.114), (9.115), and (9.116) give

$$\alpha_\omega = n_e \int \eta_{\omega A}(v) \, f(v) \, 4\pi \, v^2 \, dv \left(1 - \exp\left(-\frac{\hbar \omega}{k_B T}\right)\right) \quad (9.117)$$

Quantum mechanical determinations of $\eta_{\omega A}$ have been done using different approximations. In case of electron–atom collisions HOLSTEIN [9] derived the following relation:

$$\eta_{\omega A} \simeq \frac{n_a \, e^2}{3 \, \varepsilon_0 \, c \, \hbar \, \omega^3} v'^3 \left(1 - \frac{1}{2} \frac{\hbar \omega}{\frac{1}{2} m_e v'^2}\right) \sigma_{m,a} \left(\frac{m_e}{4} (v^2 + v'^2)\right) \quad (9.118)$$

Correlations among collisions are neglected. The momentum transfer cross section for electron–atom collisions $\sigma_{m,a}$ may only depend weakly on the electron energy. With the assumption

$$n \, v' \, \sigma_{m,a} \left(\frac{m_e}{4}(v^2 + v'^2)\right) \simeq v_{m,a} \simeq \text{const.} \quad (9.119)$$

and a MAXWELLian velocity distribution the absorption coefficient is given by

$$\alpha \approx \frac{\omega_p^2 \, v_{m,a}}{c \, \omega^2} \left(\frac{k_B T}{\hbar \omega} + \frac{1}{3}\right) \left[1 - \exp\left(-\frac{\hbar \omega}{k_B T}\right)\right] \quad (9.120)$$

With $\hbar \omega \ll k_B T$ Eq. (9.110) follows again (except a factor that describes the correlations among collisions).

In case of electron–ion collisions the calculations become much more involved. Some results can be found in BEKEFI [1]. Inserting the electron–ion collision frequency Eq. (9.97) instead of the electron–atom collision frequency Eq. (9.120) approximately gives the absorption coefficient in case of electron–ion collisions. Figure 9.6 shows the sum of the electron–atom and electron–ion collision frequencies of a Fe-plasma as a function of the temperature according to the values given in Fig. 9.5a–f with a circular frequency of 1.88×10^{14} s^{-1} (CO_2 laser). The parameter is the total pressure without COULOMB corrections:

$$p = (n_e + n_a + \sum n_i) k_B T \quad (9.121)$$

Figure 9.7 shows the corresponding electron plasma frequency and Fig. 9.8 the absorption coefficient Eq. (9.109).

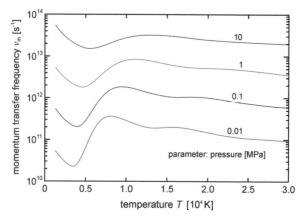

Fig. 9.6 Sum of electron–atom and electron–ion collision frequencies in a Fe-plasma as a function of the temperature computed using the values given in Fig. 9.5a–f and Eq. (9.80) and (9.97)

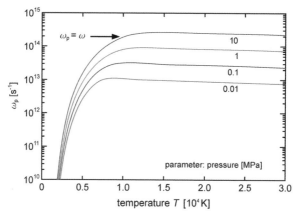

Fig. 9.7 Electron plasma frequency in a Fe-plasma as a function of the temperature computed using the values given in Fig. 9.5a–f and Eq. (9.111). The plasma frequency reaches at $p = 10$ MPa and $T \gg 1$ eV the circular frequency of the CO_2 laser radiation of $\omega = 1.87 \times 10^{14}$ s^{-1}

Besides inverse bremsstrahlung other absorption mechanisms as well as scattering can occur. Multiphoton ionization preferably occurs in the visible and ultraviolet regions and at high radiation intensities. The principal incoherent scattering process in laser-induced plasmas is RAYLEIGH scattering at small metal droplets[8] [13]. THOMSON scattering at free electrons can be neglected. Besides these incoherent processes laser radiation can be absorbed by exciting plasma waves and there can also be coherent scattering processes like stimulated BRILLOUIN scattering and stimulated RAMAN scattering. But these effects are normally unimportant at the laser intensities that are used during laser material processing.

---

[8] When the size of the droplets approaches the laser wavelength MIE scattering takes place.

9 Plasma Physics

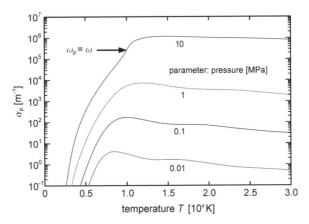

Fig. 9.8 Absorption coefficient of a Fe-plasma as a function of the temperature. Computed according to the values given in Fig. 9.5a–f with the circular frequency ω of CO_2 laser radiation. $\mathrm{Im}(n) = \kappa$ is the imaginary part of the index of refraction of a plasma (Eq. (3.47)). The strong increase of the absorption coefficient around $\omega_p = \omega$ is due to the fact that an electromagnetic wave in a collision-free plasma cannot propagate any more if the plasma frequency exceeds the circular frequency of the wave and is reflected totally (see Fig. 9.5a–f). Due to collisions this behavior is modified but as long as the collision frequency is not too high the principal characteristic is maintained

A further effect is the modification of the collision cross sections induced by electromagnetic radiation. Of special interest is the reduction of the ionization energy in a strong laser field. The determination of the collisional ionization cross section in the presence of a strong laser field has to be performed within the frame of quantum mechanics [3]. A simple estimate of the upper limit of the reduction of the ionization energy due to the laser field can be obtained by investigating the DC case. The potential that acts upon an electron far away from the kernel and the other hull electrons is approximately given by the COULOMB potential. In the presence of a DC field the total potential is given by the superposition of both potentials:

$$\phi = -\frac{e}{4\pi\varepsilon_0 r} - E_{\mathrm{DC}}\, r \tag{9.122}$$

with

E_{DC} – DC electric field.

The potential has its maximum value at

$$r(\phi_{\max}) = \sqrt{\frac{e}{4\pi\varepsilon_0 E_{\mathrm{DC}}}} \tag{9.123}$$

Inserting into Eq. (9.122) yields

$$\phi_{\max} = -2\sqrt{\frac{eE_{DC}}{4\pi\varepsilon_0}} \tag{9.124}$$

The continuum limit of the COULOMB potential is at zero energy. With this it follows for the reduction of the ionization energy that

$$\Delta E_{\text{ion}} = 2e\sqrt{\frac{eE_{DC}}{4\pi\varepsilon_0}} \tag{9.125}$$

The electric field strength of an electromagnetic wave of intensity I is given by

$$E = \sqrt{Z_0 I} \tag{9.126}$$

with

$Z_0 = 377\,\Omega$ – vacuum wave resistance.

Equation (9.126) only applies in vacuum. But as long as the plasma does not significantly change the wave propagation Eq. (9.126) can approximately also be used within plasmas. At an intensity $10^{10}\,\text{W}\,\text{m}^{-2}$ the field strength is $E = 1.9\times 10^6\,\text{Vm}^{-1}$ and the reduction of the ionization energy is $\Delta E_{\text{ion}} = 0.1$ eV. Because Eq. (9.126) was derived in the case of a DC field this value has to be understood as an upper bound. A more precise determination is, as already mentioned above, much more involved.

9.5 Non-equilibrium Processes

Up to now it was assumed that the plasma is in a state of at least local thermodynamic equilibrium. This implies that the velocity distribution function of the particles is a MAXWELLian, that all occupation numbers of atomic and ionic levels are given by BOLTZMANN factors and the ionization degree is governed by SAHA equations,[9] and that a single temperature can be applied. But thermodynamic equilibrium is more or less a special case, lasers, e.g., are an example of highly non-equilibrium systems. In the following the conditions for thermodynamic equilibrium to exist in laser-induced plasmas will be discussed and the limits of its validity will be estimated.

The strength of coupling between subsystems and other heat baths is of principal importance. To illustrate this a simple model is adopted (Fig. 9.9). Energy is fed into a system and extracted from another one. Both systems are assumed to be in thermal

---

[9] This applies accordingly to the dissociation equilibrium of molecules.

9 Plasma Physics

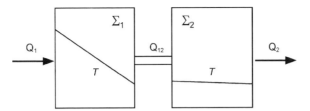

Fig. 9.9 Schematic of the thermal coupling between two systems. In system 1 energy is fed into whereas energy is extracted out of system 2. If the heat conductivity is large the temperature within the system is almost constant. This is assumed here for system 2. When the heat contact between the systems is large then the temperature in both systems at the point of contact will be equal

contact. When the heat conductivity within one of the systems is large, which means that the coupling between distinct regions of the system is large, the temperature will be almost constant everywhere within the system. The state of this system can then be characterized by a single temperature. When the heat contact between the two systems is large too the temperatures of both systems will also be equal. The condition for homogeneous temperatures depends not only on the strength of the coupling but also on the values of the energy fluxes that are fed in and extracted and the time scales of the processes involved. If at time $t = 0$ the energy flux into system 1 is switched on it will take some characteristic time to reach a state of homogeneous and equal temperatures.

The subsystems within a plasma are the free electrons, the translational degrees of freedom of atoms and ions, the electronic levels of atoms and ions, and the plasma radiation. Furthermore, electron–ion pairs are created and annihilated (ionization and recombination). The laser field energy is mainly fed into the plasma by inverse bremsstrahlung. The electrons collide with other electrons and with the heavy plasma particles. Due to the equal masses of the collision partners electron–electron collisions are accompanied by large energy exchanges. With degrees of ionization above 10^{-4} the electron–electron collisions drive the electron distribution function toward a MAXWELLian distribution. This applies accordingly to the heavy particles though the temperatures in general do not have the same values.

The mean electron energy is determined by the absorbed laser energy and by the energy transfer to other particle species. During elastic collisions between electrons and heavy particles there is a large momentum transfer but only a very small energy transfer. In case of inelastic collisions during which a heavy particle is excited or ionized the amount of energy transfer is much larger. In order to excite an atom the electron energy must exceed a threshold value, namely the excitation energy. Because of this threshold the energy losses due to inelastic collisions increase strongly with increasing mean electron energy. The energy conservation is given by

$$\frac{d(\varepsilon\, n_e)}{dt} = \alpha\, I - P_{\text{elast}} - P_{\text{ex}} - P_{\text{ion}} - P_{\text{rad}} + P_{\text{rec}} + P_{\text{superelast}} \qquad (9.127)$$

The mean energy ε of the electrons is increased by the interaction with the laser field. The first term on the right-hand side of Eq. (9.127) is the time-averaged value of the absorbed power density. The electrons lose energy by elastic collisions with heavy particles (P_{elast}), by excitation and ionization collisions (P_{ex}, P_{ion}), and by radiation (P_{rad}). The radiation losses of the electrons which are mainly due to bremsstrahlung can be neglected at the electron energies that exist in laser-induced plasmas during laser materials processing. The electron gas gains energy during electron–ion recombination (P_{rec}) and during superelastic collisions ($P_{superelast}$). The time constant of equilibration of the electron energy in laser-induced plasmas during laser material processing is in the range of 1 ns to several 100 ns. This time constant is mainly determined by the inelastic collisions, while the time constant for energy transfer from the electrons to the translational degrees of freedom of the heavy particles is determined by the elastic collisions. Due to the small mass ratio of electron and ion or atom mass the energy transfer per collision is quite low. The mean energy transfer is given by

$$\dot{\varepsilon} = \frac{2\,m_e}{M}\,v_m\,\varepsilon \qquad (9.128)$$

with

M – mass of the heavy particles (ions, atoms)
m_e – electron mass.

In a Fe-plasma ($M \approx 10^5\,m_e$) with a collision frequency of 10^{11} s^{-1} the electron relaxation time constant approximately amounts to $\tau_\varepsilon \approx 500$ ns.

The number of electrons initially increases avalanche like by collisional ionization of atoms and ions until the electron creation is compensated for by recombination and other loss processes like diffusion. The electron density conservation reads

$$\frac{dn_e}{dt} = R_{ion} - R_{diff} - R_{rec} \qquad (9.129)$$

Electrons are produced by ionizing collisions (R_{ion}), losses are due to diffusion (R_{diff}) and recombination (R_{rec}). These processes and especially ionization depend on the mean electron energy. For simplification the transport processes will not be treated explicitly, the diffusion losses can be accounted for by approximate expressions. The electrons transfer energy to the heavy particles by excitation, ionization, and by energy transfer to the translational degrees of freedom of the heavy particles. When the mean energies of these energy reservoirs reach the mean electron energy the electrons gain on the average as much energy during superelastic collisions as they lose during inelastic and elastic collisions. In case of a steady power absorption from the laser field the mean electron energy and thus the energy of the degrees of freedom of the heavy particles continue to increase. For a stationary plasma state to exist the plasma has to lose energy. This can be due to radiation, energy transfer to the workpiece, or by transformation of the internal energy of the plasma into directional translation energy. The latter mechanism implies acceleration of the plasma. The energy transfer to the workpiece can be due to heat conduction (energy of the translational degrees of freedom) or by transfer of excitation and recombination energy which is more effective. The contribution of the radiation to the energy conservation is difficult to estimate. In case of black body radiation the energy density of the radiation in the frequency interval $(v, v + dv)$ is given by PLANCK's law:

$$P_v \, dv = \frac{8 \pi v^2}{c^3} \frac{h v}{\exp\left(\dfrac{h v}{k_B T}\right) - 1} \, dv \qquad (9.130)$$

The first factor on the right-hand side is the mode density in the frequency interval $(v, v + dv)$, the second factor is the mean energy of a mode of frequency v. The intensity results by multiplying the energy density of the radiation with the velocity of light c. Integration of the intensity over all frequencies results in the STEFAN–BOLTZMANN law:

$$S = \sigma \, Z^4 \qquad (9.131)$$

with

$$\sigma = 5.7 \times 10^{-8} \, \text{W m}^{-2} \, \text{K}^{-4}.$$

With a temperature of 11,000 K and a surface of 10^{-6} m², $S = 834$ W. Equation (9.130) and (9.131) only apply in case of complete thermodynamic equilibrium including radiation. The PLANCK radiation spectrum is continuous. When PLANCK's law applies the microscopic nature of the radiating system is unimportant. The region between spectral lines is filled by line broadening and by continuum radiation (bremsstrahlung and recombination radiation). The contribution of continuum radiation processes becomes comparable to the line radiation only at very high densities, which do not exist in many laser-induced plasmas that are created during laser materials processing. Because of this the radiation spectra of these plasmas are mainly dominated by line radiation. Whereas thermodynamic equilibrium in many cases of interest exists, at least approximately, among the plasma particles, this holds much less often for the radiation. The important quantity is the mean free path of the photons. If the photon mean free path is small compared to the plasma dimension the photons are absorbed and re-emitted quite often before they can leave the plasma. The radiation that leaves the plasma is then mainly emitted from within a thin layer at the plasma surface which is a basic assumption of Eq. (9.131).

The quantitative description of the radiation of plasmas is quite involved in general. There are mainly two models that are used. In the LTE model[10] it is assumed that the states of the atoms and ions are populated according to the BOLTZMANN

---

[10] LTE = **L**ocal **T**hermodynamic **E**quilibrium.

law. The prerequisite for this to apply is that there are much more transitions due to superelastic collisions than radiation transitions. In this case the radiation does not have a significant impact on the population densities. In the CR model[11] on the other hand deexcitation by collisions are neglected. The excitation results from electron collisions, the deexcitation from radiation transitions. In the following the plasma radiation in the LTE model is discussed.

9.6 Plasma Radiation in the LTE Model

In laser-induced plasmas several distinct elemental species can be present each with several ionization stages. With the assumption that the particles are in local thermodynamic equilibrium, the states of atoms and ions are occupied according to the BOLTZMANN law. For the description of such systems the following definitions are used:

a — designates the element

z — ionization stage $z = 0, 1, ...$

n — energy state (main quantum number)

The densities of atoms, ions, and electrons are governed by a system of SAHA equations

$$\frac{N_e N_{a,z+1}}{N_{a,z}} = \frac{2}{\Lambda_e^3} \frac{U_{a,z+1}(T, \Delta E_{a,z+1}^{ion})}{U_{a,z}(T, \Delta E_{a,z}^{ion})} \exp\left(-\frac{E_{a,z}^{ion} - \Delta E_{a,z}^{ion}}{k_B T}\right) \quad (9.132)$$

$$\Lambda_e = \frac{h}{\sqrt{2\pi m_e k_B T}} \quad (9.133)$$

subject to the constraint of constant total number of the heavy particles

$$N_a = \sum_{z=0}^{n_{z_a}-1} N_{a,z} \qquad : z = 0, ..., n_{z_a} - 1 \quad (9.134)$$

and of charge neutrality:

$$N_e = \sum_{a=0}^{n_a-1} \sum_{z=0}^{n_{z_a}-1} z N_{a,z} \qquad : a = 0, ..., n_a - 1 \quad (9.135)$$

n_{z_a} is the number of relevant ionization stages of element a. If temperature and the total density of the heavy particles are given this system of equations has a

---

[11] CR = Collisional Radiative.

unique solution, although it can in general only be found by numerical methods. The partition functions are

$$U_{a,z}\left(T, \Delta E_{a,z}^{\text{ion}}\right) = \sum_{n=1}^{n_{\text{max}}} g_{a,z;n} \exp\left(-\frac{E_{a,z;n}}{k_B T}\right) \tag{9.136}$$

with

n_{max} – highest excited but still bound state

and the reduction of the ionization potential of the ith ionization stage[12]

$$\Delta E_{a,z}^{\text{ion}} = (z+1)\frac{e^2}{4\pi\epsilon_0 r_D} \tag{9.137}$$

with the DEBYE radius

$$r_D = \sqrt{\frac{\epsilon_0 k_B T}{e^2\left(n_e + \sum_{a,z} z^2 N_{a,z}\right)}} \tag{9.138}$$

The occupation numbers of ionization stage z in the excited state n is given by

$$N_{a,z;n} = N_{a,z}\frac{g_{a,z;n}}{U_{a,z}\left(T, \Delta E_{a,z}^{\text{ion}}\right)} \exp\left(-E_{a,z;n}/k_B T\right) \tag{9.139}$$

9.6.1 Line Radiation

The radiation of a plasma consists of line radiation, recombination radiation, and bremsstrahlung. The line radiation results from transitions between an energetically higher level n to an energetically lower level m in which a photon is emitted whose energy equals the energy difference of the two involved levels. The number of spontaneous transitions between the levels n and m, respectively, per volume, time, and frequency interval is

$$\frac{dN_{sp}}{dt} = A_{nm} f_{nm}^L(\nu) N_n \tag{9.140}$$

A_{nm} is the EINSTEIN coefficient of spontaneous emission, N_n is the density of the particles in the state n, $f_{nm}^L(\nu)$ is the line shape of the transition, and ν the frequency of the emitted radiation. The line shape $f_{nm}^L(\nu)$ depends on the dominant line broadening mechanism and is normalized as follows:

---

[12] For multiple ionized ions see Appendix D.1.

$$\int_0^\infty f_{nm}^L(\nu)d\nu = 1 \tag{9.141}$$

The line shape function as a function of the wavelength is with

$$\lambda = \frac{c}{\nu} \tag{9.142}$$

and

$$f(\nu)d\nu = f(\lambda)d\lambda \tag{9.143}$$

$$f(\nu) = f(\lambda)\frac{d\lambda}{d\nu} \tag{9.144}$$

given by

$$f(\nu) = f(\lambda)\frac{\lambda^2}{c} \tag{9.145}$$

Besides spontaneous emission there is also absorption and stimulated emission. The number of absorbing transitions $m \to n$ per volume, time, and frequency interval is:

$$\frac{dN_{\text{abs}}}{dt} = B_{mn} f_{nm}^L(\nu)\rho(\nu)n_m \tag{9.146}$$

$$\rho(\nu) = \frac{I(\nu)}{c}, \ \rho(\nu) = \left[\frac{Js}{m^3}\right], \ I(\nu) = \left[\frac{Ws}{m^2}\right] \tag{9.147}$$

$\rho(\nu)$ is the photon energy density per frequency interval, $I(\nu)$ the intensity per frequency interval. B_{mn} is the EINSTEIN coefficient of absorption and is related to A_{nm} by

$$A_{nm} = \frac{8\pi\nu^2}{c^3} h\nu B_{nm} \tag{9.148}$$

$$B_{nm} = A_{nm}\frac{c^3}{8\pi h\nu^3} \tag{9.149}$$

$$B_{nm} = \left[\frac{m^3}{Js^2}\right] A_{nm} = [1/s]$$

The number of stimulated emissions $n \to m$ per volume, time, and frequency interval is

$$\frac{dN_{\text{stim}}}{dt} = B_{nm} f_{nm}^L(\nu)\rho(\nu)N_n \tag{9.150}$$

$$B_{mn} = \frac{g_n}{g_m} B_{nm} \tag{9.151}$$

9.6.2 Radiation Transport

The intensity change per frequency interval and solid angle along the path element dz is given by the radiation transport equation:

$$dI(\nu) = \left(\frac{1}{4\pi} \frac{dN_{sp}}{dt} + \frac{dN_{stim}}{dt} - \frac{dN_{abs}}{dt} \right) h\nu \, dz \tag{9.152}$$

With Eqs. (9.140) and (9.146) it follows that

$$\frac{dI(\nu)}{dz} = S(\nu) - \kappa(\nu) I(\nu) \tag{9.153}$$

$$S(\nu) = \frac{1}{4\pi} A_{nm} N_n h\nu f_{nm}(\nu) \tag{9.154}$$

$$\kappa(\nu) = (B_{mn} N_m - B_{nm} N_n) \frac{h\nu}{c} f_{nm}(\nu) \tag{9.155}$$

In case of spatially constant $S(\nu)$ and $\kappa(\nu)$ Eq. (9.153) has the following solution:

$$I(\nu) = \frac{S}{\kappa} (1 - \exp(-\kappa z)) \tag{9.156}$$

In the limiting case $\kappa z \ll 1$ the plasma is optically thin and the intensity increases linearly with the path length z:

$$I(\nu) = S(\nu) z = \frac{1}{4\pi} A_{nm} N_n h\nu f_{nm}(\nu) z \tag{9.157}$$

whereas in the limiting case $\kappa z \gg 1$ the plasma is optically thick and the intensity becomes independent of the path length z:

$$I(\nu) = \frac{S(\nu)}{\kappa(\nu)} \tag{9.158}$$

Using the equilibrium densities PLANCK's law follows[13]:

$$I(\nu) = \frac{1}{4\pi} \frac{8\pi \nu^2}{c^2} \frac{h\nu}{\exp\left(\frac{h\nu}{k_B T}\right) - 1} \tag{9.159}$$

---

[13] P_ν in Eq. (9.130) is the radiation energy density whereas $I = Pc$ in Eq. (9.159) is the intensity per solid angle.

9.6.3 Radiation Power of Line Radiation

The radiation power per volume and frequency interval of the transition $n \to m$ of the element a in ionization stage z is given by [6]

$$p_{a,z;nm}(\nu) = A_{a,z;nm} h \nu_{a,z,nm} N_{a,z;n} f^L_{a,z;nm}(\nu) \qquad (9.160)$$

$$N_{a,z;n} = N_{a,z} \frac{g_{a,z;n}}{U_{a,z}(T, \Delta E^{ion}_{a,z})} \exp\left(-E_{a,z;n}/k_B T\right) \qquad (9.161)$$

with

$E^{ion}_{a,z}$ – excitation energy with respect to the ground state.

Using

$$p_{a,z;nm}(\nu) d\nu = p_{a,z;nm}(\lambda) d\lambda \qquad (9.162)$$

$$p_{a,z;nm}(\lambda) d\lambda = A_{a,z;nm} h \frac{c}{\lambda} N_{a,z;n} f^L_{a,z;nm}(\nu) d\nu \qquad (9.163)$$

$$p_{a,z;nm}(\lambda) d\lambda = A_{a,z;nm} h \frac{c}{\lambda} N_{a,z;n} f^L_{a,z;nm}(\lambda) d\lambda \qquad (9.164)$$

yields the power per volume and frequency interval of the transition $n \to m$ of the element a in the ionization stage z [6]:

$$p_{a,z;nm}(\lambda) = A_{a,z;nm} h \frac{c}{\lambda} N_{a,z;n} f^L_{a,z;nm}(\lambda) \qquad (9.165)$$

9.6.4 Line Shapes

The line shapes depend on the line broadening mechanism. All transitions are subject to the natural line broadening which is determined by the finite lifetime of the respective states. The line shape due to the natural line broadening is given by a LORENTZ function

$$f^L_{a,z;nm}(\nu) = \frac{2}{\pi} \frac{\Delta \nu^L_{nm}}{4(\nu - \nu_{nm})^2 + \left(\Delta \nu^L_{nm}\right)^2} \qquad (9.166)$$

$$\int_0^\infty f^L_{nm}(\nu) d\nu = 1 \qquad (9.167)$$

or as a function of the wavelength:

$$f^L_{a,z;nm}(\lambda) = \frac{2}{\pi} \frac{\Delta \lambda^L_{nm}}{4(\lambda - \lambda_{nm})^2 + \left(\frac{\lambda}{\lambda_{nm}} \Delta \lambda^L_{nm}\right)^2} \qquad (9.168)$$

9 Plasma Physics

$$\int_0^\infty f_{nm}^L(\lambda)d\lambda = 1 \tag{9.169}$$

$\Delta\nu_{nm}^L$ and $\Delta\lambda_{nm}^L$, respectively, are the frequency and wavelength FWHM (full width at half maximum).

Due to the thermal movement of the light-emitting particles the radiation is DOPPLER shifted. The superposition of many emitted photons with different DOPPLER-shifted wavelengths results in a DOPPLER shape function:

$$f^L(\lambda) = \frac{2\sqrt{\ln 2}}{\sqrt{\pi}} \frac{1}{\Delta\lambda_{nm}^D} \exp\left[-4\ln 2 \frac{(\lambda - \lambda_{nm})^2}{\left(\Delta\lambda_{nm}^D\right)^2}\right] \tag{9.170}$$

$$\Delta\lambda_{nm}^D = \sqrt{\frac{8k_B T \ln 2}{Mc^2}} \lambda_{nm} \tag{9.171}$$

with

$\Delta\lambda_{nm}^D$ – FWHM (full width at half maximum).

STARK broadening is caused by collisions of the electrons with the light-emitting particles, during which the potential of the impacting electrons shifts the energy levels which on the average leads to a broadening of the line. Besides line broadening the STARK effect also causes a line shift:

$$\Delta\lambda_{\text{Stark}} = w_m \frac{n_e}{n_{e,0}} \quad : \quad [w_m] = m \tag{9.172}$$

$$\Delta\lambda_{\text{Shift}} = d_m \frac{n_e}{n_{e,0}} \quad : \quad [d_m] = m \tag{9.173}$$

with

w_m – FWHM (full width at half maximum) at the electron density $n_{e,0}$
n_e – actual electron density.

Often several broadening mechanisms exist simultaneously. If there are two broadening mechanisms that both have a LORENTZian shape the resulting line shape again is a LORENTZian shape with the total width

$$\Delta\lambda = \Delta\lambda_1 + \Delta\lambda_2 \tag{9.174}$$

In case of two DOPPLER broadening mechanisms the combined line width is given by

$$\Delta\lambda = \sqrt{\Delta\lambda_1^2 + \Delta\lambda_2^2} \tag{9.175}$$

In case of a LORENTZ shape and a DOPPLER shape the combined line shape is given by the convolution of both functions:

$$f^V(\lambda) = \int_{-\infty}^{\infty} f^D(\lambda') f^L(\lambda' - \lambda) \lambda' \tag{9.176}$$

This combined line shape is called VOIGT profile. The resulting line width is approximately given by

$$\Delta \lambda_V = \frac{\Delta \lambda_L}{2} + \sqrt{\frac{\Delta \lambda_L^2}{4} + \Delta \lambda_D^2} \tag{9.177}$$

9.6.5 Bremsstrahlung

Besides line radiation, in which bound electrons are involved, in plasmas there can also be radiation emitted by free electrons. The emitted power due to bremsstrahlung per volume and frequency interval in the hydrogen approximation is given by [6]

$$\epsilon_{ff}(\nu) = 4\pi 2\pi \, 16 \frac{(\alpha a_0)^3 E_H}{3 (3\pi)^{1/2}} \sqrt{\frac{E_H}{k_B T}} \exp\left(-\frac{h\nu}{k_B T}\right) N_e \sum_{az} z^2 N_{a,z} \tag{9.178}$$

With

$$\epsilon(\nu) d\nu = \epsilon(\lambda) d\lambda \tag{9.179}$$

$$\epsilon(\lambda) = \epsilon(\nu) \frac{d\nu}{d\lambda} = \epsilon(\nu) \frac{c}{\lambda^2} \tag{9.180}$$

The bremsstrahlung power per volume and wavelength interval follows [6]:

$$\epsilon_{ff}(\lambda) = 4\pi 2\pi \, \frac{c}{\lambda^2} 16 \frac{(\alpha a_0)^3 E_H}{3 (3\pi)^{1/2}} \sqrt{\frac{E_H}{k_B T}} \exp\left(-\frac{hc}{\lambda k_B T}\right) N_e \sum_{az} z^2 N_{a,z} \tag{9.181}$$

9.6.6 Recombination Radiation

During recombination of an electron and an ion the excess energy can be taken by a photon that is emitted. The power of recombination radiation per volume and wavelength interval in the hydrogen approximation is given by [6]

9 Plasma Physics

$$\epsilon_{fb}(\nu) = C \exp\left(-\frac{h\nu}{k_B T}\right) N_e \sum_{a=0}^{n_a-1} \sum_{z=1}^{n_{za}-1} \sum_{n^*}^{n^*+\Delta n_{max}} \frac{z^4}{n^3} \exp\left(\frac{z^2 E_H}{n^2 k_B T}\right) N_{a,z} \quad (9.182)$$

$$C = 4\pi 2\pi \frac{32}{3} \frac{(\alpha a_0)^3 E_H}{3(3\pi)^{1/2}} \left(\frac{E_H}{k_B T}\right)^{3/2}$$

$$n \geq n^* = \sqrt{\frac{z^2 E_H \lambda}{hc}}$$

$\Delta n_{max} = 30$

$$\epsilon_{fb}(\lambda) = C \frac{c}{\lambda^2} \exp\left(-\frac{hc}{\lambda k_B T}\right) N_e \sum_{a=0}^{n_a-1} \sum_{z=1}^{n_{za}-1} \sum_{n^*}^{n^*+\Delta n_{max}} \frac{z^4}{n^3} \exp\left(\frac{z^2 E_H}{n^2 k_B T}\right) N_{a,z}$$
$$(9.183)$$

9.6.7 Influence of the Apparatus on Measured Spectra

When measuring the radiation spectrum emitted by a plasma the spectrum is modified by the measuring system. Two principal mechanisms can be distinguished:

- the transmission of the radiation through optical components and the detector sensitivity dependent on the wavelength
- line broadening due to the finite measuring accuracy

The influence of the wavelength-dependent transmission and detector sensitivity can be taken into account by a wavelength-dependent multiplication factor and the line broadening by the apparatus can be accounted for by convolving the spectrum that is emitted by the plasma by a function that describes the apparatus broadening called apparatus function:

$$I_m(\lambda) = S_{\text{Apparat}}(\lambda) \int_0^\infty f(\lambda, \lambda', \Delta\lambda_{\text{Apparat}}) I_p(\lambda') d\lambda' \quad (9.184)$$

with

$f(\lambda, \lambda', \Delta\lambda_{\text{Apparat}})$ – apparatus function
$\Delta\lambda_{\text{Apparat}}$ – full width at half maximum of the apparatus
$S_{\text{Apparat}}(\lambda)$ – spectral characteristic of the apparatus
$I_p(\lambda)$ – spectrum emitted by the plasma
$I_m(\lambda)$ – spectrum with the impact of the measuring apparatus taken into account.

References

1. G. Bekefi, Radiation Processes in Plasmas, Wiley, New York, 1966
2. E. Beyer, K. Behler, U. Petschke, and W. Sokolowski, Schweißen mit CO_2-Lasern, Laser und Optoelektronik, Vol. 18, pp. 35–46, 1986
3. P. Cavaliere, G. Ferrante, and C. Leone, Particle-atom ionising collisions in the presence of a laser radiation field, J. Phys. B, Vol. 13, p. 4495, 1980
4. H.W. Drawin and P. Felenbok, Data for Plasmas in Local Thermodynamic Equilibrium, Gauthier-Villars, Paris, 1965
5. W. Ebeling, W.D. Kraeft, and D. Kremp, Theory of Bound States and Ionization Equilibrium in Plasmas and Solids, Akademie-Verlag, Berlin, 1976
6. H.R. Griem, Plasma Spectroscopy, Mc-Graw-Hill, 1964
7. K. Günther, S. Lang, and R. Radtke, Electrical conductivity and charge carrier screening in weakly non-ideal plasmas, J. Phys. D: Appl. Phys., Vol. 16, p. 1235, 1983
8. K. Günther and R. Radtke, Electrical Properties of Weakly Nonideal Plasmas, Birkhäuser, Basel, 1984
9. T. Holstein, Low Frequency Approximation to Free-Free Transition Probabilities, Pittsburgh, 1965
10. S. Ichimaru, Basic Principles of Plasma Physics, Benjamin/Cummings Publishing Company, London, 1973
11. J. D. Jackson, Classical Electrodynamics, John Wiley & Sons, New York, 1975
12. E. Kocher and others, Dynamics of Laser Processing in Transparent Media, IEEE J-QE-8, 1972
13. A. Matsunawa, Beam – Plasma Interaction in Laser Materials Processing by Different Wavelength, 10.ter Internationaler Kongress LASER 91, 1991
14. W. Peschko, Abtragung fester Targets durch Laserstrahlung, 1981
15. A.N. Pirri, Plasma Energy Transfer to Metal Surfaces Irradiated by Pulsed Lasers, IAAJ., 16, 1979
16. Yu. P. Raizer, Laser-Induced Discharge Phenomena, Consultants Bureau, New York, 1977
17. F. Reif, Fundamentals of Statistical and Thermal Physics, McGraw Hill, 1965
18. E. Beyer, Eunfluss des laserindizierten Plasmas beim Schweissen mit CO_2-Lasern, Dissertation, TH Darmstadt, 1985

Chapter 10
Laser Beam Sources

Torsten Mans

10.1 CO$_2$ Laser

Torsten Mans

10.1.1 Principles

CO$_2$ lasers emit radiation at a wavelength of 10.6 μm. This is enabled via transition between two vibronical states of the linear CO$_2$ molecule. Excitation in the upper laser level takes place via collision with an electron or, much more likely, via collision with a vibronically excited nitrogen molecule. Thus nitrogen is added to the laser gas to enhance this pumping process. In addition helium is added for efficient cooling and depletion of the lower laser level via collisions. The pressure of the gas mixture between the laser mirrors is usually below normal conditions (100–250 hPa) in order to achieve a homogeneous discharge. Commercially available CO$_2$ lasers achieve electro-optic efficiencies of 5–15%.

10.1.2 Types of Construction

Different types of CO$_2$ lasers can be categorized in flowing and stationary gas mixtures. The electrical excitation can take place via alternating or direct current.

The DC excitation is realized with two electrodes inside the laser gas mixture. The direction of the discharge can be along or perpendicular to the resonator axis (e.g., TEA laser: *T*ransversely *E*xcited *A*tmospheric *P*ressure). Advantage of the DC excitation is the comparably easy setup and the high overall efficiency. Disadvantage is the burn-off of the electrodes at high average powers and the resulting service expenses.

T. Mans (✉)
Laser und Laseroptik, Fraunhofer-Institut für Lasertechnik,
52074 Aachen, Germany
e-mail: torsten.mans@ilt.fraunhofer.de

High average powers (>2 kW) are therefore usually realized with high-frequency excitation schemes. Two plates of a condensator ("Slab"- or "Tube"-configuration) are placed around the laser gas vessel and a high-frequency alternating electrical field is applied. A typical electrical frequency for this excitation is 27 MHz [7].

CO_2 lasers with more than 500 W average power usually have folded resonator cavities to obtain a large mode volume with moderate length of the whole setup.

In transversal flow CO_2 lasers, the laser gas is streaming perpendicular to the optical axis of the resonator (Fig. 10.1). After flowing through the resonator where it is heated by the electrical discharge the laser gas is sent through a heat exchanger to cool it down. Lasers with transversal flow are available up to powers of 10,000 W. Beam quality changes if it is a slow- or fast-flowing laser. Most CO_2 lasers with transversal flow are operated in the multimode regime ($M^2 \sim 5$).

Better beam qualities are achieved with axially streamed CO_2 lasers with gas flow parallel to the optical axis (Fig. 10.2) after cooling and leaving the circulation pump. Fast-flow CO_2 lasers are available from several hundred watts up to 30 kW. In comparison to transversal flow CO_2 lasers gas and energy consumption and therefore cost of ownership are usually higher for axial flow CO_2 lasers.

Significantly reduced cost of operation concerning gas consumption can be facilitated with diffusion-cooled laser setups. This concept is established for average powers up to 50 Wm^{-1} resonator length with a hermetically sealed housing for the laser gas. The housing can be sealed by the resonator mirrors or with Brewster windows inside the external resonator cavity (Fig. 10.3). The cooling takes place via heat conduction from the laser gas to the housing.

To achieve higher power levels in CO_2 lasers with diffusion cooling the slab geometry has been applied to the laser gas (Fig. 10.4). The CO_2 is located between two rectangular-shaped electrodes. The achievement is an enlargement of the

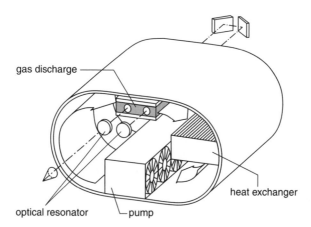

Fig. 10.1 Sketch of transversal flow gas laser [7]

10 Laser Beam Sources

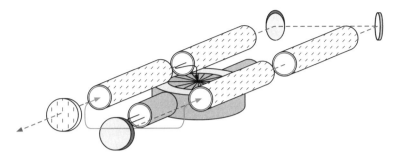

Fig. 10.2 Schematic of axial flow gas laser [7]

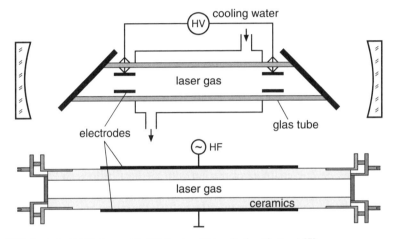

Fig. 10.3 Sealed-off CO_2 laser with DC (*top*) or HF excitation (*bottom*) [7]

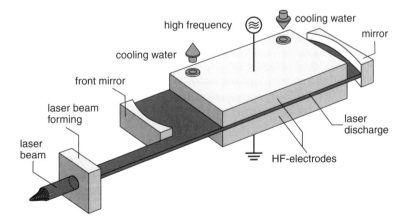

Fig. 10.4 Sketch of a CO_2 slab laser [5]

cooling area which leads to a good heat removal by conduction. CO_2 slab lasers are available up to powers of 5 kW with very good beam qualities ($M^2 < 1.1$).

10.2 Solid-State Lasers

Torsten Mans

10.2.1 Principles

Solid-state lasers have a laser-active medium which consists of a host material with ions as dopants. Within these ions the laser process of excitation and stimulated emission takes place. In the following the emphasis is on solid-state lasers with high powers (>1 W) in continuous wave (cw) operation as usually required for applications in materials processing. For these types of lasers mostly neodymium (Nd) or ytterbium (Yb) is used as laser-active ion. Depending on the geometry of the medium the most common host materials are fused silica (fiber geometry) or a crystal called yttrium aluminum garnet (YAG; disk/rod/slab geometry).

Excitation of the ions is facilitated by absorption of radiation. The electronic transition from the upper to the lower laser level of an excited ion takes place under emission of light of 1,064 nm wavelength for neodymium and 1,030 nm for ytterbium ions. Strong coupling of the ion with the host lattice can lead to longer wavelengths. This is especially the case for ytterbium doped in fused silica.

These emission wavelengths around 1 μm have the advantage that compared to the wavelength of the CO_2 laser standard optical materials (BK7, fused silica) can be used for optics and that the radiation can be transported by flexible glass fibers [8, 9].

10.2.2 Types of Construction

Solid-state lasers can be distinguished in the way they are optically excited (lamps, diode lasers) and the geometry of the active medium (rod, disk, fiber, slab). They can be operated in continuous wave (cw) and pulsed mode.

Pulsed operation can be achieved by gain switching, quality switching, or mode locking.

Gain switching is facilitated by modulation of the pumping source. Technically this is mostly done with flash lamps with the laser pulse duration following the pump pulse duration which are typically in the range of 50–2, 000 μs at repetition rates up to 4 kHz. But modulation of diode lasers is also possible.

T. Mans (✉)
Laser und Laseroptik, Fraunhofer-Institut für Lasertechnik, 52074 Aachen, Germany
e-mail: torsten.mans@ilt.fraunhofer.de

10 Laser Beam Sources

Quality switching enables the generation of shorter laser pulses with typical pulse durations from 10 to 500 ns with repetition rates of 100 kHz. The quality of the optical resonator is reduced with an optical switch (mechanical shutter, acousto-optic modulator, electro-optic modulator) during pumping, i.e., the resonator is virtually blocked. Opening the resonator again releases the stored energy in a short light pulse.

Mode locking occurs when the axial modes of a resonator are coupled in phase, i.e., they start oscillating at the same time. In this case they interfere in time to a very short pulse with pulse durations from several picoseconds down to several femtoseconds if the spectral bandwidth of the active medium supports enough axial modes. Technically, locking of the modes is realized by modulating the resonator losses with the resonator roundtrip time and caring for dispersion effects. Typical repetition rates of an oscillator are in the range of 100 MHz. Modulation can be actively controlled (e.g., electro-optic modulator) or passively achieved (Kerr lens, saturable absorber).

Lamp-pumped systems usually have solid-state media in rod geometry. The rods have typical radii of 1–4 mm and lengths of 20–200 mm and are often situated in the focal line of an elliptical reflector (Fig. 10.5). In the remaining focal line(s) is the lamp.

Laser rod and lamp are imbedded in a flow tube and are cooled by circumfluent water. The optical axis of the resonator is perpendicular on the polished circular facets of the rod. This is called a transversal pumping scheme.

Lamp-pumped cw lasers are commercially available with powers from 50 to 800 W per laser rod and beam propagation parameters M^2 from 15 to 150. Power scaling is possible by a serial arrangement of several pumped laser rods and is available up to 4 kW with beam propagation parameter $M^2 = 70$.

The flash lamps used for gain switching have typical average powers from 20 to 500 W with pulse peak powers from 5 to 20 kW.

Gas discharge lamps have a broad spectral emission compared to the absorption spectrum of doped neodymium or ytterbium ions. This leads to a poor pumping efficiency. The emission wavelength of AlGaAs diode lasers can be adjusted to the absorption spectrum of the laser-active ion. This enhances pumping efficiency

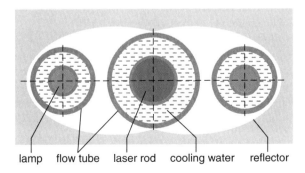

Fig. 10.5 Cavity of a lamp-pumped rod laser [7]

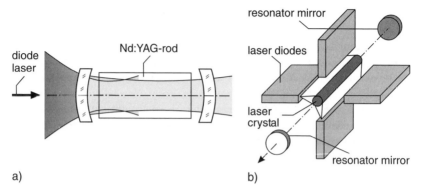

Fig. 10.6 (a) Longitudinally pumped rod laser; (b) transversally pumped rod laser [7]

significantly and in addition offers the advantage that the intensity distribution of this pump source can be tailored to the geometry of the active medium. The practically achievable electro-optic efficiency is between 10 and 25% which is an order of magnitude higher compared to the several percent of lamp-pumped laser systems.

Diode laser-pumped solid-state lasers (DPSSL) are realized in all common geometries for the active medium:

- Rod
- Disk
- Fiber
- Slab

For the rod geometry transversal and longitudinal pumping schemes are possible with diode lasers (Fig. 10.6). In the longitudinal pumping scheme the rod is pumped through a polished and coated facet of the rod in the direction of the optical axis of the resonator. The pump radiation is inserted through a dichroic mirror which is highly transmissive for the pump wavelength and highly reflective for the laser wavelength. With this setup average powers up to several tens of watts with diffracted-limited beam quality ($M^2 = 20$) are available.

Transversally diode-pumped solid state lasers (Fig. 10.6b) can be scaled with a series of pumped rods in the resonator and are available in average powers up to 8 kW. As the heat load is significantly reduced by the usage of diode lasers instead of lamps the thermally induced lens possesses less aberrations resulting in a two times better beam quality ($M^2 = 35$) compared to lamp-pumped systems. In quality switch operation pulse peak powers of 10 MW at repetition rates of 100 kHz are attainable.

To increase beam quality further different geometries of the active medium have been proposed. The three mentioned above besides the rod are the ones commercially available today. A major aim of these new developments is to generate better

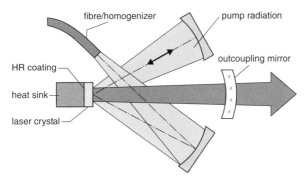

Fig. 10.7 Scheme of a disk laser [13]

cooling conditions which leads to less aberrations of the thermal lens and an increase of applicable power.

In a disk laser the active medium has a diameter of several millimeters as typical rods do but a thickness of the order of 100 μm. This leads to a large surface to volume ratio of the medium. The generated heat is efficiently removed over one of the large circular surfaces (Fig. 10.7) and the temperature offset to the heat sink can only reach low values over the small thickness of the disk. Thermal effects are efficiently reduced with this setup and very large pumping powers can be applied to the medium. A disadvantage of the small thickness is that the medium has to be heavily doped and that many passes of the pump light are necessary (typical number of passes is 16), leading to a more complex pumping setup. Disk lasers are available with average powers exceeding 8 kW with beam propagation factors $M^2 < 10$. This radiation can be coupled to transport fibers with a core diameter of 200 μm.

The active core of a single mode fiber laser has a typical diameter of 10 μm. The surrounding cladding with lower refraction index than the core to achieve total internal reflection has a typical diameter of 125 μm. The fiber is typically several meters in length. This geometry leads to a huge fraction of surface to volume from which the heat has to be removed, which makes it possible to remove this heat without active water cooling. The guidance of the light by the fiber is different compared to all other common setups of solid-state lasers which have a free beam propagation within the laser resonator. This leads to almost diffraction-limited beam qualities ($M^2 < 1.5$) at average powers as high as 3 kW. Power scaling is achieved by coupling of these diffraction-limited fiber lasers to a passive transport fiber with larger core diameter. Powers up to 50 kW are available.

Pumping of fiber lasers can be facilitated by wide stripe emitters which are directly coupled into pumping fibers. The pumping fibers are spliced to the cladding of the laser-active fiber (Fig. 10.8a). This has the advantage that the overall heat generated by the pumping diodes can be spread to a large area, and failure of a single pumping diode does not have a fatal effect on the performance. On the other hand many single packages are required. Another pumping scheme is focusing the

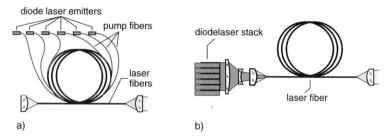

Fig. 10.8 Scheme of a slab laser: (**a**) stripe emitter pumping; (**b**) stack pumping [13]

light of a laser diode stack to the end facet of the laser-active fiber (Fig. 10.8b). This setup is more simple in principle but heat dissipation has to be more elaborate.

In slab lasers the active medium has a height (∼1 mm) much smaller than the other rectangular dimensions (>10 mm). Cooling takes place via the two largest areas of the slab (Fig. 10.9). The pump radiation is formed into a thin line which is adapted to the rectangular geometry of diode laser bars. This geometry of heat source and heat sink results in a thermal lens similar to a cylindrical lens with very low aberrations. Longitudinal ('Innoslab') and transversal (e.g., 'Zig-Zag-Slab') pumping schemes have been demonstrated but only longitudinally pumped slabs are commercially available. Average powers of 1 kW with good beam quality have been demonstrated.

The ratio of volume to cooling area of the slab geometry is smaller than in disk and fiber lasers. Therefore, applicable average pumping powers are lower compared to these lasers. The advantages of the slab geometry lie in the pulsed laser regime. Fiber lasers have in principle a very small amplifying cross section and a long amplification length which result in very high single-pass amplification but a limitation of maximum pulse power by nonlinear effects due to the low cross section. For disk lasers the opposite holds: they have a large amplification cross section and a very low amplification length resulting in low single-pass amplification but high applicable peak powers. The slab geometry has an amplification cross section which is more than three orders of magnitude larger than in fibers and somewhat lower compared to disks. The amplification length in turn is two to three orders of magnitude smaller

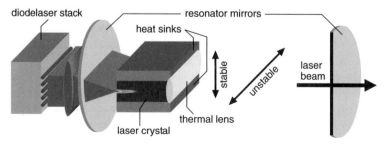

Fig. 10.9 Scheme of a slab laser

10 Laser Beam Sources

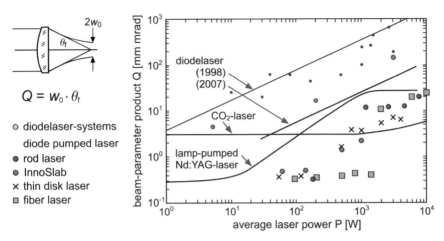

Fig. 10.10 Comparison of DPSSL: rod, disk, fiber, and slab lasers [4, 5]

than in fiber lasers but one to two orders of magnitude larger than in disks. Consequently, the slab geometry is applied effectively when the pulse regime is of interest.

Figure 10.10 shows a comparison of beam qualities versus average power for the different solid-state laser concepts. Beam quality is quantified by the beam parameter product.

For kilowatt average powers plain diode laser systems are now slowly approaching the beam quality of lamp-pumped solid-state lasers. But beam parameter products are still orders of magnitude larger than those of the new diode-pumped solid-state laser concepts.

10.3 Diode Lasers

Konstantin Boucke

10.3.1 Fundamentals

Diode lasers – or semiconductor lasers – are distinguished from all other types of solid-state lasers by their stimulation principle: While other solid-state laser media have to be pumped optically, semiconductor lasers are directly pumped by supplying an electrical current. The direct electrical pumping makes use of the specific prop-

---

K. Boucke (✉)
Laser und Laseroptik, Fraunhofer-Institut für Lasertechnik, 5207 Aachen, Germany; Oclaro Photonics Inc. Tucson 85706, USA
e-mail: konstantin.boucke@spectraphysics.com; konstantin.boucke@oclaro.com

erties of so-called direct semiconductors such as gallium arsenide (GaAs) or indium phosphide (InP) in conjunction with different dopings (n-doping and p-doping) to generate radiating transitions between different energy bands (valence band and conduction band) of the semiconductor material in the junction area between n- and p-doped semiconductor material [3].

The principal electrical characteristics of a laser diode are the same as of any non-radiating diode. To pump the laser diode an electrical voltage is applied in the conducting direction of the diode. Since the intermediate step of converting the electrical pump power in optical pump power – as required by other solid-state laser media – is not necessary, diode lasers achieve comparable high conversion efficiencies of 50–70% and high average powers can be extracted from a small volume of active material. On the other hand, physical and technological principles limit the size of the active volume to a thickness of a few micrometers, a length of a few millimeters, and a lateral extension of some $100\,\mu$m (Fig. 10.11). Therefore, despite the very high-power densities achieved in the active medium, only absolute output powers of a few watts can be extracted from a single laser diode.

For this reason typically laser diode arrays – also called laser diode bars – are employed whenever high output powers are required. A laser diode bar is a single semiconductor chip integrating approx. 20–25 laser diodes that are arranged and electrically connected in parallel (Fig. 10.12). All laser diodes emit in the same direction, forming a combined output beam with, respectively, increased power. Laser diode bars have typical dimensions of approx. 10 mm (width) $\times$ 1 mm (resonator length) $\times$ 0.1 mm (thickness).

For operation, such laser diode bars require electrical contacts on the top and bottom side of the chip (n- and p-side of the diode) and have to be mounted on a heat sink for efficient cooling (Fig. 10.13). This mounting process requires specially developed high-precision assembly and soldering procedures.

State-of-the-art laser diode bars achieve an optical output power in the range of 60–120 W with an electrical to optical efficiency of 60–70%. Laser diode bars are mainly available in the near-infrared spectral region with wavelengths between 790 and 1,080 nm.

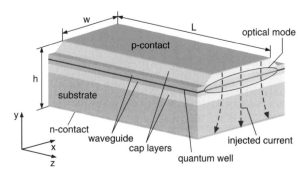

Fig. 10.11 Schematic of a laser diode. In this example a broad area laser diode is shown. Typical dimensions are $L = 1$ mm, $b = 200\,\mu$m, $h = 110\,\mu$m

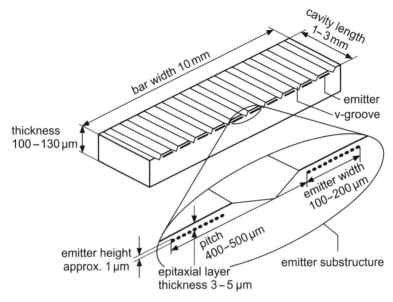

Fig. 10.12 Schematic of a laser diode bar [12]

10.3.2 Configurations and Characteristics

In Fig. 10.14 typical electro-optical and spectral characteristics of a diode laser bar are presented. Beyond the laser threshold – marked by the threshold current – the output power increases linearly with the driving current. Also the operating voltage increases with the current, due to the series resistance of the laser bar. Power losses result in a self-heating of the laser bar. Since the spectral gain distribution and the refractive index of the semiconductor material depend on the temperature the self-heating leads to a thermal wavelength shift of approx. 0.3 nm/K. In general, the output power of the laser bar is thermally limited; with rising temperature due to increased power dissipation the electro-optical efficiency of the laser diode decreases, and thus also the slope of the output power is diminished.

As a consequence of the selection rules for radiating transitions between the energy bands in semiconductors, the radiation generated by laser diodes is in general linearly polarized. The orientation of the polarization (TE – electrical field vector parallel to the epitaxial layers; TM – electrical field vector perpendicular to the epitaxial layers) depends on the specific material and epitaxial layer design of the laser diode.

As a consequence of their geometry and typical dimensions, the beam properties of laser diodes distinguish them significantly from other laser types. In the vertical direction the design of the waveguide leads to a diffraction-limited beam ($M^2 = 1$). However, due to the small waveguide thickness of only a few micrometers the diffraction at the narrow exit from the waveguide results in a large divergence angle: The divergence angle of near-IR laser diodes in the vertical axis is typically in the

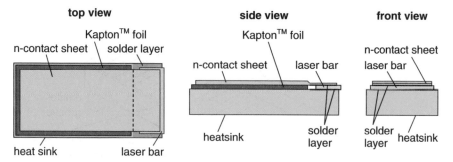

Fig. 10.13 Schematic of a packaged laser diode bar. The laser diode bar is soldered to a heat sink, also serving as a p-contact. The n-contact is provided by a contact sheet soldered to the top side of the bar

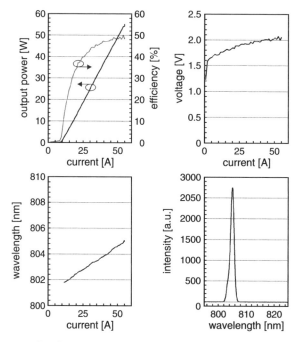

Fig. 10.14 Representative electro-optical and spectral characteristics of a laser diode bar

order of 45–60° (full angle with 95% enclosed power). Because of the 'fast' beam widening, the vertical beam axis is also referred to as 'fast axis' (Fig. 10.15).

In the lateral direction, the typical width of each emitter of the laser bar is 100–200 μm. Due to the large dimension compared to the wavelength and strong nonlinearities in the active semiconductor medium, in this axis a multimode beam with only a poor beam quality is generated ($M^2 \sim 20$–30). Nevertheless, the divergence angle in the lateral direction is typically only 6–7° (full angle with 95%

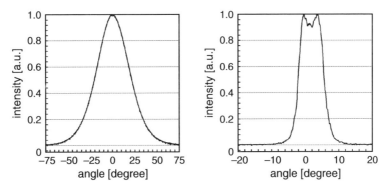

Fig. 10.15 Representative far-field intensity profiles of a laser diode bar in fast axis (*left*) and slow axis (*right*)

enclosed power) due to the large width of the emitter. Because of the comparably 'slow' widening of the beam the lateral axis is also referred to as 'slow axis'.

Since the laser bar comprises 20–25 emitters arranged in parallel, its output beam also consists of 20–25 partial beams arranged in a fixed lateral distance to each other. Thus, the overall beam quality of the laser bar in slow axis is reduced further to typical values of $M^2 = 1,000-1,500$.

These beam properties call for special collimation and transformation techniques, in order to shape the beam for its direct application or for fiber coupling. The large divergence angle in fast axis necessitates a collimation immediately in front of the diode laser bar. Therefore, cylindrical micro-lenses with a large numerical aperture (NA) of 0.6–0.8 are employed that have to be positioned and fixed with high-precision mounting processes to the diode laser bar.

Most applications – and especially fiber coupling – require at least an approximately rotational symmetric beam. The line-shaped and with respect to beam quality in fast and slow axis very asymmetric beam of the diode laser bar therefore needs to be symmetrized using special beam-shaping optics. One example for such an optical system is the so-called step mirror; its mode of operation is illustrated in Fig. 10.16. Based on the same or similar functional principles a variety of different optical systems are available for the symmetrization of diode laser bar beams. These symmetrization optics change the intensity distribution of the beam, but due to principal rules of optics it is not possible to increase the optical power density by any kind of conventional optics.

Three technical approaches are employed to increase the optical power of diode laser systems:

1. Stacking: Several laser bars – mounted on heat sinks – are densely stacked together in vertical or horizontal direction. This arrangement is called a diode laser stack. Typical stacks comprise of 6–15 laser bars, but also larger stacks and two-dimensional stacks – with stacked bars in horizontal and vertical direction – are possible. Stacking increases the output power of the diode laser system nearly

linear with the number of employed bars. However, the beam quality of the system decreases in the same manner, because the power enhancement is achieved simply by increasing the dimension of the beam source.
2. Polarization multiplexing: Since the radiation of laser diodes is linearly polarized, the beams of two diode laser bars or stacks can be superimposed using a polarization beam combiner. Hereby the output power is nearly doubled and the beam quality is conserved.
3. Wavelength multiplexing: Dichroic mirrors with specified wavelength transmission or reflection bands allow the superpositioning of beams from diode laser bars or stacks with different wavelengths. In principle a large number of beams with a wavelength spacing of only slightly more than their respective spectral width can be superimposed. For technical and economical reasons only combinations of up to four wavelengths have been realized in high-power diode laser systems so far, resulting in a nearly quadrupled output power with a nearly unchanged beam quality.[1]

Based on the described techniques diode laser systems with output powers ranging in the multi-kilowatt regime have been realized [10]. However, especially in the upper power range the beam quality of diode lasers is not comparable to the one of diode-pumped solid-state lasers or fiber lasers: While the beam of a 4 kW diode-pumped thin-disc laser can be easily coupled into a 100 μm fiber core and fiber lasers have been realized with more than 1 kW output power and nearly diffraction-limited beam quality, a 4 kW diode laser already requires a 400–600 μm fiber core (Fig. 10.17).

In general, diode lasers are – due to their specific properties – well suited for direct applications as well as for fiber coupling. On the one hand even diode laser systems in the kilowatt range can still be compact and light-weighted (volume <10 l,

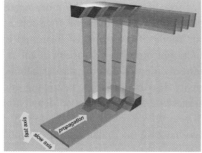

Fig. 10.16 *Left*: functional principle of the step mirror. *Right*: fiber-coupled diode laser module based on the step mirror for beam symmetrization [14]

---

[1] As a consequence of additional optical elements required for polarization and wavelength multiplexing, losses in optical power and beam quality are technically unavoidable. The mentioned values have to be seen as theoretical limits.

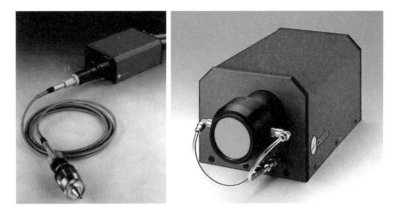

Fig. 10.17 *Left*: Fiber-coupled high-power diode laser system. *Right*: High-power diode laser system for direct application (pictures: Laserline GmbH, Germany)

weight <25 kg) and can be mounted on translation axes or robot arms. On the other hand the typical wavelengths of high-power diode lasers in the near-infrared spectral range are compatible with optical fibers allowing a nearly lossless transport of high optical powers over a long distance.

10.4 Excimer Laser

Torsten Mans

10.4.1 Principles

Excimer lasers emit light in the ultraviolet spectral range (UV). Excimer is an abbreviation for excited dimer. A dimer is a molecule consisting of two subunits. In the case of an excimer laser the subunits are two different atoms. One atom is a noble gas (argon, krypton, xenon) and the other a halogen (fluorine, chlorine). Depending on different combinations of the binding partners different central wavelengths of the excimer laser result (Table 10.1).

Table 10.1 Wavelengths of different excimer lasers [11]

Excimer	F_2**	ArF	KrF	XeC	XeF
Wavelength [nm]	157	193	248	308	351

** excited

T. Mans (✉)
Laser und Laseroptik, Fraunhofer-Institut für Lasertechnik, 52074 Aachen, Germany
e-mail: torsten.mans@ilt.fraunhofer.de

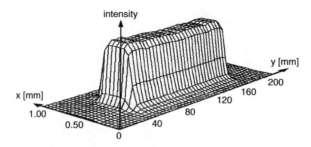

Fig. 10.18 Intensity profile of an excimer laser [2]

Table 10.2 Typical parameters of commercial excimer lasers

Pulse Energy [mJ]	Repetition Rate [Hz]	Average Power [W]	Pulse Duration [ns]
1–1,200	10–4,000	0.5–300	10–40
Beam Dimensions [mm]	Beam Divergence [mrad]	Gas Lifetime [10^6 Pulses]	Chamber Lifetime [10^6 Pulses]
$(1 \times 3) - (2 \times 6)$	$(6 \times 2) - (3 \times 1)$	20–50	500–2,000

Excitation is facilitated by a gas discharge. The predominant reaction path occurs via ionization of halogen and noble gas by electron collision. The excited ions join in a three-body collision to the excimer. The third collision partner is needed to carry away excessive energy and momentum. Very often neon or helium gas is added for this purpose. To let this collision process occur often enough the gas mixture is under pressure (2–5 bar).

As the excimer has an average lifetime of 10 ns high pumping rates and current densities are mandatory. Typical setups have breakdown voltages for the electrical discharge of 50 kV and peak currents of 100 kA. These high currents are only manageable in pulsed operation [1].

The resulting amplification during operation is quite high, which leads to many higher order modes inside the resonator. This leads to a beam quality which is quite bad compared to CO_2 or solid-state lasers. The intensity profile of excimer laser is determined by the geometry of the laser gas tube and the discharge electrodes (Fig. 10.18).

10.4.2 Setup

The technical setup to realize the noble gas supply of the laser gas tube is comparatively simple. The halogens (fluorine, chlorine) on the other hand require tight safety precautions as they are highly reactive. Their aggressive chemical properties limit the lifetime of the laser gas and the gas tube (Table 10.2). Therefore, dust separators

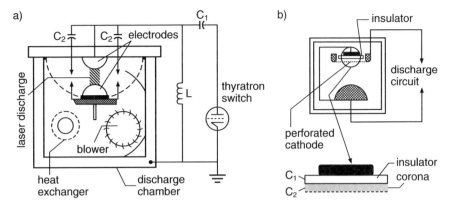

Fig. 10.19 Scheme of (**a**) excimer laser; (**b**) preionization (from [6])

and cryogenic gas cleaning techniques are integrated in the cooling circuit to remove solid and gaseous products of the laser gas and the electrodes.

For electrical switching usually thyratron switches which support high peak currents and fast rise times (∼10 ns) combined with saturable inductances are applied (Fig. 10.19).

Before the glow discharge under the high pressure starts the gas mixture is ionized in order to obtain a constant and homogeneous discharge with a lowered electrical voltage for the breakdown. Preionization can be realized with UV radiation or a corona discharge (Fig. 10.19).

References

1. D. Basting, G. Marowsky, "Excimer Laser Technology", Springer Verlag Berlin-Heidelberg, 2005, ISBN-13 978-540-20056-7
2. D. Basting, K. Pippert, U. Stamm, "History and Future Prospects of Excimer Laser Technology", 2nd International Symposium on Laser Precision Microfabrication, LPM2001; RIKEN Review No.43; Jan. 2002
3. L.A. Coldren, S.W. Corzine, "Diode Lasers and Photonic Integrated Circuits", John Wiley & Sons Inc., 1995, ISBN 0-471-11875-3
4. R. Diehl (ed.), "High-power Diode Lasers – Fundamentals, Technology, Applications", Springer, 2001, ISBN 3-540-66693-1
5. "Einführung in die industrielle Materialbearbeitung", Rofin-Sinar Laser GmbH, 10/2002
6. D.J. Elliot, "Ultraviolet Laser, Technology and Applications", Academic Press, Inc., 1995, ISBN 0-12-237070-8
7. G. Herziger, R. Poprawe, "Lasertechnik I", Chair for Laser Technology, RWTH Aachen, 1997, Lecture Notes
8. F.K. Kneubühl, M.W. Sigrist, "Laser", Teubner, 1991, ISBN 3-519-23032-1
9. W. Koechner, "Solid State Laser Engineering", Springer Verlag, 2003, ISBN 3-540-65064-4
10. R. Poprawe, P. Loosen, F. Bachmann (eds.), "High Power Diode Lasers – Technology and Applications", Springer, 2007, ISBN 0-387-34453-5

11. VDI-Technologiezentrum, "Materialbearbeitung mit Excimerlasern", Laser in der Materialbearbeitung Band 11, 1998, ISBN 3-00-003443-9
12. P. Loosen, "Beam Quality Limits and Comparison to Coherent Coupling", in *High Power Diode Lasers*, ed. F. Bachmann, P. Loosen, R. Poprawe, Springer, 2007, pp. 175
13. P. Loosen, Lehrstuhl für Technologie Optischer Systeme der RWTH Aachen, Lecture Notes
14. P. Loosen, "Incoherent Beam Combining", in *High Power Diode Lasers* (Fundamentals, Technology, Applications), ed. R. Diehl, Springer, pp. 309

Chapter 11
Surface Treatment

11.1 Transformation Hardening[1]

Konrad Wissenbach

11.1.1 Motivation

Laser beam hardening produces hard, wear-resistant surface layers on tools and components made from steel and cast iron. Laser beam hardening offers technically rewarding solutions especially for small and/or complex-shaped components where conventional methods of hardening cause problems in regard to distortion or make compliance with preset geometries of the hardness penetration layer difficult.

Comparable to conventional methods of hardening (e.g., induction hardening, flame hardening, and furnace hardening), transformation hardening with laser radiation is characterized by a well-defined temperature–time sequence in the solidus range [1–5]. In comparison to other methods of hardening, the entire temperature cycle runs in relatively short time, i.e., a few tenths of a second up to several seconds. Thus, laser beam hardening is considered a short-time hardening process.

In comparison to most of the conventional procedures, laser beam hardening generally does not require the use of external cooling agents. The high thermal conductivity of metals causes rapid self-quenching which means that the laser radiation heats up near-surface layers only. Absorbed optical energy in the surface layers is thermalized and transported into the bulk of the part via thermal conduction. The bulk material remains almost unchanged due to the low amount of total energy used during the process. This enables surface hardening even on small components.

The most important advantages of laser beam hardening are as follows:

K. Wissenbach (✉)
Oberflächentechnik, Fraunhofer-Institut für Lasertechnik, 5207 Aachen, Germany
e-mail: konrad.wissenbach@ilt.fraunhofer.de

[1] Definitions in these notes: transformation hardening equals martensitic surface layer hardening equals laser hardening.

- Minimal distortion of the processed parts
- Option of hardening partial surface areas and generation of hardening patterns, respectively
- Usually no external cooling agents required
- Easy integration into automated flexible manufacturing
- Suitable for online process monitoring and control

For numerous applications laser beam hardening is in competition with induction hardening. Although diode lasers have lower investment and operational costs compared to CO_2 and Nd:YAG lasers, the capital investment for a laser beam hardening system is often higher than the one required for an induction system.

11.1.2 Process Description

Figure 11.1 shows a diagram of laser beam hardening in case of relative movement between laser beam and component. Areas 1–3 represent a spatial allocation of the transformation processes during laser beam hardening. Area 1 comprises austenite formation, area 2 martensite formation, and area 3 the hardened track. As pictured in Fig. 11.2, the resulting temperature–time cycle of the volume element of a component is determined by the following factors: the optical and thermo-physical material properties, the material volume available for heat conduction out of the surface layer, the geometry of the workpiece, the power density distribution (PDD) on the component surface, and the relative speed, V_H, of the component surface relative to the optical axis. The temperature–time cycle of transformation hardening is divided into three phases:

- Heating-up the workpiece above the A_{c3}[2]-temperature
- Dwell time to achieve the austenitic microstructure
- Cooling down with a material-specific minimum cooling rate

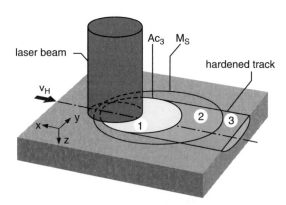

Fig. 11.1 Diagram of transformation hardening with laser radiation

---

[2] A_{c1} temperature: start of austenite formation. A_{c3} temperature: end of austenite formation.

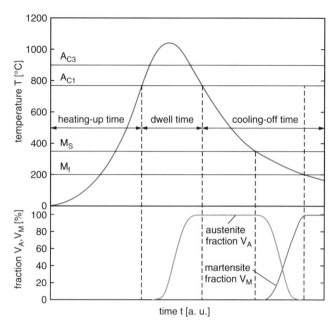

Fig. 11.2 Temperature–time sequence during transformation hardening with laser radiation

Figure 11.2 shows a diagram of the transformation of the microstructure components during transformation hardening with a predefined temperature–time cycle [3]. After exceeding the A_{c1}[2] temperature, the austenite formation sets in (see Chap. 6). The volume to be hardened has to stay above the transformation temperature (A_{c3} temperature) for a certain time in order to ensure the transformation into an austenitic microstructure is as complete as possible. The actual time necessary does not only depend on the individual process parameters but also on the microstructure. Subsequent self-quenching by means of heat conduction into the surrounding material with a minimum quenching rate initiates martensite formation as soon as the temperature falls below the M_s[3] temperature (see Sect. 6.1). Martensite formation ends when the M_f[3] temperature is reached.

In general, all materials that can be hardened with conventional methods are suitable for transformation hardening with laser radiation, i.e., steel and cast iron with a carbon content $\geq 0.3\%$.

Due to the short temperature–time sequences (0.1–10 s), materials with small-grained structure, a high proportion of pearlite or tempered materials are better suited for laser hardening than materials with a high proportion of ferrite and stable carbides [1]. The time–temperature–austenitization diagrams (TTA) and the time–temperature–transformation diagrams (TTT) [6, 7] provide important indications for the layout of the process as well as for the austenitization process and

---

[3] M_s temperature: start of martensite formation. M_f temperature: end of martensite formation

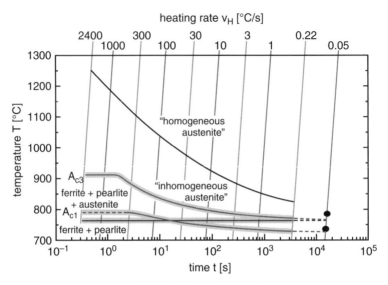

Fig. 11.3 Time–temperature–austenitization diagram (TTA) for DIN Ck45 (AISI 1045) [6, 7]

for the required minimum cooling rate. The TTA diagrams allow to determine the transformation temperatures for various material conditions at delivery depending on the heating-up speed as well as the dwell time necessary for the generation of a homogeneous austenite. As an example, Fig. 11.3 shows the TTA diagram for steel DIN Ck45 (AISI 1045). The pearlite–austenite transformation runs above the A_{c1} temperature while the homogenization of the austenite runs above the A_{c3} temperature. The progression of the A_{c1} temperature and the A_{c3} temperature in relation to the heating-up rate can be seen in Fig. 11.3.

The austenitization temperature depends on the heating-up rate, the state of heat treatment, the carbon content, and the alloying elements. Since the carbon content and concentration of alloying elements vary for different materials, it is necessary to determine individual TTA diagrams for each material which display the dependence of the austenitization temperature and time on the heating-up rate. Figure 11.3 comprises the temperatures necessary for complete austenitization. Typical heating-up rates for laser beam hardening are between 300 and 3000 °C/s. Consequently, steel DIN Ck45 (AISI 1045) has an A_{c1} temperature of 790 °C and an A_{c3} temperature of 911 °C.

The minimum cooling rate necessary for the generation of a complete martensite microstructure can be found in the TTT diagrams.

Figure 11.4 shows the TTT diagrams for two austenitization temperatures of steel DIN Ck45 (AISI 1045). The cooling curves and the domains of the different microstructure types are presented. Depending on its cooling curve, the microstructure contains ferritic, pearlitic, bainitic, and martensitic fractions that are proportionate to the percentages displayed on the individual cooling curves. Predictions about transformations of microstructures are only possible on the basis of their cooling

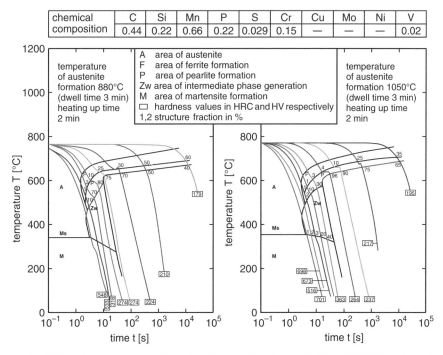

Fig. 11.4 Time–temperature–transformation diagrams for steel DIN Ck45 (AISI 1045). Austenitization temperatures 880 and 1050 °C [6, 7]

curves. The cooling-down time starts as soon as the temperature falls below the A_{c1} temperature. TTT diagrams are solely exact valid for a given austenite grain size and a predefined heating-up and dwell time during the stated austenitization temperature.

Depending on the cooling curve, the austenite transforms at certain temperatures into the structures listed in Fig. 11.4. In order to achieve a martensitic microstructure, it is necessary to select a cooling rate that avoids passing through regions of ferrite, pearlite, and bainite structures. These structures possess a considerably lower hardness than martensite. During martensite formation the carbon dissolved in the austenite is unable to move in the lattice structure. The undercooled austenite starts turning into the tetragonal body-centered martensite at the M_s temperature. The formation of the martensite structure continues when the temperature is reduced further until the entire austenite is transformed upon reaching the M_f temperature.

11.1.3 Physical Background

The physical processes relevant for transformation hardening are as follows:

The *absorption* of laser radiation on technical steel surfaces (see Sects. 3.4, 3.5, 3.6, and 3.7).

The absorbed optical energy is transformed into thermal energy and transported into the component through *heat conduction* (see Chap. 4). Within the component, the temperature distribution depends on time and location in accordance with the process parameters and the thermo-physical material characteristics.

Transformation kinetics (see Chap. 6) deals essentially with austenite and martensite formation and depends on the microstructure and the temperature–time cycle. Austenite formation is a diffusion process largely determined by nucleation and grain growth. The diffusion of carbon (which depends on temperature and time) is crucial for the process (see basic principles, Sect. 6.2). In contrast to austenite formation, martensite formation is *not* a diffusion process. Martensite formation sets in above the critical cooling rate at the M_s temperature and only progresses when the temperature is reduced further. Martensite formation sets in abruptly and with continued cooling it occurs as a cascade within split seconds. The formation of the martensite is triggered by a collective change of location of atomic groups in a coordinated movement. The face-centered cubic austenite lattice (FCC) turns into the body-centered cubic martensite lattice (BCC). Martensite formation ends in the cooling phase when the M_f temperature is reached.

Depending on the component geometry, the thermo-physical material properties, the temperature distribution, and the resulting transformation processes a 3D transient *tension* field is formed that is decisive in determining the deformation and the residual stress behavior of the component after the hardening process [8] (see Chap. 5).

11.1.4 Experimental Results

Basically, CO_2 as well as Nd:YAG diode and fiber lasers are suitable for transformation hardening. Since CO_2 laser radiation has a low absorptivity A on technical steel ($A<10\%$), the application of coatings that enhance the absorption (e.g., graphite, phosphate, metallic oxides) becomes necessary [3, 4, 9]. This requires additional work steps for the application and removal of coatings. Above all, the low reproducibility of the properties of the applied coatings poses a problem. The fundamental advantages of hardening with Nd:YAG, diode and fiber laser radiation versus application of CO_2 laser radiation are as follows:

- Superior absorptivity on technical steel surfaces (30–35%) compared to CO_2 lasers renders the use of absorption-enhancing coatings unnecessary
- Potential transmission of laser radiation using optical fibers
- Simplified process control as there is no need for temperature measurements on coated surfaces

Transformation hardening often requires a homogeneous power density distribution (PDD) in order to avoid local remelting of the component surface and also to achieve a uniform hardness penetration depth across the track width.

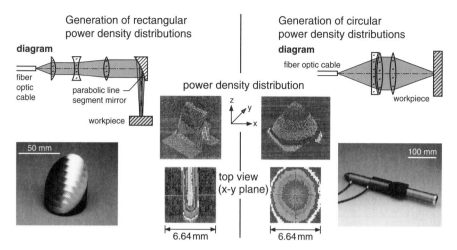

Fig. 11.5 Beam guidance and shaping systems

Homogenization of the PDD is achieved when optical fibers are utilized. Additional options for beam transformation include scanner systems, integrating mirrors for the generation of line shaped and quadratic PDDs, as well as transmittive optics (Fig. 11.5). The desired hardness penetration depth, the utilized laser source as well as the component configuration significantly influence the appropriate beam guidance and shaping system selection.

Typical characteristics of laser beam hardening are as follows:

Laser output: several 100 W–5 kW
Beam cross sections: $1\,\text{mm}^2 (1 \times 1\,\text{mm}^2) - 500\,\text{mm}^2 (100 \times 5\,\text{mm}^2)$
Incident power density: $1 \times 10^3 – 10^4\,\text{W/cm}^2$
Hardness penetration depth: 0.1–1 (2) mm (depending on material)
Hardness penetration width: $< 1\,\text{mm} - \geq 100\,\text{mm}$
Process velocity: 0.1–4 m/min

Process diagrams (Fig. 11.6) for materials DIN 90 MnCrV 8 (AISI O2) and DIN Ck45 (AISI 1045) are exemplified below [9]. The diagrams are based on experimental results and theoretical models, with the models being checked for consistency with the experimental results. The model calculates the 3D temperature field, and the hardness penetration depth is determined by the profile of the A_{c1} temperature. The model calculations make the following assumptions:

The absorptivity A for hardening of graphite-coated surfaces with CO_2 laser radiation amounts to 65%.
The power density distribution corresponds to a rectangular distribution with the dimensions: $F_L = 10 \times 22\,\text{mm}^2$.

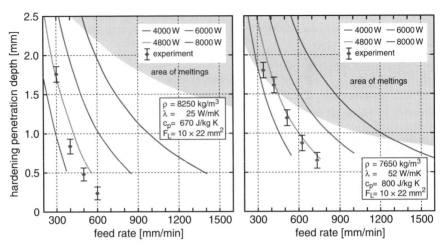

Fig. 11.6 Process diagrams for materials DIN 90 MnCrV 8 (AISI O2) (**a**) and Ck45 (AISI 1045) (**b**)

During the model calculation, the thermal conductivity and specific heat capacity are varied until the calculated hardness penetration depth and width align with the experimental hardness penetration depth and width of a reference-hardened sample. This is the proper method to establish the temperature dependency of these thermo-physical parameters. It is important that the adjusted values stay within the ranges published in pertinent literature. The deviation of the theoretical values from the experimental results should remain below 50 μm. For each of the two materials, the iteration is carried out once. The established values are retained for all parameter settings.

Figure 11.6a and b shows calculated process diagrams for materials DIN 90 MnCrV 8 (AISI O2) and DIN Ck45 (AISI 1045). The hardness penetration depths that have been established experimentally are recorded for a laser power of $P_L = 4.8$ kW. When the two materials DIN Ck45 (AISI 1045) and DIN 90 MnCrV 8 (AISI O2) are compared under identical parameter settings, a considerably greater hardness penetration depth can be achieved in DIN 90 MnCrV 8 (AISI O2). This is caused by the different thermo-physical properties of the two materials. The determining factor affecting the hardness penetration depth is the thermal conductivity (for a given P, v, r). As is the case with DIN Ck45 (AISI 1045), a low hardness penetration depth might be caused by high thermal conductivity, the heat transmitted into the material is distributed fast and the overall temperature in the material stays low. For the same reason materials with low thermal conductivity (see DIN 90 MnCrV 8 (AISI O2)) have a limited parameter range due to the possibility of local remelting of the material surface. Experiments show that local remelting sets in at an even lower laser output level than the theoretical model predicts because the model is based on an ideal rectangular power density distribution that does not take into account any inhomogeneities of the power density distribution. Parameter combinations of high laser power and high feed rates have a very limited applicability because they bear the risk of local remelting.

Overall, there is a good consistency between the experimental and theoretically predicted hardness penetration depths for a given parameter range. In every case, the hardness penetration depths obtained experimentally did match the theoretical model with an average deviation of 0.1 mm. This way, the diagrams are easy-to-handle resources for the user and allow the preselection of process parameters. However, the user should always take into account specific characteristics of the material to be processed (e.g., carbide precipitation, overheating, formation of residual austenite). The process diagrams shown here reflect only a small portion of the variable process parameters. In case of altered power density distributions, different beam dimensions, materials, or finite, i.e., thin wall thickness of the components to be hardened, these process diagrams are not applicable anymore and need to be adopted. Approaches to combine the multitude of process parameters in one useable diagram are mostly futile due to the complexity of the presentation and fail to make reasonable predictions for individual hardening applications.

Figure 11.7 shows in the upper portion a cross-section of a laser beam hardened track on a hollow profile made from DIN Ck 45 (AISI 1045). The hardness penetration depth is uniform across the entire track width. On the left, the related pictures

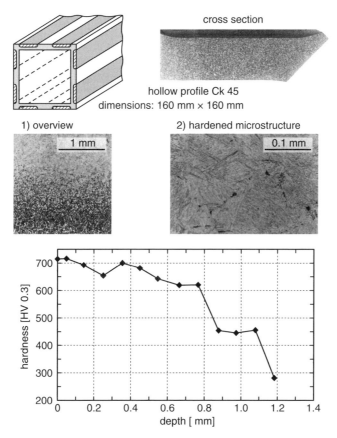

Fig. 11.7 Example of laser beam hardening

of the microstructure show an overview including the hardening geometry and the heat-affected zone (HAZ); to the right they show the martensitic microstructure in the hardened zone. The hardness as a function of depth shows maximum values of 700 HV 0.3 at a hardness penetration depth of 0.8 mm.

11.1.5 Applications

11.1.5.1 Hardening of Ball Grooves on Ring Joints [5]

Ring joints are a component of the slip joint assembly group. The ball groove areas of the ring joints made from DIN Cf 53 (AISI 1050) are to be hardened (Fig. 11.8). Induction hardening hardens the entire interior of the joint with a maximum hardness penetration depth of about 1.7 mm. During the process, large distortions are created in the area of the drillings and ball grooves. A subsequent finishing (grinding) is required to eliminate these distortions. Nd:YAG laser radiation hardens only the area of the ball grooves with specified hardening penetration depths of about 1 mm, hardness penetration widths of about 3–6 mm, and a length of the treatment area of about 20 mm. This way, distortions are minimized and subsequent finishing can be avoided.

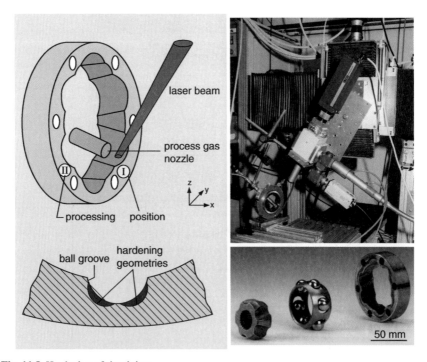

Fig. 11.8 Hardening of ring joints

The diagram in Fig. 11.8 shows the treatment with laser radiation. Due to the given geometry of the ring, the minimum angle of incidence of the laser beam is 30° (angle between optical axis and ring joint radius on the xz-plane) when transmitted into the area to be hardened (ball groove). The process gas nozzle is a circular nozzle with a diameter of 10 mm that has been positioned under an angle of 30° in the xz-plane onto the ball groove relative to the interaction zone.

In order to harden the 12 ball grooves of a ring, two processing positions (I and II) and a 180° rotation of the joint ring become necessary. This is due to the ball grooves being tilted toward each other under the angle α (xy-plane). Figure 11.8 shows on the right-hand side a cross-section of the positions of the two zones of the ball track to be hardened. Figure 11.8, center, displays the experimental setup with ring joint fixture, process gas nozzle, as well as beam-guiding and beam-shaping optics.

A laser power of 1950 W and a feed rate of about 600 mm/min produces the required hardness penetration depths of about 1 mm and hardness penetration widths of about 6.3 mm. Based on these feed rates, the process takes about twice as long as induction hardening, excluding the setup time for a ring joint. Figure 11.9 shows a fully hardened ring joint in overview (a) and a detail of the ball groove (b). On the right-hand side, a cross-section as well as a longitudinal section are presented. The ring joint geometry before and after laser hardening is measured and translated into a distortion coefficient. The evaluation of the distortion coefficients demonstrates that the distortion during laser beam hardening of ring joints (distortion coefficient 8) is half the amount of the distortion that takes place during induction hardening (distortion coefficients 15–17).

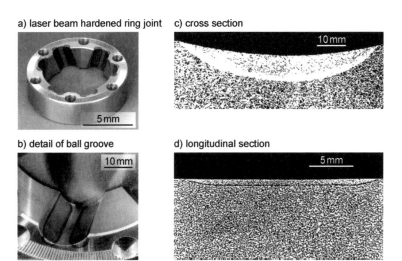

Fig. 11.9 Laser beam hardened ring joint

11.1.5.2 Hardening of Torsion Springs [10]

Car doors are being attached to the car body using door hinges that are equipped with a sintered reel and torsion spring. Wear of the torsion spring occurs on some types of these door hinges in the area of contact between the torsion spring and the sintered reel. In order to reduce the wear, the contact area is to be hardened in a locally confined area by means of martensitic hardening with laser radiation, without affecting the mechanical properties of the torsion spring in the remaining regions. The contact area extends over a perimeter section of about 170° with a length of 10–12 mm. The diagram of the required hardening geometry is presented in Fig. 11.10. Eight different types of torsion springs are to be hardened using this procedure.

In order to achieve the hardening geometry depicted in Fig. 11.10, the setup illustrated in Fig. 11.11 is utilized for processing with two-beam technique. The hardening is conducted using two high-power diode lasers (HPDL) systems with a maximum output power of 1.5 kW each and a central wavelength of 940 nm.

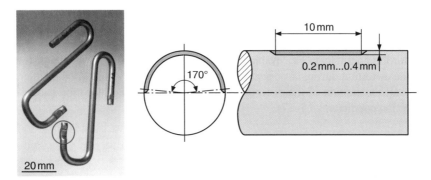

Fig. 11.10 Hardening of torsion springs

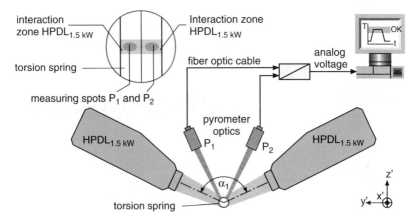

Fig. 11.11 Diagram of the setup for hardening of torsion springs

11 Surface Treatment

The power density distributions of the two lasers are aligned on the surface in a way that ensures a nearly homogeneous heating of the torsion spring over a perimeter section of about 180°. The torsion spring is fixed in a support fixture in order to ensure exact positioning between laser beam and workpiece geometry.

While the HPDL systems remain stationary, the torsion spring is moved in linear motion along the negative x'-direction. The hardening geometry is determined by the following process parameters: beam power of the individual beams, feed rate, angle between the two laser beams (see Fig. 11.11), and the position of the laser beams relative to the surface of the torsion spring. The position of the laser beams along the z'-direction affects the hardening geometry in particular. For values of z' that are too large, the hardening zone does not extend over the entire angle of 170° anymore, for values of z' that are too small, the hardening geometry is constricted or even interrupted in the center of the spring wire (0°-Position).

On the right-hand side of Fig. 11.12 a photograph of the torsion spring during the hardening process is shown. A homogeneous glow over approximately half of the perimeter is clearly visible. The corresponding cross-section of the resulting hardening geometry shows a uniform hardness penetration depth over a 170° wide section of the perimeter with hardness values of 900 ± 50 HV0.1. The procedure for the different types of torsion springs was configured in a way that all of them could be hardened using the same set of parameters while meeting all specifications of hardening geometry and hardness value.

In order to realize a process control method that fulfills all requirements mentioned above, the thermal radiation originating from the interaction zone is measured using two pyrometers (Fig. 11.11). The recorded temperatures provide information

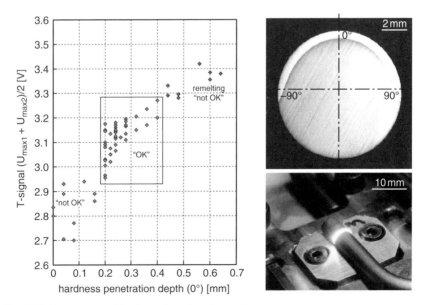

Fig. 11.12 Process control for hardening of torsion springs

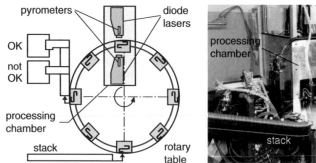

Fig. 11.13 Plant for hardening of torsion springs

about the hardness penetration depth. Two characteristic values of the temperature–time sequences are the basis for evaluating the hardening result. Through correlation between these two characteristic values and the hardness penetration depth a process window is defined (Fig. 11.12). When the range of this process window is exceeded, the hardness penetration depth is either too small or remelting of the surface occurs. For both characteristic values the corresponding limits are determined.

Based on the results of process setup and process control, two plants for hardening of torsion springs were assembled using two-beam technique (Fig. 11.13).

After transfer from the stack, a torsion spring is inserted into the support fixture and fastened. The support fixture is moved parallel to the laser beams in a linear motion. While the hardened torsion spring is transported out of the process chamber, the temperature–time sequences that have been recorded by the process control system are evaluated. The evaluation results in a signal indicating whether the processing was "OK" or "not OK". The hardened torsion spring is unloaded and characterized as a finished part or a part for quality check.

The two plants described here are operational since March 1999 and are hardening about 4 million springs each year per system. The uptime of the system remains close to 100%. Random, destructive samples have confirmed the functionality of the process control.

11.2 Remelting

Andreas Weisheit

During remelting with laser radiation, a small volume of a material melts rapidly and when the laser beam is shut off solidifies rapidly [11–13]. The high solidification rate leads to a fine dendritic or cellular microstructure. Depending on the material also metastable phases can occur.

A. Weisheit (✉)
Oberflächentechnik, Fraunhofer-Institut für Lasertechnik, 5207 Aachen, Germany
e-mail: andreas.weisheit@ilt.fraunhofer.de

Remelting can improve corrosion or wear properties. The corrosion resistance is increased by the homogenization of the microstructure or the dilution of corrosion inhibiting elements (e.g., Cr in steel). The wear resistance is increased by grain refinement or formation of hard phases.

Remelting is most effective for cast materials, e.g., cast iron, aluminum, magnesium, and copper-based alloys. The refinement after remelting can be in the order of magnitudes and metastable phases can be formed (e.g., ledeburite in cast iron) which leads to significant changes in properties.

Remelting with laser radiation features a very high cooling rate which cannot be achieved by other heat sources like plasma, arc, or flame. Comparable cooling rates can only be achieved by electron beam melting. However, EB melting has to be performed in a high vacuum.

11.2.1 Physical Fundamentals

The main physical fundamentals of laser remelting are as follows:

- Absorption of the laser radiation
- Heat conduction
- Melt pool dynamic
- Rapid solidification

Below the melt pool dynamic and the rapid solidification will be explained in more detail. Absorption of laser radiation, heat conduction, and theoretical background of melt pool dynamic are discussed in Chaps. 4, 5, and 8.

11.2.1.1 Melt Pool Dynamic [14]

When a material is melted due to the absorbed energy of the laser radiation, convective flows are created which are related to temperature gradients on the surface of the melt pool. Assuming an intensity distribution with a maximum in the center of the laser beam a temperature gradient forms in radial direction since the temperature in the center of the melt pool is higher than at the edges. (Fig. 11.14a).

The temperature gradient causes a surface tension gradient σ since the surface tension is dependent on temperature. The gradient of σ in y-direction (Fig. 11.14) can be written as follows:

$$\frac{\partial \sigma}{\partial y} = \frac{\partial \sigma \cdot \partial T}{\partial T \cdot \partial y} \tag{11.1}$$

For $d\sigma/dT < 0$ a tension field is formed with low tension in the center and high tension at the edges of the melt pool. This gradient causes a melt flow from the center to the edges. Sheer stresses cause a flow of material along the solid–liquid line at the edges of the melt pool to the bottom. In the center of the melt pool the material flow rises to the surface again. The flow of the material causes a deformation of the melt pool (Fig. 11.14b). The convective flow driven by the surface tension gradient is

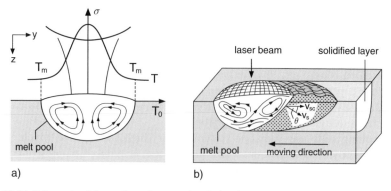

Fig. 11.14 Schematic of the melt pool convection during laser remelting for $d\sigma/dT < 0$: (**a**) cross section of melt pool; (**b**) longitudinal section of the melt pool

known as Marangoni convection. The flow velocity is in the range of several meters per second which is 2–3 orders of magnitudes higher than the typical velocity during laser remelting (0.005–0.05 m/s).

The surface tension and its temperature dependence are determined by the alloying elements. For example, oxygen and sulfur cause a significant reduction of surface tension in steel. In some cases even a positive gradient can form which leads to a reverse of the convective flow but for metals normally $d\sigma/dT < 0$ is true.

The precondition for Marangoni convection is a free surface on the melt pool. A solid layer (e.g., oxide layer) prevents the formation of the convective flows. Therefore the shielding of the melt pool is of great significance in this respect.

11.2.1.2 Rapid Cooling [15]

The solidification morphology of a metal can be planar, cellular, or dendritic. A planar solidification requires the flow of the latent heat released on solidification into the solidified volume. In this case the real temperature of the melt is always greater than the liquidus temperature of the material. If the planar solidification forms, a stochastic bulge would grow into a region of increased temperature which causes the break down of the bulge. Planar solidification can be achieved for high-purity metals in a wide parameter range. If the metal contains impurities or alloying elements the constitutional supercooling plays an important role. Supercooling requires that no concentration balance at the solidification front driven by diffusion or convection occurs. This is the case for laser remelting because the melt exists only for a very short time (typically 0.5–2 s) and the time-dependent diffusion will be incomplete. Convection is impeded at the solid–liquid interface due to friction. Under these conditions the melt enriches with impurity and alloying elements during crystallization. The melting temperature, T_{liq}, of the enriched layer decreases toward the solidification front and drops below the real temperature of the melt. In this case a non-planar solidification is possible even if the latent heat released on solidification is completely flowing through the solidified volume. The growth of stochastic bulges now

11 Surface Treatment

will be stable because they grow into a supercooled melt. The planar solidification front breaks down and a dendritic solidification occurs. The growth morphology can be cellular-dendritic (also known as cellular) or dendritic. A cellular structure is generated when the crystals grow in columns without formation of secondary dendrite arms. If secondary or even ternary dendrite arms form then the structure is called dendritic. Which morphology forms depends mainly on the solidification rate R. If R is small the growth velocity of the dendrites is also small. The dendrites form a great tip radius. The volume of the melt between the dendrites becomes too small to form secondary arms. A high solidification rate leads to rapid growth of the dendrites with a small tip radius. Between the dendrites, enough melted volume is left to form secondary arms.

For metals with low purity and alloys cellular or dendritic growth can be suppressed only in a very small parameter range.

The relevant parameters for the solidification morphology are the temperature gradient,

$$G = \frac{\partial T}{\partial x} \tag{11.2}$$

the solidification rate,

$$v_s = \frac{\partial x}{\partial t} \tag{11.3}$$

and the cooling rate.

$$\dot{T} = \frac{\partial T}{\partial t} \tag{11.4}$$

The parameters are linked in the equation

$$\dot{T} = v_s \cdot G \tag{11.5}$$

Figure 11.15 shows the correlation between G and v_s. The limits of cellular and dendritic growth are determined by the constitutional supercooling and the absolute stability which is characterized by a critical solidification rate which leads to a planar solidification independently from the temperature gradient.

The refinement of the microstructure is determined mainly by the dissipated heat. With increasing cooling rate and increasing temperature gradient the dissipated heat per time unit increases too. This reduces the time for crystal growth and as a result the solidified structure is fine. The growth morphology (planar, cellular, dendritic) is determined by the value G/v_s, because the dimension of the constitutional supercooled region depends on v_s and G. With increasing v_s the solidified volume per time unit increases which results in an increased volume of the supersaturated melt at the solidification front. This volume will also increase with decreasing temperature gradient.

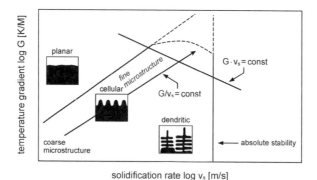

Fig. 11.15 Schematic of the solidification morphology as a function of solidification rate and temperature gradient

The width of the melt pool generated during laser remelting is typically 0.1–4 mm, the melted depth lies in the range of 0.1–2 mm. Therefore the volume of the melt pool is small compared to the whole volume of the workpiece. When the laser beam moves on or is shut down the heat is flowing rapidly into the cold bulk volume (self-quenching). The melt solidified with high cooling rates (10^2–10^7 K/s) which are determined by the process parameters, the thermo-physical properties of the material and the geometry of the workpiece. The most important process parameters are velocity and intensity. With increasing velocity the solidification rate increases. However, the solidification rate is not constant as shown in Fig. 11.16 which illustrates a longitudinal section of the melt pool (x–z-plane). Along the melting front (line A–B) the material is melted and solidified again along the solidification front (line B–C). The local solidification rate v_s is determined by the projection of the velocity v_v onto the normal at the solidification front:

$$v_s = v_v \cdot \cos \Theta \qquad (11.6)$$

The solidification rate at point B is zero and increases along the line B–C reaching its maximum at C ($v_{s,\max} = v_v$).

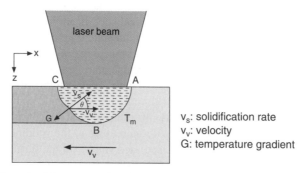

Fig. 11.16 Schematic of the cross section of the melt pool

The intensity has a strong impact on the temperature gradient in the melt and the solidified volume. With increasing laser beam intensity the temperature gradient also increases. With increasing velocity the time for heat flow is reduced which increases the temperature gradient. However, the impact is much lower than that of the laser beam intensity.

The cooling rate is also determined by the thermal conductivity of the material. For materials with a low thermal conductivity (e.g., titanium alloys, $\lambda_{Ti} = 21.9\,\text{W}/(\text{mK})$), the heat flow is low and the cooling rate decrease whereas for materials with a high thermal conductivity (e.g., aluminum alloys, $\lambda_{Al} = 37\,\text{W}/(\text{mK})$) the effect is reverse.

The workpiece geometry is only of significance if the volume of the melt is big compared to the surrounding bulk volume, e.g., when thin sheets or small edges of a workpiece have to be remelted. In these cases the heat flow is restricted and the cooling rate decreases.

Cooling rates up to 10^4 K/s lead to a cellular or dendritic solidification which is much finer than for conventional casting. The formation of metastable phases depends on the material. In case of very high cooling rates ($> 10^6$ K/s) a planar solidification occurs. For some alloys even an amorphous solidification can be achieved.

11.2.2 Process

Figure 11.17 shows the scheme of the laser remelting process. Larger areas are remelted by scanning the surface. To protect the melt pool from oxidation an off-axis nozzle provides an inert gas flow. The required flow rate is determined by the oxidation sensitivity of the material and the size of the melt pool. Aluminum and titanium alloys require a bigger flow rate than steel or cast iron. Typical values are 3–20 l/min. Higher flow rates can cause turbulences inside the gas flow which lead to a contamination with the surrounding air. Furthermore, a strong gas flow can cause turbulences in the melt pool, which can lead to a rough surface after solidification. To protect the solidified volume from surface oxidation after the laser has moved on a shroud or a second shielding gas nozzle has to be used. If any contamination of the melt or oxidation has to be avoided remelting has to be done in a shielding gas chamber.

For melted depths greater than 100 μm, continuous wave (cw) Nd:YAG, diode, fiber, and CO_2 lasers with an output power of minimum 500 W are suitable. Thin layers (several micrometers of thickness) are best remelted with a pulsed laser (e.g., Nd:YAG or Excimer) because the heat flow into the material can be controlled on a finer scale.

As inert gases, helium or argon can be used. Helium has a higher cooling effect due to its higher heat conductivity. Because helium has a lower density than air it tends to flow upwards away from the process region. Argon has a lower cooling effect (low heat conductivity) but is more dense than air and tends to flow downwards. For reactive metals and alloys (e.g., Al, Ti, and their alloys) helium has

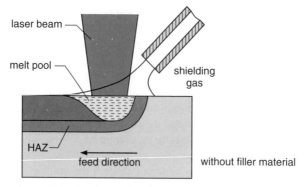

Fig. 11.17 Schematic of the laser remelting process

proven to be the best choice. For metals and alloys with less sensitivity to oxidation (e.g., Fe, Co, Cu, and their alloys) argon is sufficient.

Below the melted zone the material is heat treated. This area is therefore called the heat-affected zone (HAZ, Fig. 11.17). Whether the heat affection causes significant changes in the microstructure depends on the material. In hardenable steels and cast iron martensite can form accompanied by a significant increase in hardness. In cast aluminium alloys only a slight grain growth occurs which has no significant effect on the properties.

11.2.3 Examples for Laser Remelting

In view of potential applications, laser remelting is only of interest for those materials and initial microstructures where a rapid cooling causes significant effects on the microstructure and properties. This is true for cast alloys which have solidified slowly (e.g., by sand casting or chill casting) and form a coarse cast microstructure. Remelting can lead to a refinement and homogenization. Furthermore materials are of interest which form metastable phases during solidification or cooling in the solid state. Two examples are given below.

The effect of grain refinement and homogenization is shown in Fig. 11.18 for the aluminum alloy AlSi10Mg. A scan velocity of 0.5 m/min leads to a refinement

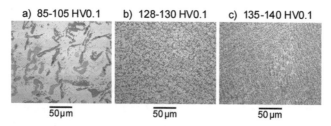

Fig. 11.18 Grain refinement in the alloy AlSi10Mg after laser remelting: $P = 1\,\text{kW}$, $d_L = 1\,\text{mm}$, and $v_{He} = 5\,\text{l/min}$; (**a**) conventional cast; (**b**) $v_v = 0.5\,\text{m/min}$; (**c**) $v_v = 5\,\text{m/min}$

in orders of magnitude. For a velocity of $v_v = 5\,\mathrm{m/min}$ a further refinement is observed. The rapid cooling leads to a very fine precipitation of the Si phase which causes an increase in hardness. Refinement and hardness increase can improve the wear as well as the corrosion resistance.

Cast iron solidifies according to the stable Fe–C diagram ("gray" cast iron). The microstructure is composed of ferrite and/or pearlite and graphite which precipitates as lamellae or spheres (Fig. 11.19). During laser remelting the graphite is completely dissolved in the iron melt. During rapid cooling the melt is supercooled and the solidification is governed by the metastable Fe–Fe$_3$C diagram. The microstructure is composed of primary solidified austenite (which transforms into pearlite on further cooling) and the eutectic ledeburite (Fig. 11.19). This so-called "white" solidification is accompanied by a significant hardness increase (Fig. 11.20) which leads to a superior wear resistance compared to the gray cast iron.

At very high supercooling the crystallization can be suppressed completely and the melt is frozen in an amorphous structure (metallic glass, Fig. 11.21). Metallic glasses feature some interesting properties:

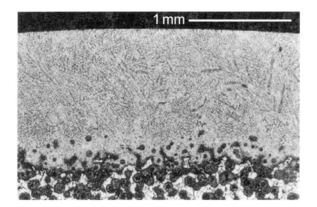

Fig. 11.19 Ledeburitic microstructure after laser remelting GGG60; $P_\mathrm{L} = 1900\,\mathrm{W}$; $A_\mathrm{L} = 0.6\,\mathrm{mm}^2$; $v_v = 0.24\,\mathrm{m/min}$

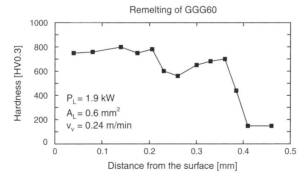

Fig. 11.20 Hardness profile across a remelted layer of GGG60

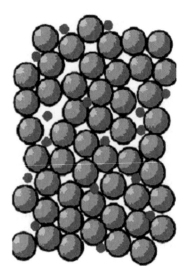

Fig. 11.21 Amorphous structure of a metallic glass with interstitial dissolved atoms of another element

- High strength, hardness, wear resistance, and fatigue strength
- Ductility
- Excellent corrosion resistance because of the homogeneous structure and the absence of grain boundaries
- Good electric and thermal conductivity comparable to metallic melts

Especially the ductility and the electric and thermal conductivity are different to conventional glasses.

Amorphous structures in pure metals require cooling rates of 10^9 K/s which are difficult to achieve. However, for certain alloys cooling rates of 10^6–10^8 K/s are sufficient which can also be achieved by laser remelting. The most important metallic glasses are transition metal–metal alloys (e.g., Cu–Zr, Cu–Ti) and transition metal–metalloid alloys (e.g., Fe–B, Ni–P). The composition should be in the region of eutectics with a low melting point. Glazing by laser remelting has been documented in numerous publications [16–18]. However, only a very thin layer of several micrometers can be glazed. In deeper regions the cooling rate is not sufficient.

11.2.4 Application

In many car engines camshafts made from cast iron are used. Casting allows near-net-shape manufacturing which minimizes machining. Strength and ductility of cast iron are sufficient for the mechanical stresses during operation. However, in the contact area with the cam follower a high wear resistance is required. This can be improved, e.g., by remelting a thin layer followed by rapid cooling which forces the material to solidify according to the metastable Fe–Fe$_3$C phase diagram. For laser remelting a line focus is used which covers the whole width of the cam [1]. Only a single pass is necessary to remelt the required surface area. Figure 11.22a shows the process of remelting and Fig. 11.22c shows a longitudinal section of a cam. The depth of the remelted zone is 0.5–0.8 mm. State of the art is remelting with TIG. Table 11.1 lists some features of TIG and laser remelting.

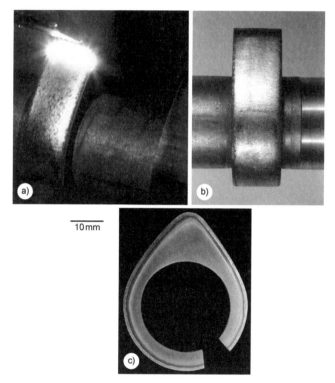

Fig. 11.22 Laser remelting of a cam with a line focus beam: (**a**) process; (**b**) remelted surface of a cam; (**c**) longitudinal section of a remelted cam

Table 11.1 Comparison of laser remelting and TIG for camshaft remelting

TIG	Laser beam
Huge heat input = big distortion	Low heat input = small distortion
High roughness = big oversize required	Low roughness = small oversize required
Big melt pool = significant deviation from the original geometry especially at the cam tip	Small melt pool = low deviation from the original geometry especially at the cam tip
Low investment costs	High investment costs

11.3 Polishing with Laser Radiation

Edgar Willenborg

The surface roughness of a part or product strongly influences its properties and functions. Among these can be counted abrasion and corrosion resistance, tribological properties, optical properties, haptics as well as the visual impression the customer desires. Therefore, in industrial manufacturing grinding and polishing techniques are widely used to reduce the roughness of surfaces.

A new method to attain such high-quality surfaces is polishing with laser radiation. In principle there are three different process variants [19]. At the top of Fig. 11.23 a sketch of a cross-section of a milled surface is shown. With *polishing by large-area ablation* material is ablated over the whole surface. Thereby, the smoothing is achieved by increased ablation of the peaks of the surface and decreased ablation in the valleys. In contrast, *polishing by localized ablation* requires a precise measurement device for measuring the initial surface profile. After a nominal/actual value comparison only the peaks of the profile are ablated by a controlled laser pattern. The third process variant is *polishing by remelting*. A thin surface layer is molten and the surface tension leads to a material flow from the peaks to the valleys. No material is removed but reallocated while molten. In the following, further details are shown for all three process variants [63].

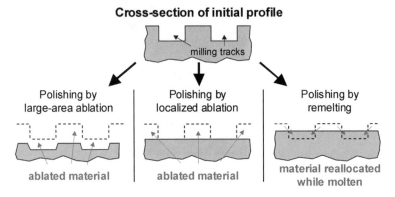

Fig. 11.23 Process variants of polishing with lasers

11.3.1 Polishing by Large-Area Ablation

Polishing by large-area ablation is predominantly used for CVD diamond films and plates [20–22]. For the most part, excimer lasers are used (ArF, KrF, XeCl). For thick

---

E. Willenborg (✉)
Oberflächentechnik, Fraunhofer-Institut für Lasertechnik, 52074 Aachen, Germany
e-mail: edgar.willenborg@ilt.fraunhofer.de

11 Surface Treatment

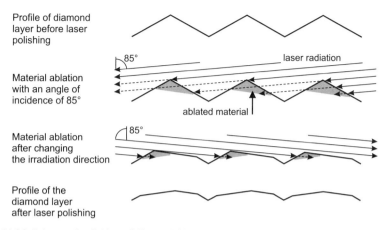

Fig. 11.24 Scheme of polishing of diamond films and plates with lasers

films and plates ($>100\,\mu$m) argon ion lasers, copper vapor lasers, and frequency-doubled Nd:YAG lasers are also suitable, often combined with a finishing step, again with an excimer laser. In order to ensure increased material removal at the profile peaks, laser polishing is carried out with an angle of incidence of up to 85° to the normal of the aimed surface (Fig. 11.24). A further reduction of the roughness can be achieved by rotating the sample during processing.

The roughness of thin films ($<100\,\mu$m) with initially $Ra = 0.1$–$1\,\mu$m can be reduced by a factor of 2 to 4. For thick films with initially $Ra = 20$–$30\,\mu$m, even higher reductions are possible. The processing time is between a few minutes up to several hours per square centimeter depending on the laser source, one or two step processing, and initial roughness.

11.3.2 Polishing by Localized Ablation

Polishing by localized ablation is based on the controlled ablation of profile peaks with pulsed laser radiation. To locate the position of the profile peaks, an elaborate and costly profile measurement system is required. The principle of this process variant is described in the patent specification [23], but no details are given concerning process parameters, achievable surface roughness and processing time. Polishing by localized ablation has not been investigated in much detail to date. In the industry localized ablation has been mainly used to structure surfaces for the manufacturing of micro tools.

11.3.3 Polishing by Remelting – Metals

Polishing of metals by remelting with laser radiation is a new method for the automatic polishing of 3D surfaces. The main characteristics are as follows [19]:

- A high level of automation.
- Short machining times especially in comparison to manual polishing.
- No pollutive impact from grinding and polishing wastes and chemicals.
- Polishing of grained and microstructured surfaces without damaging the structures.
- The generation of user-definable and localized surface roughness.
- No changing of the form of the workpieces. Deviations of the form are not corrected but otherwise an already perfect form is not damaged.
- Small micro roughness as the surface results from the liquid phase.

For metals for the process-variant polishing by remelting, two subvariants exist: macro polishing and micro polishing.

Macro polishing is carried out with cw laser radiation. Milled, turned, or EDM-processed surfaces with a roughness Ra up to several micrometers can be polished [24, 19, 25]. Remelting depths between 20 and 200 μm are used. Beam diameter and remelting depth have to be chosen according to the material and the initial surface roughness. Normally, fiber-coupled Nd:YAG lasers are used with laser powers of 70–300 W. The processing time is between 10 and 200 s/cm^2 depending on the initial surface roughness, the material and the desired roughness after laser polishing. The achievable roughness depends on several influencing variables [24]:

- Initial surface roughness and especially the lateral dimensions of the surface structures.
- Thermo-physical material properties, e.g., heat conductivity and capacity, absorption coefficient, viscosity, surface tension, and melting and evaporation temperature.
- Homogeneity of the material: segregations and inclusions especially downgrade the surface quality.
- Medium grain size and statistical distribution, a small grain size is preferable.

Theoretically, the surface tension smoothes the surface of the melt pool, and a "perfect smooth" surface should be achievable. But during the remelting and solidification process, several mechanisms produce new surface structures resulting in a process-induced roughness. Figure 11.25 shows the most important mechanisms. The challenge of laser polishing is to minimize the sum of these mechanisms. This can lead, however, to competitive objectives for the process parameters. The main field of application for macro polishing is the substitution of the time- and cost-consuming manual polishing, e.g., in tool and mold manufacturing.

In contrast to macro polishing, *micro polishing* is carried out with pulsed laser radiation [19, 27–31]. The pulse duration normally is in the range of 20–1000 ns and the remelting depth in the range of 0.5–5 μm. With the micro polishing process variant, only fine pre-processed surfaces (e.g., grinded, micro milled) can be polished. Due to the small remelting depth, larger surface structures remain unaffected and can, therefore, not be eliminated. The most important process parameters are pulse duration and intensity. Longer pulses can eliminate laterally larger surface structures. The intensity has to be chosen according to the pulse duration and the

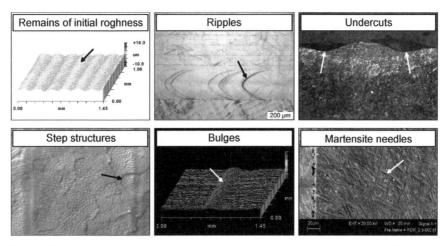

Fig. 11.25 Structures preventing a "perfect smooth" surface after laser polishing [24, 26]

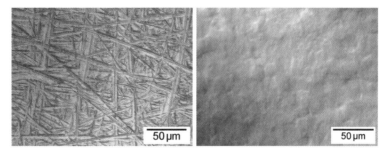

Fig. 11.26 TiAl6V4 surface milled (*left*) and micro polished (*right*)

material to be polished. A top-hat intensity distribution is preferable to generate a homogenous remelting depth. Fiber-coupled Nd:YAG and excimer lasers are used. Processing times less than 3 s/cm^2 can be achieved. Figure 11.26 shows a TiAl6V4 surface before (left) and after micro polishing (right). Due to the fine micro roughness, micro polishing is especially suitable for tribological and medical applications. Furthermore, the localized polishing and adjustment of the gloss level allows the generation of design surfaces [32].

The micro polishing process variant is limited to smooth surface structures with lateral dimensions of up to 40 μm. Larger surface structures can only be removed with the macro polishing process. But, in contrast to the macro polishing process, the micro polishing process often leads to a finer micro roughness and, therefore, to a higher gloss level. As a consequence for some applications, a combination of both variants is used: first macro polishing to eliminate the tracks from milling or turning, then micro polishing to enhance the gloss level.

Figure 11.27 shows a glass form before and after laser polishing. The processing of 3D surfaces is still under investigation, but this example already shows that laser polishing can be applicable even for free form surfaces [64].

Fig. 11.27 Glass form for the manufacturing of shafts and feet of wine glasses grinded (*left*) and laser polished (*front*) [24]

Examples for polishing results are shown in Table 11.2. Copper, gold, and aluminium alloys show hitherto predominantly unsatisfying results, but otherwise laser polishing of these metals has not yet been investigated as closely as the polishing of steels and titanium alloys has.

Table 11.2 Polishing results for selected metals [19, 24]

Metal	Subvariant	Initial roughness, Ra (μm)	Roughness after laser polishing Ra (μm)	Processing time (s/cm^2)
Tool steels, e.g., 1.2343, 1.2344, 1.2316, 1.2365	Macro	1–4	0.07–0.15	60–180
1.3344	Micro	0.5–1	0.3	3
Titanium, TiAl6V4	Macro	3	0.5	10
	Micro	0.3–0.5	0.1	3
Bronze	Macro	10	1	10
Stainless steel 1.4435, 1.4571	Macro	1–3	0.2–1	60–120

11.3.4 Polishing by Remelting – Glass

Laser polishing of glass by remelting is similar to the conventional and widely used fire polishing with manual or automatic guided burners. But as a laser can be controlled more precisely than a flame, laser polishing allows higher surface qualities. Due to the high absorption coefficient, cw CO_2 lasers with a wavelength

11 Surface Treatment

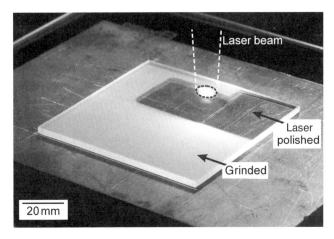

Fig. 11.28 Polishing of fused silica with CO_2 laser radiation [33]

of $\lambda = 10.6\,\mu m$ are used (Fig. 11.28). With the laser radiation the surface of the glass is heated. Evaporation has to be avoided, because otherwise material would be removed and a dent would occur. But the temperature must also be high enough to reduce the viscosity of the glass sufficiently in order to allow effective material flow from the peaks to the valleys. So the surface should be heated slightly below evaporation temperature.

The most important process parameters are the interaction time and the intensity. The longer the interaction time, the lower the roughness after laser polishing and the required intensity, but the higher the processing time. Usually, the laser power is between 30 and 4000 W, the intensity on the workpiece between 70 and 500 W/cm^2 and the feed rate between 2 and 80 mm/s [33–39]. Typical processing times are 1–10 s/cm^2. To avoid cracks preheating may be necessary, especially for glass with high thermal expansion coefficients. For preheating various heating devices or ovens can be used. For good polishing results, homogeneous preheating is a prerequisite.

In particular the micro roughness is smoothed. This and the avoidance of polishing lubricants can lead to the reduction of scattering losses of optics as well as to the enhancement of the destruction threshold. Form correction of optics through polishing by remelting has not been investigated yet. Examples for polishing results are shown in Table 11.3.

Table 11.3 Polishing results for different glass types [34, 33, 36, 38, 39]

Glass type	Initial roughness	Roughness after laser polishing
Lead glass	$R_z = 13.3\,\mu m$	$R_z = 2.5\,\mu m$
Fused silica	$R_{max} = 2\,\mu m$	$R_{max} = 50\,nm$
	$R_a = 150\,nm$	$R_a = 10\,nm$
TRC-33	$R_q = 500\,nm$	$R_q = 1\,nm$

11.3.5 Polishing by Remelting – Thermoplastics

Polishing of thermoplastics by remelting is similar to the polishing of glass. The main differences result from the lower process temperatures needed as a result of the low melting temperatures of thermoplastics. With a CO_2 laser with 500 W laser power, processing times less than $0.1\,\text{s/cm}^2$ are possible. Thermosetting plastics cannot be polished by remelting due to the lack of a liquid phase.

11.3.6 Summary of the Three Process Variants

Most common are the process variants polishing by large-area ablation and polishing by remelting. Polishing by localized ablation has only been slightly investigated. Table 11.4 shows a rough comparison of the three process variants. Polishing by remelting is divided into the subvariants, micro and macro polishing.

Table 11.4 Comparison of the process variants for laser polishing

	Polishing by large-area ablation	Polishing by localized ablation	Polishing by remelting	
			Micro	Macro
Materials	CVD diamond	Metals	Titanium, steel, nickel, cast iron	Titanium, steel, cast iron, glass, thermoplastics
Initial surfaces	Crystal structure from CVD process	n/a	Grinded, diamond milled	Milled, turned, EDM
Processing time	Minutes to hours per square centimeter	n/a	$1-10\,\text{s/cm}^2$	$0.5-3\,\text{min/cm}^2$
Laser sources	Excimer lasers and frequency converted Nd:YAG lasers	Pulsed lasers with nanosecond or subnanosecond pulse duration	Fiber-coupled Q-switch Nd:YAG lasers or excimer lasers with 20 ns to 1 µs pulse duration	Fiber-coupled cw Nd:YAG lasers for metals and cw CO_2 lasers for glass
Required machinery precision	Medium	High	Low	Medium
Achievable roughness	Strongly dependent on the material and initial surface roughness (see text)			
Special features	Flat angle of incidence required	Elaborate and costly profile measurement system required	Very homogeneous materials required	Homogeneous materials required

11.4 Structuring by Remelting

Andre Temmler

As already mentioned in Sect. 11.3, the surface of a part or product strongly influences its properties and functions. Therefore, many plastic parts have structured surfaces such as leather textures on car dashboards. The manufacture of the structures in injection molds is normally done by photochemical etching of the steel surface. Extensive studies have been carried out in order to substitute photochemical etching by an environmentally friendly laser-based process. But in spite of all this research work, the processing time for the structuring by lasers, e.g., ablation could not be reduced to an economical level suitable for the structuring of large tools. Usual processing times for structuring by laser ablation are about 5–10 min/cm^2 for a 100 µm deep structure. In addition, the quality of the surfaces manufactured by laser ablation normally demands extensive post-processing steps to remove the residues from the ablation process.

A new approach to structuring metallic surfaces with laser radiation is structuring by remelting. In this process no material is removed but reallocated while molten (Fig. 11.29).

This structuring process is based on the new active principle of remelting in comparison to the conventional structuring by photochemical etching or the structuring by laser ablation, which are based on removal. The surface structure and the micro roughness result from a laser-controlled self-organization of the melt pool due to surface tension. The structuring process is based on reallocation of material instead of ablation [65].

11.4.1 Active Principle

The process of structuring by laser remelting is based upon the physical interrelationship between the variation of the melt pool volume and the movement of the three phase line [26]. This movement determines the resulting surface topography

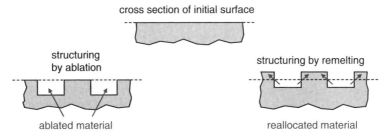

Fig. 11.29 Scheme of surface structuring by removing or reallocating of material

---

A. Temmler (✉)
Oberflächentechnik, Fraunhofer-Institut für Lasertechnik, 5207 Aachen, Germany
e-mail: andre.temmler@ilt.fraunhofer.de

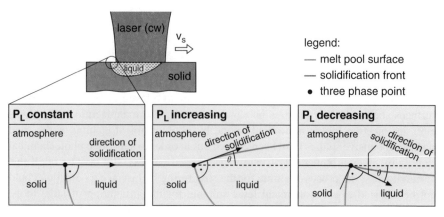

Fig. 11.30 Scheme of the active principle of the structuring by remelting process

as a consequence of the variation of the melt pool volume. In case of structuring by remelting, the melt pool volume can be precisely modulated, for example, by a modulation of the laser power. The resulting surface topography, as a consequence of the induced melt pool volume modulation, is generated by the same mechanism as shown for the ripple formation during laser polishing. So the FEM in-house program for the free boundary value problem for surface remelting can be used to predict the surface topography of structuring by remelting and forms the theoretical basis to understand the structuring process.

In practice the control of the melt pool volume is carried out by modulating the laser power of a cw laser. Figure 11.30 shows the principle of the structuring process. A thin surface layer ($< 100\,\mu m$) is molten and solidifies afterward. The direction of the solidification follows the melt pool surface [26]. With constant laser power the melt pool surface is approximately flat and no structuring appears. With increasing laser power the volume of the melt pool increases, and the melt pool surface is bulged outward due to the density change from solid to liquid and, therefore, the bigger volume of the molten material. The solidification now follows the bulged surface and structuring is achieved. With decreasing laser power the process works exactly the other way around. Therefore, with a modulation of laser power while remelting a thin surface layer of the material, structuring can be achieved [65].

11.4.2 Process and Relevant Procedural Parameters

The controlled modulation of the melt pool volume is essential for the process of structuring by remelting. The height, size, and form of the produced surface topography are directly depending on the average melt pool volume, and the absolute change and the time-dependant alteration rate of the melt pool volume. While the average size of the melt pool is responsible for the structural resolution of the produced structures, the absolute change and the time-dependant alteration rate of the melt pool volume are responsible for the height and symmetry of the produced structures.

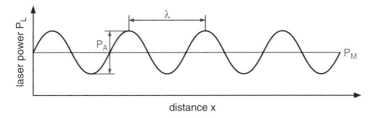

Fig. 11.31 Scheme for the controlled modulation of laser power

In order to achieve periodic structures the laser power is modulated sinusoidally at the average laser power P_M with an amplitude of P_A and a wavelength λ (Fig. 11.31). A defined average laser power and laser beam diameter are chosen in order to adjust the average size of the melt pool volume. The amplitude of the modulated laser power is responsible for the absolute change of the melt pool volume while the combination of wavelength and amplitude defines the alteration rate of the changing melt pool volume.

While the laser power is modulated, the laser beam is moved unidirectionally over the surface by a 2D laser scanner with a defined scanning velocity and laser beam diameter. In this case the wavelength of the modulated laser power is equivalent to the wavelength of the remelted structures. In order to obtain compact areas structured by remelting instead of single tracks, an overlap of the remelted tracks is necessary.

Consequentially, the relevant procedural parameters are the average laser power, the scanning velocity, the amplitude of modulated laser power, the wavelength of modulation, and the diameter of the laser beam (Fig. 11.32).

Depending on the wavelength used, the height of a remelted track varies, for example, from 5 μm for a wavelength of 0.5 mm up to 20 μm for a wavelength of 4 mm. In order to achieve greater heights of the surfaces' structures, the process has to be repeated several times.

Every additional time the surface is remelted using the same procedural parameters, the height of the structures grows. At a well-defined wavelength the machining time strongly depends on the demanded structure height, and therefore, on how many times the surface has to be remelted [65].

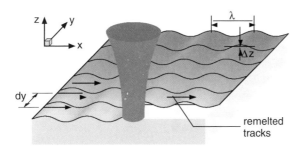

Fig. 11.32 Scheme of the process and the procedural parameters

11.4.3 Achieved Structures and Perspective

Periodic and aperiodic structures have been attained so far. The periodically structures achieved have a 1D sinusoidal periodicity like waves or 2D sinusoidal periodicity such as burl structures. The periodical structures achieved have a wavelength from 0.5 to 4 mm and a structure height of 5 µm up to 1 mm (Fig. 11.33 left).

However, the production of symmetrical structures is a special case and needs a precise adjustment of wavelength and the amplitude of the laser power modulation. In normal cases, this means that the structured surface topography is skewed without adjusting the laser power amplitude dependent on the wavelength used of the modulation (Fig. 11.34 left).

On its right side Fig. 11.34 shows a non-periodic structure similar to an artificial leather texture. This artificial leather texture was build up with structuring by remelting using non-periodic laser power jumps instead of a sinusoid modulation. Despite the quantitative differences in height and dimensions of the mountains and valleys in comparison to the original, this structure qualitatively shows already a good similarity to the original texture.

Figure 11.35 shows the latest result of a large-scale structure ($120 \times 150\,\text{mm}^2$) with a structural height of about 200 µm based on the new principle of structuring by remelting.

Up to now the achieved results for the structuring process by remelting of a thin surface layer, while modulating the process parameters, have been very promising. The machining time of the structuring process strongly depends on the laser beam

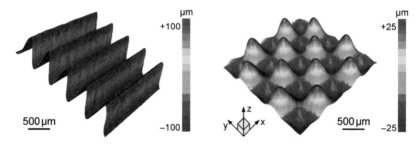

Fig. 11.33 Attained periodic structures

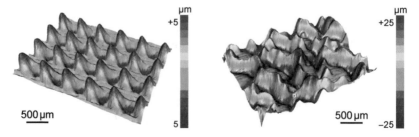

Fig. 11.34 Attained skew and aperiodic structures

Fig. 11.35 Picture of a periodically remelted surface topography

diameter used, and therefore, on the wavelength chosen and the desired height of the surface structures. The machining time varies, e.g., from 10 min/cm^2 for a structural wavelength of 0.5 mm and a height of 100 μm down to 30 s/cm^2 for a structural wavelength of 4 mm and a height of 50 μm.

Up to now these machining times have not yet been suitable for application in an industrial environment, especially one producing fine structures with wavelengths lower than 2 mm. In addition the structuring process has been attained on flat 2D samples so far and has to be adapted to 3D free form surfaces in order to be an even more attractive process for multiple industrial production lines.

Design structures and light path forming surfaces are just two potential applications for this technology [65].

11.5 Alloying and Dispersing

Andreas Weisheit and Gerhard Backes

11.5.1 Motivation

During laser alloying a thin surface layer is melted (typical melting depth: 0.1–2 mm) and a filler material is simultaneously added which is melted completely and dissolved in the melt of the material. Convective melt flows lead to mixing of both materials. When the laser beam moves on the melt pool is rapidly cooled by heat flow into the surrounding bulk volume. When the temperature falls below the liquidus temperature the melt solidifies rapidly and a fine cast structure is formed.

A. Weisheit (✉)
Oberflächentechnik, Fraunhofer-Institut für Lasertechnik, 5207 Aachen, Germany
e-mail: andreas.weisheit@ilt.fraunhofer.de

Which phases are present depends on nature and amount of the filler material and on the process parameters.

During dispersing, the filler material is not or at least only partly melted. The convective flows in the melt pool lead to the distribution of the inserted particles. After solidification, a dispersion layer, normally composed of a soft matrix material (substrate) and embedded hard particles (filler), has formed.

Alloying can improve surface properties like corrosion and wear resistance. Unlike remelting, alloying offers more options to achieve the desired properties because the chemical composition of a thin surface layer can be changed. An increased corrosion resistance can be achieved, e.g., by adding elements like Al or Cr which form dense oxide layers. Wear properties are improved by adding elements which form hard phases (e.g., W or V in steel to form carbides) or intermetallics (e.g., Ni in Al to form nickel aluminides).

Dispersing aims at the improvement of wear properties only. The addition of hard particles like carbides, nitrides, or oxides increases the resistance against (sliding) abrasion.

Alloying and dispersing are interesting processes for light metal alloys on the basis of aluminum, magnesium and titanium [42, 43].

11.5.2 Physical Fundamentals

The most important physical fundamentals of laser alloying and dispersing are

- absorption of the laser beam,
- heat conduction,
- melt pool dynamic, and
- rapid solidification

The absorption of the laser beam, the heat conduction and the theoretical fundamentals of the melt pool dynamic are described in Chaps. 4, 5, and 8. Melt pool dynamics and rapid solidification are discussed in detail in Sect. 11.2 "Remelting". The melt pool dynamic is important for alloying and dispersing. Due to the short existence of the melt pool (seconds) the filler material cannot be homogeneous distributed by diffusion only. The flow velocity of the melt which is orders of magnitude greater than the scan velocity during laser alloying/dispersing ensures a sufficient mixing. Figure 11.36 shows this effect for a single volume element which was traced in FEM simulation.

11.5.3 Process

Figure 11.37 shows the process of laser alloying and dispersing. Larger areas can be treated by scanning the surface. Powder is used as filler material which is fed through special powder feed nozzles. Ar or He is used as carrier and shielding gas. Some alloys can also be alloyed with gas (e.g., titanium with N_2).

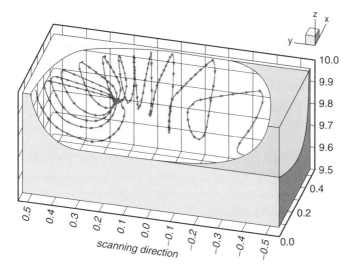

Fig. 11.36 Schematic demonstration of the trace of a volume element in the melt pool (FEM simulation)

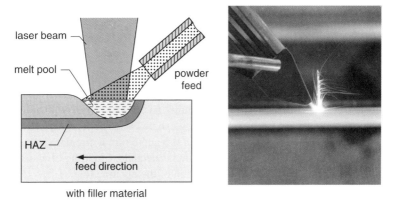

Fig. 11.37 Process of alloying/dispersing with powder as filler material; scheme (*left*) and process (*right*)

11.5.4 Powder Injection Nozzles [40]

Powder injection is a key factor in laser alloying/dispersing and cladding as well. The design of the injection nozzle determines the powder efficiency and the oxidation of the melted material. Three different concepts of powder injection can be used (Fig. 11.38):

- Off-axis powder injection (a single powder gas stream is fed lateral into the laser beam)
- Continuous coaxial powder injection (a powder gas stream cone is produced which encloses the laser beam)

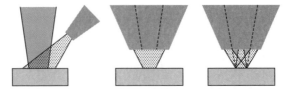

Fig. 11.38 Different powder injection concepts: *left*, off-axis; *center*, continuous coaxial; *right*, discontinuous coaxial

- Discontinuous coaxial powder injection (three or more powder gas streams are fed coaxial to the laser beam)

11.5.4.1 Off-Axis Powder Injection

For the off-axis powder injection the nozzle is positioned lateral to the laser beam. The position is determined by the angle between nozzle and workpiece and the distance between nozzle tip and workpiece. Since the powder stream diverges when leaving the nozzle (Fig. 11.39) the distance should be as small as possible to achieve a high powder efficiency (= weight of clad track per unit length versus powder amount fed per unit length). Typical values range from 8 to 12 mm. A parameter determining the efficiency is the direction of injection. When the flow direction of the particles is the same as the moving direction of the workpiece (dragging injection) a higher efficiency is achieved as for the opposite situation (stinging injection). In case of dragging injection more powder is injected directly into the melt than during stinging injection. From this result it is clear that off-axis powder injection is not suitable for 3D cladding. Figure 11.39 shows an off-axis nozzle with adjustment unit. Size and geometry of the nozzle opening can be varied depending on the width of the clad track. For a track width between 0.5 and 5 mm a circular cross section of the opening with a diameter of 1.5–3.5 mm is adequate. Cladding of wider tracks (5–25 mm) requires a rectangular cross section of the opening (e.g., 1.5×15 mm^2).

Fig. 11.39 Off-axis powder injection nozzle (*left*) and powder gas stream of an off-axis powder injection nozzle (*right*)

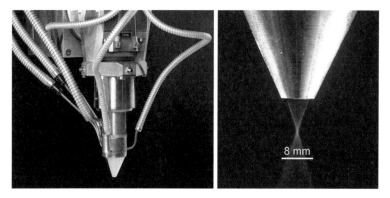

Fig. 11.40 Continuous coaxial powder injection nozzle mounted on a laser beam optic (*left*) and powder gas stream (*right*)

Since the nozzle is positioned close to the melt pool it is exposed to laser beam reflections. Therefore the nozzle should be water-cooled, to ensure long-term operation without any damage.

11.5.4.2 Continuous Coaxial Powder Injection

In continuous coaxial powder injection (below named "coaxial powder injection") a powder stream cone encloses the laser beam. Figure 11.40 shows a coaxial powder injection nozzle. For long-term operation it is essential that the nozzle is water-cooled. The powder stream cone is produced as follows: The powder stream of the powder feed unit is split into three identical streams which are fed into a ring-shaped expansion chamber inside the nozzle. In this chamber a homogenous "powder cloud" forms which is then fed into a cone-shaped slit. Eventually the powder leaves the nozzle in the form of a hollow cone. The diameter of the powder stream focus can be adapted to the laser beam area on the workpiece. Depending on process parameters such as particle size, gas flow rate, and powder mass flow, powder stream diameters below 500 μm can be achieved.

The major advantage of coaxial powder injection compared to off-axis powder injection is the potential for 3D cladding. However, tilting of the nozzle is restricted. Since the homogeneity of the powder stream cone depends on the powder distribution inside the expansion chamber of the nozzle, gravity will affect the powder stream when the nozzle is tilted. Experiments have shown that a maximum tilt angle of approximately 20° can be accepted without significant effects on the geometry of the clad layer.

11.5.4.3 Discontinuous Coaxial Powder Injection

In discontinuous coaxial powder injection several individual powder streams are distributed around the laser beam, thus forming a powder stream focus. Figure 11.41 shows a nozzle with three individual powder streams (below named as "three way

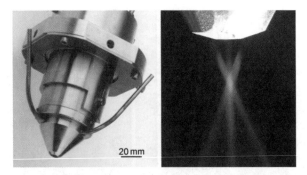

Fig. 11.41 Discontinuous powder injection nozzle (*left*) and powder gas streams (*right*)

nozzle"). The core diameter of the powder stream focus depends on the angle between the individual powder streams, the diameter of the nozzle holes, the distance between nozzle tip and powder stream focus, the powder feed rate, and the particle size. The major advantage of the three way nozzle is the potential for 3D cladding including the potential to tilt the nozzle up to 180°.

For reactive materials an additional shroud can be used to improve the shielding of the melt pool from oxidation. Examples are shown in Fig. 11.42.

Typical lasers used for alloying/dispersing are Nd:YAG, diode, fiber, or CO_2 lasers with an output power greater than 500 W.

11.5.5 Material Combinations for Alloying and Dispersing

Alloying/dispersing is most interesting for light metal alloys on the basis of aluminum, magnesium and titanium. The wear resistance of these alloys is not sufficient for many applications. Alloying/dispersing can lead to an improvement leaving the bulk properties unchanged. Below three examples are described.

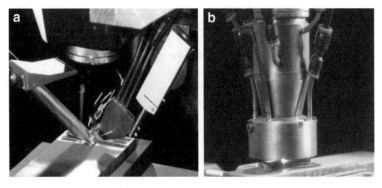

Fig. 11.42 Additional shrouds for an improved shielding from oxidation during laser melting; off-axis shroud (*left*) and coaxial shroud (*right*)

11.5.5.1 Alloying of AlSi10Mg with Nickel

Ni and Al form the aluminum-rich intermetallic phases Al_3Ni and Al_3Ni_2, which exhibit a high hardness due to a complex crystal structure. Figure 11.43 shows an alloyed layer with a content of 15 wt.% Ni in AlSi10Mg. The dark phases are primary solidified dendrites of Al_3Ni. The remaining melt solidifies as ternary eutectic $\alpha - Al + Si + Al_3Ni$. In the bottom area of the layer the dendrites have grown almost perpendicular to the surface which means that the growth velocity is faster than the movement of the solidification front (see also Sect. 11.2 "Remelting"). Since the velocity of the solidification front increases from bottom to top the dendrites are "catched" in the top region leading to a new growth orientation almost parallel to the surface according to the curvature of the solidification front. The alloyed zone has a hardness around 300 HV0.3. The base material has a hardness of approximately 60 HV0.3.

11.5.5.2 Gas alloying of TiAl6V4 with N_2

Titanium has a high affinity for nitrogen. If gaseous nitrogen is fed into the melt pool TiN is formed in an exothermic reaction. Pure TiN has a hardness of 2500 HV. Gas alloying with nitrogen is therefore an effective method to improve the hardness of titanium alloys. Figure 11.44 shows the microstructure of an alloyed layer on TiAl6V4. TiN solidifies primary as dendrites embedded in a solid solution matrix. On the surface a dense layer of TiN is formed with a thickness of several micrometers with a characteristic golden color. The amount of alloyed nitrogen determines the hardness of the layer. Figure 11.45 shows hardness profiles for various gas mixtures of Ar/N_2. Because TiN is a brittle phase cracking occurs for layers with a hardness higher than approximately 600 HV. Cracks are due to thermal stresses during cooling. Cracking can be avoided by preheating ($> 400°C$).

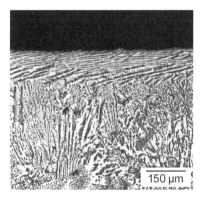

Fig. 11.43 AlSi10Mg alloyed with pure Ni; primary solidified dendrites of Al3Ni (*dark*) in a eutectic matrix (*light*) of $Al + Al_3Ni + Si$, Laser beam source: CO_2, $P = 4.5$ kW, $I = 4.4 \times 10^6$ W/cm², $v = 0.2$ m/min, and $m_P = 3$ g/min

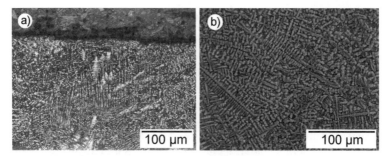

Fig. 11.44 TiAl6V4 alloyed with N_2; primary solidified dendrites of TiN (*light*) in a matrix of α-Ti solid solution (*dark*)

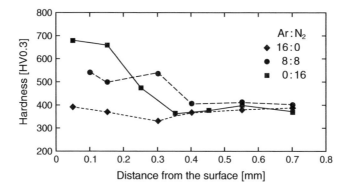

Fig. 11.45 Hardness profiles of TiAl6V4 alloyed with various gas mixtures of Ar/N_2

11.5.5.3 Dispersing of AlSi10Mg with TiC

Precondition for dispersing is a high difference in the melting point between dispersoid and substrate. Therefore only high melting point materials are suited as dispersoids. Another important property is the density. Dispersing a material with a high density (e.g., WC, $\delta = 15.6\,\text{g/cm}^3$) in a metal with low density (e.g., Al, $\delta = 2.7\,\text{g/cm}^3$) leads to gravity segregation which cannot be completely compensated by the convection flows. Typical materials for dispersion are carbides (WC, TiC, etc.) nitrides (TiN, etc.) or borides (Ti_2B, etc.). By laser dispersing volume fractions of the dispersoid up to 60–70% can be achieved. Higher contents lead to the formation of cavities because the melted volume is too small to embed all particles. Furthermore the brittleness of such layers is extremely high which causes cracking. Figure 11.46a shows a layer of TiC dispersed in AlSi10Mg. Such layers increase the resistance against abrasion which is shown in Fig. 11.46b for a pin-on-disk test.

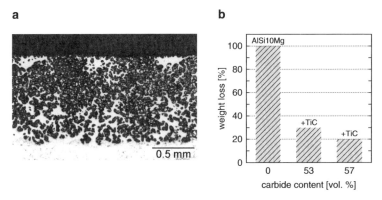

Fig. 11.46 AlSi10Mg dispersed with TiC; *Left*: cross section, TiC particles (*dark*) embedded in Al matrix (*light*). *Right*: improved abrasion resistance for dispersed layers (pin-on-disk test)

11.5.5.4 Application

Cylinder blocks for cars are nowadays cast mainly from eutectic Al–Si alloys. To improve the wear resistance gray cast iron liners are integrated. However, this causes thermal stresses during operation due to the different thermal expansion of gray cast iron and Al. Furthermore, the heat flow into the cylinder block is restricted because the liner and the block have no metallurgical bonding. Laser alloying with Si can be an alternative to form a wear-resistant layer on the surface. As base material, a hypoeutectic cast alloy can be used (e.g., AlSi8Cu3) which is alloyed with Si forming a hypereutectic layer with good wear behavior.

The metallurgical bonding between layer and substrate ensures an optimum heat flow into the bulk. Figure 11.47 shows the process known as TRIBOSIL [41]. The cylinder block is fixed and a head for inside treatment is rotating. Alloyed cylinder liners have been successfully tested in engine tests.

Fig. 11.47 Laser beam alloying of a cylinder liner with Si [41]

11.6 Laser Metal Deposition

Andres Gasser

11.6.1 Motivation

During the last decades **L**aser **M**etal **D**eposition LMD (also known as laser cladding) has become an established technique in many companies for applying wear and corrosion protection layers on metallic surfaces as well as for the repair of high added value components. Application fields are dye and tool makings, turbine components for aero engines and power generation, and machine components such as axes, gears, and oil-drilling components. Continuous wave lasers with power ranges up to 18 kW are used on automated machines with three or more axes, enabling 3D laser metal deposition (see 3D laser metal deposition). Mainly powder additive materials are used which implies a high diversity of available materials. Laser metal deposition with powder additive materials has unique properties like [44]

- Metallurgical bonding between deposited material and substrate material
- Very low porosity and no bonding defects and undercuts
- No oxidation, even for oxidation-sensitive materials
- Layer thickness ranging between 0.1 and 1.5 mm for a single layer
- Minimized heat input and distortion
- Suitable for iron (Fe), cobalt (Co), nickel (Ni), titanium (Ti), aluminium (Al), copper (Cu), and other metallic base alloys
- Highly automated production of high precision layers

11.6.2 Process Description

CO_2, solid state lasers SSL (either lamp-pumped Nd:YAG, fiber, or disc lasers) and high-power diode lasers (HPDL) are used for laser metal deposition. Due to the higher absorptivity and the higher process efficiency of the shorter wavelengths of the SLL and HPDL, this lasers are entering more and more into the LMD market. Furthermore, these lasers can be guided through fibers which offer higher flexibility in the beam-guiding system. In addition, fiber lasers, disc lasers, and HPDL possess a high wall-plug efficiency leading to an improved economic efficiency. Figure 11.48 shows a schematic of LMD process with a lateral powder feeding nozzle.

Typical powder grain sizes used for LMD range between 20 and 100 μm. The powder is fed into the interaction area with a carrier gas, usually argon or helium.

A. Gasser (✉)
Oberflächentechnik, Fraunhofer-Institut für Lasertechnik, 52074 Aachen, Germany
e-mail: andres.gasser@ilt.fraunhofer.de

11 Surface Treatment

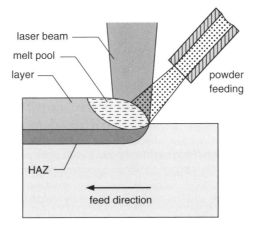

Fig. 11.48 Schematic of the LMD process

The carrier gas also acts as shielding gas. The laser melts a thin layer of the surface (some tenths of millimeters) and the incoming powder leading to a layer with high density (100%) and metallurgical bonding. Due to the local heat input given by the focussed beam a low thermal load is achieved, resulting in a small heat-affected zone. Typical values for the heat-affected zone range from some tenths of millimeters up to 1 mm.

For the repair of tools and molds also small pulsed lasers using manual wire feeding are established. The manual wire feeding offers high flexibility without the need of complex programming. These machines have limited deposition rates due to the low average power of the lasers (up to 200 W).

Figure 11.49 shows the manual pulsed LMD process (left side) and the automated LMD process with cw laser radiation and continuous powder feeding (right side).

Conventional powder feeders used for thermal spraying are used for the powder feeding in LMD processes. Depending on the complexity of the part to be treated different powder feeding concepts may be used (see Sect. 11.5).

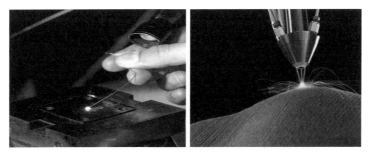

Fig. 11.49 Pulsed laser LMD process with manual wire feeding and cw LDM process with continuous powder feeding (source: TRUMPF)

Table 11.5 Most important parameters in LMD and typical values

Parameter	Typical values for LMD
Laser power	200–4000 W
Beam diameter	0.6–8 mm
Speed	200–2000 mm/min
Powder feeding rate	0.5–30 g/min
Feeding gas rate	2–15 l/min Ar or He
Shielding gas rate	2–15 l/min Ar or He

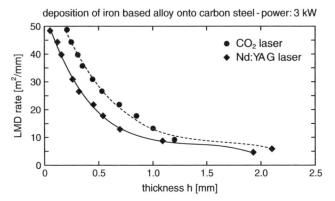

Fig. 11.50 LMD rates (cm^2/min) at different layer thickness h (mm) achieved with CO_2 laser and Nd:YAG laser radiation (power: 3 kW)

The most important parameters in LMD are given in Table 11.5.

Figure 11.50 shows the LMD rate versus the achieved layer thickness for two different wavelengths (CO_2 laser 10.6 μm and Nd:YAG 1,064 nm) at 3 kW laser power. A low carbon steel was chosen as a base material, and the additive material is a Fe-based alloy. The higher absorptivity of the Nd:YAG laser leads to higher LMD rates (about 50% higher rates).

Figure 11.51 shows the overall absorptivity in LMD process with CO_2 laser radiation compared to diode and solid state laser radiation. On the right side, two LMD results (cross sections) are shown, which were produced with CO_2 laser and Nd:YAG laser radiation. For achieving the same geometrical results (layer width and thickness) about 2.5 times more laser power is required for the CO_2-laser. The overall energetic efficiency is given by η

$$\eta = \text{abs} \times \eta_{\text{wall-plug}} \tag{11.7}$$

Introducing typical values for the wall-plug efficiency of about 10% for the CO_2 laser and of about 30% for solid state lasers (not lamp pumped) and diode lasers result in an energetic efficiency of

$\eta_{CO2} = 2.5$–3.5%

$\eta_{SSL/diode} = 18$–21%

11 Surface Treatment

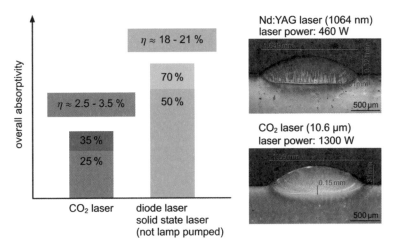

Fig. 11.51 Overall absorptivity and energetic efficiency achieved for LMD with different wavelengths

11.6.3 Materials

Figure 11.52 shows different layers achieved by LMD. Single layers up to 1–1.5 mm thickness as well as multiple layers up to several centimeters thickness are feasible. On the right side of the figure typical material combinations are shown. Depending on the application different additive materials can be applied resulting in different hardness ranges.

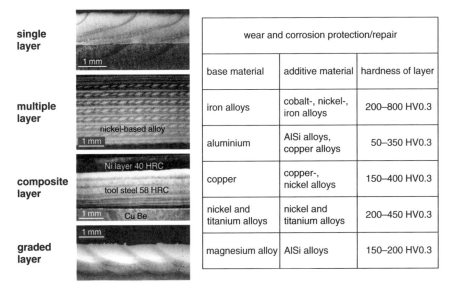

wear and corrosion protection/repair		
base material	additive material	hardness of layer
iron alloys	cobalt-, nickel-, iron alloys	200–800 HV0.3
aluminium	AlSi alloys, copper alloys	50–350 HV0.3
copper	copper-, nickel alloys	150–400 HV0.3
nickel and titanium alloys	nickel and titanium alloys	200–450 HV0.3
magnesium alloy	AlSi alloys	150–200 HV0.3

Fig. 11.52 Different layers produced by LMD

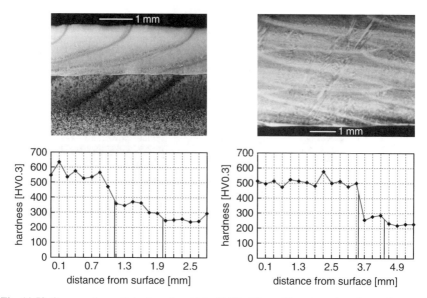

Fig. 11.53 Cross sections of single and multiple Stellite 6 layer (three layers) on low carbon steel and corresponding hardness profiles

Figure 11.53 shows the hardness distribution of two Stellite 6 (cobalt-based alloy) layers, one single layer and one layer out of three single layers. The layers were deposited on a low carbon steel (0.45% C). The hardness values achieved range between 500 HV 0.3 and 650 HV 0.3, the thickness from 1.2 up to 3.7 mm. The heat-affected zone (HAZ) is in the range of 1 mm. Usually the HAZ ranges in the measure of the layer thickness and can be reduced by reducing the thickness of the deposited layer (e.g., by applying less laser power during the LMD process).

11.6.4 Applications

Today's most important applications for LMD are the repair, wear, and corrosion protection of high added value and high precision machine components. One of the reasons is the high investment costs for the LMD machines. For serial production LMD has been used for the cladding of valves and valve seats at some car manufacturers and the wear protection of oil-drilling components [45–47].

11.6.4.1 Molds and Dyes

LMD is a well-established technique for the repair and modification of plastic injection molds [48, 49]. Also the repair and modification of dyes for metal sheet forming is getting more and more established [50].

11 Surface Treatment

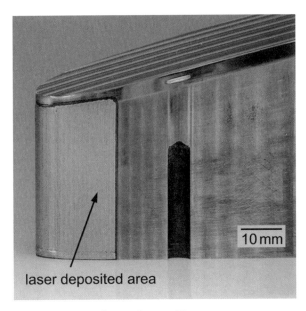

Fig. 11.54 Repair of worn-out areas in a car lamp mold

LMD allows the repair of critical areas also near polished or grained (chemically etched) surfaces. Figure 11.54 shows the repair of a car lamp mold. This mold is worn out in the front area and a layer of 0.15 mm thickness has been applied. No damage was observed even in the proximity of the polished surface. This type of repair cannot be performed with any other welding technique (e.g., TIG welding).

11.6.4.2 Modification of Molds

LMD offers the unique property to add material on sensitive surfaces and materials without damaging them. This can be used to add material to high added valued components during the manufacturing process, especially when design changes are required. Short-term changes during the manufacturing of molds usually lead to time-consuming repair processes. With LMD the missing material can be added precisely. Also molds and inserts can be modified when the plastic infection process requires a change in the design of the mold. Figure 11.55 shows an example for the modification of a car mold. A several millimeters thick layer has been applied onto the top of the mold.

11.6.4.3 Turbine Components for Aero Engines and Power Generation

The refurbishment of turbine components is performed with conventional techniques like TIG welding (tungsten inert gas welding), PTA welding, and thermal spraying. In the last years, LMD has been established for the repair of aero engine components at several facilities and offers several advantages like higher process

Fig. 11.55 Modification of a car lamp mold

speeds, near-net-shape deposition, and low heat input [51]. Figure 11.56 shows the repair of rotating seals on a cooling plate from a helicopter. Multiple layers are applied on the sealing geometry with a width of 0.6 mm.

11.6.4.4 Gears and Axes

In the areas of petrochemical industry, offshore drilling, shipping, storage, and transhipment, sugar industry, steel and energy industry machine components like shafts, gears, gearboxes, and drilling components are being repaired or protected by means of LMD. Figure 11.57 shows a repaired axis of a centrifuge and Fig. 11.58 the repair process during LMD of a gear component.

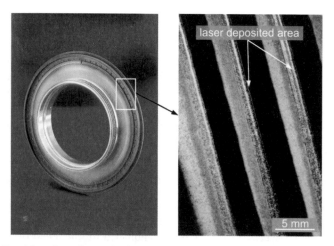

Fig. 11.56 Repairing of rotating seals of a cooling plate

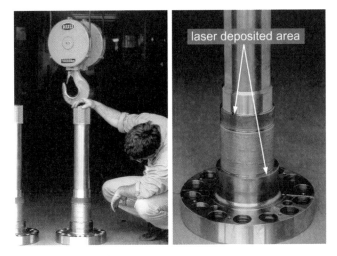

Fig. 11.57 Repair of centrifuge axis

Fig. 11.58 Repair of gear components with LMD (source: Stork Gears & Services)

11.6.4.5 Micro Laser Metal Depostion

New laser sources like fiber and disk lasers offer the possibility to produce LMD structures below 10 μm thickness and widths below 100 μm. In the last years much effort has been made to downscale the LMD process. For such structures, powder grain sizes in the range of 10 μm have to be fed and suited powder nozzles have been built. Possible application areas for micro LMD are medical science, electronic

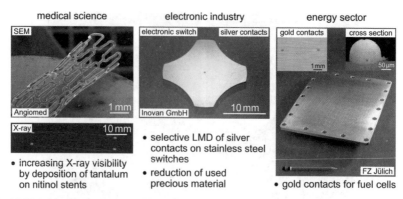

Fig. 11.59 Micro LMD applications

industry, and energy sector as shown in Fig. 11.59. These applications include the increasing of the X-ray visibility of stents and the selective deposition of silver and gold contacts for switches and fuel cells.

11.6.4.6 3D Laser Metal Deposition

By applying single deposited tracks or layers one on top of the other, 3D structures can be achieved. This can be used, for example, for repairing turbine blades (tip-repair, trailing edge repair) or even to build up complete new blades. Figure 11.60 shows blades that have been deposited on a shaft directly from CAD data. Recent development include the implementation of suited tools into LMD machines for closed chain repair and buildup based on CAD data or data acquired from digitalization.

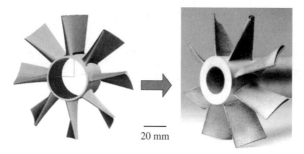

Fig. 11.60 Deposition of blades on a shaft starting from CAD data (3D LMD)

11.7 Pulsed Laser Deposition

Jens Gottmann

Pulsed laser deposition (PLD) [52, 53] is a physical vapor deposition (PVD) process. The material is ablated by intense pulsed laser radiation in a vacuum recipient with reactive or inert processing gas atmosphere. A laser-induced plasma is formed, the exited material expands through the processing gas atmosphere and is deposited as thin film on a substrate (Fig. 11.61).

Complex compositions of target material (e.g., superconducting $YBa_2Cu_3O_7$ and ferroelectric $PbZr_{0.52}Ti_{0.48}O_3$) can be reproduced as a thin film due to congruent material removal without alternation of the composition. The range of materials to be deposited using PLD is wide because the material removal is independent of special material properties like electrical, magnetic, or chemical properties, if laser radiation of suitable wavelength, intensity, and pulse duration is applied.

The variability of the process originates from the freedom to choose an appropriate processing gas (vacuum, reactive, non-reactive processing gas, or mixtures of different gases), the straightforward implementation of hybride processes (e.g., additional ion- or photon bombardment during deposition) and the possibility to adjust a multitude of processing variables independent from each other. By variation of, e.g., laser power, pulse repetition rate, processing gas pressure, and substrate temperature processing parameters like deposition rate, kinetic energy of the particles, and the resulting film structure can be tailored for applications. For instance, the kinetic energy of the film-forming particles can be controlled in a typical range

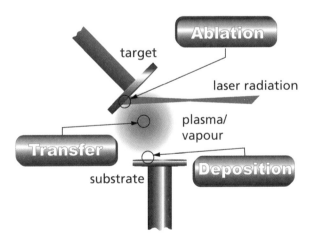

Fig. 11.61 Scheme of the processes during PLD

J. Gottmann (✉)
Fraunhofer-Institut für Lasertechnik, 5207 Aachen, Germany
e-mail: jens.gottmann@ilt.fraunhofer.de

of 1–100 eV and the deposition rate can be varied between 0.05 and 50 nm/s independent of each other.

Because of the pulsed nature of the deposition, a unique extreme instantaneous deposition rate of 10–100 μm/s is reached during several microseconds resulting in exceptional conditions during the film formation process. Therefore, multilayer film systems and gradient films can be fabricated and novel materials can be synthesized combinatorial far from thermal equilibrium as a thin film.

Since the first deposition of the high-temperature superconductor $YBa_2Cu_3O_7$ the PLD process has been developed to an established laboratory tool. For the deposition of diverse materials, lasers with large pulse power are applied. The selection of the suitable laser source is carried out by means of the absorption spectrum of the to-be-deposited material, the desired deposition rate, and the appropriate properties of the laser-induced plasma necessary for the formation of the desired film structure. For example metals, semiconductors, ceramics, and polymers have been deposited as thin films using excimer lasers, q-switched Nd:YAG lasers, ruby lasers, as well as ps and fs lasers. The PLD process has been successfully extend to a hybrid deposition process. The spectrum of new materials and film structures to be synthesized has been extended exploiting additional ion bombardment from an ion gun or a low-pressure gas discharge and additional photon bombardment of the growing thin film.

The PLD process is now about to be used for industrial applications and has to compete with mature established deposition processes. e.g., various chemical vapor deposition (CVD) technologies and other physical vapor deposition technologies like sputtering and thermal evaporation. Mainly due to the still high cost of high-power laser beam sources, the industrial application of PLD will be initially limited for the deposition of thin films of multicomponent materials not possible to be deposited with established technologies, for metastable film structures or for crystalline thin films to be deposited at low substrate temperatures in applications with a constrained thermal budget. Examples are high-temperature superconductors, diamond-like carbon, and laser-active materials to be integrated with electronic components.

11.7.1 Fundamentals

The processes taking place during PLD are subsumed to the three stages: removal, transfer, and deposition of the material (Fig. 11.61). The laser radiation is absorbed in the target material which is heated, melted, evaporated, photolytic decomposed, and/or electronically ablated resulting in a vapor or plasma (see Chaps. 8 and 9).

In general, above the ablation threshold the ablation rate increases initially linearly with fluence (Fig. 11.62). However, a fraction of the laser radiation is absorbed in the vapor/plasma of the ablated material. Therefore, on the one hand the efficiency of the material removal process is reduced and on the other hand the optical energy is fed to the vapor/plasma in the form of internal excitation (electronic excitation, ionization) and kinetic energy of the particles. In general, only a small fraction of

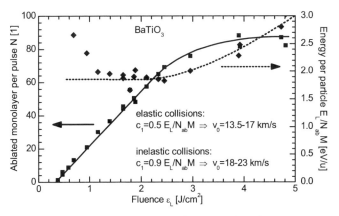

Fig. 11.62 Ablation rate and energy per ablated particle versus laser fluence

the optical energy of the laser radiation results in the removal of target material ($\sim 1\%$) while losses due to heat conduction into the target and excitation of the vapor/plasma are dominant as a function of the laser fluence (Fig. 11.63). With increasing fluence the fraction of optical energy converted to energy deposited into the plasma increases, resulting in an increasing energy of the later film-forming particles (Fig. 11.62, right scale).

The choice of the fluence used for pulsed laser deposition is hence a compromise between efficiency of the material removal (and hence the deposition rate) and the optimal energy of the film-forming particles governing the desired film structure. The efficiency of the material removal rate is maximal at the end of the linear increase of the ablation rate (e.g. $3\,\mathrm{J/cm^2}$ in the case of PLD of $BaTiO_3$ using KrF excimer laser radiation; Fig. 11.62). The ablation threshold ($\sim 0.3\,\mathrm{J/cm^2}$ in this example) is negligible low and losses due to the calculated heat conduction are small ($\sim 10\%$, Fig. 11.63). The optimal energy of the film-forming particles

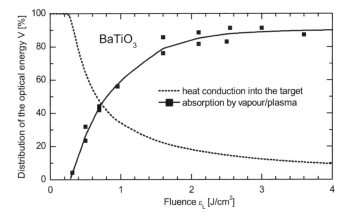

Fig. 11.63 Distribution of the optical energy to heat conduction into the target and to the plasma versus laser fluence

is depending on the desired film quality and the required film structure for it. In general, the optimal kinetic energy is in the range of 10 eV (e.g. for crystalline thin films) and up to 100 eV (e.g. for metastable super hard thin films) [54].

Impinging on the film growing on the substrate, the kinetic energy and the internal energy of the film-forming particles are released locally in a small volume (several nanometers in diameter). At moderate kinetic energies ($\sim$ 10 eV) the ad-atom diffusion is increased resulting in larger mobility and improvement of the crystal quality of the growing film (Fig. 11.64). At kinetic energies > 30–50 eV, the impinging particles may penetrate deeper than the first monolayer resulting in the creation of subsurface defects. Such bulk defects reduce the crystal quality and have to be annealed in a thermal treatment at elevated temperatures typical for sintering or solid phase epitaxy. In general, the activation energy for diffusion of bulk atoms is twice the activation energy for surface atoms due to the ratio of the number of surrounding atoms.

For the synthesis of new materials (e.g. diamond-like carbon, DLC, or cubic boron nitride) a localized energy deposition by impinging energetic particles ($\sim$ 100 eV) in a volume of $\sim$ 1 nm^3 is necessary to grow metastable materials far from the thermodynamic equilibrium. The metastable phase is quenched due to the very rapid cooling after the thermal spike induced by the energetic particle.

11.7.2 Kinetic Energy of the Film-Forming Particles

The kinetic energy of the film-forming particles is controlled by the processing variables fluence ε_L, the processing gas pressure p, the distance between target and substrate r, and the substrate temperature T, as well as material parameters like molecular mass M and the atomic masses m_i of the different atoms i constituting

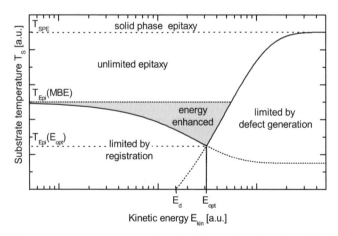

Fig. 11.64 Substrate temperature necessary for epitaxial growth as a function of kinetic energy of the film-forming particles [66]

the material. For the calculation of kinetic energy, the physical processes such as absorption of the laser radiation, heat diffusion, phase transitions, absorption of laser radiation in the vapor/plasma, and the expansion of the laser-induced plasma in the processing gas atmosphere have to be modeled, coupled to a system of differential equations and solved numerically [55]. Using some assumptions taking into account either elastic or inelastic collision between the ablated particles and the processing gas particles simple functional correlations between the average kinetic energy of the film-forming particles and the processing variables are obtained [56].

Firstly, taking only elastic collisions between the ablated particles and the processing gas particles into account a model has been derived with the assumptions, that the fraction f_E of the laser energy E_L is distributed to the kinetic energy of the ablated particles and to the processing gas particles in the volume between target and substrate and that after several collisions the velocity distributions of the particles have been averaged independent of their mass [57, 58]. For the mean kinetic energy $\langle E_{\text{kin}} \rangle_i$ of the impinging particle i with mass m_i holds

$$\langle E_{\text{kin}} \rangle_i = f_E \frac{E_L}{N_{ab}} \frac{m_i}{M + M_p N_p / N_{ab}}, \qquad (11.8)$$

with the N_{ab} the number of ablated target molecules with mass M, the mass of the processing gas particles M_p, and the number of processing gas particles N_p which are accelerated by the energetic ablated particles. To find an expression for the ratio between the number of accelerated processing gas particles and the number of ablated molecules local spherical expansion within the small volume angle element $\delta\Omega$ of a pyramid is considered. The base area of the pyramid is the square of the unit cell a^2 of the growing film material, located at the distance d_S from the target with an angle α between the line from target-to-substrate and the substrate normal, while the tip of the pyramid is where the material is ablated from the target, resulting in

$$\delta\Omega = a^2 \cos\alpha / d_S^2 \qquad (11.9)$$

The number of ablated particles entering the volume element $\delta\Omega$ is determined by measuring the thickness D_S of the film grown using N_L laser pulses to

$$N_{\text{dep}} = D_S / a N_L = c_2' N_{ab} / d_S^2 \qquad (11.10)$$

which is proportional to the number of ablated particles and, due to conservation of the number of particles, proportional to the inverse square of the distance. Using Eq. (11.10) the proportional constant c_2' is related to the ablation rate and the deposition rate and is therefore experimentally easy accessible. For the number of processing gas particles on the way of the ablated particles through the volume element $\delta\Omega$ holds

$$N_P = \frac{pV}{kT} = \frac{p}{kT} \frac{1}{3} r^3 \delta\Omega = \frac{pr^3}{3kT d_S^2} a^2 \cos\alpha \qquad (11.11)$$

resulting in the expression for the kinetic energy of the film-forming particles

$$\langle E_{\text{kin}}\rangle_i = \underbrace{\frac{f_E E_L}{N_{ab}} \frac{m_i}{M}}_{E_{\text{kin}}(p=0)} \underbrace{\frac{1}{1 + \frac{M_P}{M} \frac{pr^3 a^2 \cos\alpha}{3kT} \frac{1}{c_2' N_{ab}}}}_{\text{deceleration by processing gas}} . \tag{11.12}$$

Using the approximations

$$N_{ab} = a_1 E_L = a_1 \varepsilon_L A_L \quad \text{and} \quad f_E = 1/2 = \text{const.} \tag{11.13}$$

the mean kinetic energy of the impinging film-forming particles can be expressed generally as

$$\langle E_{\text{kin}}\rangle_i = \frac{1}{2a_1 M} \frac{m_i}{1 + \frac{pr^3}{\varepsilon_L T} \frac{M_P}{M} \frac{a^2 \cos\alpha}{3k A_L c_2' a_1}} = c_1 \frac{m_i}{1 + c_2 \frac{pr^3}{\varepsilon_L T}}, \tag{11.14}$$

where the square of the initial plasma velocity, the constant

$$c_1 = f_E \frac{E_L}{N_{ab} M} = \frac{1}{2} \frac{E_L}{N_{ab} M} = \frac{1}{2} v_0^2, \text{ in units } \left[\frac{m^2}{s^2}\right] \text{ or } \left[\frac{eV}{u}\right] \tag{11.15}$$

and the deceleration constant

$$c_2 = \frac{M_P}{kT} \frac{a^3 \cos\alpha}{3k} \frac{N_L \varepsilon_L}{D_S d_s^2}, \text{ in units } \left[\frac{K}{m^2}\right] \tag{11.16}$$

are to be computed for a each material, laser, and PLD setup after measurement of the resulting film thickness D_S and ablation rate N_{ab}, obtained using average processing variables (angle between target normal and substrate normal α, temperature of the processing gas T, energy of the laser radiation E_L, fluence ε_L, number of laser pulses N_L, and target-to-substrate distance d_S).

Besides elastic collisions between the atoms and ions in the laser-induced plasma and/or the processing gas particles inelastic collisions are possible, if the inelastic energy transfer during the collision is large enough to excite internal degrees of freedom of the particles. The maximum inelastic energy transfer during the collision of two particles A and B with masses m_A and m_B and velocities v_A and v_B is

$$E_{\text{inel}} = \frac{m_A m_B}{2(m_A + m_B)} (v_A - v_B)^2 \tag{11.17}$$

and becomes only mass dependent, if one of the particles is resting, e.g., the processing gas particle B,

$$E_{\text{inel}} = E_{\text{kin}} \frac{m_A}{m_A + m_B} \tag{11.18}$$

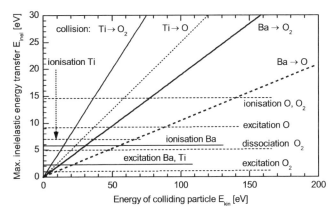

Fig. 11.65 Maximum inelastic energy transfer during the collision of energetic Ba and Ti particles with processing gas particles

and is the largest for equal masses. An inelastic collision is only possible, if the inelastic energy transfer exceeds the energy to excite an internal degree of freedom (Fig. 11.65). Cross sections of inelastic collisions have been measured for some noble gases and hydrogen-type particles. Cross sections of inelastic collisions of K^+, Na^+, Ca^+, and Mg^+ with molecular N_2 increase from $\sigma = 10^{-16}$ cm^2 at $E_{kin} = 400$ eV to $\sigma = 40 \times 10^{-16}$ cm^2 at several kiloelectron volts and are comparable to the cross sections for elastic collisions [59].

Because the cross sections for the inelastic collisions are unknown for most collision pairs a quantitative microscopic description is not possible. Therefore, a global inelastic collision of the ablated material with the whole processing gas between target and substrate is considered. Using energy and momentum conservation including a term for energy dissipation E_{dis}

$$\frac{1}{2} M N_{ab} v_0^2 = \frac{1}{2} (M N_{ab} + M_P N_P) v^2 + E_{dis} \quad (11.19)$$
$$M N_{ab} v_0 = (M N_{ab} + M_P N_P) v \quad (11.20)$$

results in the kinetic energy of the plasma E_{kin} and the dissipation E_{dis}

$$E_{kin} = E_0 \frac{1}{1 + \frac{M_P}{M} \frac{N_P}{N_{ab}}}, \quad E_{dis} = E_0 \frac{\frac{M_P}{M} \frac{N_P}{N_{ab}}}{1 + \frac{M_P}{M} \frac{N_P}{N_{ab}}}, \quad (11.21)$$

with E_0 being the kinetic energy at the beginning of the expansion. Analogous to Eq. (11.14) the mean kinetic energy of the particle kind i is deduced to

$$\langle E_{kin} \rangle_i = c_1 \frac{m_i}{\left(1 + c_2 \frac{pr^3}{\varepsilon_L T}\right)^2} \quad [56]. \quad (11.22)$$

Because the energy dissipation into internal degrees of freedom of the plasma is already considered, resulting, e.g., in plasma emission, the fraction of energy deposited initially in the plasma should be increased to $f_E = 0.7$–0.9, e.g.,

$$c_1 = 0.9 \frac{E_L}{N_{ab} M} = 0.9 v_0^2 \qquad (11.23)$$

while the rest of the energy (10%, see also Fig. 11.63) is dissipated via heat conduction in the target material.

The optical energy absorbed by the laser-induced plasma/vapor is measured and has been determined experimentally by measuring the transmission of the laser radiation through the laser-induced plasma above a thin film of $BaTiO_3$ with a small hole 30 μm in diameter [60]. Up to 90% of the optical energy is absorbed in the laser-induced plasma during ablation of $BaTiO_3$ in air (Fig. 11.63). The other portion of the optical energy is dissipated as heat into the target volume while the fraction of the energy for the ablation of material is less than 1% of the optical energy and can be neglected.

For the calculation of the initial plasma velocity (Eq. (11.16)) the ablation rate of $BaTiO_3$ is measured using different fluences (Fig. 11.62). The threshold for ablation of $BaTiO_3$ is $0.3\,J/cm^2$. The ablation rate increases linearly with increasing fluence up to 80 monolayers per pulse at $3\,J/cm^2$ and above $3\,J/cm^2$ the slope of the ablation rate is strongly reduced. From the ablation rate as a function of fluence the initial kinetic energy of the particles is computed to $c_1 = 0.9 \pm 0.1\,eV/u$ assuming elastic collisions ($f_E = 0.5$) and $c_1 = 1.6 \pm 0.15\,eV/u$ assuming inelastic collisions ($f_E = 0.9$) and is fairly constant using fluences of 1–3 J/cm^2. At further increasing fluences the energy per particle increases resulting in an increased initial kinetic energy (e.g., $c_1 = 2.1 \pm 0.15\,eV/u$ at the rate of $3.6\,J/cm^2$ assuming inelastic collisions).

To determine the deceleration constant, $BaTiO_3$ thin films have been deposited using different fluences at a substrate temperature of 550 °C. To prevent the plume narrowing at oxygen processing gas pressures $p > 20\,Pa$, resulting in a significant increase of the deposition rate in the film center of a factor of 2–4 [57, 61] and therefore a decreasing deceleration "constant" with increasing pressure, the processing gas pressure was 10 Pa. The deceleration constant is determined to $c_2 = 8 \pm 1 \times 10^{10}\,K/m^2$ (or $c_2/M_P = 2.5 \pm 0.8 \times 10^9\,K/m^2 u$ for the use of other processing gases) and quite constant at fluences of 1–5 J/cm^2 (Fig. 11.66).

The calculated mean kinetic energy of the barium particles after ablation of $BaTiO_3$ in oxygen processing gas environment is decreasing with increasing processing gas pressure and increasing distance from the target (Fig. 11.67). The decrease of the kinetic energy is much steeper in the inelastic collision model, because the energy dissipation as well as the number of particles to which the residual kinetic energy is distributed increases with distance and processing gas pressure (Eq. (11.22)). The maximum kinetic energy at low distances and low processing gas pressures is predicted higher in the inelastic collision model than when assuming elastic collision corresponding to the ratio between Eqs. (11.23) and (11.15).

11 Surface Treatment

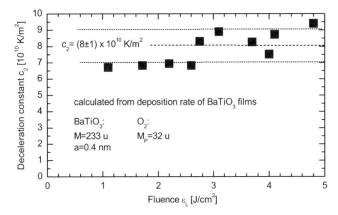

Fig. 11.66 Deceleration constant c_2 as calculated from the thickness of $BaTiO_3$ films deposited using different fluences

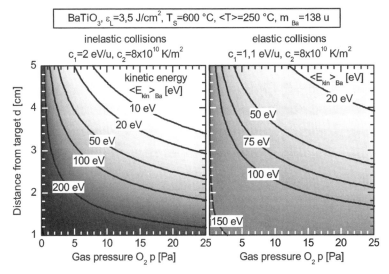

Fig. 11.67 Lines of constant mean kinetic energy of Ba particles during ablation of $BaTiO_3$ calculated as a function of distance and gas pressure using the inelastic (*left*) and the elastic collision models (*right*)

11.7.3 Plasma and Thin Film Properties

The dynamics of the laser-induced plasma is experimentally investigated to test the theoretical models and to correlate the thin film properties with the processing variables. In general, time-resolved techniques like ion-probe, time-resolved mass spectrometry, high-speed photography, and optical spectroscopy are used. For example, during the pulsed laser deposition of $BaTiO_3$ thin films the emission of excited barium atoms is investigated using a gated intensified CCD camera (exposure time

> 1 ns) and an interference filter with a central wavelength of $\lambda = 554$ nm. By variation of the time between laser pulse and exposure, the dynamic of the laser-induced plasma is investigated as a function of the processing gas pressure. The fluence is $\varepsilon_L = 3$ J/cm^2, the angle between target normal and substrate normal is $\alpha = 45°$, and the target-to-substrate distance is $d_S = 3$ cm. At a processing gas pressure of $p = 1$ Pa the first excited Ba atoms arrive at the substrate (located at the lower edge of each picture) after $t = 1.2$ μs and with increasing processing gas pressure the arrival time is delayed (Fig. 11.68). At a high processing gas pressure of $p = 50$ Pa focussing of the plasma caused by inelastic collisions between ablated particles and the processing gas particles is observed, resulting in an increased deposition rate on a smaller area (< 1 cm^2). Above the substrate, a film of intense emission is observed, due to the compression of the accelerated processing gas particles (accelerated by the snow plough effect). After more than 10 μs, emission from Ba atoms reflected from the substrate is detected. Reflection of film-forming particles from the substrate is not observed at lower processing gas pressures, therefore the sticking coefficient is considered as 1 (Fig. 11.68).

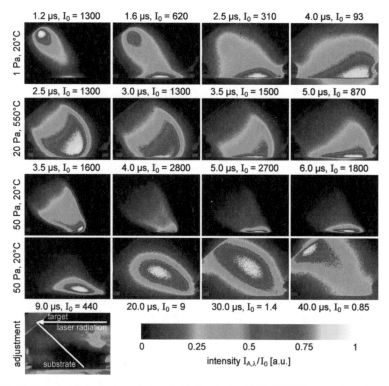

Fig. 11.68 Emission from the laser-induced ablation during PLD of BaTiO$_3$ at various times after the laser pulse. The O$_2$ processing gas pressure is $p = 1$ Pa (*top*), 20 Pa (*middle*), and 50 Pa (*bottom*). The angle between target and substrate normal is $\alpha = 45°$ and the target-to-substrate distance is $d_S = 3$ cm

Using high-speed photography (e.g., using a gated intensified CCD camera) the plasma emission during laser ablation of $BaTiO_3$ at different fluences is recorded. Using subsequent Abel inversion of the image the 3D radiation density is obtained. From this data the center of mass of the radiation density within the angle of $5°$ to the target normal is computed.

The calculated velocity of the expansion of the plasma using the inelastic collision model is in agreement to experimentally determined velocities of the center of mass of the radiation density for distances $R > 0.5–1$ cm (Fig. 11.69). The experimentally observed acceleration of the center of mass of the radiation density at low distances $R < 0.5–1$ cm from the target is due to the high-excited plasma core, where the high radiation density is caused by the laser-induced excitation of the plasma, therefore shifting the center of mass of the radiation density toward the target surface. The calculated velocity of the expansion of the plasma using the elastic collision model is not in agreement with experimentally determined velocities (Fig. 11.69).

The dielectric constant of films deposited using a substrate temperature of $T_S = 550\,°C$ and a target-to-substrate distance of $d_S = 2.6$ cm increases with increasing processing gas pressure, while at a target-to-substrate distance of $d_S = 4$ cm the dielectric constant exhibits a maximum of $\varepsilon_r = 1,000–1,300$ at a pressure of $p = 10–20$ Pa (Fig. 11.70, [62]). From the processing variables, the mean kinetic energy of the film-forming barium particles is computed using the inelastic collision model equation (11.21). The dielectric constant of the $BaTiO_3$ films deposited at $T_S = 550\,°C$ increases with increasing kinetic energy up to $\varepsilon_r = 1,100–1,300$ at $E_{kin} \sim 15–20$ eV. A further increase of the kinetic energy results in a sharp decrease in the dielectric properties (Fig. 11.71). Below the optimum mean kinetic energy $<E_{opt}> \sim 15–20$ eV an increase in the kinetic energy leads to an increased surface activation by energy transfer to the surface and enhanced surface diffusion by induced surface defects. The crystal quality is limited by insufficient registration

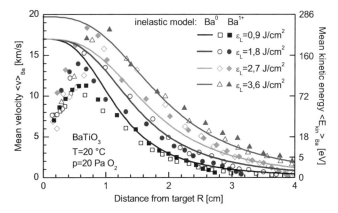

Fig. 11.69 Comparison of the calculated plasma expansion velocities versus the velocities of the measured center of mass of the radiation density as a function of the distance from the target

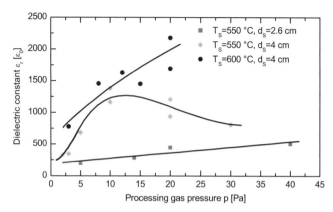

Fig. 11.70 Dielectric constant of BaTiO$_3$ films as a function of processing gas pressure, substrate temperature, and target-to-substrate distance

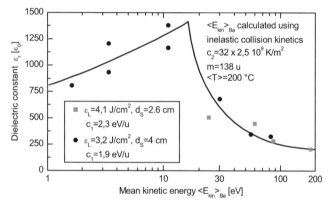

Fig. 11.71 Dielectric constant of BaTiO$_3$ films as a function of the calculated mean kinetic energy of the film-forming barium particles

[66]. Above the optimum mean kinetic energy, an increasing kinetic energy results in bulk displacements beneath the first monolayer, needing a higher annealing temperature. The crystal quality is limited by defect generation. The optimum kinetic energy of the film-forming particles is suggested to be 30–50 eV, because for the creation of a bulk defect 10–20 eV are necessary to be deposited within a monolayer. This seeming discrepancy is due to the spread of the real kinetic energy distribution around the calculated mean kinetic energy: At a mean kinetic energy of $E_{kin} \sim$ 15–20 eV a sufficient high proportion of the barium particles exhibits a kinetic energy $>$ 30–50 eV to limit the crystal quality by defect generation.

For a wide range of other thin film materials, thin film structures fabricated by pulsed laser deposition and their applications, see [52] and [53].

References

1. W. Amende, Oberflächenbehandlung mit Laserstrahlung, in H. Treiber "Der Laser in der Fertigungstechnik", Hoppenstedt Technik, Darmstadt 1990
2. E. Beyer, K. Wissenbach, Oberflächenbehandlung mit Laserstrahlung, Springer Verlag, ISBN 3-540-63224-7, 1998
3. C. Schmitz-Justen, Einordnung des Laserstrahlhärtens in die fertigungstechnische Praxis, Dissertation RWTH Aachen 1986
4. H. Willerscheid, Prozessüberwachung und Konzepte zur Prozessoptimierung des Laserstrahlhärtens, Dissertation RWTH Aachen 1990
5. G. Vitr, B. Ollier, F. Küpper, K. Wissenbach, Umwandlungshärten komplizierter Werkstückgeometrien mit Nd:YAG-Laserstrahlung, in Laser in der Materialbearbeitung, Band 4, Präzisionsbearbeitung mit Festkörperlasern, VDI Verlag 1995
6. J. Ohrlich, A. Rose., P. Wiest, Atlas zur Wärmebehandlung der Stähle, Bd. 3, Stahleisen, 1973
7. J. Ohrlich, H.-J. Pietrzeniuk, Atlas zur Wärmebehandlung der Stähle, Bd. 4, Stahleisen, 1976
8. K. Müller, C. Körner, H.-W. Bergmann HTM 51(1), 19, 1996
9. F. Küpper, Entwicklung eines industriellen Verfahrens zur Fest-Phasen-Härtung von Langprodukten mit CO_2-Laserstrahlung, Dissertation RWTH Aachen, Shaker Verlag 1997
10. B. Schürmann, Umwandlungshärten mit Dioden-Laserstrahlung in Mehrstrahltechnik, Dissertation RWTH Aachen, Shaker Verlag 2000
11. S. Mordike, Grundlagen und Anwendungen der Laseroberflächenbehandlung von Metallen, Dissertation, TU Clausthal, 1991
12. A. Gasser, Oberflächenbehandlung metallischer Werkstoffe mit CO_2-Laserstrahlung in der flüssigen Phase; Dissertation, RWTH Aachen, 1993
13. G. Herziger, P. Loosen, Werkstoffbearbeitung mit Laserstrahlung Grundlagen – Systeme – Verfahren, Hanser Verlag, ISBN 3-446-15915-0, 1993
14. N. Pirch, E.W. Kreutz, B. Ollier, X. He, The modelling of heat and solute transport in surface processing with laser radiation, in: NATO Advanced Study Institute, Laser Processing: Surface Treatment and Film Deposition, Sesimbra, Portugal, 3–16 July, 1994
15. B.L. Mordike, A. Weisheit, Werkstoffkundliche Aspekte der schmelzmetallurgischen Lasermaterialbearbeitung, in: Laser – Von der Wissenschaft zur Anwendung, Ed. W. Jüptner, Strahltechnik Bd. 10, BIAS-Verlag, ISBN 3-9805011-4-0, 1996
16. Y. Yan, Y. Yang, C. Zhao, W. Wu, M. Wang, M. Lu, Laser glazing of NI–Nb and Ni–Ni–Cr alloys and corrosion resistance of amorphous layers. Acta Met. Sinica 6(1), 52, 1993
17. H.W. Bergman, B.L. Mordike, Laser and electron-beam melted amorphous layers. Journal of Materials Science 16, 863, 1981
18. M. Miglierini et al., Laser-induced structural modifications of FeMoCuB metallic glasses before and after transformation into a nanocrystalline state. Journal of Physics: Condensed Matter 13, 10359–10369, 2001
19. E. Willenborg, Polieren von Werkzeugstählen mit Laserstrahlung, Dissertation RWTH Aachen, Shaker Verlag Aachen 2006
20. S.M. Pimenov, V.V. Kononenko, V.G. Ralchenko, V.I. Konov, S. Gloor, W. Lüthy, H.P. Weber, A.V. Khomich, Laser polishing of diamond plates, Appl. Phys. A, Mai 1999, Band 69, S. 81–88
21. R.K. Singh, D.G. Lee, Excimer laser assisted planarization of thick diamond films, Journal of Electronic Materials, 1996, Band 25, Heft 1, S. 137–142
22. A.P. Malshe, B.S. Park, W.D. Brown, N.A. Naseem, A review of techniques for polishing and planarizing chemically vapor-deposited (CVD) diamond films and substrates, Diamond and related materials, 1999, Band 8, Heft 7, S. 1198–1213
23. A. Bestenlehrer, Verfahren und Vorrichtung zum Bearbeiten von beliebigen 3D-Formflächen mittels Laser, Europäische Patentschrift EP 0 819 036 B1, 1996
24. T. Kiedrowski, Oberflächenstrukturbildung beim Laserstrahlpolieren von Stahlwerkstoffen, to be published at the end of 2009

25. J.A. Ramos, J. Murphy, K. Wood, D.L. Bourell, J.J. Beaman, Surface roughness enhancement of indirect-SLS metal parts by laser surface polishing, Konferenz-Einzelbericht: Solid Freeform Fabrication Proceedings, Proceedings of the SFF Symp., 2001, S. 28–38
26. N. Pirch, S. Höges, K. Wissenbach, Mechanisms of surface rippling during laser polishing, Proc. 8 International Seminar on Numerical Analysis of Weldability, Graz-Seggau, Austria, 25–27, 2006
27. Laser polishing of Nickel underplating, AMP Inc., Oktober 2002, http://rf.rfglobalnet.com/library/applicationnotes/files/5/laser.htm
28. K.Richter, G. Barton, Verfahren zur Bearbeitung von durch Reibung hochbeanspruchten Flächen in Brennkraftmaschinen, Europäische Patentschrift EP 0 419 999 B1, 1990
29. H.W. Bergmann, H. Lindner, L. Zacherl, C. Brandenstein, Verfahren zum Herstellen von Zylinderlaufbahnen von Hubkolbenmaschinen, Offenlegungsschrift DE 197 06 833 A 1, 1997
30. G.P. Singh, M. Suk, T.R. Albrecht, W.J. Kozlovsky, Laser smoothing of the load/unload tabs of magnetic recording head gimbal assemblies, Journal of Tribology, Oktober 2002, Band 124, S. 863–865
31. T.A. Mai, G.C. Lim, Micromelting and its effects on surface topography and properties in laser polishing of stainless steel, Journal of Laser Applications, 2004
32. T. Kiedrowski, E. Willenborg, K. Wissenbasch, S. Hack, Generation of design structures by selective polishing of metals with laser radiation, LIM, International WLT-Conference on Lasers in Manufacturing 2005, München
33. E. Willenborg, K. Wissenbach, T. Kiedrowski, A. Gasser, Polieren von Glas und Glasformen mit Laserstrahlung Tagungsband: Lasertechnologie für die Glasbearbeitung, glasstec 2004, Düsseldorf
34. E. Willenborg, Polieren von Quarzglas, anual report 2003, Fraunhofer-Institut für Lasertechnik, Aachen, 2004
35. P.-O. Wichell, U. Stute, J.Z. Zänkert, W. Wilke, A. Horn, M. Eschler, M. Mallah, Abschlussbericht zum Forschungsvorhaben InProGlas – Innovatives Produktionsverfahren zur Politur von Glasoberflächen, Laser-Zentrum Hannover, 2005
36. F. Vega, N. Lupón, J.A. Cebrian, F. Laguarta, Laser application for optical glass polishing, Optical Engineering, January 1998, Band 37, Heft 1, S. 272–279
37. F.-S. Ludwig, Thermodynamische Untersuchungen zur Politur von Glas mittels Laser, Dissertation TU München, 1998
38. R. Jaschek, A. Geith, H.W. Bergmann, Umweltfreundliche Politur von geschliffenen Bleikristallgläsern mit CO_2-Lasern, Zeitschriftenaufsatz: Sprechsaal, 1992, Band 125, Heft 5, S. 278–283
39. H. Wang, D. Bourell, J.J. Beaman, Laser polishing of silica rods, Proceedings of the Solid Freeform Fabrication Symposium, 1998, S. 37–45
40. A. Weisheit, G. Backes, R. Stromeyer, A. Gasser, K. Wissenbach, R. Poprawe, Powder Injection: The Key to Reconditioning and Generating Components Using Laser Cladding, Proceedings of Materials Week 2001
41. F.-J. Feikus, A. Fischer, Laserlegieren von Al-Zylinderkurbelgehäusen, Konferenz Zylinderlauffläche, Hochleistungskolben, Pleuel – Innnovative System im Vergleich, VDI-Ges. Werkstofftechnik, Stuttgart, D, 25–26 Sept., 2001, Bd 1621, 83–96
42. X. He, L. Mordike, N. Pirch, E. W. Kreutz, Laser surface alloying of metallic materials, Lasers in Engineering 4, Overseas Publishers Association, 1995, 291–316
43. A. Weisheit, Laserlegieren von Titanwerkstoffen mit Kobalt, Nickel, Silizium und Molybdändislizid, Dissertation, TU Clausthal, 1993
44. A. Gasser, G. Backes, K. Wissenbach, E. Hoffmann, R. Poprawe, Maßgeschneiderte Oberflächen durch Laserstrahl-Oberflächenbehandlung mit Zusatzwerkstoffen – eine Übersicht, Laser und Optoelektronik 3/97.
45. M. Kawasaki et al., The Laser Cladding Application at Automotive Parts in Toyota, Proceedings ISATA '92, 1992.
46. Schutzauftrag, LaserCommunity, Das Laser-Magazin von TRUMPF, 03/07

47. http://www.technogenia.com/fr/services/services-lasercarb.php
48. K. Eimann, M. Drach, Instandsetzung von Werkzeugen mit Laserstrahlung, Tagungsband Aachener Kolloquium Lasertechnik AKL 2000, 29–31 Mai 2000.
49. K. Eimann, M. Drach, K. Wissenbach, A. Gasser, Lasereinsatz im Werkzeug- und Formenbau, Proceedings Stuttgarter Lasertage, 25–26, Sep. 2003
50. J. Nagel, Laserhärten und Laser-Pulver-Auftragschweißen von Umformwerkzeugen, Tagungsband Aachener Kolloquium Lasertechnik AKL 2004, 295–305, 2004.
51. A. Gasser, K. Wissenbach, I. Kelbassa, G. Backes, Aero engine repair, Industrial Laser Solutions, Sep. 2007, 15–20.
52. D.B. Chrisey, G.K. Huber, Pulsed laser deposition of thin films, John Wiley & Sons, New York, 1994
53. R. Eason, Pulsed laser deposition of thin films, John Wiley & Sons, New York, 2006
54. M. Ohring, The materials science of thin films, Academic Press, San Diego, 1992
55. M. Aden, Plasmadynamik beim laserinduzierten Verdampfungsprocess einer ebenen Metalloberfläche, Dissertation RWTH Aachen, Shaker Verlag, Aachen, 1993
56. J. Gottmann, Dynamik der Schichtabscheidung von Keramiken mit KrF Excimer-Laserstrahlung, Thesis RWTH-Aachen 2001, http://darwin.bth.rwth-aachen.de/opus3/volltexte/2001/223/
57. J. Gottmann, T. Klotzbücher, E.W. Kreutz, Growth of ceramic thin films by pulsed laser deposition, the role of the kinetic energy of the film-forming particles, ALT'97 Laser Surface Processing, Proc. SPIE 3404, 8, 1998
58. E.W. Kreutz, J. Gottmann, Dynamics in pulsed laser deposition of ceramics: experimental, theoretical and numerical studies, Phys. Stat. Sol. (a) 166, 569, 1998
59. S.H. Neff, N.P. Carleton, "Excitation and change of charge in ion-molecule collisions in the adiabatic region" in M.R.C. McDowell Atomic collision processes, North-Holland publishing company, Amsterdam, 1964
60. M. Aden, E.W. Kreutz, H. Schlüter, K. Wissenbach, The applicability of the Sedov-Taylor scaling during material removal of metals and oxide layers with pulsed CO_2 and excimer radiation, Journal of Physics D 30, 980, 1997
61. D.J. Lichtenwalner, O. Auciello, R. Dat, A.I. Kingon, Investigation of the ablated flux characteristics during pulsed laser ablation of multicomponent oxides, Journal of Applied Physics 74, 7497, 1993
62. J. Gottmann, E.W. Kreutz, Controlling crystal quality and orientation of pulsed-laser-deposited $BaTiO_3$ thin films by kinetic energy of the film-forming particles, Applied Physics A 70, 275–281, 2000
63. E. Willenborg, Polishing with Laser Radiation, http://www.ilt.fraunhofer.de
64. R. Ostholt, E. Willenborg, K. Wissenbach, Laser polishing of metallic freeform surfaces, Proceedings of the fifth International WLT-Conference on Lasers in Manufacturing, Munich, June 2009
65. A. Temmler, E. Willenborg, N. Pirch, K. Wissenbach, *Structuring by Remelting*, Proceedings of the 5th International WLT-Conference on Lasers in Manufacturing, 2009
66. K.J. Boyd, D. Marton, J.W. Rabalais, S. Uhlmann, T. Frauenheim, Semiquantitative subplantation model for low energy ion interactions with solid surfaces, J. Vac. Sci. Technol. A 16, 463, 1998

Chapter 12
Forming

Alexander Olowinsky

12.1 Bending

12.1.1 Introduction

Smaller, more intelligent, and more complex – this is the aim of multifunctional components which introduce more autonomy into machines and tools. Indicators such as functionality, quality, and production costs are heavily influenced by these components. Thus, microsystem technology plays an important role for the technological development of the 21st century [1]. Micromechanical sensors such as gear rate sensors, acceleration sensors, or RADAR distance sensors with highly integrated electronics claim new safety standards in automotive [2]. Micro-optical components in growth markets such as the ICT are gaining enormous importance [3, 4]. Miniaturized endoscopes and catheter systems, as well as intelligent implants are demonstrating the potential of micro-technological solutions in the field of medical devices [1]. All these products have one thing in common: they are all based on miniaturization and integrated intelligence, asking for microtechnology – actuators, sensors, intelligence, and communication – and reflect the character of a cross-sectional technology.

Due to the high requirements for accuracy and the sensitivity to damage of the fragile parts the assembly of MEMS is challenging for a lot of producers in terms of manufacturing problems, especially tools and methods for handling high accuracy positioning and joining are required. The increasing level of integration raises new requirements for the assembly technology. In particular in automation the conventional methods for adjustment are limited. The stroke of electrical relays is indeed adjusted by screws nowadays, but by reducing the production cycle times this procedure can no longer be applied. Here a new method has been established. By means of pulsed laser irradiation thermal stress is induced inside the relay spring, which leads to a movement of the spring's end and thus to an adjustment of the stroke [5].

A. Olowinsky (✉)
Mikrotechnik, Fraunhofer-Institut für Lasertechnik, 5207 Aachen, Germany
e-mail: alexander.olowinsky@ilt.fraunhofer.de

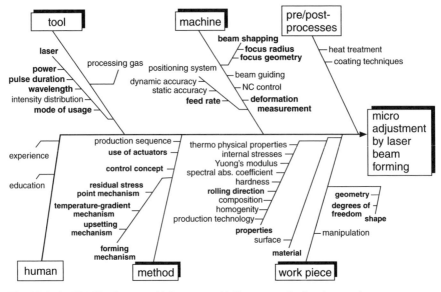

Fig. 12.1 ISHIKAWA diagram with impacts and influences on the forming result

The forming by laser radiation is influenced by several parameters which partly depend on each other. A part of the relevant process parameters is displayed in an ISHIKAWA diagram (see Fig. 12.1).

The laser radiation as a dominant process parameter is described primarily by wavelength, power, and pulse duration. In combination with the machine parameters such as beam shaping and feed rate the method of *forming* is developed.

Depending on the used laser beam parameters two fundamental types of forming can be distinguished (see Fig. 12.2).

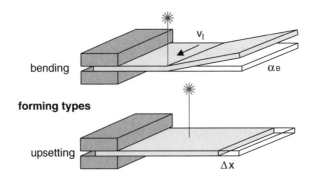

Fig. 12.2 Principal differentiation on the basis of directions of movement

12.1.2 Process Models

In literature several mechanisms for laser beam forming are mentioned [10–12]. The main difference between them lies in the temperature development during the laser treatment. The mechanisms can be divided into several groups (see Fig. 12.3):

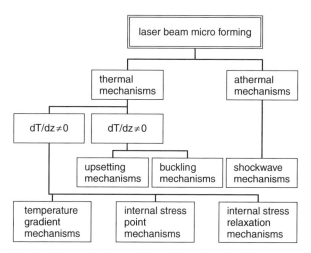

Fig. 12.3 Classification of the mechanisms of laser-induced forming [6]. dT/dz describes the temperature distribution over the sheet thickness

In the following, three methods among the thermal mechanisms will be investigated exemplarily.

12.1.2.1 Internal Stress Point Mechanism

The easiest case of simple bending by laser irradiation can be performed by single point irradiation without relative movement between laser beam and part. The part is locally heated in the region of the zone of interaction between beam and matter and it expands. The developing compressive stresses are converted into plastic compressive strain due to the decreasing yield stress as the surrounding material can be seen as rigid. According to the ratio between hindering section modulus and the induced compressive stresses a counter bending away from the laser beam may occur. As soon as the tensile strength is reached all additional thermal stresses are converted into plastic deformations, so that the counter bending angle remains constant. After the irradiation stops, the heated and compressed area cools down and is combined with the thermal contraction. The residual compressive stresses are relieved and residual tensile stresses develop. These tensile stresses lead to a bending toward the laser beam.

These stresses also reach the maximum yield stress which in the cooling phase is higher due to the lower temperature compared to the heating phase. The residual stresses increase in the same degree as the yield stress with decreasing temperature. The thermal contraction is compensated by local plastic strain. After complete cooling, only a small residual stress source remains which creates a slight bending. The bending edge is determined by the component itself and runs through the point of irradiation. The orientation of the bending edge follows the minimum geometrical moment of inertia. Therefore the position of the point of irradiation relative to the edges of the part does not play an important role.

Multiple irradiations at the same position do not lead to an increase of the bending angle as at first the elastic deformations are relieved due to the residual stresses in the upper layer of the part before the above-described process phases are run through. Larger bending angles can be achieved by placing several spots side by side.

12.1.2.2 Temperature Gradient Mechanism

The transition from residual stress point mechanism to TGM is caused by the relative movement between the laser beam and the component. Reference [11] shows the elementary model: A symmetrical two-layer model calculates the bending angle via the difference of the thermal expansion of the two layers.

The laser power is absorbed in the upper layer and is transformed into heat. The thermal expansion is hindered by the lower layer and is transformed into plastic compression. The lower layer stays thermally and geometrically unchanged.

$$\alpha_B = 4 \cdot \frac{\alpha_{th}}{\rho \cdot c_p} \cdot \frac{A \cdot P_L}{v_f} \cdot \frac{1}{s_0^2} (\text{rad}) \tag{12.1}$$

The bending angle increases linearly with the energy introduced into the part and decreases with the reciprocal of the square of the sheet thickness. The following section tries to explain the sequence of the process steps which are illustrated in the figure (see Fig. 12.4).

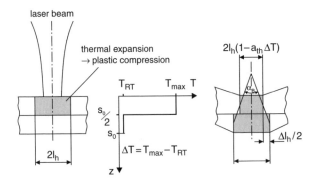

Fig. 12.4 Elementary model of laser beam bending [11]

12 Forming

The sheet is irradiated by the laser beam. The upper layer facing the laser beam is heated and it expands. During heating the melting temperature can be increased but this is not a precondition for the process.

The thermal expansion causes a counter bending. This counter bending depends on the width b of the sheet and is pure elastic deformation [12].

The yield strength decreases due to the increasing temperature in this area. The section modulus of the non-heated lower layer and the area ahead of the laser beam in feed direction work against this counter bending. The developing compressive stresses increase up to the yield stress. Beyond this point a plastic deformation occurs. As the temperature still increases the elastic stresses are partially converted into plastic compression.

After the irradiation the cooling phase starts. The heat is dissipated in the surrounding material. By heat conduction the temperature in the lower layer increases. The thermal contraction of the upper layer which was plastically compressed during the heating phase leads to bending toward the laser beam. The lower layer expands and it is plastically deformed. As the temperature is uniformly distributed the cooling takes place without further changes in the geometry. The single steps are displayed in Fig. 12.5.

For the TGM a model has been developed giving the bending angle α_B which depends on the laser power and the sheet thickness. This model is given in the following equation:

$$\alpha_B = 3 \cdot \frac{\alpha_{th}}{\rho \cdot c_p} \cdot \frac{A \cdot P_L}{v_f} \cdot \frac{1}{s_0^2} \quad (\text{rad}) \tag{12.2}$$

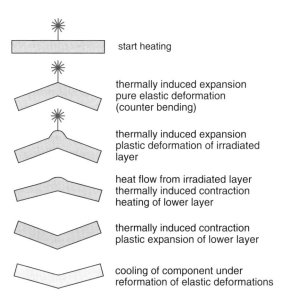

Fig. 12.5 Sequence of forming states

α_B bending angle,
α_{th} coefficient of thermal expansion, CTE,
ρ density,
c_p specific heat capacity,
A absorption,
P_L laser power,
v_f feed rate,
s_0 sheet thickness.

The main differences compared to the residual stress point mechanism are the following:

- The bending edge is not given by the minimum section modulus but follows the line of irradiation. Therefore non-linear deformations can also be realized.
- The deformation can be increased by multiple irradiations of the same line.

12.1.2.3 Upsetting Mechanism

The upsetting mechanism is based on a nearly homogeneous temperature distribution over the sheet thickness. Thus the thermal expansion is transformed uniformly into plastic deformation without bending of the part. As mentioned in the previous section the asymmetric compression leads to an out-of-plane deformation. Using an appropriate process parameter set and an adapted temperature profile a stress distribution can be created which is symmetric to the neutral fiber of the component. The deformation runs in three phases. These phases are illustrated in Fig. 12.6 showing a cross section of a sheet which is not clamped.

In a first step the component is heated locally. The diameter of the laser beam on the surface is smaller than the width of the component. Heat conduction creates a homogeneous temperature field over the thickness of the sheet. The thermal expansion in-plane is hindered by the surrounding non-heated material. Only the expansion perpendicular to the surface remains as degree of freedom.

With increasing temperature the yield strength of the material decreases. The thermally induced stresses exceed the yield stress and the material is plastically deformed.

During cooling after the laser pulse the plastically compressed area contracts. But due to the plastic compression the initial state cannot be reached. The surrounding

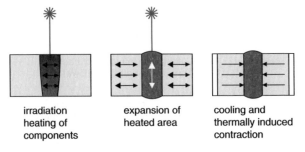

Fig. 12.6 Schematic of the upsetting mechanism

material hinders the shrinkage during cooling and the expansion during heating. The again decreasing temperature makes the yield stress rise. The increasing shrinkage forces lead to compressive stresses in the non-irradiated area. The stress state is not homogeneous and ranges just below the yield strength. Therefore mechanical loads provoke deformation of the component.

12.1.3 Forming Results

On the basis of comb-like structures the bending behavior of narrow parts is investigated. The influence of the sheet thickness plays an important role besides the number of irradiations. The two-layer model shown in Fig. 12.4 will be used for explanation. The development of the temperature gradient over the sheet thickness and the thicknesses of the upper and the lower layers influence the achieved bending angle. The thicker the material the smaller the upper heated layer and the connected tensile stresses. Accordingly the resulting bending angle will be smaller [7, 8] (see Fig. 12.7).

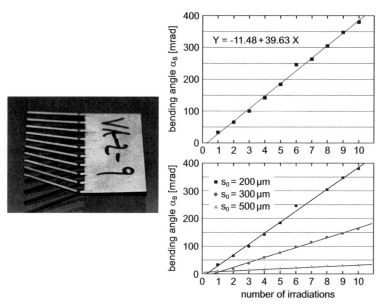

Fig. 12.7 Bending behavior of stainless steel with repeated irradiation. Thickness $s_0 = 200\,\mu\mathrm{m}$; laser parameters: $P_P = 260\,\mathrm{W}$, $\tau_H = 0,3\,\mathrm{ms}$, $f_P = 20\,\mathrm{Hz}$, v_L $100\,\mathrm{mm/min}$, and $2w = 100\,\mu\mathrm{m}$

12.1.4 Applications of Laser Beam Forming for Actuators

The mechanisms of laser beam forming can be used for manufacturing various actuators. Here the different mechanisms lead to different actuator types (see Fig. 12.8).

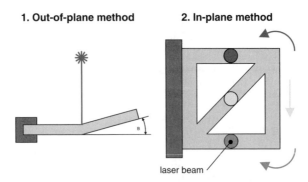

Fig. 12.8 Basics of the laser beam microforming as adjustment tool

In principle two kinds of actuators can be distinguished according to the mechanisms. The actual bending that occurs with the out-of-plane method and the actual bending that occurs with the in-plane method where all deformations take place in the plane of the component.

The out-of-plane method characterizes a movement in the direction of the incident laser beam, and the deformation or displacement occurs out of the principal plane of the actuator. However, in-plane indicates a movement perpendicular to the laser beam but in the principal plane.

12.1.4.1 Framework Actuator

The framework actuator consists of a square with a crossbeam. With an appropriate irradiation of the actuator several movements can be realized. The upsetting mechanism is used as the forming mechanism. The laser parameters have to be chosen in such a way that a movement out of the plane caused by the residual stress point mechanism is suppressed. The framework actuator shown in Fig. 12.9 gives the unit cell to assemble quite complex adjustment structures. The different movements can be assigned to the different irradiation points:

- An irradiation at point a causes a lateral movement of the right-hand side in positive y direction because the crossbar is compressed (Fig. 12.1).
- A simultaneous irradiation of points b and c leads to a shortening of the upper and lower bars of the square. The crossbeam suppresses the displacement in negative × direction and a movement of the right-hand side in negative y direction occurs (Fig. 12.2).
- Irradiation of all points a, b, and c simultaneously leads to a shortening of the complete actuator (Fig. 12.3).
- An angular change α at the right-hand side can be achieved by irradiating points a and c as shown in Fig. 12.8 or alternatively a and b for a rotation in the other direction (Fig. 12.4).

12 Forming

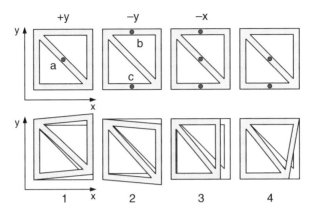

Fig. 12.9 Framework actuator. The result of the irradiation at different points in the upper row is shown below

12.1.4.2 Two Bridges Actuator

For the two bridges actuator (TBA) the upsetting mechanism is used to change the length of the actuator. The shortening of the actuator can be divided into different phases (see Fig. 12.10).

In the first step the laser is used to heat position 1. The thermally induced compressive stresses are balanced by tensile stresses at position 2. During the transition to the molten state at position 1 all stresses at this position as well as at position 2 are released. The molten pool can only expand in the direction perpendicular to the part.

The procedure can be repeated until the desired displacement is reached. The repetitions are only limited by the fact that the two shoulders of the actuators are touching each other and that the actuator cannot be compressed anymore. This actuator can be seen as a basic cell to realize more complex movements [7, 9].

The tube actuator with three two bridges actuators gives a simple structure to adjust mirrors and lenses in optical instruments. The rigid construction and the compact arrangement of the double bridges allow very small units that can exactly be

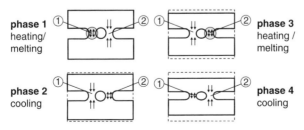

Fig. 12.10 Two bridges actuator

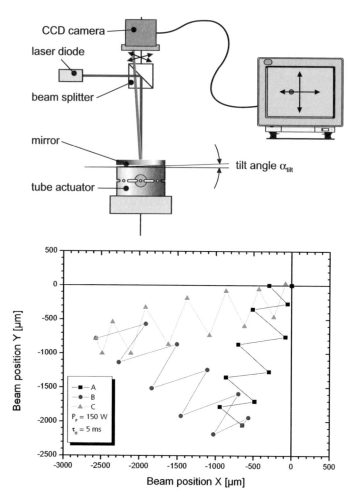

Fig. 12.11 *Top*: Measurement setup. *Bottom*: Movement of the center of gravity of the measurement beam [8]

adapted for the application. Figure 12.11 shows the measurement setup that displays the behavior of the actuator and the movement of the center of gravity of a deflected laser beam. Treating the first double bridge causes the COG to move toward the lower left corner. The irradiation of the second TBA results in an upward movement, whereas the last TBA causes a nearly horizontal displacement to initial point.

Hence all points within the measurement area can be reached by a combination of all three TBAs. This means that within the maximum deflection any tilt angle can be realized.

Table 12.1 Phases of treatment of the unidirectional two bridges actuator

	Position "1"	Position "2"
Phase 1: heating	Heating Thermal expansion (compressive stresses) Reaching the melting point: relaxation of compressive stresses	Obstruction of the thermal expansion (tensile stresses) Relaxation of tensile stresses
Phase 2: cooling	Thermal contraction (tensile stresses)	Obstruction of the thermal contraction (compressive stresses)
	⇨ Shortening of the component	
Phase 3: heating	Obstruction of the thermal expansion (tensile stresses) Relaxation of tensile stresses	Heating Thermal expansion (compressive stresses) Reaching the melting point: relaxation of compressive stresses
	⇨ Shortening of the component	
Phase 4: cooling	Obstruction of the thermal contraction (compressive stresses)	Thermal contraction (tensile stresses)
	⇨ Shortening of the component	

12.1.5 Conclusion

Laser beam forming offers a variety of tools and methods to be used in adjusting microcomponents. The combination of different actuators gives room for the realization of very complex adjustment structures. The stroke and the accuracy of the adjustment can be adapted for the application. As forming depends mainly on the laser parameters, automated adjustment procedures can be set up.

References

1. Miller, F.: "Mikrotechniken", Fraunhofer Magazin, Zeitschrift für Forschung, Technik und Innovation, 4.1998, Fraunhofer Gesellschaft, München (1998)
2. Robert Bosch AG: Geschäftsbericht 1998, Pressebild-Nr. 2-K8-10516
3. Kopka, P., Hoffmann, M., Voges, E.: "Bistable 2 × 2 and multistable 1 × 4 micromechanical fibre-optic switches on silicon", Proceedings 3rd Int. Conf. on MOEMS, Mainz, (1999)
4. Bishop, D.J., Giles, R.: "Silicon micromechanics takes on light-wave networks", The industrial physicist, 1998, S. 39–40
5. Hamann, Chr., Rosen, H.-G.: "Relaisfederjustierung mittels gepulster Nd:YAG-Laser", Laser/Optoelektronik in der Technik, Hrsg. W. Waidelich, Berlin: Springer Verlag (1990)
6. Vollersten, F.: "Laserstrahlumformen, lasergestützte Formgebung: Verfahren, Mechanismen, Modellierung", Bamberg: Meisenbach Verlag (1996)

7. Olowinsky, A., Gillner, A., Poprawe, R.: "Mikrojustage durch Laserstrahlumformen", in Proceedings of the Sensor'97 Vol.4 (1997), 133–137
8. Olowinsky, A.: "Laserstrahlmikroumformen – neues Justageverfahren in der Mikrotechnik", Dissertation RWTH Aachen, Shaker, 2003
9. Huber, A.: "Justieren vormontierter Systeme mit dem Nd:YAG-Laser unter Einsatz von Aktoren" aus der Reihe Fertigungstechnik – Erlangen (114), Meisenbachverlag Bamberg, 2001
10. Vollertsen, F.: "Mechanisms and models for laser forming", Proceedings LANE 94, Bamberg: Meisenbach Verlag (1994), S. 345–359
11. Vollertsen, F., Holzer, S.: "Laserstrahlumformen – Grundlagen und Anwendungsmöglichkeiten", VDI-Z 136 (1994) Nr 1\2 Januar/Februar
12. Geiger, M., Becker, W., Rebhan, T., Hutfless, J.: "Microbending and precision straightening by the aid of excimer laser radiation", Proceedings LANE 95, Bamberg: Meisenbach Verlag 1995 S. 306–309

Chapter 13
Rapid Prototyping and Rapid Tooling

13.1 Selective Laser Sintering (SLS)

Christoph Over

13.1.1 Introduction

Sintering is a term in the field of powder metallurgy and describes a process which takes place under a certain pressure and temperature over a period of time [1]. During sintering particles of a powder material are bound together in a mold to a solid part. In selective laser sintering (SLS) the crucial elements pressure and time are obsolete and the powder particles are only heated for a short period of time. SLS uses the fact that every physical system tends to achieve a condition of minimum energy. In the case of powder the partially melted particles aim to minimize their in comparison to a solid block of material enormous surface area through fusing their outer skins.

Like all generative manufacturing processes laser sintering gains the geometrical information out of a 3D CAD model. This model is subdivided into slices or layers of a certain layer thickness. Following this is a revolving process which consists of three basic process steps: recoating, exposure, and lowering of the build platform until the part is finished completely. The different process variations of SLS are based on the materials used (polymer powders, multicomponent metal–polymer powder mixtures, multicomponent metal–metal powder mixtures, single component metal powder) and subsequent post-treatments.

C. Over (✉)
Inno-shape c/o C.F.K. GmbH, 65830 Kriftel, Germany
e-mail: over@inno-shape.de

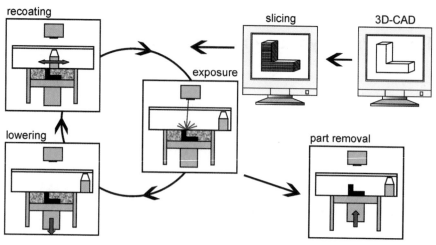

Fig. 13.1 Schematic of the process steps of selective laser sintering [5]

13.1.2 Selective Laser Sintering of Polymer Powders

The relatively low-temperature range up to 200 °C used in polymer processing promotes the sintering of polymer powders. Therefore the process chamber of an SLS machine is designed to be pre-heated to almost melting temperature of the material to be sintered. In most cases, radiant heaters heat up the powder bed. The laser just contributes a small amount of differential energy in order to melt the powder particles. In general CO_2 lasers with a power up to 50 W are used. The most commonly used materials are polyamide as a crystalline and polycarbonate as well as polystyrene as amorphous materials. Depending on the material and the application, subsequent processes such as manual grinding, sand blasting, or infiltration for surface improvement are necessary.

Table 13.1 Mechanical properties of laser-sintered polyamide [2]

Property	Polyamide	Polyamide glass filled
Tensile strength	44 N/mm^2	38 N/mm^2
Young's modulus	1600 N/mm^2	5910 N/mm^2
Breaking elongation	9%	2%
Bending modulus	1285 N/mm^2	3300 N/mm^2

Polymer SLS is mainly used for the direct manufacturing of design and functional prototypes. The use as master models for casting processes such as vacuum or fine casting is also possible but usually creates the necessity for post-processing of the surface.

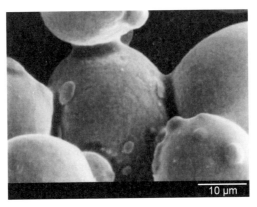

Fig. 13.2 Polymer-coated metal particles connected by SLS particle size: 50 μm. Thickness of the polymer layer: approx. 10 μm [3]

13.1.3 Indirect Selective Laser Sintering of Metals

Indirect selective laser sintering uses the process technology of SLS of polymers to manufacture metallic parts [3]. This is made possible through coating of the metal particles with a polymer layer. Every single powder particle (size approx. 50 μm) of the metal powder is coated with a polymer layer of thickness 5–10 μm.

Through irradiation with a laser beam (CO_2 laser up to 100 W) the relatively low melting (in comparison to the metallic material) polymer coating is melted and connects the metal particles after solidification (Fig. 13.2). The resulting green part shows significantly low mechanical properties. In a subsequent oven process in a reducing hydrogen atmosphere the part is post-sintered and the polymer is evaporated. As a finish a low melting metal, usually copper or bronze, is infiltrated into the voids of the porous material. This takes place using capillary forces in the same oven process at the corresponding melting temperature of the infiltrating material. During the oven process a geometry-dependent shrinkage occurs. After cooling down to room temperature a part with a density of 100% consisting of 60% steel and 40% of low melting metal such as copper can be achieved. Table 13.2 shows some of the characteristic mechanical properties of parts manufactured by indirect selective laser sintering.

The main field of application for this technology is the manufacturing of inserts for prototype tooling (rapid tooling), plastic injection molding, and light-metal die

Table 13.2 Mechanical properties of indirect SLS parts [2]

Property	Value
Yield strength (0.2%)	305 N/mm^2
Tensile strength	510 N/mm^2
Breaking elongation	10%
Hardness	87 HRB

casting. The lifetime of the tools depends on the geometry and the material processed.

A variation is the processing of polymer-bound forming sand in order to achieve cores for sand casting.

13.1.4 Direct Selective Laser Sintering of Metals

Direct selective laser sintering of metals includes the use of special multicomponent powder systems. The components are metal powder materials of which one serves as a binder material. Therefore high and low-melting materials are used and in general they are mixed mechanically. In some cases the high-melting component is coated with the low-melting one. During the process only the low-melting component is melted by a laser beam and binds the high-melting particles together which stay in solid phase. According to the principle of liquid phase sintering the densification of the powder layer takes place through the rearrangement of the high-melting particles in the liquid phase of the low-melting component (Fig. 13.3).

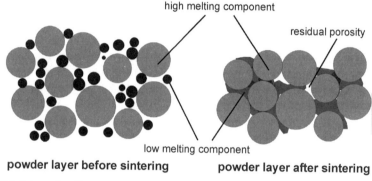

Fig. 13.3 Schematic of liquid phase sintering [3]

Several different powder systems have been investigated so far. The objective of these investigations is the development of a powder system with good wetting characteristics in order to cover the high-melting particles completely with the liquid phase of the low-melting component and achieve a good material composite. Mainly copper-based powder systems, such as Cu–(Sn/Pb), Cu–Sn, Cu–Fe, Bronze–Ni, and pre-alloyed Bronze powder, are investigated [3]. To improve the useful properties of parts manufactured by direct SLS (Table 13.3) more and more powder systems are developed, which exhibit higher strength and hardness such as bronze–steel, WC–Co, and WC/Co–Ni.

In general CO_2 laser systems with an output power of up to 200 W are used. The main application of direct SLS is also the manufacturing of small lot or pre-series tools. The advantage in comparison to indirect SLS is the avoidance of the complex oven process.

13.1.5 Selective Laser Melting (SLM)

Selective laser melting, developed by the Fraunhofer ILT, is an enhancement of direct selective laser sintering. The main differences are the materials used and the actual melting process. The materials used are stainless steel, tool steel, nickel-based alloys, titanium, and aluminum alloys without any additives. These materials are completely molten in the process (just like laser welding) through the use of solid state lasers (diode pumped Nd:YAG) with laser powers up to 500 W. Through the combination of the process parameters the actual powder layer is melted completely as well as the already solidified material underneath is melted partially.

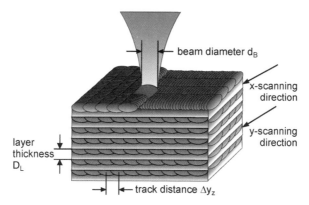

Fig. 13.4 Schematic of the layer-wise structure of SLM parts with the process parameters laser beam diameter, layer thickness, and overlap (without surrounding powder bed) [5]

Through this measure a metallurgical connection between neighboring tracks and underlying layers can be achieved. The choice of the right combination of processing parameters regarding the material used results in a part density of approx. 100%. Subsequent processing steps needed in sintering in order to achieve higher densities can be avoided completely.

Figure 13.5 shows a cross section of a part manufactured out of tool steel X38CrMoV5-1 by SLM parallel to the build direction. The single tracks and layers are clearly visible through the characteristic melt structure. Due to the rapid solidification characteristic of this process a dendritic cast-like structure consisting of martensite, residual austenite, and carbides results.

Table 13.3 Mechanical properties of direct SLS parts [4]

Property	Bronze-based powder	Steel-based powder
Residual porosity	20–25%	2–5%
Tensile strength (N/mm^2)	120–180	500–550
Breaking elongation	4.2%	7%
Roughness R_z (μm)	50–60	50

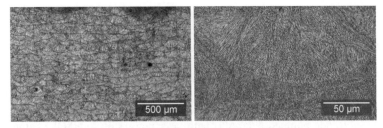

Fig. 13.5 Cross section parallel to build direction. Material: X38CrMoV5-1

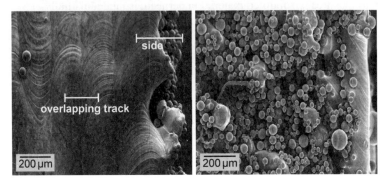

Fig. 13.6 SEM pictures of a SLM part. *Left*: top surface and *right*: side surface

The last molten layer (top surface) of a SLM part shows a characteristic structure with overlapping tracks exhibiting an appearance known from welding (Fig. 13.6 left). The width of the single tracks varies significantly and subsequently leads to a roughness of the side surfaces (right outline of part in Fig. 13.6). In addition molten particles are attached to the surface and increase the roughness (Fig. 13.6). This surface structure usually requires post-machining but can also be used as an additional functionality, e.g., for medical implants.

As opposed to parts resulting from indirect or direct selective laser sintering SLM parts exhibit mechanical properties comparable to cast parts out of the same material.

The use of serial metallic materials in combination with the geometric freedom opens up a broad variety of applications.

In injection molding SLM enables the integration of conformal cooling channels which are not manufacturable by conventional machining. Through this cycle times

Table 13.4 Properties of SLM parts out of different materials [3, 5]

	Stainless steel X2CrNiMo17-13-2	Tool steel X38CrMoV5-1	Titanium TiAl6V4
Tensile strength (N/mm^2)	550	1720	1140
Yield strength (N/mm^2)	450	1000	1040
Breaking elongation (%)	15	2.3	6
Hardness	240 HV0,1	52 HRC	435 HV10

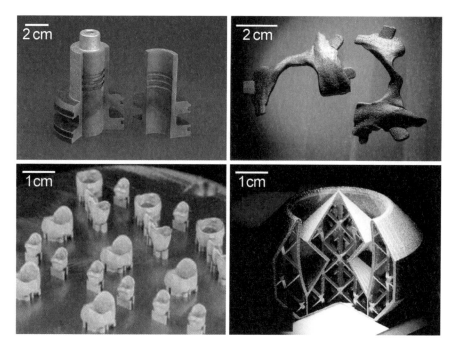

Fig. 13.7 Applications for SLM. *Top left*: injection mold tool (X38CrMoV5-1). *Top right*: facial implant (TiAl6V4). *Bottom left*: dental restoration (pure Titanium). *Bottom right*: lightweight part with internal structure (TiAl6V4) [5, 6]

can be reduced as well as the part's quality can be improved through lower warpage. In the medical field SLM opens up the possibility of simultaneous manufacturing of individual implants directly out of CT data. The first industrial serial production is applied for dental restorations. The combination of a lightweight material such as titanium with internal hollow structures gives the possibility of ultra lightweight parts, e.g., for space or racing applications.

13.2 Stereolithography

Wilhelm Meiners

13.2.1 Description of the Process

Stereolithography is one of the first rapid prototyping technologies applied industrially, in which an UV light hardens a liquid monomer by spontaneous polymerization.

W. Meiners (✉)
Oberflächentechnik, Fraunhofer-Institut für Lasertechnik, 5207 Aachen, Germany
e-mail: wilhelm.meiners@ilt.fraunhofer.de

The mechanical component for stereolithography is a container, used as a reservoir and a building room at the same time. On a vertically movable platform which is installed in this container, the part will be built up. This is done by moving the laser beam in x- and y-axes through a scanner. The energy of the laser beam initiates a chemical reaction which hardens the liquid monomer. After completion of one layer the building platform is lowered about the amount which equals the thickness of one layer (Fig. 13.8). This procedure will be repeated until completion of the whole part.

The parameters which decide on the quality of the part are

- laser power;
- beam characteristics eg. beam diameter, beam shape;
- scanning velocity;
- material characteristics of the used material;
- layer thickness.

Usable materials are

- epoxy resins;
- acrylates;
- polymer resins.

Depending on the manufacturer, different strategies for the exposure to light are applied. Generally, liquid layers are outlined first to generate the white borders. Afterward the inner area is exposed the laser beam moves in hatchings. To achieve the exact dimensions of the part, the beam diameter has to be compensated. This means the path of the beam will be misaligned by half of the diameter of the laser beam.

The volume of the liquid resin shrinks due to polymerization. By using resins instead of acrylates the shrinkage can be reduced. The linear shrinkage of epoxy resin is 0.06%, whereas acrylates shrink by 0.6%. The disadvantage of epoxy to acrylates is its exposure to light time which is three times the amount which is

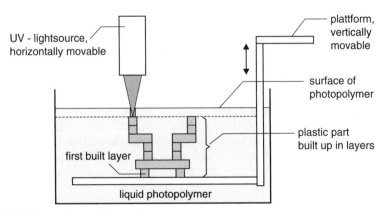

Fig. 13.8 Scheme of stereolithography process

needed to process acrylates. To compensate shrinkage, different scanning strategies and optimized processing parameters are used.

During building the stability of the parts is low (comparable to galantine). Therefore supports are used. Overhanging structures which exceed a certain angle have to be backed up by supports. These supports can easily be removed after completion of the part.

After complete hardening of the last layer, the building process is finished. The part is moved up and can be taken off from the platform. Surplus resin remains in the container and can be used for further parts. To flush the model completely, hollow structures need to be endued with holes.

After the process, polymerization of the parts reaches up to 96%. Complete 100% hardening is achieved by after heat treatment in an UV oven. To enhance the quality of the part furthermore, mechanical treatments are applied. This can be sandblasting, grinding, milling, etc.

Compared to different RP methods, stereolithography offers the highest grade of accuracy and best surface quality. Common usage of this technology is in casting processes which require high detail resolution, for instance, vacuum casting.

Used for

- geometry prototypes;
- functional prototypes;
- technical prototypes (Fig. 13.9).

Fig. 13.9 Examples of models fabricated by stereolithography

13.3 Laminated Object Manufacturing (LOM)

Christoph Over

The term laminated object manufacturing (LOM) which has been primarily used by the developer and manufacturer "Helisys" did displace the broader term layer laminate manufacturing (LLM).

C. Over (✉)
Inno-shape c/o C.F.K. GmbH, 65830 Kriftel, Germany
e-mail: over@inno-shape.de

LOM describes an additive manufacturing technique where parts are built up layer by layer. The material is coated with glue and pressed together with a preheated roll. Its temperature depends on the layer material and its thickness. After pressing, the layers are cut by a CO_2 laser according to the layer outline (Fig. 13.10). Up to four layers can be cut simultaneously. Thereby building time will be increased but accuracy will be decreased since chamfers are reproduced by steps. The thinnest a layer can be is up to 0.25 mm.

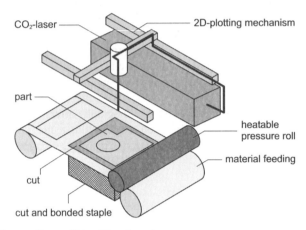

Fig. 13.10 Scheme of Layer Object Manufacturing process

A wide range of materials can be processed, and the following itemization shows one of the most widely used (Fig. 13.11):

- plastics;
- ceramics;
- paper coated with PE;
- fiber-glass-reinforced composites.

When the building process is finished, the part is taken out of the machine and excrescent material is taken off. Afterward it is essential to clean and to varnish the model to reduce the risk of unwanted growth. This growth can be 1–2% in

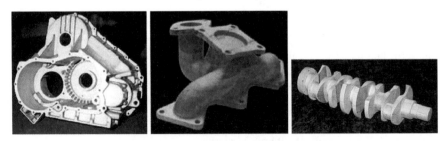

Fig. 13.11 Examples of parts manufactured by Layer Object Manufacturing

vertical direction. For varnishing a special lacquer is used which allows penetration up to 5 mm deep. By this treatment the stability of the models will be enhanced as well. Models which are prepared in this way can be treated afterwards by grinding, milling, cutting, etc.

Compared to raster scanner processes, LOM allows much larger massive models with big differences in their walls' thickness. Furthermore the created models are capable of withstanding high pressures.

Due to the construction in layers, the model possesses anisotropic mechanical properties. This means the tensile strength is higher in x- and y-axes, while z-axis is the building direction.

References

1. A. Gebhardt, Rapid Prototyping: Werkzeuge für die schnelle Produktentstehung, Carl Hanser Verlag München Wien, 2000
2. www.3dsystems.com
3. W. Meiners, Direktes Selektives Laser Sintern einkomponentiger metallischer Werkstoffe, Dissertation RWTH Aachen, April 1999
4. www.eos-gmbh.de
5. C. Over, Generative Fertigung von Bauteilen aus Werkzeugstahl X38CrMoV5-1 und Titan TiAl6V4 mit "Selective Laser Melting", Dissertation RWTH Aachen, June 2003
6. Fraunhofer ILT, 52074 Aachen, Germany

Chapter 14
Joining

14.1 Heat Conduction Welding

Norbert Wolf

14.1.1 Introduction

In the heat conduction welding process the material is heated to above the melting point through the energy of a laser beam, but only so high, so that no measurable evaporation will occur. The shape of the melt pool and the welding depth are dependent on the heat conduction of the material. Important factors which influence the heat conduction are the material and the geometry and temperature of the work piece. Figure 14.1 shows the principle of the heat conduction welding process. The theoretical foundation of energy transportation and the heat conduction mechanism are explained in the chapter "Fundamentals of Material" (p. 41ff).

With steel and aluminum the heat conduction welding process is a quiet and stable process. The seam quality from heat conduction welds is normally quite good; in particular with austenitic stainless steel the visible seam quality is very good. On the basis of achievable welding speeds, heat conduction welding is best applied to foils and thin sheets up to 1.5 mm thick (for example auto movie sheets, armatures) as well as for welding wires and tubes from interest. For various joining configurations (e.g., butt or lap joint) and some kinds of seams (e.g., square butt or lap joint) the process is useful. Additional material cannot normally be used.

Compared to other welding methods, heat conduction welding delivers better results by requiring lower energy input into the workpiece, resulting in lower distortion in the workpiece and higher process efficiency.

N. Wolf (✉)
Trenn- und Fügeverfahren, Fraunhofer-Institut für Lasertechnik,
5207 Aachen, Germany
e-mail: norbert.wolf@ilt.fraunhofer.de

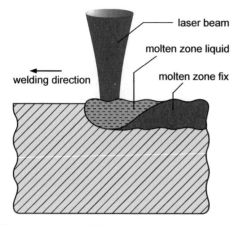

Fig. 14.1 Principle of heat conduction welding process

14.1.2 Principle and Analysis of the Heat Conduction Welding Processes

Heat conduction welding can be performed with each of three commercial beam sources: gas laser, solid-state laser and high-power diode laser. The intensity I of the laser radiation must be high enough to heat up the workpiece to its melting temperature but lower than the border intensity I_C, or "threshold intensity" (in the literature named "Plasma threshold" but this is not absolutely correct, while inevitably achieve the border intensity I_c develop a laser plasma). If the intensity is higher than the threshold intensity, the heat conduction welding process changes into the deep penetration welding process.

Below this threshold intensity the welding depth increases a bit with increasing intensity, and the welding depth depends exclusively on heat conduction. When the threshold intensity is reached, the welding depth increases rapidly.

The threshold intensity depends on the heat conduction, the absorptivity of the useable laser radiation, and the laser beam (kind of beam source, wavelength, power, geometry of beam spot). Accurate data for threshold intensities are only conditional possibility and not generally available. In aluminum material the transition to deep welding is the result of the higher heat conduction and usually lower absorption grad at higher intensities as compared to steel material. With the maximum intensity of high–power diode lasers sources currently available, the threshold intensity for aluminum is not achieved, which means that all welds in aluminum material with these beam sources are heat conduction welds.

In contrast to gas- and solid-state lasers, high-power diode lasers with optical output power in the kilowatt range are preferred for heat conduction welding applications, because of their lower beam qualities and intensities.

14 Joining

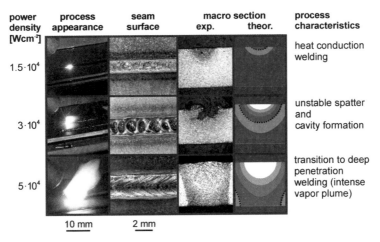

Fig. 14.2 3 process areas of heat conduction welding

Looking at the process behavior as well as experimental and theoretical analysis of the seam properties as a function of the intensity (Fig. 14.2) in welding 2 mm thick mild steel with a welding speed v_w of 1.8 m/min, a continuously transition of three different process areas observably. The focus dimensions of the high power-diode laser which is used are 1.7 mm × 3.8 mm with an output power of 2.5 kW. The simulation with "LaserWeld3D" [1] is based on the Rosenthal-solution of the heat conduction equation to take account of finite workpiece thickness.

With a pure heat conduction welding process with low intensity, a small ember illuminated appearance as well as a quiet, stable and soundless process can be observed. Small welds will be generated with a scaled upper seam. The aspect proportion (proportion of welding depth to seam witness) of the dome-shaped weld penetration is always lower than 1.

When the intensity increases the process gets unstable, which is visible in the form of spatter and the development of a very irregular upper seam. The simulation obtains, that also theoretical on the melting surface the boiling point arrives (temperature values above the melting temperature painted white in Fig. 14.2).

When the intensity increases the vapor plume will be greater and the process is stabilized. In the case of CrNi-steel the heat conduction welding process already approximates by an intensity of ca. 5×10^4 W/cm², the border intensity.

14.1.3 Characteristic Curves for Welds with High–Power Diode Laser and Different Materials

In the double logarithmic diagram (Fig. 14.3) the experimental and theoretically calculated function for the welding depth s_w of 2 mm thick sheets depends on the welding speed v_w.

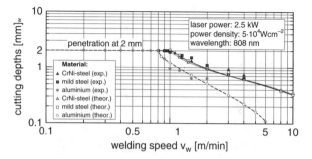

Fig. 14.3 Welding depth as a function of welding speed v_w [2]

With a lower welding speed and all other parameters held constant, the welding depth increases. The influence of the heat conductivity K of different materials is also shown in the diagram Fig. 14.3. When the heat conductivity is low the energy (heat) moves more slowly from the input point and the interaction zone between the laser radiation and workpiece, into the surrounding areas of the workpiece and results in increased heat accumulation, making possible a greater welding depth or a higher welding speed. CrNi-steel has the lowest heat conductivity with 27.5 W/(m · K) and the highest welding speed with 1 m/min, so that full penetration welds can be developed.

The heat conductivity of mild steel is greater than 40 W/(m·K). This is the reason that full penetration welds can only be achieved by a welding speed of 0.9 m/min. The heat conductivity of aluminum 99.5 at 235 W/(m · K) is significantly higher, which manifests the plunge down of the curve to higher welding speeds. Full penetration in aluminum 99.5 may be achieved at up to 0.8 m/min. Aluminum alloys have a heat conductivity range from 100 W/(m · K) to 160 W/(m · K) and transforms the input energy effectively.

In addition to the metal's heat conductivity, the temperature and geometry of the workpiece influence the rate of heat propagation. In workpieces which are pre-heated for example through welds, heat accumulation develops faster than in a cooler workpiece, which allows either a greater depth and wider seam shape or higher welding speed. Heat accumulation occurs if the heat flow is close to the workpiece pages. Injecting sufficient energy, due to the finite sheet thickness, the welding depth increases by developing heat accumulation on the sheet underneath disproportionately. Full penetration welds in thin sheets already achieve at certain welding speeds, during in thick sheets from the same materiel at the same speed do not develop heat accumulation and therefore explicit lower welding depth.

The results described here are generally available for heat conduction welding processes; quantitative data's only obtain for the used high-power- diode laser systems and the used process parameters.

14.1.4 Example of Use

Figure 14.4 illustrates the heat conduction welding process with a high-power diode laser (year 2000). The whole laser system is mounted on a six axis articulated robot and is guided with the welding speed over the workpiece; here CrNi-steel sheet.

Figure 14.5 shows some examples of cross sections of heat conduction welding. The intensity of the laser radiation was approximately $5 \times 10^4\,\text{W/cm}^2$.

High welding speed up to 10 m/min and good seam quality at 0.5 mm thick CrNi-steel sheets are suitable for this process with such applications. With thicker steel, sheets up to 1 to 2 mm, typical welding speeds range from 1 to 2 m/min. Furthermore, the joining geometry has an influence on the possible welding speed. Particularly via the welding of CrNi-steels through a suitable supply of inert gas, normally Argon, Helium or inert gas mixture, metallic blank and industry requirement welding seams can be developed. The gas, which is delivered through a suitable nozzle, protects the interaction zone from the oxygen in the atmosphere; this disables oxidation processes and prevents the accumulation of oxide films on the upper seam.

Fig. 14.4 Heat conduction welding process of CrNi-steel with robot guided high-power diode laser

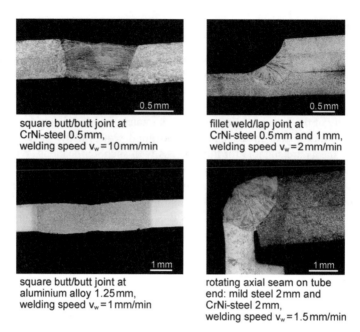

Fig. 14.5 Examples of heat conduction welding. Joints in steel and aluminum with high-power diode laser

The required seam geometry and the low welding speed of 1 m/min at 1.25 mm thick aluminum sheet are the result of the influence of higher heat conductivity, compared to steel material, with essential higher heat conduction and the typical high melting oxide film for aluminum material, clear. For penetration welding, the welding speed must be reduced, so that there is a higher heat flow into the work piece, to melt the oxide film and the backside of the sheet. Through vertical heat transportation from the fusion zone, wide welding seams are produced. In the example in Fig. 14.5 (left side down) the weld seam is five times wider than it is deep. As has been mentioned, it is not currently possible to achieve the threshold intensity for aluminum and his alloys with a high-power diode laser.

One example of industrial heat conduction welding is the joining of stainless steel sinks. Since 1998 a robot equipped with a direct working high-power diode laser system and a maximum laser output power of 1.5 kW has been installed in a company in Baden-Württemberg, Gemany, Fig. 14.6. Stainless steel parts from 0.6 to 1.25 mm thick are welded as a fillet weld at a lap joint with welding speeds up to 1 m/min. For seam protection a special shielding gas device was developed, so that it not necessary to finish the seam on the upper side after welding. Process-specific disadvantages are the necessity for very accurate clamping techniques to hold the tolerance and the cooling measures for the gas nozzle, which is heated up by the back reflection of laser radiation [3].

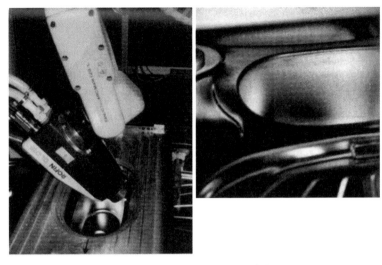

Fig. 14.6 Heat conduction welding process for stainless steel sinks

14.2 Deep Penetration Welding

Norbert Wolf

14.2.1 Introduction

In contrast to heat conduction welding, in the deep penetration welding regime the energy of the laser radiation heats up the material to over its evaporation temperature. Because of the generated pressure and the flow of metal vapor, a vapor channel (keyhole) in the melt pool is established. This effect is named the deep penetration welding effect DIN 32511 [4] and allows aspect ratios (seam depth/width approximately z_R/w_0) greater than 10:1. Figure 14.7 illustrates the process schematically.

The welding process is directed by moving the laser beam or the workpiece and therewith the vapor channel. In front of the vapor channel the material is completely molten. The molten material flows to surround the vapor channel and solidifies behind the keyhole at the weld seam. To achieve this effect, a specific laser beam intensity (Fig. 14.8) must be provided for each kind of laser, for example, CO_2-laser, Nd:YAG-laser (fiber/disk-laser) or diode laser. Characteristic of this process is the vapor plume (plasma).

N. Wolf (✉)
Trenn- und Fügeverfahren, Fraunhofer-Institut für Lasertechnik, 5207 Aachen, Germany
e-mail: norbert.wolf@ilt.fraunhofer.de

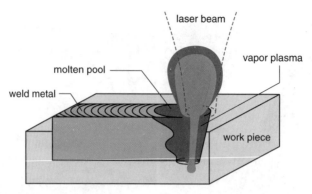

Fig. 14.7 Principle of deep penetration welding process

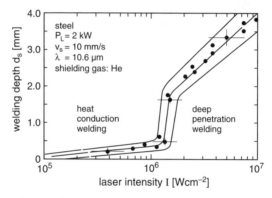

Fig. 14.8 Difference between laser deep welding process and heat conduction welding versus welding depth

The application spectrum for this process is extensive. This process is useful for materials such as steel, aluminum, titanium, and their alloys. With the application of a solid-state-laser the area of non-metallic material is for the deep penetration welding worth mentioning. The material thickness that can be welded is on a scale from some tenths of millimeters to 20 mm, in some cases more than 20 mm [5].

Deep penetration welding can be used for various joining configurations such as butt-joint, t-joint, lap joint and fillet-weld, in different welding positions. The application areas extend to the micro techniques, the automotive industry, off shore applications and shipping and tank construction. Deep penetration welding is used by manufacturing shops for welding simple geometric parts as well as for 3D-maching. The suitability of filler materials and laser-hybrid-technology (Sect. 14.3) considerably extend the spectrum of laser deep penetration welding, while the small tolerance zone of laser beam welding will be major.

With regard to conventional welding techniques such as MIG/MAG/TIG/UP etc., laser beam welding has a higher efficiency. The relatively low electric power efficiency of the lasers compared to that of conventional sources is more than compensated for by much better process efficiency (speed, welding depth). This is the result of the high welding speed and precision made possible by using CNC handling systems. More advantages are provided by the low distortion of the welded parts, cost reduction, and truing as well as contactless and non-wearing machining with a laser tool. These are characteristic advantages of deep penetration welding.

To use this process with all its advantages, it is necessary that the design and construction of parts to be welded are suitable for laser technology [6]. Accessibility for the laser beam, e.g., the optical components, and the material-specific properties of the parts must be considered.

14.2.2 Principle of Deep Penetration Welding and Physical Foundations

Above a laser intensity I, which is able to vaporize the material, a vapor capillary is created (Fundamentals Chap. 4, pp. 41ff). This is associated with metal vapor plasma, which is developed from the absorption of the laser irradiation and by ionization of the metal vapor (plasma). This phenomenon assists the input coupling of the laser beam into the workpiece, but it also interferes with the laser beam before it strikes the workpiece. The energy transfer is carried out by a wrench of ions at the capillary wall. In the Foundation Chap. 9 the development of a plasma is described. The absorption of the laser irradiation in front of the vapor capillary can be described by the Fresnel formula (foundation Sect. 2.1). By suitable geometrical conditions of the vapor capillary, the absorption rate may be increased by multi-reflection.

When the laser radiation rate increases, the vaporization rate and the metal vapor density increase. If the laser intensity becomes higher than a threshold value, the density of the plasma will increase, so that energy is absorbed in the plasma, isolating the laser beam from the working area. When the energy input is too high, the plasma separates from the workpiece and interrupts the welding process [7].

The vaporized metal (Chap. 8) streams up in a vertical direction to the capillary wall, so that one part of the vapor condenses by hitting the back wall. On zero-penetration welds, another part of the vaporized metal streams through the upper capillary aperture. On full penetration welds an additional part of the vapor streams through the lower aperture. The generated pressure inside the capillary is in balance with the pressure outside, so that a vapor capillary forms (Chap. 8):

$$p_i = p_a$$

Inside the vapor capillary the gas pressure and ablation pressure are added. On the other hand, from the outer side, there is the inertia of the melt, the curvature pressure and the solid pressure. The pressure balance in the vapor capillary is regulated through the plasma pressure $p = pkT$ and the ablation pressure. The geometry of the vapor capillary depends on laser and process parameters. Its size is in the range of the laser beam diameter.

The plasma plume develops when there is sufficient metal vapor with adequate density. The required threshold intensity depends on different process parameters. Assist gases (e.g., Helium, Argon) have a great influence on the vapor plume. Their size depends on the kind of gas and escape proportion.

The plasma threshold for a welding process, for example for steel material with a gas- or solid-state-laser, is in the range of 1×10^6 W/cm^2 to 2×10^6 W/cm^2 (Fig. 14.8) in addition to the wavelength of lasers and material. In comparison with heat conduction welding, the energy input is usually higher with deep penetration welding, as shown in Fig. 14.8.

14.2.3 Function of Vapor Capillary (Keyhole)

The vapor capillary allows the great penetration depth of the laser beam into the workpiece. The penetration depth is approximately the Rayleigh length z_r.

Because of multiple reflections in the vapor capillary it is possible to reach high absorption rates for highly reflective materials. The vapor capillary allows uniform heating of the welding seam over the whole sheet thickness. In addition, the vapor capillary is an out-gassing channel for developing gases, reducing porosity in the melt.

Interaction processes inside the vapor capillary provide signals for process monitoring of the welding process.

14.2.4 Significant Parameters for Laser Beam Deep Penetration Welding

For laser beam deep penetration welding, many parameters are significant:

- laser beam: wavelength, power, polarization, focal diameter, focusing number, intensity distribution
- process media: assist gas, gas rate, kind of nozzle, nozzle geometry, additional wire, wire diameter, wire speed
- workpiece: material, thickness, structure, surface condition, edge preparation
- handling: welding speed, weld direction, focus position, acceleration
- In addition: work piece geometry (2D, 3D), weld position, accessibility

In Fig. 14.9 the values for the laser deep welding process are shown schematically.

14 Joining

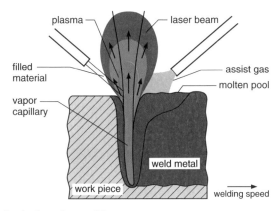

Fig. 14.9 Values for the laser deep welding process

14.2.5 Example of Use

Example: Welding of thin sheets (Figs. 14.10 and 14.11)
Example: laser welding for Off-Shore Industry (Fig. 14.12)
Example of laser welding of heavy metal section (Fig. 14.13)
Example of welding with disk laser (Fig. 14.14)

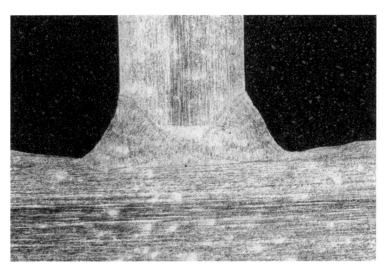

Fig. 14.10 One sided laser welded T-joint in high-strength stainless steel. CO_2-laser; P_1 = 3kW; V_s = 2.5 m/min; s = 3.8 mm

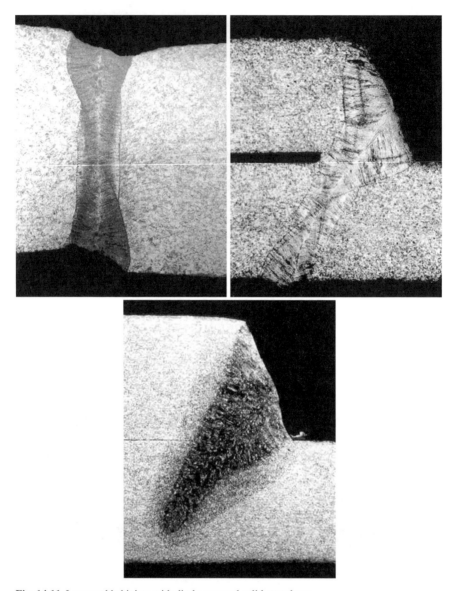

Fig. 14.11 Laser welded joints with diode-pumped solid-state laser

14 Joining

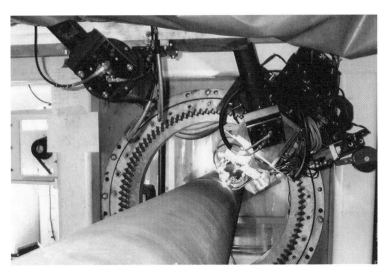

Fig. 14.12 Laser-orbital welding with CO_2-Laser for Off-Shore-Industry (Company: Bouygues Offshore, France)

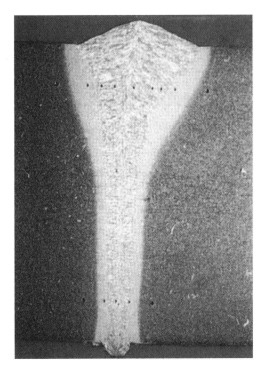

Fig. 14.13 Welding with 20 kW CO_2-laser. Mild steel s = 15 mm

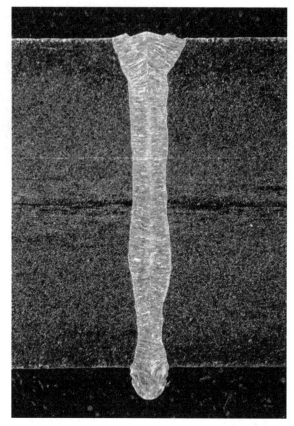

Fig. 14.14 Welding with disk laser bead on plate welding stainless steel s = 10 mm, $P_l = 8$ kw, $v_s = 3$ m/min

14.3 Hybrid Welding

Dirk Petring

The expression "laser hybrid welding" describes welding processes which use—besides the laser beam—a second welding source, acting in the same melt pool as the laser beam. The second source provides additional heat and—as the case may be—also filler material.

Besides the combination of different laser beams, the best-established hybrid welding variant is the support of lasers by arc power. Laser-arc hybrid welding processes are recognized for their robustness, efficiency, and flexibility. In par-

---

D. Petring (✉)
Trenn- und Fügeverfahren, Fraunhofer-Institut für Lasertechnik, 5207 Aachen, Germany
e-mail: dirk.petring@ilt.fraunhofer.de

ticular, the coupling of a deep-penetrating laser beam with the heat and molten metal-supplying gas metal arc (GMA) is a proven hybrid technique. It significantly expands the original welding application range of lasers.

Laser-arc hybrid welding has been investigated since 1978, when Prof. William Steen and co-workers in the UK published their first paper about TIG (Tungsten Inert Gas) augmented laser welding [8]. By replacing the tungsten electrode by a wire fed electrode, the heat of the arc as well as the molten filler material of the wire are transferred to the welding zone by the so-called MIG/MAG (Metal Inert/Active Gas) process. After the first industrial laser-MIG hybrid system was put into operation by Fraunhofer ILT in the year 2000 at a German oil tank manufacturer [9], a hybrid boom could be observed and various new installations followed. Laser hybrid welding systems have been installed in shipbuilding [10, 11], as well as in the automotive industry [12].

14.3.1 Fundamentals

The laser-arc hybrid process is characterized by the simultaneous application of a focused laser beam and an arc, creating and moving a common melt pool along the weld pass (see Fig. 14.15). The combination offers an increased number of parameters compared to the individual processes, thus allowing flexible control of a welding process adapted to the demands of material, design, and manufacturing conditions.

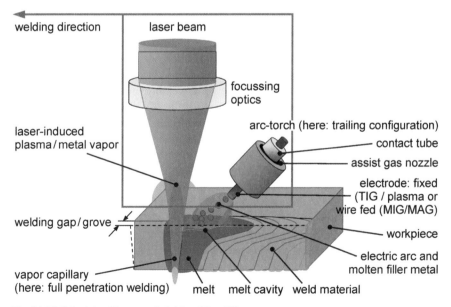

Fig. 14.15 Principle of laser-arc hybrid welding [9]

The main benefits of the hybrid technique compared to laser beam welding, with or without filler wire are:

- better gap bridging capability at lower laser beam power.
- better leveling of edge offset.
- lower demands on edge preparation and clamping.
- improved adjustment of the thermal cycle.

And there are also significant benefits compared to arc welding techniques:

- higher welding speed.
- weldability of zero-gap and I-seam at lap joint.
- single-pass full-penetration welding even at high welding depth.
- lower heat input.
- lower distortion.
- at T-joints or in corners: smaller fillet, more clearance.

The result is a more flexible and robust welding process which provides higher productivity and quality.

The arrangement of the arc relative to the laser beam axis (leading, trailing or coaxial type, inclination and distance between laser and arc) depends on the material to be welded and its surface properties. Also important are the type of joint, edge preparation, and welding position. The laser and arc power, the laser focusing parameters, and the laser wavelength, as well as the chosen metal-transfer mode (MIG/MAG) and special boundary conditions, e.g., accessibility for a seam tracking sensor, are other factors with a strong influence on the design of hybrid processes and equipment.

Normally, the smallest possible arc inclination is desired. Angles in the range of $15°$ to $30°$ relative to the laser axis work with technically acceptable effort. Nd:YAG, disk and fiber laser radiation, due to a lower interaction with the arc's plasma, allows a closer approach to the arc than CO_2 laser radiation, as long as the laser induced vapor jet does not lead to detrimental interaction effects.

Currently the most preferred laser-arc hybrid welding process uses MIG/MAG. The process can be controlled in such a way that the MIG/MAG part provides the appropriate amount of molten filler material to bridge the gap or fill the groove while the laser is generating a vapor capillary within the molten pool to ensure the desired welding depth at high speed. The combination of processes increases the welding speed above the sum of the single speeds, and produces an enhanced regularity of the weld bead. Improvements of metallurgical properties associated with hardness and toughness, as well as diminished porosity due to the promoted escaping of gas out of the enlarged melt pool, may be noted.

The correct setting of gas parameters is an important factor in hybrid welding, where laser and arc-specific criteria have to be taken into account simultaneously. Using Nd:YAG, disk or fiber laser radiation, beam absorption within the plasma and resulting plasma shielding can be neglected due to the shorter wavelength as compared to CO_2 lasers. Thus, with these lasers the selection of process gas can be determined according to arc stability demands and bead shielding properties. In this case argon is the dominant portion of the gas used. Also, for MIG/MAG welding, droplet detachment and spatter-free metal-transfer have to be considered. A small addition of oxygen promotes droplet detachment and reduces spatter. Admixtures of helium lead to higher arc voltage and the corresponding power increase results in wider seams, but these also lead to destabilizing the arc. Nevertheless, using CO_2 lasers, a helium mixture is necessary to avoid plasma shielding. Fortunately, the presence of the laser beam enables acceptable arc stability even with a significant helium flow.

14.3.2 Integrated Hybrid Welding Nozzle

With the standard approach of combining a discrete arc-torch in off-axis configuration with a laser focusing head, there are certain limitations on the possible position and orientation of the arc. In order to prevent the torch nozzle from interfering with the laser beam, it has to be positioned at a sufficient distance and inclination. Another problem with this off-axis configuration approach is that it promotes entrainment of air into the weld by the Venturi effect.

To address these problems, a more sophisticated approach uses a welding head in which the laser beam and arc are surrounded by a common water-cooled nozzle device with an integrated contact tube for the stable guiding of the wire electrode (see Fig. 14.16) [9]. This arrangement provides the closest laser and arc proximity at the steepest arc inclination. The assist gas flows out of an annular channel coaxially with the laser beam. A diffusing aperture within the channel enables a homogeneously distributed stream of the assist gas into the welding zone. Thus, a transverse suction of air by the Venturi effect is avoided and effective protection of the weld bead is ensured (Fig. 14.16). Moreover, a minimal but sufficient leak gas flow in the upward direction avoids process gas contamination by air entrainment via the laser beam entrance.

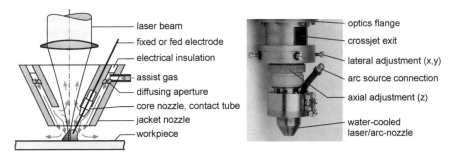

Fig. 14.16 Principle (*left*) and practical device (*right*) of the "Integrated Hybrid Welding Nozzle" [13]

14.3.3 Welding of Steel and Aluminum

To demonstrate the general capabilities of laser-MIG/MAG hybrid welding, typical macro sections of seams in aluminum and steel are shown in Fig. 14.17. The

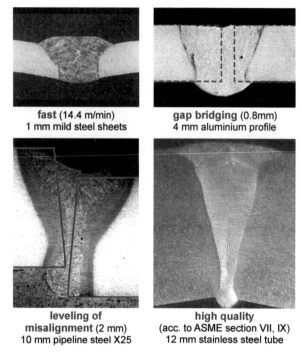

Fig. 14.17 Cross sections and related parameters showing benefits of hybrid welding [13]

examples provide evidence of fast speed, wide gap bridging, smooth leveling of misalignment and high weld quality.

Further examples will illustrate some hybrid welding features in more detail. In Fig. 14.18 the gap variation at a butt joint in 4 mm aluminum alloy sheets could be

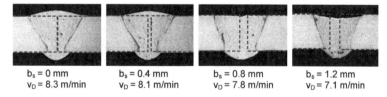

$b_s = 0$ mm $b_s = 0.4$ mm $b_s = 0.8$ mm $b_s = 1.2$ mm
$v_D = 8.3$ m/min $v_D = 8.1$ m/min $v_D = 7.8$ m/min $v_D = 7.1$ m/min

- Aluminum 6000 series alloy
- sheet thickness 4 mm
- square butt weld in flat position **without backing bar or root protection**
- gap width b_s
- welding speed 2.5 m/min
- Nd:YAG-laser 2.7 kW
- MIG impulse arc in trailing configuration
- wire material S-AlSi12, diameter 1.2 mm, wire feed rate v_D
- assist gas argon

Fig. 14.18 Gap bridging capability in an aluminum alloy [9]

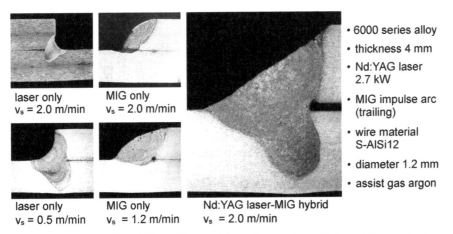

Fig. 14.19 Hybrid welding of fillet weld at lap joint and comparison with the welding result using the single processes alone [13]

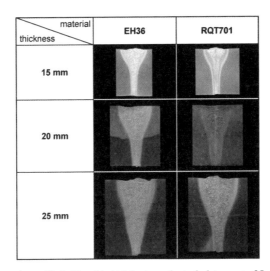

Fig. 14.20 Cross sections of hybrid welded high-strength steel plates up to 25 mm thickness using a 20 kW CO_2 laser at adapted power level [14]

bridged without any backing melt support or root protection. Fillet welds at a lap joint (Fig. 14.19) get effective reinforcement by the contribution of the MIG process. Furthermore, thick section welds can be produced, up to 25 mm in a single pass in structural steel plates (Fig. 14.20). The respective laser beam powers, welding speeds, and gap bridging capabilities can be read from Fig. 14.21. Corresponding fatigue results are shown in Fig. 14.22.

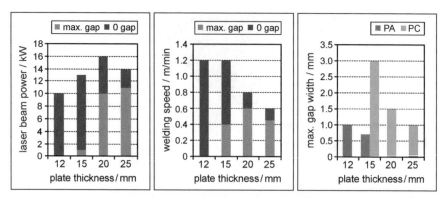

Fig. 14.21 CO_2-Laser-MAG hybrid welding parameters versus thickness, gap size and welding position for butt joints in structural steel [14]

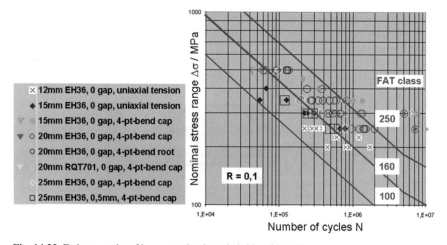

Fig. 14.22 Fatigue results of heavy section laser hybrid welds [14]

14.4 Laser Beam Welding of Thermoplastics

A.L. Boglea, Andreas Roesner, and Ulrich Andreas Russek

14.4.1 Motivation

Plastics play an important role in almost every facet of our lives and are used in a wide variety of products, from everyday products such as food and beverage pack-

---

A.L. Boglea (✉)
Mikrotechnik, Fraunhofer-Institut für Lasertechnik, 5207 Aachen, Germany
e-mail: andrei.boglea@ilt.fraunhofer.de

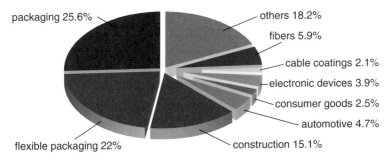

Fig. 14.23 Application fields of plastic materials in western Europe 2002

aging, to furniture and building materials, to high tech products in the automotive, electronics, aerospace, white goods, medical and other sectors (Fig. 14.23). Today, advanced plastic polymers have already replaced other materials (such as metals for rigid components, glass in the optical industry and even silicon in low cost electronics) and have made drastic changes in the design and setup of almost all products in terms of weight reduction, cost saving, and environmental footprint [15, 16]. These are crucial factors in manufacturing and use for the success of both disposable and long lasting products.

Successful (plastic) products are based on an optimal balance, including:

- careful selection of the plastic material,
- proper construction (Design for Manufacturing),
- appropriate manufacturing and joining methods

Although plastics provide previously-unattained manufacturing capabilities (e.g. large parts can be produced by injection molding with high reproducibility) they often require assembly technologies that must enable the massive production of components, even if they are made from complex parts. In a global market driven by reduced production costs in low wage countries, industry needs to achieve higher quality and higher flexibility of product assembly with reduced costs in order to remain competitive, since assembly costs often account for more than 30% of product manufacturing costs (see Fig. 14.24, left side).

To this end, compared to conventional plastic joining (such as adhesive bonding, mechanical crimping, hot plate welding, vibrational welding and ultrasonic welding), laser plastic welding technology offers significant advantages, such as low cost, power scalability, flexible geometry and contour shaping, joining capability without additional material, and process stability combined with the possibility of on-line process control, which have already established laser plastic welding as a versatile process for high quality assembly in many industrial applications (see examples in Fig. 14.24, right side).

Through the specific characteristics of all laser processes such as contactless and locally defined energy deposition, thermal damage to the joining partners is

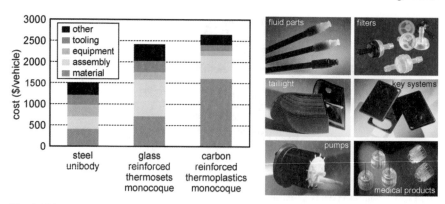

Fig. 14.24 Cost structure of Body-in-White-Manufacturing for cars, assembly makes up to 30% (*left* – Source: Dieffenbach et al., 1996); Spectra of current laser welded polymer parts (*right*)

avoided. Mechanical stress is also reduced or avoided since during the joining process the components experience no movement relative to each other. For optimal plastic material combinations the weld strength is close to that of the base materials. Furthermore, tight, porosity-free and optically high quality weld seams can be achieved through laser beam welding of plastic materials. The idea of laser beam welding of polymers was developed in the 1960s. However, this process gained importance mainly due to the development of high-power diode lasers in the early 1990s. The previously used laser systems based on CO_2 and Nd:YAG lasers were replaced by diode lasers due to their significantly reduced price and increased efficiency [17, 18].

The quality of the welding as well as the various applications that can be covered by the laser beam welding of plastics reveal its great market potential. At present laser beam welding can be considered for approximately 20% of the plastic processing market in fields such as electronics, automobile or packaging.

14.4.2 Process Basics

14.4.2.1 The Joint Configuration

For the laser welding of plastics with high-power diode lasers (HPDL) the overlap joint (Fig. 14.25, left) enjoys greater success than the butt joint (Fig. 14.25, right) and has become the most used joint configuration.

The principle of the laser beam welding of plastic materials is based on the absorption of the electro-magnetic radiation into the material.

In the case of an overlap joint transmission welding, one of the joining partners is transparent to the wavelength of the laser beam so the beam can pass through the first joining partner, depositing the necessary energy for the welding process in the joining area. The joining area is represented by the contact surface between the two

14 Joining

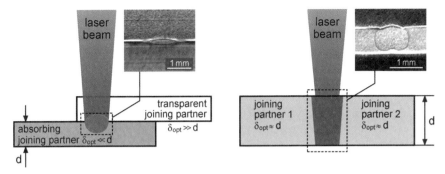

Fig. 14.25 Typical joint configurations for the laser beam welding of plastics

joining partners. Here, the electro-magnetic energy is converted by the absorbing joining partner into thermal energy with almost no loss. Through heat conduction the wavelength-transparent joining partner is also melted. To guarantee efficient heat transfer, firm contact between the joining partners is required. In the case of transmission welding a thin film of molten material in the range of 30–300 μm develops in the welding zone.

14.4.2.2 Irradiation Methods

To benefit from the specific advantages of the laser process, it is important to observe or generate boundary conditions that are conclusive to the process. This includes having a part design and a weld geometry that are suitably tailored to the laser process. Due to the optical properties of plastics, diode lasers, Nd:YAG lasers and fiber lasers can be used for laser transmission welding. A suitable laser should be selected on the basis of the welding configuration employed. There are four main irradiation methods (Fig. 14.26).

The standard method is called "vector welding" or "contour welding." In this method, the focused laser beam is moved over the material's surface, following the

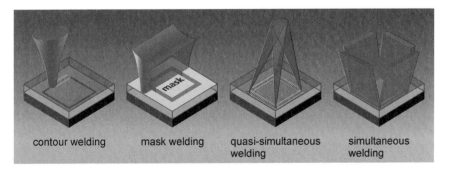

Fig. 14.26 Process variants based on different irradiation strategies

weld seam geometry. The laser source might be a diode or YAG laser, either with fiber coupling or without, dependent on the application, and the typical laser power is in the range of 10–100 W. The movement along the desired welding contour can be performed using an x–y handling system to move the joining partners under a fixed laser beam or by the use of galvanometer scanners to move the laser beam while holding the joining partners fixed. Welding speeds typically range from 2 to 5 m/min, in some cases up to 25 m/min. This method provides a simple, very flexible, easy controllable and cost efficient method for the laser transmission welding of polymers.

Guiding the laser beam by means of a fast galvano-scanning mirror system over an arbitrary 2-dimensional weld seam contour within a limited scanning field is called quasi-simultaneous welding. An F-Theta-optic ensures that the focal spot of the laser beam lying at each point of the scanning field plane will be within this plane. If the power is sufficiently high and the scanning speed is fast enough, in fact so fast that heat loss is so small that no re-solidification occurs, the entire seam will be softened as well, so that a quasi-simultaneous weld procedure is realized. Because of the plastification of the entire weld seam, part tolerances may be compensated for, depending on the weld seam and part design.

To generate narrow weld seams ($\approx 100 \mu m$) and complex seam designs in micro technology, biology and the life sciences, mask welding is preferred. A mask between the laser beam source and joining partners guarantees the irradiation only of areas to be welded. This mask is conformal with the weld seam, so those places that must not be exposed to laser radiation are protected. High-power diode lasers with a collimated, homogeneous, line-shaped intensity distribution are preferred for this irradiation method [19]. The resolution or the smallness, respectively, depends on the thickness of the transparent joining partner, the optical and thermal properties of both joining partners, and the intensity distribution as well as on the arrangement of the laser source, clamping device and mask. The process efficiency, however, is reduced because a portion of the laser power is blocked by the mask and thus is not used for the process.

The compact and modular setup of high-power diode lasers permits simultaneous laser welding. By means of appropriate beam forming and guiding, a homogeneous intensity distribution conforming to the entire weld seam geometry may be generated [20, 21]. There are different strategies possible to realize simultaneous welding, such as direct irradiation of the weld seam contour, using classic optical components, employing mask technology or the application of fiber bundles. During the welding process the entire weld seam is irradiated and welded with one single laser pulse. Apart from short process times, which are of interest for big lot production, simultaneous welding allows a more moderate irradiation of the parts to be welded. It offers wider process windows and higher bridging capability as well as enhanced weld strength compared to contour welding. Because of the plastification of the entire weld seam, part tolerances may be compensated for, depending on the weld seam and environmental design.

14.4.2.3 Process Management

Process management is essential in order to achieve a high quality joint. A detailed process analysis reveals that the weld strength has a typical evolution for different laser energy levels [22–24]. Therefore, the graphical representation of the weld seam strength versus laser energy represents the so-called "characteristic curve" of the laser beam welding process (Fig. 14.27). On this characteristic curve the different stages of a laser transmission welding process can be indentified. In the first stage (Fig. 14.28-1), a reduced energy leads to the melting of the absorbing joining partner and just to the heating of the transparent joining partner. In this case the weld seam has a poor strength. With an increase of the energy (Fig. 14.28-2) a sufficient molten volume will be produced in both partners and a strong weld seam can be achieved. The optimal process parameters for achieving the strongest weld seam are within this domain of the characteristic curve. A further increase in the energy will lead to thermal decomposition of the absorbent joining partner and reduce the strength of the weld (Fig. 14.28-3). However, in particular cases an increase in the weld strength can be noticed, but process instability will rise, too. One of the most common consequences of such a situation is the formation of pores in the weld seam, which may eventually cause the failure of the weld seam or compromise the tightness of the assembly.

In addition to the laser energy absorbed into the material, the weld strength also depends on the time of interaction between energy and material. A short interaction

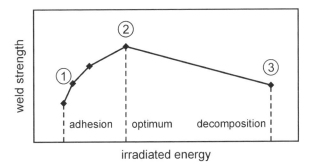

Fig. 14.27 The characteristic curve for a laser beam transmission welding process

Fig. 14.28 Different states of the welding process (heat affected zone)

time prevents the plastification of the transparent joining partner. With increasing irradiated energy, the absorbent joining partner decomposes while the heat flow into the transparent joining partner is still too small to plastify the transparent material sufficiently. If the interaction time between laser beam and material is too long the heat affected zone will reach the surface of the joining partner and a surface distortion will be observed [25].

14.4.2.4 Light-Material Interaction

As previously mentioned, for polymeric materials to be joined, a heat source is generated in the joining zone by the absorption of laser radiation. Thus, the parts can fuse together by means of the generated melt that forms in the heat-affected zone. Therefore, the basis for machining or processing materials with laser light depends on the way in which the laser reacts with the specific material involved [26, 27]. The different light-material interactions are displayed in Fig. 14.29 (neglect the second boundary reflection).

If light hits a material surface it either is reflected or penetrates into the material. Surface reflection is represented as R_G in Fig. 14.29. The part of the light that is not reflected at the surface enters the material. As the light passes through the medium, its intensity decreases according to Lambert-Beer's law:

$$I(x) = I_0 \cdot e^{C_{ex} \cdot x}$$

$I(x)$ is the intensity at the point x, I_0 denotes the intensity of the entering light and C_{ex} represents the extinction coefficient, which depends on the material, wavelength, additives and other parameters. C_{ex} consists of two parts. The first one is the absorption coefficient C_{abs}, which indicates the absorbed energy A. In fact, the absorption

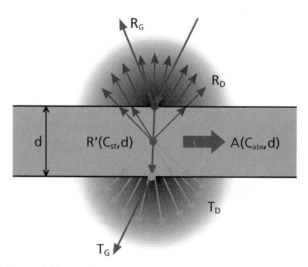

Fig. 14.29 Light-material interaction

coefficient C_{abs} can cover values from nearly zero (transparent) over a medium range (volume absorber) up to high numbers (surface absorber). The absorbed energy is transformed into heat and, thus, leads to a temperature increase, which may cause melting or even thermal destruction [28].

The second part of the extinction coefficient C_{ex} is the scattering coefficient C_{st}, which indicates the deflection of the light. Scattering (R′) deflects the light out of its original propagation direction. This can be caused, for example, by particles, crystals or grain boundaries, but whatever the cause the overall energy of the light is not transferred into the material and, thus does not affect the state of the material. The deflected light that leaves the material can be measured as diffuse reflectance (R_D) above the material or diffuse transmittance (T_D) below the material.

The light that is propagated through the material is indicated by the transmittance (T). T consists of two components. The first component is the directed transmittance (T_G). This component has the same direction as the entering light. The second component is the diffusive transmittance (T_D), which is characterized by all directions except the one of the directed light.

14.4.2.5 Optical and Thermal Properties of Thermoplastics

Optical Properties

The optical properties of the material significantly influence the thermoplastic laser transmission welding process. The top layer has to be sufficiently transparent to the wavelength used. The more transparent the material is, the more energy can be irradiated into the welding area. For the bottom layer the absorption is the important parameter for the welding process. The interaction of material and radiation only takes place in the upper surface of the bottom welding partner.

The reduction of the original radiation flux is caused by scattering and absorption. Since both physical phenomena appear at the same time it is difficult to separate them. Loss of laser power due to scattering and absorption within the transparent joining partner results in an increased laser power requirement to guarantee sufficient laser energy for plastification within the joining area. Furthermore, the more laser energy is absorbed per length (volume) within the transparent joining partner, the larger the risk of melting or burning the transparent thermoplastic. Due to these factors, the process window becomes smaller. Therefore, knowledge about influencing the optical properties of the transparent partner, while adapting them to the laser beam process, is a crucial point in the laser beam welding of thermoplastics. Dyes, crystallinity, additives and other ingredients added to the basic thermoplastic influence the optical properties. Therefore, they allow adapting the material's optical properties to the technical needs of the process.

From the physical point of view, absorption is the damping of electro-magnetic waves while they are propagating through a spatially extended media. In plastic welding, absorption takes place because of the interaction of electro-magnetic laser radiation and matter (thermoplastic). Due to dielectric and magnetic losses of the media, absorbed energy is transformed into heat energy. Absorption of the laser power in the infrared (IR) range occurs by excitation of vibrations in molecules. The light is absorbed in the ultraviolet range (UV) by excitation of electrons. Polymers show characteristically strong absorption in the deep IR or deep UV range. The strength and structure of the absorption in the IR range is governed by molecular groups, with absorption coefficients for radiation at 10 μm wavelength typically lying in the range 100–$1,000\,\text{cm}^{-1}$.

The optical penetration depth is the reciprocal value of the absorption coefficient and depends on the actual thermoplastic used (chemical compound, chemical ingredients), the morphology (e.g., the degree of crystallinity) and processing of the thermoplastic, and the selected laser wavelength, and it furthermore depends on the nature, quantity (concentration), size and distribution of the added substances, such as color additives as well as flame retardants, fillers and reinforcement materials.

In the IR wavelength range most materials show adequate absorption without additives at a special wavelength that correlates with the chemical structure of the polymer. In Fig. 14.30 this behavior can be seen for different polymers. At several wavelengths (1,200; 1,450; 1,700; 1,900; and > 2,400 nm) the absorption reaches a value sufficient for melting the joining partners. If the laser wavelength is chosen according to the results of the spectra, no further IR absorber has to be added.

In the visible and near IR range most non-pigmented thermoplastics are transparent or show an opaque or translucent behavior. Therefore, the extent to which the laser radiation is absorbed is usually adjusted by IR absorbers such as carbon black, pigments or dyes and not by the polymer matrix itself. Due to its broad-band absorbing abilities and its low price, carbon black is the standard IR absorber for applications in which the optical properties of the welding partners are of minor interest. By changing the amount of carbon black added, the optical penetration depth can be altered. In Fig. 14.31 the optical penetration depth is depicted over the amount of added carbon black for a few standard plastics. The optical penetration depth governs the molten volume created in the welding area. Higher molten volume

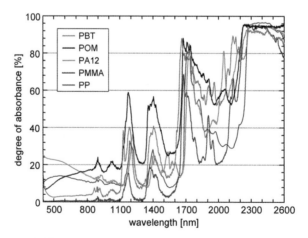

Fig. 14.30 Spectra of different polymers (PA 12, PBT, PMMA, POM and PP)

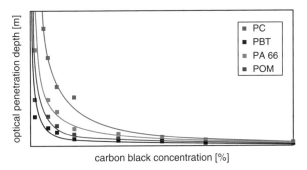

Fig. 14.31 Optical penetration depth over the amount of added carbon black

allows higher gap bridging capabilities but the process speed is lowered. Therefore, a trade-off for the desired application has to be found.

Combinations with a higher degree of complexity could be realized with special laser additives. But as such absorbing additives usually have an intrinsic color of their own in the visible range (e.g., a slight green for an IR absorber), a compromise has to be found between technical design and the demands of the marketing department with regard to coloring.

In contrast to absorption, scattering causes a change of the original intensity distribution due to an interaction of laser radiation and matter. Scattering means a change in the original beam propagation behavior of the laser radiation. While scattering is negligible within amorphous thermoplastics, it is of relevance within partially crystalline thermoplastics. Crystalline superstructures (spherulites) cause the scattering (a sort of multiple reflection). The original optical path becomes changed, while the optical path within the scattering medium becomes longer. This again causes increased absorption. The absorption increases either with an increasing degree of crystallinity or at the same degree of crystallinity but with smaller (and therefore more) spherulites. Together both effects (absorption and scattering) cause a broadening and a decrease of the original intensity. Due to scattering, broadened intensity distribution may cause broader weld seams. The more transparent the material is, the more energy can be irradiated into the welding area. The higher the amount of additives (e.g. glass fibers, etc.), the degree of crystallinity and the thickness of the top layer, the lower is the amount of direct laser energy transmitted into the welding area. Due to scattering of light the laser beam gets broader and the intensity in the welding area is reduced. In Fig. 14.32 the behavior for a natural PA is shown for different thicknesses.

Thermal Properties

Considering the thermal properties of thermoplastics, such as melting and decomposition behavior, mechanical strength vs. temperature, phase transition, heat conductivity, and thermal expansion coefficient as well as viscosity, welding thermoplastics is quite different from welding metals. The temperature within the joining area has

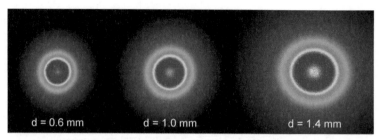

Fig. 14.32 Increase of the laser beam diameter due to scattering for Polyamide (PA 6) and different material thicknesses

to be kept above the melting and below the decomposition temperature. Depending on the material, this temperature range may be pretty small. Mechanical strength vs. temperature for thermoplastics is different for amorphous and partially crystalline thermoplastics, especially in that there is no distinct phase transition for amorphous thermoplastics as there is for metals and for partially crystalline thermoplastics. Considering the short interaction times of laser radiation with the thermoplastic as well as the low heat conductivity of polymers, on the order of 0.1–1 W/(m.K), the intensity distribution is imaged into an energy density profile corresponding to a temperature profile within a layer with the dimensions of the optical penetration depth. The large thermal expansion coefficients of thermoplastics not only support the mixing and diffusion of the thermoplastics to be welded, but also may cause internal stresses because of shrinkage, even after long periods of time after solidification. The viscosity of plastified (molten) thermoplastics is much larger than that for molten metals, therefore, molten thermoplastics do not flow and mix sufficiently within the available interaction time without any external force, such as inherent pressure (see Sect. 2.1.1).

In summary, fulfilling optical and thermal process requirements is mandatory for obtaining high-quality welds.

Further Physical Properties

Alongside the physical properties, such as transmission, absorption and melting point, there are also other factors to be borne in mind with respect to weldability. On the one hand, the polymer compatibility (Fig. 14.33) of the parts to be joined must be ensured if dissimilar thermoplastics are to be welded. On the other hand, it must also be considered that the modification of a single type of polymer generates new polymer types that can show very different joining behavior. The joining of identical-type thermoplastics has become established in industry over the past few years. In future, development work will focus especially on the joining of dissimilar material types, since this will permit the targeted selection and alignment of the material to the specific conditions that prevail in the application. This will go hand in hand with increased functionality and enhanced added value for new, innovative products.

14 Joining

material-matrix		ABS	Pa6 (tk)	Pa66 (tk)	PC	PE-HD (tk)	PE-LD (tk)	PMMA	POM (tk)	PP (tk)	PS	PBT (tk)	SAN	TPE	PPS
ABS						0	0			0	0				
Pa6	27.8					~	~			~					
Pa66	27.8				~	~	~			~					
PC	19.4			~		0	0			0	0				
PE-HD	16.6	0	~	~	0			~	~	0	0	0	0		
PE-LD	16.6	0	~	~	0			~	~		0	0	0		
PMMA	19.1				~	~			~						
POM	21.7				~	~				~	0				
PP	19.0	0	~	~	0	0		~	~		0	0	0		
PS	18.1	0			0	0	0		0	0		0	0	0	
PBT	21.9					0	0			0	0				
SAN						0	0			0	0				
TPE	20.5														
PPS															

	acceptable welding	0	impossible welding
	good welding		same material = allways good welding
~	unacceptable welding	tk	partially crystaline polymer

Fig. 14.33 Polymer compatibility matrix [Franck, A.: Kunststoff-Kompendium, 4. Aufl. Vogel Buchverlag, Würzburg 1996]

Plastics contain a large number of additives that are incorporated in order to achieve specific product properties for the application in question. A number of these are visually perceptible additives, while others are not. Problems that impair or even prevent laser welding and which have their origins in the additives employed, are not always recognized immediately on account of the complex way in which the additives interact.

14.4.2.6 Laser Beam Welding Adapted Design

In obtaining high-quality laser welds, different design and technical factors have to be considered compared to conventional welding technologies, and certain requirements have to be fulfilled, such as:

- constructional or laser adapted design of the joining partners
- demands on the product reconciled with laser adapted design of the joining partners
- guaranteeing the clamping and positioning of joining partners during the irradiation process
- avoiding or minimizing gaps between the joining partners
- ensuring accessibility, avoiding beam obstructing parts between laser and joining area
- specific polymer design of the joining partners

- guaranteeing welding suitability of the joining partners (polymer compatibility)
- considering the influence of processing conditions on crucial demands for laser process
- matching the physical properties to the laser beam process
- considering type, concentration, distribution and size of additives
- avoiding impurities and moisture (especially for PA)
- irradiation method and joint geometry
- laser beam source (i.e., wavelength, power)
- joining parameters as well as the clamping strategy and parts supply and removal

If laser beam welding is being considered for an existing or new product, a laser beam adapted design has to be considered early, in order to use the process's technical opportunities and advantages.

14.4.2.7 Laser Sources for Thermoplastic Welding

In principle, for the laser welding of thermoplastics the usual laser sources are represented by CO_2, Nd:YAG and diode lasers. The wavelength of the CO_2 laser leads to very short optical penetration depths, while in the case of a diode laser the material would be transparent. Therefore. CO_2 lasers are used more for the welding of thin foils than for the overlap welding of thicker plastic components [29]. Due to the lower investment cost and the possibility of fiber delivery for the laser beam, diode lasers have enjoyed an increasing industrial acceptance. However, the low beam quality of these laser systems and the high thermal expansion of the welding area are some drawbacks to their use. The optical penetration depth can be adjusted for the relevant process window through carbon black pigments, for example.

The laser beam welding of thermoplastics with high-power diode lasers generates optically high value weld seams and presents several process-specific advantages such as contactless deposition of the joining energy, no process induced vibrations, no thermal stress of the entire joining component, no particle release during welding, and high industrial integration and automation potential. Compared to alternative processes like adhesives, the impact on the working environment via the development of vapors is, for laser transmission welding, negligible. The dimensions of the possible joining partners extend from macro-components, e.g., hermetically sealed liquid containers for the automotive industry to micro-components, e.g., housings for electronic components or medical products.

For such applications diode lasers, due to their compactness and reliability (lifetime of over 10,000 hours under optimal working conditions), are the most used laser sources. The emission wavelengths for these laser systems are in the range of approximately 800 to 1,000 nm. For this domain the laser beam can be guided through optical fibers allowing the easy integration of the laser source into processing heads or robot systems. Furthermore, the optical properties of the polymers in this wavelength range make high-power diode lasers most suitable for laser beam welding.

Laser source	CO_2	Diode	Nd:YAG	Fiber
Wavelength [μm]	9.6–10.3	0.8–1.0	1.06	1.06–1.09
Beam quality M^2	1.1	47	1.1	1.05
Efficiency [%]	10	30	3–10	20
Beam delivery	free beam	optical fiber	optical fiber	optical fiber
Compactness	+	+	−	++
Scanner beam manipulation	+	(+)	+	++
Spot diameter [μm]	50	∼200	20	10

14.4.3 New Approaches for Plastic Welding

Recent technological advances have enabled the development of miniaturized plastic components, starting from simple geometries up to high levels of complexity. The low cost of these materials and their cost-effective manufacturing have simultaneously brought opportunities and challenges for the development of hybrid microsystems or larger products containing plastic materials [30, 31]. Since such devices usually involve the assembly of multiple components with different functions, one of the main challenges for their complete realization is represented by the joining process. The real growth of the laser transmission welding of plastics and the increase in its industrial acceptance started with the development of high-power diode lasers (HPDL). The development of such laser sources allowed a considerable improvement in the initial investment costs, easier integration into manufacturing systems due to their reduced dimensions and an increase of the wall-plug efficiency (∼30%) compared to the alternative laser sources, represented at that time by Nd:YAG and CO_2 lasers.

Nowadays, due to the development of new laser sources for materials processing with an almost ideal beam quality, e.g, fiber or disk lasers, new possibilities are open in the welding of plastics. The high beam quality enables the achievement of a laser spot of only a few micrometers or working with a long focusing length while still having a small laser spot. New concepts like the remote welding of plastics for joining large components or maskless achievement of weld seams as narrow as 100 μm can now be considered. The attractive investment costs, extremely reduced dimensions and attractive wall-plug efficiency of fiber lasers as well as the availability of new wavelengths including 1.55 μm or 1.9 μm have raised the interest of the research and industrial community and subsequently, their performance for material processing is currently the subject of intensive research. If in the case of metallic materials, concrete positive results are already demonstrating the high potential of these new lasers, for the case of plastic materials the investigations are still at their infancy. Therefore, the current research is focused on evaluating the applicability of fiber lasers for the transmission welding of thermoplastics. In the following sections two of the latest research results related to this topic are presented in detail.

14.4.3.1 TWIST® – A New Approach for the High Speed Welding of Plastics

The high beam quality of single mode fiber lasers ($M^2 \approx 1.05$) offers the advantage that very small laser focal spots can be achieved, fulfilling the theoretical prerequisites to producing very narrow weld seams. However, when using such a high quality laser beam focused into a very small focal spot, the power density is on the order of $I = 10^5 - 10^6 \text{ W/cm}^2$. Such laser intensities might be ideal for the cutting or welding of metals but in the case of polymer contour welding, where the commonly used laser intensities are in the range of $I = 10^2 - 10^3 \text{ W/cm}^2$, they would lead to overheating and instantaneous material degradation even at low laser power levels.

In order to overcome these problems and to achieve a thermally optimized laser welding process, an innovative strategy for the coupling of the laser energy into the materials was developed. According to this approach the laser beam has to be moved with a high dynamic (at a high rate) to avoid overheating the material, but simultaneously, through adequate overlap of the laser spot passing over the material for consecutive increments of the weld seam, to ensure the required time for the heat transfer between joining partners and the sufficient diffusion of the polymeric chains to obtain a high quality weld seam. The new approach was called TWIST® – Transmission Welding by an Incremental Scanning Technique because it keeps the incremental weld seam forming characteristic of contour welding while the high dynamic movement and the effect of this movement are typical to the quasi-simultaneous welding technique. Furthermore, such complex, accurate and fast movement of the laser beam can be realized only by elements from the field of high speed scanning technology like galvanometric, polygon or resonance scanners [32, 33].

There are multiple possibilities for performing such a weld seam. In a simplified manner, as shown in Fig. 14.34a,b,c, we can consider, for example, a circle, a line or even the Lemniscate of Bernoulli as individual elements of the weld seam. If the laser beam is moved with high speed along such an individual element and if at the same time the element is shifted at the desired feed rate from one increment of the weld contour to the consecutive one, the weld seam can be achieved. Nevertheless, a corresponding correlation between the laser spot, feed rate and the high speed movement of the laser beam has to be carefully taken into consideration in order to get a homogeneous and high quality weld seam. Such an approach has been already applied in the past to the electron beam welding of metals for achieving different properties of the weld seam or to achieve wider weld seams. Nevertheless, here—

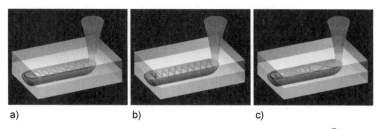

Fig. 14.34 Different possibilities for the high dynamic movement of the TWIST® approach

independent of the chosen geometry for the individual elements of the weld seam—the aim is not to broaden the weld seam but to achieve a thermally optimized welding process.

For a better understanding of the advantages of such an approach, computer simulations were performed for the case of circular beam movement and for the movement along a Lemniscate. The results are shown in Fig. 14.35. The first simulation results reveal that for the circular beam movement an overheating at both sides of the weld seam can occur. A more homogeneous temperature distribution without overheating areas can be achieved for the high dynamic movement of the laser beam along a Lemniscate.

Considering the case of the circular geometry for the high dynamic movement of the laser beam, for the first experiments the process parameters were selected in order to achieve a 15, 30, and 45% overlapping grade for the laser spot (considered for two consecutive increments of the weld seam). The results of the tensile test for the achieved weld seams are shown in Fig. 14.36. The pulling force was applied with a speed of 20 mm/s perpendicularly to the 15 mm weld seam.

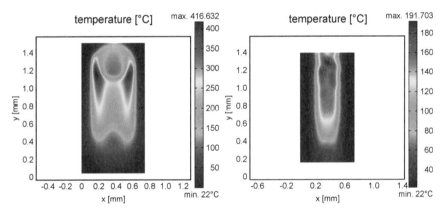

Fig. 14.35 Top view of the welding zone for a computer simulation of the temperature gradients developed for circular beam movement (*left*) and for movement along a Lemniscate (*right*)

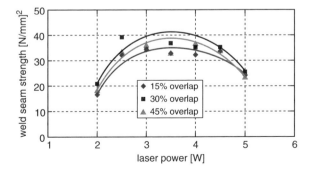

Fig. 14.36 Tensile test results for weld seams with a 500 μm width; feed rate v = 50 mm/s

The strength of the weld seams for a 15% overlapping of the laser spot reaches a maximum of 34.9 N/mm² for a laser power of 3 W. As expected, with the increase of the overlapping the maximum weld seam strength increases. For the same laser power and feed rate the amount of the deposited energy into the joining area increases due to the increased overlapping. Therefore, a higher volume of molten material is produced and subsequently a stronger weld seam is achieved. However, for a 45% overlapping, due to the higher energy input, in the microtome cuts the first signs of material degradation start to be noticeable from a power of 3.5 W while for a 30% overlapping these signs appear starting from 4 W. Therefore, the overall trend of the tensile test results shows the highest weld seam strength for an overlapping of 30%. Another result noticed in the experiments is that there is a slight tendency to reach maximum weld seam strength for a lower laser power with an increase of the overlapping rate. This tendency is also the result of the higher energy input through a greater overlap.

The micrographs of the microtome cuts show the appearance of the Heat Affected Zone (HAZ) and the increase of its width and depth along with the varied process parameters (Fig. 14.37). The thermal decomposition of the material due to laser power that is too high can be easily observed in Fig. 14.37d. However, the most significant result revealed by the analysis of the weld seam cross-section is the flat appearance of the HAZ. In opposition to the HAZ usually achieved with diode lasers, which has a spherical lens shape, the HAZ obtained through the TWIST® approach has an elongated shape, being very narrow, and its depth remaining almost constant across the entire weld seam. The high speed movement considered within the irradiation strategy leads to an uniform laser energy deposition in the material across the weld seam.

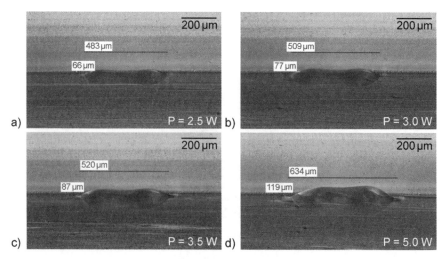

Fig. 14.37 Micrographs of the microtome cuts of the weld seams realized for $\eta = 30\%$, $v = 50$ mm/s. (**a**) P = 2.5 W; (**b**) P = 3 W; (**c**) P = 3.5 W; (**d**) P = 5 W

Reducing the radius of the circular movement in order to obtain weld seams with a width of 250 μm, a different appearance of the HAZ was achieved for the same process parameters as in the case of the 500 μm weld seams. It was noticed that by reducing the diameter of the circular movement for the overlapping rates considered, the energy deposition in the absorbing material was higher, causing thermal degradation for laser powers starting with 3.5 W. Furthermore, for the weld seams where no material degradation was observed, the appearance of the HAZ was closer to the spherical lens shape of the HAZ usually achieved with the diode laser. For minimized weld seams it was determined that higher feed rates are possible even if there is no overlap between two consecutive increments of the weld seam. Since the laser beam is passing over the same place on the material with a period equal to the time required to cover the diameter of the circular movement with the defined feed rate, it seems to be sufficient to reach a firm joining in the case of thinner weld seams. Therefore, detailed investigations have to be carried out for a better understanding of the energy deposition in the case of narrower weld seams. Weld seams as narrow as 130 μm with a good optical appearance were obtained for feed rates up to 18 m/min.

By increasing the feed rate and adjusting the parameters accordingly for circular movement, an optimized energy deposition can be achieved. Figure 14.38 shows the micrograph of a 250 μm thin weld seam achieved with a laser power of 2.5 W, where the tendency to a flat appearance of the HAZ can be identified.

The experimental results showed the viability of the TWIST® approach and subsequently, the possibility to weld thermoplastics with fiber lasers even when high laser intensities are involved. The high dynamic movement considered by the TWIST® approach has the benefit that a homogeneous laser energy distribution takes place in the absorbing joining partner. The effect is similar to the result followed by the top-hat shaping of the laser intensity distribution, but without necessitating any beam shaping operation. Compared to the results achieved for the conventional laser welding of polypropylene, where the maximum weld strengths were in the range of 25–30 N/mm^2, the weld strength achieved through TWIST®

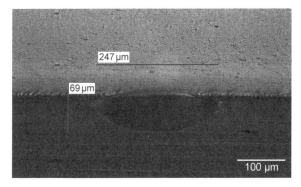

Fig. 14.38 Micrographs of the microtome cuts of the weld seams realized for v = 100 mm/s; P = 2.5 W

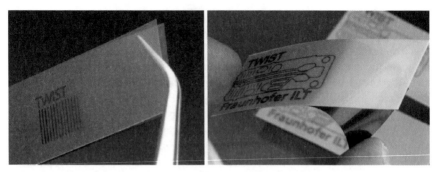

Fig. 14.39 TWIST® welding of polypropylene foils with a thickness of 100 μm

reached values up to 35 N/mm². The higher strength achieved through the proposed method may be the result of reduced surface tensions at the weld seam's interfaces with the joining partners.

The reduced HAZ achieved might represent a successful solution for the welding of plastic components where no distortion on the back side of the components is allowed, or where a minimum distortion of the welded structures is required. For the last situation an eloquent example is given by the recently developed microfluidic chips, where channels with a width of less then 50 μm have to be sealed. Through "conventional" laser welding the HAZ is still too high and causes the collapse of the channel walls, closing them and compromising the chip's functionality. Also, the welding of thin foils or membranes could be considered. Figure 14.39 shows the welding, using the TWIST® approach, of 100 μm thick polypropylene foils with weld seams having a thickness from 500 μm down to 130 μm. The feed rate was varied, up to 12 m/min.

The next steps for the presented work have to be taken in the direction of investigating the phenomenon occurring when reducing the weld seam width down to 100 μm. Furthermore, for the proposed approach (TWIST®) the limits concerning maximum feed rate, minimum HAZ and therefore distortion on the backside of the joining components have to be determined.

14.4.3.2 Absorber-Free Laser Beam Welding of Transparent Thermoplastics

The optical properties of the joining partners play an important role in the welding process. In particular, one problem is the energy deposition in the welding area, as soon as the material does not match the desired optical requirements; that is, when materials of the same optical properties are to be welded. To overcome these difficulties carbon black is usually used for the absorbing joining partner but the black color is not always desired. One solution is the use of a color that is IR-transparent for the transparent joining partner or a visually transparent IR absorber for the absorbing joining partner. This IR absorber can be an additive that is incorporated into the plastic mold or a coating applied directly to the surface layer between the joining partners.

14 Joining

In this field several developments have been made, so that nearly all different colors are weldable. From the cost-efficiency point of view, there is the problem of high cost for the absorbing additives. To reduce costs, especially for large scale parts, a two-component injection molding process can be used. With this process the absorber is only placed in the welding area of the absorbing part.

The use of an IR absorber is not always a desired method, especially when medical applications are concerned. The use of an additional chemical product can lead to problems in the approval of the product in terms of biological compatibility.

A new approach is the use of tailored lasers. The wavelength of these lasers is adapted to the absorption spectra of the material. In combination with optics with high numerical apertures the welding of similar thermoplastics without an IR absorber is feasible.

To weld similar thermoplastics with an overlap geometry, several circumstances have to be considered. First of all, there has to be an intrinsic absorption by the polymer at the chosen wavelength, or the wavelength of the laser system used can be chosen by the absorption spectra of the material. If the absorption coefficient is high enough, the corresponding wavelength is suitable. In Fig. 14.40 a spectrum of PA 6-3T is given. A peak in the absorption line at 1,500 nm can be identified. Therefore the material is suitable for a laser welding process at this wavelength.

A further step is the use of a working head with an optic with a high numerical aperture to keep the intensity on the sample surface below the threshold at which the material starts to melt. In the welding area the diameter of the laser beam is small enough to deliver an appropriate intensity to meld the material within the joining area (Fig. 14.41).

However, both the optics and the absorption coefficient must be aligned to each other to allow the welding of similar materials. Figure 14.42 shows the calculated intensity in combination with both factors for a material without scattering behavior ($C_{st} \sim 0$). On the abscissa the z-coordinate is displayed, while 0 is the welding area,

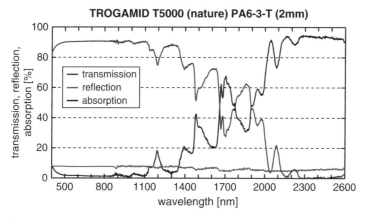

Fig. 14.40 Measured transmission and reflection and absorption from a 2 mm-thick test specimen of PA 6 3 T

Fig. 14.41 Experimental set-up for transparent-transparent welding

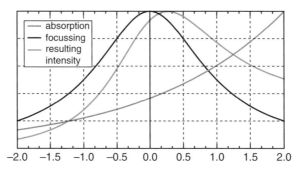

Fig. 14.42 Effects on the intensity due to absorption and focusing

2 is the surface of the upper sample and −2 is the bottom of the lower sample. In this case the intensity is always standardized to 1 at the point of maximum intensity. By the means of absorption (blue) the intensity follows the Lambert-Beer Law. Therefore, it has its maximum value on the surface (2) and falls with increasing material depth. By means of focusing (green) the intensity reaches a maximum at the focal point (0), which is located within the joining area [34].

It can be recognized that the maximum intensity and by implication, the highest temperature is not in the welding area, but 0.2 mm above, in the upper joining partner. And for that reason, the focal position has to be adjusted. Furthermore, it becomes clear that for this material and optical setup, the depth of the heat-affected zone will be around 1 mm, if the required dose is considered to be 0.8 of the maximum intensity. Though this seems to be quite high, this value is in the range of those for welding with IR absorbers.

In several experiments the feasibility of the absorber free welding has been shown for both thin foils and thicker bulk material. Whereas for the foil material higher feed rates are possible, the applications demand injection molded parts. Especially for medical applications like microfluidic devices, where transparent materials have

Fig. 14.43 Microfluidic mixer (PA PACM)

to be used to allow spectroscopy and IR absorber is no choice because of the costs, this welding technique shows remarkable opportunities. In Fig. 14.43 a microfluidic mixer is shown, where as a first step the microfluidic channel is made by shaping with an excimer laser source and afterwards the channels are sealed with a diode laser at 1.5 μm. As this method is used for prototypes a different way has to be used for mass production: to allow high welding feed rates, the closure of the microfluidic structure can be made of a thin foil material and the container of the microfluidic structure can be made by the injection molding process. But still there are challenges to face concerning the light material interaction and the adaptation of the process to every application.

14.4.4 Applications and Further Prospects

The first industrial application of laser beam transmission welding of plastic was developed by Marquardt GmbH with the collaboration of the Fraunhofer Institute for Laser Technology in 1998. The new technique was applied to an electronic car key. The challenge for welding an electronic car access key with integrated radio remote control electronics was to set up a waterproof junction between the plastic housing and plastic flexible cover foil (Fig. 14.44 right) and not to damage the electronics inside the key during the joining process. Compared to most of the conventional plastic welding techniques (ultrasonic, friction, hot plate), laser beam welding is well suited to such an application. Both plastic parts have to appear black to the human eye, which is implemented for the absorbing housing by adding carbon black during the polymer fabrication process. The upper flexible plastic foil is pigmented with a special additive that leads to a "black" visible appearance but keeps transmission high in the near-infrared laser wavelength range (see spectra in Fig. 14.44 left).

The seam width of the laser weld is in the 1 mm range, and the contour welding time is approximately 3 s. The seam is airtight and free of hazardous pollution, and the key is contour welded in a batch production process.

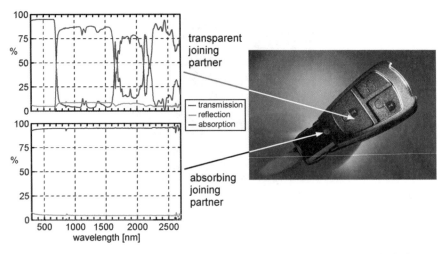

Fig. 14.44 Right: Electronic car access key. Lower housing color is black by carbon black pigmentation, upper flexible foil's color is black as well, but transparent at near-infrared laser wavelength. *Left*: Transmission (*red curve*) of lower and upper polymers

To demonstrate the state of laser welding process understanding and the potential of industrial implementation, a partly automated laser plastic welding machine with an integrated high-power diode laser was set up (Fig. 14.45).

The assembly cell consists of a horizontal moving robot, a pneumatically driven picking unit, a diode laser source including power supply and optical fiber with laser processing head, a component adapted clamping unit, and process control means.

The robot consecutively picks up a black and a transparent part out of the two 25-part magazines and puts them into the clamping device. Within this unit, both parts are fixed in their final position and clamped together for optimum heat flow. The robot moves the laser processing head along the circular contour at the black part's outer circumference. For a laser output power of 10 W, the overlap weld takes

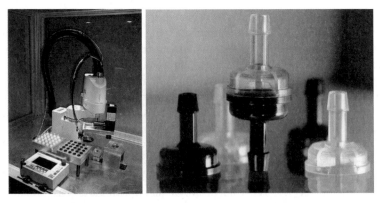

Fig. 14.45 Partly automated assembly cell for laser welding of thermoplastic polymers (Polyamide PA black and PA transparent, not reinforced)

place within 2 s. The weld seam has a width of 0.8 mm according to the focused laser beam diameter; the circular contour diameter (equivalent to the part's crank, see Fig. 14.45, right) is 14 mm.

The laser system, consisting of high-power diode laser source (maximum laser power 50 Watt, wavelength 940 nm), electric power supply (maximum DC current: 80 Amps) and laser chiller unit is contained in the bottom part of the assembly cell. Laser radiation is fed into an optical fiber of 600 μm core diameter and numerical aperture NA = 0.22 (equivalent to a divergence angle of 12.7°). This fiber guides the radiation to the focusing head, which is mounted on the robot's vertical axis. The focusing head also contents an infrared sensor, which detects heat radiation in the wavelength range 1.5–2.5 μm emitted from the welding zone and transmitted through the upper polymer part and the clamping device's transparent plate. This signal, which is a measure of the joining section's temperature, is used for closed-loop process control, i.e., to adjust the laser power to the appropriate level. The welding process can therefore be monitored and controlled to avoid high temperatures and hence polymer decomposition, to guarantee reproducible welding performance.

The laser welding of plastics is not a means to replace all conventional plastic welding processes, but stands as an additional welding process with special features. To benefit from the laser's specific advantages (small, well defined spatial and precise temporal energy deposition), some process-adapted surrounding conditions have to be fulfilled, mainly adapted component and weld seam design and material and additive choice. The last years' technological progress, increasing market demands and higher acceptance of laser plastic welding technology today leads to promising process applications. Future development trends are specific modification of polymer optical properties, polymer compatibility and the further reduction of productive welding time. For contour welding, the welding time might be high if there is a long welding seam length. Time reduction is feasible by applying higher laser output power or optimizing simultaneous welding strategies which are favorable for large-scale production purposes.

14.5 Laser Transmission Bonding

Fahri Sari

14.5.1 Introduction

The miniaturization of electronic chips shows that the packaging of front-end processed wafers for the production of those micro-chips containing sensitive electronic

---

F. Sari (✉)
Mikrotechnik, Fraunhofer-Institut für Lasertechnik, 5207 Aachen, Germany
e-mail: fahri.sari@ilt.fraunhofer.de

devices, e.g., sensors is becoming more important. The high temperature stress during the conventional Wafer Level Packaging (back-end process) can cause damage to those microelectronic components. The state-of-the-art for Wafer Level Packaging is the sealing of front-end processed wafers with cap-wafers in bond stations. A bond station consists mostly of a wet chemical cleaning device, a bond aligner and a bonding device. In a standard procedure the wafers are cleaned with wet chemical sample preparation methods like RCA [35], then fixed and pressed (pre-bonding) to each other in a bond aligner and in a last step mostly bonded in a bond-chamber through a heat treatment (tempering process) supported with a diverse gas (inert or active) atmosphere. The tempering process (increasing the temperature) has an influence on the bonding energy which in turn influences, for example, the mechanical strength. On the other hand, the limits on the thermal load of the components integrated in a chip have a decisive influence on the maximum temperature that can be applied to ensure successful packaging.

Silicon Direct Bonding [36–38] is one of the most-applied bonding techniques. Depending on the temperature limits and wafer treatment it is differentiated between high temperature bonding (> 800 °C) and low temperature bonding (< 450 °C) [38]. Figure 14.46 gives an overview of current bonding methods. Laser Transmission Bonding is shown as an alternative method to conventional bonding techniques.

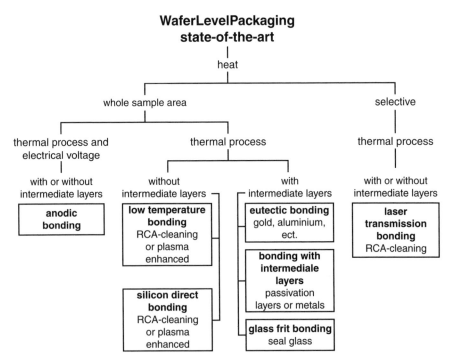

Fig. 14.46 Bonding methods

The disadvantage of Silicon Direct Bonding is the high thermal input ($> 800\,°C$) which, as mentioned, has a negative influence on the functions of the sensitive integrated sensors. Low Temperature Bonding is an alternative to conventional Silicon Direct Bonding. In Low Temperature Bonding, the wafers can be prepared with RCA Cleaning [35], applying plasma enhanced surface treatment with various gases like O_2 [39] or using a spin-on glass [40] as an intermediate layer. This preparation helps to bond under $450\,°C$ but with a long process period. Applying metallic intermediate layers like gold or aluminum between two silicon wafers leads to low temperature bonding by formation of low temperature eutectic bonds [41]. The eutectic point for silicon-gold is at $363\,°C$ and for silicon-aluminum at $577\,°C$ which are suitable processing temperatures to achieve good bonding strength. Glass frit bonding is based on the application of a seal glass with low softening temperature between two wafers. The bonding temperature for this thermo-compressive process is between 400 and $450\,°C$ [42].

A new trend is chip-to-wafer packaging. This process enables three-dimensional chips with multiple functions by packaging of two front-end processed chips without wire bonding.

14.5.2 Thermochemistry of Bonding

The process of bonding needs flat and clean surfaces to activate the chemical reactions at an atomic level between two surfaces. Silicon at room temperature has an oxide layer. This layer reacts with water (air humidity) and forms silanol groups (Si-OH) on the surface. It is important to use a surface preparation method to achieve a homogenous and reproducible surface condition free from any organic or metallic contamination. There are some preparation methods which can be applied. The most common method is the application of the wet chemical surface cleaning method RCA. Through RCA cleaning the generation of hydrophilic surfaces can be achieved by applying two RCA cleaning steps: SC1 (NH_4OH : H_2O_2 : $DI-H_2O$ with a vol. composition 1:1:5; 10 min. at $75-85\,°C$) and SC2 (HCl : H_2O_2 : $DI-H_2O$; with a vol. composition 1:1:6; 10 min at $75-85\,°C$). The result is a surface with a homogeneous chemical oxide layer and silanol groups (Fig. 14.47). Some bonding methods such as eutectic bonding need a hydrophobic surface. A hydrophobic surface can be formed by applying a third cleaning step with hydrofluoric acid (HF dip).

After cleaning, the wafers are dried and pressed against each other and annealed. The thermochemistry of a hydrophilic wafer bonding procedure can be explained in four stages as described in [43]:

Room temperature to $110\,°C$: at room temperature two wafers bond to each other through the formation of hydrogen bonds (prebonding). The rearrangement of water molecules and formation of new silanol groups by the fracture of Si-O-Si (14.1) is observed up to $110\,°C$.

$$Si - O - Si + H_2O \rightarrow Si - OH + Si - OH \quad (14.1)$$

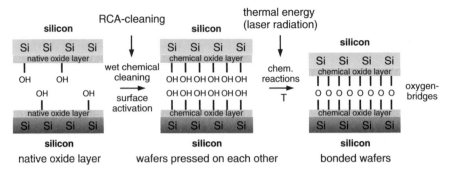

Fig. 14.47 Schematic illustration of surface activation and formation of oxygen bridges by thermochemical reactions

From 110 to 150 °C: diffusion of water molecules along the bonding interface and also through the oxide layer to the silicon atoms brings the silanol groups of both surfaces closer by the formation of siloxane (14.2) and water. The bond energy is increasing.

$$\text{Si–OH} + \text{Si–OH} \rightarrow \text{Si–O–Si} + \text{H}_2\text{O} \quad (14.2)$$

The reaction product, water, can reach the wafer edge by diffusion along the bonding interface and evaporate, or it can reach the deeper silicon atoms by diffusion through the oxide layer. This results in the formation of new SiO_2 (14.3), which in turn is involved in the thermo chemical process.

$$\text{Si} + 2\text{H}_2\text{O} \rightarrow \text{SiO}_2 + 2\text{H}_2 \quad (14.3)$$

From 150 to 800 °C: At 150 °C the saturated surface energy reaches a value of approx. $1200\,\text{mJ/m}^2$. This value is constant up to 800 °C. It is believed that all silanol groups have converted to siloxane bonds at 150 °C.

Over 800 °C: the decreased viscosity of SiO_2 allows an oxide flow. The contact area between the wafers is increased by this viscous flow.

14.5.3 Principle of Laser Transmission Bonding

Laser Transmission Bonding (LTB) is based on transmission heating of the interface of the sample pairs to be joined [44] and the subsequent thermo-chemical reactions [43] between the surfaces. The optical energy is transmitted through the upper sample (e.g., glass) and absorbed by the lower sample (e.g., silicon) at the silicon-glass interface (Fig. 14.48, left). The main part of the optical energy will be converted into heat energy and the bonding zone at the interface will be heated. This enables localized chemical reactions between the surfaces and leads to covalent bonds between silicon atoms by the formation of siloxane [43].

14 Joining

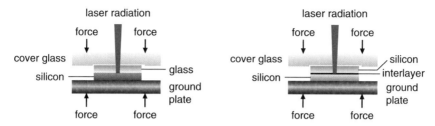

Fig. 14.48 Principle of Laser Transmission Bonding (silicon-to-glass (*left*) and silicon-to-silicon (*right*))

The application of a suitable wavelength is important for joining similar materials like silicon-to-silicon and glass-to-glass. The transmission of the laser radiation must be guaranteed by the upper sample. At the same time it is important to apply a metallic interlayer at the interface which absorbs the laser radiation to activate the thermo-chemical reactions (Fig. 14.48, right). A pressure device with integrated pneumatic cylinder ensures homogeneous physical contact between the sample pairs at an atomic level.

14.5.4 Laser Transmission Bonding of Silicon-to-Glass

A cw (continuous wave) Nd:YAG laser at $\lambda = 1,064$ nm is suitable for the bonding of silicon-to-glass along continuous lines. A pyrometric sensor, which enables thermal process control, has been installed in the laser processing head (Fig. 14.49) to ensure a constant temperature level at the bond front in real time. The evaluation of the resulting pyrometer signals during the bonding process gives information about the online temperature development at the bond front [45]. To enable dynamic thermal process control, Labview® based software with integrated closed-loop control (PID Controller) has been developed. In Fig. 14.50 a signal flow plan of the closed loop used with the transfer behavior is shown. During the bonding process the temperature development at the bond front is recorded using the pyrometric sensor. The difference between set point value (pyrometer set point value) and actual value (actual pyrometric value) is multiplied with the PID parameters in the PID Controller (proportional plus integral plus derivative controller) and an output signal (diode current control value) is sent to the laser source, so the laser power can be re-adjusted in real time. The control frequency is given with approximately 1,000 Hz.

The coefficient of thermal expansion for single crystal silicon is given with $\alpha_{20-300} = 2.5 \cdot 10^{-6}\,\mathrm{K}^{-1}$ and for Borofloat® 33 borosilicate glass $\alpha_{20-300} = 3.25 \cdot 10^{-6}\,\mathrm{K}^{-1}$. The minor difference between the thermal expansion coefficients of silicon and glass allows stress-free bonding and a low-stress cool down process after bonding.

Borofloat® 33 shows over 90% transmittance (thickness $d = 500\,\mu\mathrm{m}$) at $\lambda = 1,064$ nm (Fig. 14.51) and silicon samples (thickness of 525 μm) have an absorbing capacity of 36–38% at $\lambda = 1,064$ nm (Fig. 14.52).

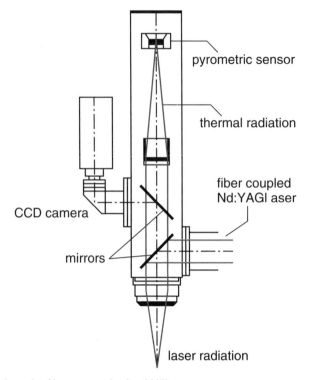

Fig. 14.49 Schematic of laser processing head [45]

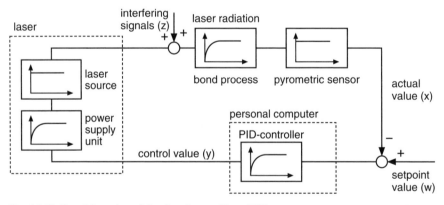

Fig. 14.50 Signal flow plan of the closed control loop [45]

The application of a laser for selective laser bonding of silicon-to-glass and the results for these using cw Nd:YAG laser sources at a wavelength of $\lambda = 1,064$ nm have been published [44, 46, 47]. As a further development to this work, the first bonding results of a current study by applying a fast oscillating galvanometric scanner combined with a fiber laser (wavelength of $\lambda = 1,090$ nm) will be introduced.

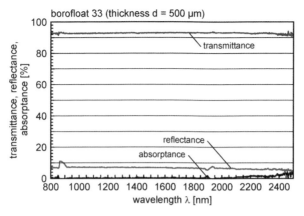

Fig. 14.51 Optical properties of a borosilicate glass (thickness 500 μm)

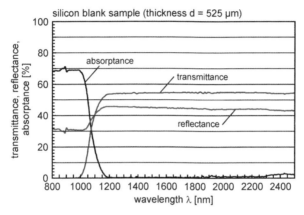

Fig. 14.52 Optical properties of single-crystal silicon (thickness 525 μm) [48]

Compared to a laser processing head which is usually integrated with a 3-axis machining system to enable flexible movement of the laser radiation over the samples, scanners with integrated mirrors also achieve flexible movement of the laser radiation. The intention of this work is to show the joinability of silicon-to-glass by using fast oscillating scanners as a function of the applied laser power and the wobble frequency of the scanner. The bond quality is determined by evaluating the bond seam using a microscope.

The experiments described below were performed on small 5×5 mm^2 silicon and glass samples; hence the wafers were diced beforehand. 4-inch Borofloat® 33 wafers (borosilicate glass, Schott Glas, polished on both sides, thickness of 500 ± 25 μm, surface roughness < 1.5 nm, total thickness variation TTV < 10 μm) and silicon wafers (4-inch [100 mm], p-doped, polished on both sides) with a thickness of 525 ± 10 μm, surface roughness < 1 nm and TTV ≤ 3 μm are generally used for LTB.

A fiber laser (SPI, SP100C, max. power 100 W, fiber diameter of 8 μm, single mode) with a wavelength of $\lambda = 1,090$ nm and a laser spot of approx. 30 μm and a scanner (Scanlab, SK1020, F-theta lens with a focal length of 80 mm) with a scan field of 35×35 mm^2 were used [49]. The application of scanners allows local power modulation through special geometries over the wobble function by quickly scanning the laser radiation perpendicular to the bond line. It is possible to generate bond seams by overlapping the feed rate with oscillating figures (Fig. 14.53, e.g., circles). The variation of the bond seam width can be also adjusted. In addition, a further advantage of the wobble function is that melting can be avoided at the heat affected zone during the joining process.

The results of the joining experiments show that the laser power and the wobble frequency have a direct influence on the bond seam quality and on the bond seam width. Figure 14.54 shows the influence of the laser power P [49]. Two sample pairs of a test series with constant process parameters ($f_W = 3,000$ Hz, $v = 1$ mm/s) were bonded with two different laser power values ($P = 41.5$ W and $P = 46.5$ W). It can be seen that the bond seam width increases with higher laser power values, from approx. 130–180 μm (Fig. 14.54 left) to 180–236 μm (Fig. 14.54 right). The variation in the bond seam width within a sample pair can be explained by applying coaxial illumination method by microscopy. The light dispersion by applying this method is not always homogeneous. The determination of the exact bond seam

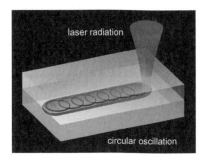

Fig. 14.53 Feed motion with overlapping oscillating circular motion [49]

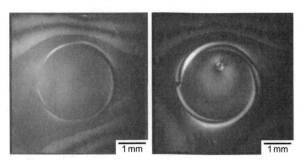

Fig. 14.54 Influence of laser power on bond quality by $f_W = 3000$ Hz (*left*: $P = 41.5$ W and *right*: $P = 46.5$ W) [49]

width can be carried out by the application of chemical etching methods like KOH for silicon.

To show the influence of the wobble frequency, the frequency was varied between $f_w = 500\,\text{Hz}$ and $f_w = 6,000\,\text{Hz}$ with constant parameters ($P = 41.5\,\text{W}$, $v = 1\,\text{mm/s}$) within a test series. The examination of the bond seams for closed, continuous and tight bonds is shown for four sample pairs with increasing wobble frequency in Figs. 14.55 and 14.56. For wobble frequency $f_W < 1,500\,\text{Hz}$ the bond seam is unstable (Fig. 14.55 left). Closed, continuous and tight bond seams could be achieved for $f_W \geq 2,000\,\text{Hz}$ (Fig. 14.56). A correlation between wobble frequency and bond seam width could not be seen. A possible reason is that the applied laser power was too low to show a relationship [49].

The achieved bond results suggest the applicability of scanners combined with fiber lasers for a silicon-glass bonding process [49]. In addition, it is possible to achieve bonding results free of cracks and free of melt and without any application of thermal process control for feed rate values up to $v = 8.33\,\text{mm/s}$ and laser power values up to $P = 52\,\text{W}$. This is possible through spatial power modulation by means of fast laser scanning.

There are some published works on selective laser transmission bonding. The common use is the application of short-pulsed laser sources. Tseng et al. reports and discusses in [50–57] the characterization of the mechanical strength of bonded

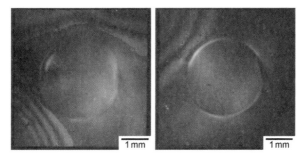

Fig. 14.55 Influence of wobble frequency on bond quality (*left*: $f_W = 500\,\text{Hz}$ and *right*: $f_W = 1,500\,\text{Hz}$) [49]

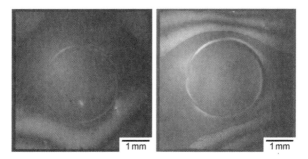

Fig. 14.56 Influence of wobble frequency on bond quality (*left*: $f_W = 2,000\,\text{Hz}$ and *right*: $f_W = 3,000\,\text{Hz}$) [49]

sample pairs and the investigation of the bond interface to identify the diffusion process. Additionally, the influence of surface roughness, oxides (several thicknesses) and contact pressure are discussed. Tseng et al. applied a nanosecond Nd:YAG laser (pulse duration 6.5 ns to 7.0 ns, repletion rate 10 Hz) at $\lambda = 532$ nm. The generated bonds are punctual. Tseng et al. also reported about generating line contours by applying overlapped punctual bonds by scanning a line through moving the laser spot. Tan et al. discuss in [58] and [59] the eutectic bonding of single crystal quartz to silicon with the application of gold and tin as intermediate layers. A Q-switched Nd:YAG laser at $\lambda = 355$ nm (repetition rate up to 20 kHz) with a laser spot of 25 μm and max. laser power of 2 W is used. The eutectic bonds are generated with a variation of the laser power between 0.08 and 0.83 W with a scanning speed of 0.1–0.5 mm/s. Mescheder et al. reports in [60] and [61] about the application of a Q-switched Nd:YAG laser at $\lambda = 1,064$ nm to bond silicon-to-glass. In [62] Thjeppakuttai reports about generated silicon – pyrex bonds by applying a Q-switched Nd:YAG laser at $\lambda = 1,064$ nm with a pulse duration of 12 ns.

In the reported works, laser sources are applied in pulsed operation mode. Application in pulse mode allows the laser to operate in thermal diffusion mode to melt a small region of the surface, e.g., of silicon as Tseng et al. reports. In this process the glass sample is indirectly melted so a diffusion bond can be generated.

In contrast to the presented works, the application of laser systems in cw-mode leads to bonding in the solid state. The application of laser radiation in cw-mode ensures continuous selective bonds along a line. Wild reports in [44] that by the application of laser sources in cw-operation mode, bonding results without any melt phase are achievable. An additional advantage is a controllable warm-up and cool down bonding process without any very intensive heat input as compared to pulsed bonding with very intensive laser power for a few milliseconds. The formation of cracks can thereby be avoided.

14.5.5 Laser Transmission Bonding of Silicon-to-Silicon

In contrast to the silicon-to-glass bonding process, the bonding of similar materials like silicon-to-silicon and glass-to-glass needs an intermediate metallic layer to absorb the laser radiation at the interface.

A cw-thulium fiber laser at $\lambda = 1,908$ nm (single mode, laser spot 41–50 μm) with max. laser power of $P = 52$ Watt is applied [48]. The collimated end of the fiber laser with focusing optics is integrated with a three-axis machining system to enable rapid, controlled movement of the laser radiation over the samples. The feed rate v of the machining system defines the speed that the focused laser moves.

A metallic absorbing intermediate layer (thickness 50 nm, sputtered) is applied. The metallic intermediate layer shows an absorbing capacity of approx. 25% between two silicon samples at $\lambda = 1,908$ nm (Fig. 14.57).

Experimental tests show that an absorbing capacity of 25% of the interlayer at $\lambda = 1,908$ nm is enough to activate thermo-chemical reactions at the interface to achieve bond results [48]. Bond seams are generated by applying spiral-shaped-

14 Joining

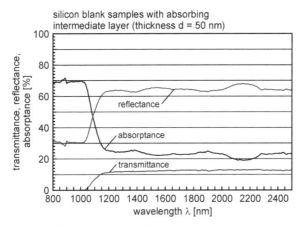

Fig. 14.57 Optical properties of a silicon sample pair with intermediate absorbing metallic layer [48]

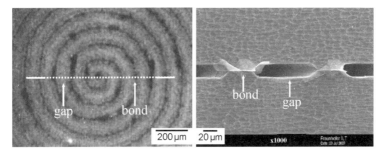

Fig. 14.58 IR-Transmission picture of a sample pair bonded with $P = 24\,\text{W}$ (*left*). SEM picture (cross-section) of a sample pair bonded with $P = 24\,\text{W}$ (*right*) [48]

contours. The IR transmission method (Fig. 14.58, left) is applied to visualize and evaluate the quality of the bond contours. To support the identification of the bond contours, SEM analyses are performed. Figure 14.58 right shows a cross-section picture of a bonded sample pair after being etched with 15% KOH for 30 min. Non-bonded areas in the interface are etched. The width of the bonded areas is measured at 35–65 μm and the width of the gaps is between 110 and 150 μm.

The mechanical strength of the silicon-silicon bonds is determined by applying micro-chevron tests [63]. For this purpose micro-chevron contours are generated on $5 \times 5\,\text{mm}^2$ sample pairs. Figure 14.59 shows a sample pair after the micro-chevron test. At the beginning the crack grows in the interface, and the growth then partly extends into the bulk material. This can be seen as evidence of high mechanical strength of the bonding result. Maximum fracture loads up to $F = 6.35\,\text{N}$ could be measured by micro-chevron-tests [48].

Selective laser bonding results can also be generated by applying intermediate absorbing layers between borosilicate glass – borosilicate glass samples.

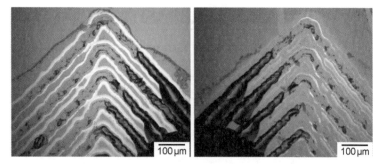

Fig. 14.59 Silicon samples after micro-chevron ($P = 24$ W). Upper sample (*left*). Lower sample with absorbing intermediate layer (*right*) [48]

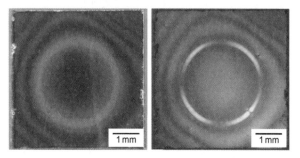

Fig. 14.60 Bonded borosilicate glass sample pair. Sample pair under light-optical microscope (*left*). Coaxial illumination of the same sample pair by microscopy (*right*)

Figure 14.60 shows a representative bonded pair made by applying an intermediate absorbing metallic layer.

14.6 Soldering

Luedger Bosse and Felix Schmitt

14.6.1 Introduction

In addition to welding and gluing, soldering is one of the most important joining processes. Soldering is a thermal process for substance-to-substance bonds and for the coating of materials wherein a liquid phase is formed by melting an added solder

---

F. Schmitt (✉)
Mikrotechnik, Fraunhofer-Institut für Lasertechnik, 52074 Aachen, Germany
e-mail: felix.schmitt@ilt.fraunhofer.de

alloy or by diffusion at the interfaces. The melting temperature of the base material is not exceeded. In principle, the joining process is based on interaction reactions between the joining partners and the melted solder. Therefore, direct, oxide- and contamination-free contact between the metal surfaces of the joining partners and the solder alloy is one of the most important process requirements. If the melting temperature of the additional material is below 450 °C (840 °F) the process is called soldering, while above 450 °C the process is called brazing.

A process-adapted heating cycle is necessary for energy input during the soldering process. This heating cycle has to enable the processes of surface activation, melting of the solder alloy, wetting of the surfaces, and the spreading of the solder and filling of the gap. The heating cycle is based on four important parameters: the heating period, with heating rate and dwell time for heating, the peak soldering temperature, the dwell time above the melting point of the solder alloy, and the cooling rate [64, 65]. In general, it is desirable to use a high heating rate but the maximum heating rate is normally constrained by the form of the energy input. By means of a laser and its high energy density it is possible to realize a maximum heating rate. The heating dwell time is necessary for the evaporation of vapors and constituents of the flux and for the uniform heating of the joining partners up to the wetting temperature. This temperature is below the melting temperature of the solder alloy. The peak soldering temperature should be such that the solder alloy is certain to melt, but at the same time the solder alloy should not be overheated so that it degrades through the loss of constituents. The peak temperature is normally set about 20–30 °C above the melting point. The minimum time that the joint geometry is held at this temperature must be sufficient to ensure that the solder alloy has melted over the entire area of the joint. Extended holding times tend to result in excessive spreading of the molten solder alloy, possible oxidation gradually taking place, and deterioration of the properties of the parent materials. The cooling stage of the cycle is not controlled by the operator but is normally governed by the thermal mass of the joint geometry. For laser processing it is very fast because of the instantaneous switch-off of the laser power, resulting in a fine-grained microstructure of the joint (Fig. 14.61).

In contrast to conventional selective soldering techniques, laser beam soldering features a contactless, temporally and spatially well-controlled energy input. Because of these characteristics laser beam soldering is ideal for joining tasks where miniaturization and reduced thermal and mechanical stresses are required. Special features of laser beam soldered joints are fine-grained microstructure and the low amount of intermetallic phases due to the fast heating and cooling rates of this process. In principle laser beam soldering is characterized by temporally and spatially selective energy input by surface absorption in the joining area, successive heat conduction and interface processes. The joining process is determined by the characteristics of the laser beam source, the chosen process parameters and thermo-physical properties of the joining partners.

A cross-section view is given in Fig. 14.62 of a simulated temperature distribution for an electrical contact during laser beam soldering. Computation of the temperature distribution is done by finite element software. The computed temperature

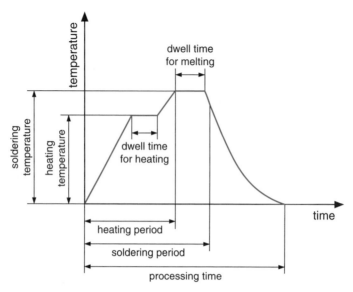

Fig. 14.61 Heating cycle for soldering

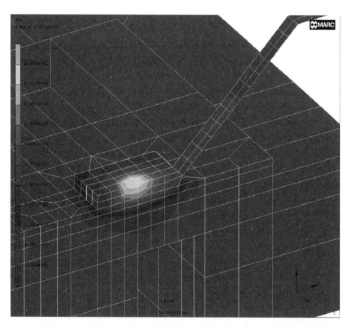

Fig. 14.62 Simulated temperature distribution of an electrical contact during laser irradiation

14 Joining

distribution is showing spatially selective energy input and furthermore, temperature gradients in the joining area.

14.6.2 Physical-Technical Fundamentals

The wetting of the surfaces is an important requirement for a good solder joint [66]. Wetting is defined as a reaction between a liquid solder droplet and the solid substrate. Wetting will occur if the liquid solder alloy has direct contact with the surface of the metal substrate. Any adhering contaminants or oxides in the joining area will form a barrier layer and reduce wetting (Fig. 14.63, left).

If the metal substrate is clean, the metal atoms are exposed at the surface, and the solder alloy wets the surface. Contact between atoms of the solder alloy and the metal substrate leads to alloying and resulting good electrical conductivity and adhesion.

Wetting of the solid metal substrate by the liquid solder alloy is an interfacial process in which the contact angle Θ provides a measure of the quality of wetting. If $\Theta < 90°$, a liquid droplet will wet the substrate. The contact angle is defined by the metal substrate, the solder alloy and the atmosphere.

Three surface tension forces σ_{12}, σ_{13} and σ_{23} are in balance as shown in Fig. 14.64. The following equation is applied between surface tension forces and contact angle Θ (Wetting or Young equation):

$$\cos \Theta = \frac{\sigma_{13} - \sigma_{23}}{\sigma_{12}} \qquad (14.4)$$

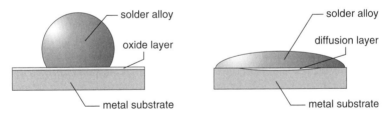

Fig. 14.63 *Left*: Liquid solder alloy on oxidized substrate. Solder alloy does not wet the surface. *Right*: Liquid solder alloy on clean substrate. The solder alloy wets the surface and a diffusion layer is formed at the interface

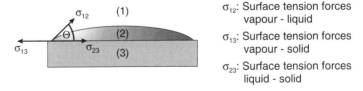

σ_{12}: Surface tension forces vapour - liquid

σ_{13}: Surface tension forces vapour - solid

σ_{23}: Surface tension forces liquid - solid

Fig. 14.64 Surface tension forces for wetting of solid substrate and liquid solder alloy

The difference between surface tension forces σ_{13} and σ_{23} is known as the adhesion tension force σ_H. The contact angle Θ is commonly used as an estimation for the adhesion tension force σ_H. The surface tension forces are the important parameters for wetting. Good wetting occurs when

$$\sigma_{13} \geq \sigma_{23} + \sigma_{12} \tag{14.5}$$

In this case, the contact angle reaches zero and the liquid solder alloy spreads to cover the entire solid substrate. Fluxing agents enhance the wetting of the substrate. In principle, fluxing agents fulfill the following three functions:

- physical: removal of oxides and other films as well as soldering reaction products from the surface to provide direct contact between solder alloy and substrate
- chemical: removal of tarnish on the surfaces and protection from reoxidation of these surfaces during the soldering process
- thermal: enhancement of heat flow between heat source and solder area.

The selection of fluxes follows two criteria, efficiency and corrosiveness, but these criteria are opposed to each other. The efficiency of a flux is rated according to its ability to wet a surface in a certain time. On the other hand, corrosion by fluxes can lead to undesirable chemical reactions on the joining partner surfaces.

The more active a flux is, the more corrosive it is, while a non-corrosive flux will not enhance—or will restrict—the wetting of solder alloy on the substrate. Because many parts in electronic industries are corrosion-sensitive the selection of flux is restricted to the less active fluxes.

14.6.3 Process Description

There are numerous different application-specific solutions available commercially for laser beam soldering machines. In principle, they are based on flexible beam shaping and guidance using galvanometric scanners or multi-axis systems. For fiber-guided systems the processing optics are moved but there are also systems in which the entire laser beam source is moved. In Fig. 14.65 a production cell and laser processing optics are shown, based on a galvanometric scanner.

The machine is designed for the laser beam soldering of an automotive microelectronic module (an alternator regulator realized in thick-film technology) with solder pads printed on an alumina substrate (Fig. 14.66). The housing has seven terminal leads to be soldered to the substrate. The entire alumina substrate is bonded by a heat conducting adhesive to an aluminum base plate (Fig. 14.66).

Major components of the production cell are a fiber coupled, continuous wave (cw) diode laser system and a processing head with integrated pyrometric and power sensors.

The diode laser system has a maximum optical output power of 250 W, which can be modulated by controlling the pump current. The collimated laser beam passes through a galvanometer scanner and is focused by an F-Theta lens on the lead/solder

14 Joining

Fig. 14.65 System design for laser beam soldering based on galvanometric scanners: production cell (*left*) and laser processing optics (*right*)

Fig. 14.66 Automotive microelectronic module in a plastic housing – alternator regulator

pad area to generate the joint. The circular focus geometry of the laser beam is aligned with the center of the semicircle at the end of the terminal lead (Fig. 14.67). The working distance between the optics and the laser beam interaction area is about 80 mm. With an image projection ratio of 1:2 the minimum focal diameter is 1.2 mm, which is double the fiber core diameter of 0.6 mm.

Thermal radiation emitted from the surface follows the beam delivery system of the galvanometer scanner and passes through a dichroic mirror, which is transparent to this wavelength range. After passing through the dichroic mirror the thermal radiation is focused by a lens on a photo detector (Ex., InGaAs, peak wavelength 2.3 μm). The output signal of the detector is conditioned by a logarithmic amplifier circuit. The integrated pyrometric sensor is conditioned for laser beam applications with process temperatures in the range of 150 °C, e.g., the welding of plastics or

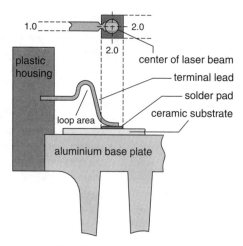

Fig. 14.67 Solder joint configuration (dimensions: mm)

soldering. The pyrometric sensor is calibrated by means of a standardized black body and the response time of the sensor is about 1 ms at 150 °C.

The surface of the lead/solder pad area is imaged onto a CCD camera via a deflecting mirror.

Apart from interconnection requirements, a high production rate has to be ensured for the process to remain attractive for mass production. For this reason the total process period, especially the irradiation time, has to be as short as possible. However, to achieve an adequate solder joint with reduced irradiation time the laser power has to be increased. To avoid the hazard of superheating, the laser power has to be limited and controlled. Therefore, the thermal radiation from the interaction zone must be detected and analyzed in more detail. In a series of experiments the following features could reproducibly be observed in the recorded pyrometric signal [67]. A typical profile is presented in Fig. 14.68, where the laser is switched on at

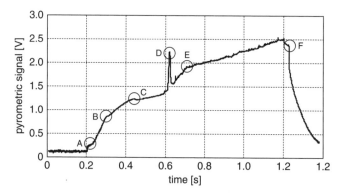

Fig. 14.68 Pyrometric signal detected during soldering; irradiation time: 1000 ms

14 Joining

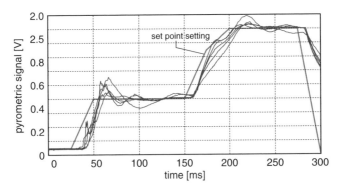

Fig. 14.69 Array of pyrometer signals recorded during different closed-loop controlled LBS processes. The gray curve represents the defined set point settings

time $t = 0.2$ s. At point [A] the reduction in the ascending slope indicates the initial activation of the applied adipic acid flux. The second change in the pyrometric signal, at point [B], is related to the onset of localized melting of the solder pad and the out-gassing of volatile components. Due to the continued energy input by the laser beam, the terminal lead reaches the wetting temperature (point [C]). In the next phase there is a sudden increase in heat dissipation due to the wetting of the terminal lead, which often results in a temperature decrease (points [C] to [E]). At point [D] a bubble consisting of volatile components leaves the molten solder. The variation of the signal curve following point [E] is induced by self-optimization of the surface tension and by superheating of the molten solder pool. After the laser beam is switched off at $t = 1.2$ s, very high cooling rates are observed. This high rate is caused by optimized heat transfer into the aluminum base plate. At point [F] the solder solidifies. The change of the descending slope in the signal curve at the crystallization point [F] is known from the thermal analysis of solidification reactions in literature (Fig. 14.68).

Based on a set of characteristic curves, benchmarks can be determined, and by changing specific process parameters a thermally and temporally optimized profile can be generated. Using these analytical profiles as set point settings for a closed-loop control system, the energy input can be controlled individually for each joining application or product (Fig. 14.69).

Figure 14.70 shows a detailed view of two solder joints and a cross-section of a laser soldered joint.

14.6.4 Applications

The application described shows that there is a huge potential for laser beam soldering in the automotive industry. Here, electronic sensors and control systems based on ceramic substrates are used in many cases. Ceramic substrates offer superior characteristics regarding the environmental conditions in the engine compartment. These

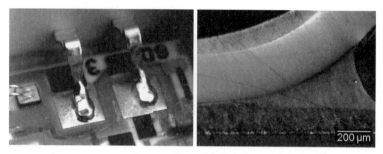

Fig. 14.70 Detailed view of two solder joints (*left*), cross-section of a laser-soldered joint (*right*)

electronic systems withstand thermal loads reliably and durably when mounted on active or passive heat sinks. Unfortunately, in combination with these heat sinks the assembly of such systems is complicated. Electronic and mechanical joints between the terminal lead and the corresponding connecting area on the ceramic substrate can only be realized by joining methods with high energy densities [68].

An innovative application for laser beam soldering is the electrical contacting of solar cells for photovoltaic module production (Fig. 14.71). Due to the decreasing thickness of solar cells, today 220 μm but in future below 150 μm, the demand for a soldering method without any mechanical contact has led to the development of the laser beam soldering process. The process is controlled by pyrometric sensors to avoid thermal damage to the thin silicon wafers.

Laser soldering is playing an increasingly important role as an alternative to the adhesive bonding or clamping of micro-optical components into metallic mountings (Fig. 14.72). In contrast to laser soldering, energy input by induction is difficult for miniaturized optics with a diameter smaller than 1 mm and mounting widths below 50 μm because of the smaller amount of material for heating. Similarly, manual soldering using a soldering iron raises problems because of the small dimensions and the resulting insufficient reproducibility. An alternative to these processes is soldering by using a high-power diode laser or a fiber laser. For these experiments the joining components consist of a gold metalized stainless steel mounting and sapphire optics, which are also metalized with gold. An AuSn solder alloy with a melting temperature of 280 °C is used. By using a fluxing agent the surfaces are

Fig. 14.71 Electrical contacting for solar cell interconnection

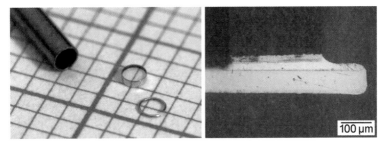

Fig. 14.72 Joining components: bushing, sapphire optics, solder preform (*left*). Cross section of a soldered joint (*right*)

cleaned of oxides before soldering and the joining area is protected from oxidation during the soldering process. The flux causes pores in the solder joint and these pores can be detected. By means of a pyrometer it is possible to establish a controlled process and a two-step temperature profile, as recommended in the literature for soldering. At the beginning of the laser soldering process the flux is activated at a lower temperature ($\sim 150\,^{\circ}$C), while in the second step the necessary energy for the melting of the solder alloy is applied. This process management reduces the number of pores within the soldered joint significantly.

The gap is filled homogeneously with solder by capillary forces. Excess solder does not wet the surfaces of the sapphire but wets the mounting on the laser-facing side. Both diode lasers focused to 1 mm and fiber lasers collimated to 1 mm diameter or lower can be used as laser sources. The advantage of the fiber laser is that the focus position does not have to be aligned because of a Rayleigh length greater than 1600 mm.

Application areas of selective laser beam soldering using high-power diode lasers are manifold and are not confined to a special branch of industry. Currently, industrial applications are focused on electronics assembly, especially for the automotive sector. Discrete mounting of critical components, soldering of cable strands, soldering and brazing of micro-electronic and micro-mechanical components and cable assemblies are industrial applications of laser beam soldering.

Because of its good focusability, accurate control of the energy input and huge power potential, laser beam soldering has the best potential to carry out these joining tasks reliably and reproducibly. By combining diode laser systems with precise positioning systems based on fast galvanometric scanners, user-defined, product specific joint configurations can be soldered at a high production rate. Here, substance-to-substance bonds are generated with high quality in combination with minimal thermal loads on the substrate. Because of its high intensities, laser beam soldering offers process durations in the range of some 100 ms per contact, depending on the joint configuration and the choice of the solder alloy. The decreasing price for diode laser systems is enabling laser beam soldering as an economically attractive alternative to other soldering methods. In combination with the possibility of controlling the temperature within the joining area, laser beam soldering is well-suited to automated production.

14.7 Laser Beam Microwelding

Kilian Klages and Alexander Olowinsky

14.7.1 Introduction

The joining processes in electronic device manufacturing are today still dominated by conventional joining techniques like press fitting, crimping and resistance welding. Laser beam joining techniques have been under intensive investigation and subsequently new processes for mass manufacturing and high accuracy assembling have been established. With the newly developed SHADOW® welding technology, technical concerns such as the tensile strength, geometry and precision of the weld have been improved. This technology provides the highest flexibility in weld geometry with a minimum welding time as well as new possibilities in using application-adapted materials. Different parts and even different metals can be joined by a non-contact process. The application of relative movement between the laser beam and the parts to be joined at feed rates of up to 60 m/min produces weld seams with a length from 0.6 to 15.7 mm using a pulsed Nd:YAG laser with a pulse duration of up to 50 ms. Due to the low energy input, typically 1 to 6 J, a weld width as small as 50 µm and a weld depth as small as 20 µm have been attained. This results in low distortion of the joined components.

In the field of micro production a variety of materials with individual product-specific dimensions are commonly used. In particular, the manufacturing of hybrid micro systems, built up of different functional groups, requires variable joining technologies tailored to the specific demands of each component or each material combination.

14.7.2 Laser Beam Microwelding

Laser beam microwelding is a versatile and flexible manufacturing technology that has found its way into various industrial applications. Electron guns for CRT displays have been produced using Nd:YAG lasers since the '70s of the last century. Such an electron gun requires more than 150 spot welds to assemble the different parts. This adds up to 15 Mio laser pulses per day. In many other industrial fields, laser beam microwelding is becoming a standard manufacturing technology for small products.

In the watch industry, gear wheels and arbors are no longer joined in a press fit process but by means of laser beam welding. In the automotive industry, more and more sensors and components such as relays and control units are mounted directly under the hood and have to undergo heavy vibrations and high temperatures.

A. Olowinsky (✉)
Mikrotechnik, Fraunhofer-Institut für Lasertechnik, 5207 Aachen, Germany
e-mail: alexander.olowinsky@ilt.fraunhofer.de

14 Joining

The joints in these components have to survive these stresses with a long predicted lifetime and a very low failure probability, as they are part of the safety equipment.

As a variety of different geometries and different accessibilities has to be joined securely, only a joining technology with high flexibility at reasonable costs and the ability to provide short cycle times can be used. Alternative joining methods often reach their limits in terms of product quality and reliability (see Table 14.1).

Laser beam microwelding is a non-contact process without any tool wear-out. The process duration is shorter than with competing techniques. The joining process may be finished within a few milliseconds, with the cycle time determined by the loading and unloading of the components to be joined as well as the specifications of the laser source.

One main advantage of laser beam microwelding is its flexibility: part geometry, material and material combinations can be changed very easily because the energy input can be controlled and the intensity and the power can be adapted to the task over a wide range. Spot welds as well as continuous weld seams can be performed. Process monitoring, as a main requirement in industrial production lines, can easily be integrated as via an inline weld monitor or offline inspection of the weld.

Laser beam microwelding requires good contact between the joining partners. To obtain good results the following joint geometries have been established (Figs. 14.73, 14.74, 14.75, 14.76, 14.77 and 14.78)

Table 14.1 Alternative joining methods

Joining method	Disadvantage in comparison to laser beam welding
Adhesive bonding	Elaborate surface pre-conditioning
	Lower bond strength
	Long process time
Swedging or border crimping	Tool wear out
	Additional forces
Resistance welding	Two-sided accessibility
	Limited material choice
Soldering	Reduced high temperature strength

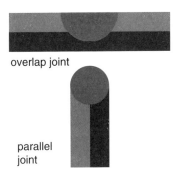
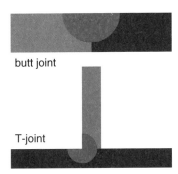

Fig. 14.73 Joint geometries for microwelding

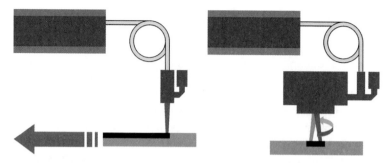

Fig. 14.74 Positioning of the laser beam for microwelding

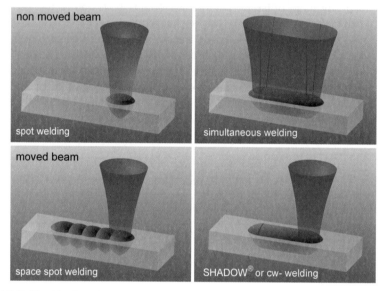

Fig. 14.75 Classification of Laser Beam Micro Welding

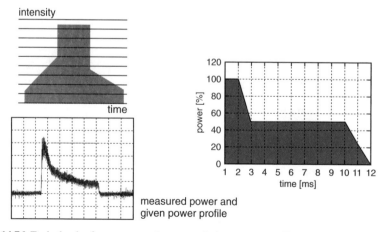

Fig. 14.76 Typical pulse form, measured power and given power profile

14 Joining

Fig. 14.77 Typical application for spaced spot welding (Source: Lasag)

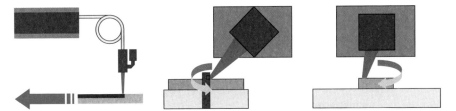

Fig. 14.78 Schematic drawing of the set-up

14.7.3 Processes and Results

The most commonly used laser source is a pulsed flashlamp pumped Nd:YAG laser at a wavelength of $l = 1,064$nm. Typical data for commercially available laser sources are listed in Table 14.2.

Pulsed flashlamp pumped Nd:YAG lasers offer some advantages compared to continuously emitting (cw) Nd:YAG lasers:

- high maximum pulse power at moderate average power
- better beam quality
- affordable investment costs and low cost-of-ownership
- lower requirements for cooling
- steeper slopes for pulse rise time
- pulse forming capability

Table 14.2 Typical Specifications of pulsed Nd:YAG laser sources and fiber lasers

		Pulsed Nd:YAG	Fiber laser
Average power	[W]	10–400	100–200
Pulse power	[kW]	1–7	–
Pulse energy	[J]	1–50	–
Pulse duration	[ms]	0.1–20	Cw
Beam quality	[mm mrad]	8–16	0.4
Beam diameter	[μm]	50–400	35
Fiber diameter	[μm]	100–500	10–50

The applicability of optical fibers to guide the laser light offers new possibilities for industrial use within manufacturing equipment. The separation of the laser source itself and the working head inside the machine or even the possibility of using one laser source for different machines by energy sharing or time sharing mechanisms makes the use of lasers more economically feasible.

The new sources, e.g., fiber lasers now combine better beam quality with reduced costs.

14.7.4 Beam Delivery

For Nd:YAG lasers, there are two possible means for beam delivery: direct beam and fiber delivery. The beam quality of a direct beam laser (BPP 8–20 mm mrad) is better than the beam quality of a fiber guided system (BPP 15–30 mm mrad). The intensity distribution of a direct beam is normally a Gaussian distribution, whereas the fiber guided system has a top-hat distribution. The Gaussian distribution can be focused to a smaller beam diameter.

However a disadvantage of Nd:YAG rod lasers is the influence of the thermal lens. Beam quality and intensity distribution depend on duty cycle, pulse duration and laser power. They also can change from pulse to pulse as well as within one pulse. A laser beam guided through optical fiber by multiple reflections is homogenized. Beam quality and intensity distribution are predetermined by the diameter of the fiber and its NA and vary only slightly. Furthermore, the maximum temperature of the weld bead using a Gaussian distribution is normally higher, so the top-hat distribution is normally more appropriate for laser beam welding.

The positioning of the beam can be performed by using a Cartesian positioning system to move the work piece or by moving the beam by means of a galvanometer scanner.

Typical applications of laser beam microwelding are dealing with wires and thin sheets ranging from 10 microns to 1 millimeter in thickness. The diameter of the laser beam should be smaller than the thickness of the parts to be welded or in certain applications it can be larger. In particular, the parts are sometimes already placed in an plastic housing (e.g., pre-molded package). Here we have to take into account that the housing material must not be influenced by diverging laser radiation or the heat created by the joining process. The main materials are steel and coated and uncoated copper alloys.

Combinations of materials often have to be joined, e.g., steel/copper or steel/brass. Here, the joint geometry determines the weldability. As the joining tasks differ very much in terms of geometry, dimensions and material, we have to pay attention to heat conduction in the parts. Sheets with a thickness below 500 μm cannot be treated as semi-infinite bodies. Heat accumulation at the back side of the sheet strongly influences the welding as well as the heat losses into the surrounding material of the parts and the clamping devices.

In microwelding technology only three different types of joining methods are applied: spot welding, spaced spot welding to create lines and continuous seam welding. Simultaneous welding with elongated line focus is very seldomly applied.

14.7.5 Spot Welding

Spot welding should be applied if only small connection cross-sections are needed or the available space is not sufficient for elongated weld seams. The diameter ranges from 100 to 800 μm depending on the beam diameter, the material and the laser power. The spot welding process can be divided into four phases: heating, melting, melt flow dynamics and cooling. Depending on the laser intensity, evaporation of material may occur.

By means of pulse forming the intensity can be adapted to the sequence of the process phases. A typical pulse form is given in Figure 14.76.

For some materials, preheating, as shown in Fig. 14.76 left is advantageous. Other materials, such as copper alloys, require high intensities at the beginning of the pulse in order to break existing oxide layers and to assure a stable incoupling of the laser energy.

Post heating with well controlled cooling conditions may reduce the risk of cracks. For this purpose a pulse form as shown in the right picture can be used. Typical pulse durations range from 1 to 15 ms.

For pure heat conduction welding the weld depth is equivalent to the radius of the weld spot diameter. Increasing the intensity leads to the evaporation of material and to establishing a capillary. This now-developing keyhole welding process, described earlier, permits greater weld depths. The line between pure heat conduction welding and keyhole welding cannot be accurately drawn for micro parts due to the facts of heat accumulation and the dimensions of the parts.

14.7.6 Spaced Spot Welding

Spaced spot welding is realized by placing several spot welds at a certain overlap in order to achieve a seam weld. The length of the seam is scalable but the heat input is very high because for each spot all process phases of spot welding have to be passed through. This may lead to distortion or thermal damage of the parts. The following picture shows the cover of a battery housing for a heart pacemaker.

Important process parameters, besides pulse power and pulse duration, are pulse repetition rate and feed rate. Together with the spot diameter, the latter two determine the overlap of two consecutive spots, which is usually in the range of 60%.

14.7.7 Continuous Welding

Continuously emitting lasers are seldom used in microtechnology to realize weld seams because of the large beam diameter.

Until now, cw laser welding has been only used for longer joints and for larger parts. A high average laser power, $P_{av} > 500\,W$, and a high processing velocity, $v > 5\,m/min$ are required for cw laser welding. Above all, cw laser sources are more expensive than pulsed laser sources. Nevertheless the joints obtained by cw laser welding show a smooth surface and an optimized microstucture nearly free of pores. The required energy per length is less for cw laser welding than for pulsed laser welding.

The idea is to apply continuous wave welding to micro parts with part dimensions less than $500\,\mu m$ and a weld width less than $100\,\mu m$. The use of pulse forming (temporal shape of the laser pulse) enables the joining of dissimilar materials like steel to copper. As the length of the weld seam will be very short, continuous wave lasers tend to be off rather than being used for welding. Therefore a new technique called SHADOW® was developed to realize extended weld seams with a pulsed laser using a single pulse and sweeping the laser beam over the surface during the duration of the pulse.

SHADOW® stands for **S**tepless **H**igh Speed **A**ccurate and **D**iscrete **O**ne Pulse **W**elding. It was invented to weld small axis-symmetrical parts which can be turned rapidly during a single laser pulse. This technique combines the advantages of continuous wave welding, such as a smooth surface and a high process speed, with the possibilities of pulsed laser systems which include lower costs and the capability of forming the temporal shape of the pulse. Since the parts are small, the latter advantage of the pulsed process enables its application in microtechnology, where the thermal load on the assembled parts has to be well controlled.

At present, pulsed laser sources are able to generate a maximum pulse duration of $\tau_{H,max} = 20\,ms$. To weld parts with a length of $l = 2\,mm$ a processing velocity of $v = 6\,m/min$ is therefore required.

Comparing the energy input ($E_{H,SHADOW} = 6\,J$) to the energy input for a similar joint using the multi-pulse technique, where ten pulses without overlap are needed ($E_{H,p} = 10 \times 2.4\,J = 24\,J$) it is less by a factor of four. Moreover the joined parts show less debris or pollution on the surface and neglecting the time needed to accelerate the parts, the processing time is dramatically reduced.

Fig. 14.79 Comparison of spaced spot welding and SHADOW® for stainless steel and brass

The effect of reduced energy input can be seen in Fig. 14.79. As the transition from solid to liquid is limited to the very first beginning of the process, no particles or melt ejections occur. The result is a smooth and even surface of the weld seam.

SHADOW® can be used for welding difficult materials and to improve weld quality. The applications are shown in the following section.

14.7.8 Applications of SHADOW® in Fine Mechanics and Electronics

The accuracy of a mechanical watch depends on the quality of the rotating spring assembly (balancer). For high end watches this balancer consists of a ring with four pins. Here SHADOW® is used to weld the pins to the ring. As the outer surface is diamond turned after the welding process, the weld seam cannot be seen. The diameter of the annular weld seam can be adjusted to achieve either a ring around the pin or by reducing the diameter, the pin can be molten in total (see Fig. 14.80).

Instead of moving the laser beam by means of a scanning head the complete part can be turned, especially for the joining of a wheel to an axle or, as shown in the following example, the inner cage of a ball bearing. The axle in the center of the part prohibits the use of a scanning head where the beam comes from the center of the field of view. This application is done with a high speed rotating workpiece with a tilted beam targeting from the outside to the center of the part at an inclination angle of 45° (see Fig. 14.81).

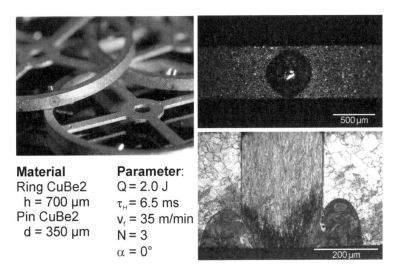

Material
Ring CuBe2
h = 700 µm
Pin CuBe2
d = 350 µm

Parameter:
Q = 2.0 J
τ_H = 6.5 ms
v_f = 35 m/min
N = 3
α = 0°

Fig. 14.80 Balancer for mechanical watches

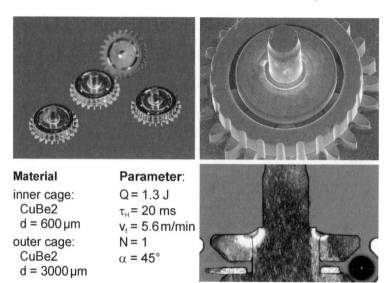

Fig. 14.81 Ball bearing

14.7.9 Comparison of Conventional Pulsed Mode Welding to SHADOW®

The already mentioned advantages of the SHADOW® technique can be discussed with another example from the watch industry. The application is the seconds hand's gear wheel. The typical combination of steel and brass with its problem of the evaporation of zinc is shown in Fig. 14.82. The two different methods are applied and the results are discussed in the following.

Fig. 14.82 Comparison of conventional pulsed mode (*left*) welding to SHADOW® (*right*) Material: Axis S20AP, Wheel CuZn37, Ø 0.3 mm

In pulsed mode, 130 pulses with a pulse energy $Q = 0.1$ J are applied. The total energy amounts to 14 J. In comparison the SHADOW® technique only uses one pulse with an energy $Q = 1.3$ J. The reduction of the energy results in a smooth surface without any material ejection or particles on the part.

14.7.10 Preconditions and Limits of Laser Beam Microwelding

To determine the applicability of laser beam microwelding to a specific joining problem there are some crucial points that have to be investigated.

Offsets in a butt joint configuration and gaps in all instances commonly create kerfs at the edges of the weld seam, which decrease the stability and strength of the joint. Furthermore the surface conditions such as oxidation or contamination with lubricants due to preceding manufacturing steps, e.g. stamping, change the energy absorption and therefore the welding result. As a consequence, more porosity may occur.

Reproducible clamping conditions are one major precondition to assure reproducible weld results. The thermal mass and the dimensions of the parts are so small that differences in the positioning within the clamping, varying gaps or clamping forces may not be compensated.

Furthermore the position of the focal plane with respect to the surface of the workpiece is crucial to the weld result. A defocusing due to misalignment will lead to broadening of the weld seam, but taking into account that the intensity distribution may seriously change the welding results might also differ substantially. Finally, the mode of operation of the laser source itself influences the weld results. At the limits of the working range the focusing conditions vary from pulse to pulse in terms of pulse power, pulse form and intensity distribution.

14.7.11 Conclusion

Laser beam joining offers the advantage of well controlled energy input into the parts with low effects on the surrounding material.

For welding micro parts, two different kinds of laser sources can be chosen: diode pumped systems like fiber or thin disk lasers for longer weld seams, and flash lamp pumped rod lasers used for the SHADOW®-technique. Fiber laser system up to 200 W are able to weld stainless steel as well as copper up to thicknesses of at least 250 mm. It is also possible to weld dissimilar metals in an overlap configuration. The beam diameter has to be small to obtain a keyhole. However, the Rayleigh length of a laser beam with a 15 µm beam diameter is smaller than 200 µm so the requirements for beam positioning are quite high.

The flash lamp pumped rod laser used for the SHADOW® technique is more economic for weld seams length up to several 5 mm. The SHADOW® technique combines the advantages of continuous welding, such as low contamination and smooth weld beads, with the lower investment costs of a pulsed solid state laser with its capability of pulse forming and good focusability due to its beam quality. As most micro parts need only short weld seams, the SHADOW® technique is often the more economical welding technique.

Fiber lasers, with their unique beam quality at reasonable cost and extremely small footprint, are at their start for industrial use. The possibility of realizing very small weld seams due to their achievable focus diameters opens up new fields of use in micromechanics and microelectronics such as the replacement of thermo sonic ribbon bonding for high power electronics.

References

1. J. Michel, M. Niessen, V. Kostrykin, W. Schulz, C. Zimmermann, D. Petring. (1999) "LaserWeld3D", ILT Software Paket
2. R. Poprawe, D. Petring, C. Benter. (2001) "Schweißen mit Diodenlasern", iASTK 2001
3. C. Brettschneider. (1998) Rofin-Sinar Laser GmbH: "Im Vordergrund steht die Ästhetik", Laser-Praxis, Oktober 1998, Carl Hanser Verlag, München
4. DIN32511 Beuth-Verlag Elektronenstrahl- und Laserschweißverfahren:Begriffe für Verfahren und Geräte
5. E. Beyer. (1987) Schweißen mit CO_2-Hochleistungslasern. Technologie Aktuell. VDI-Verlag
6. Merkblatt DVS 3203 Teil 4. Qualitätsicherung von CO_2-Laserstrahl-Schweißarbeiten. Nahtvorbereitung und konstruktive Hinweise
7. E. Beyer. (1985) Einfluß des laserinduzierten Plasmas beim Schweißen mit CO_2-Lasern. Dissertation, TH-Darmstadt
8. M. Eboo, W. M. Steen, J. Clark (1978) Arc-augmented laser welding. Proceedings of 4th Int. Conf. on Advances in Welding Processes, UK, 9–11 May 1978, pp. 257–265
9. D. Petring (2001) Hybrid laser welding. Industrial Laser Solutions, December 2001, pp. 12–16
10. H. Lembeck (2002) Laser-Hybrid-Schweißen im Schiffbau. Proceedings Aachener Kolloquium für Lasertechnik 2002, Aachen, Germany, September 2002, pp. 177–192
11. U. Jasnau, J. Hoffmann, P. Seyffarth, R. Reipa, G. Milbradt. (2003) Laser-MSG-Hybridschweißen im Schiffbau. Proceedings of European Automotive Laser Application 2003, Bad Nauheim, Germany, January 2003
12. T. Graf, H. Staufer. (2003) Laser-Hybrid-Welding Drives VW Improvements. Welding Journal, January 2003, pp. 42–48
13. D. Petring, C. Fuhrmann. (2003) Hybrid laser welding: laser and arc in concert. The Industrial Laser User, No. 33, December 2003, pp. 34–36
14. D. Petring, C. Fuhrmann, N. Wolf, R. Poprawe. (2007) Progress in Laser-MAG Hybrid Welding of High Strength Steels up to 30 mm Thickness, in Proceedings of 26th International Congress on Applications of Lasers & Electro-Optics ICALEO 2007, OR, USA, pp. 300–307
15. Frost & Sullivan, Market report: "Advances in Automotive Plastics", 11pp, (2007)
16. H. Saechtling. (1998) Kunststoff-Taschenbuch, 27. Ausgabe, Tabelle 5.6, Carl Hanser Verlag, München Wien
17. U.A. Russek, G. Otto, M. Poggel. (2001) Verbindliche Nähte – Automatisiertes Fügen von Kunststoffen mit Hochleistungs-Diodenlasern, Laser Praxis 1/2001, pp. 14–16
18. F.G Bachmann, U.A. Russek. (2002) Laser welding of polymers using high power diode lasers, Proceedings Photonics West
19. J.W. Chen. (2000) Mit der Maske in die Mikrowelt – Neues Laserschweißverfahren für Kunststoffe, Zeitschrift TAE-Aktuell, S. 2–4, Heft 12
20. F. Becker. (2003) Einsatz des Laserdurchstrahlschweißens zum Fügen von Thermoplasten, Dissertation Universität Paderborn, Institut für Kunststofftechnik

21. U.-A. Russek. (2003) Simultaneous Laser Beam Welding of Thermoplastics – Innovations and Challenges, ICALEO 2003, Jacksonville, Florida, USA, Paper ID 604, October 13th–16th
22. R. Klein, R. Poprawe, M. Wehner. (1987) Thermal Processing of Plastics by Laser Radiation; Proceedings Laser 87, Springer, Heidelberg
23. D. Hänsch, H. Pütz. (1998) Treusch: Harte und weiche Kunststoffe mit Diodenlaser verbinden, Kunststoffe 88, Carl Hanser Verlag
24. U.A. Russek, A. Palmen, H. Staub, J. Pöhler, C. Wenzlau, G. Otto, M. Poggel, A. Koeppe, H. Kind. (2003) Laser beam welding of thermoplastics, Proceedings Photonics West
25. U.A. Russek. (2003) Innovative Trends in Laser Beam Welding of Thermoplastics, Proceedings of the Second International WLT-Conference on Lasers in Manufacturing, pp. 105–112, Munich, Germany
26. P.A. Atanasov. (1995) Laser welding of plastics – theory and experiment, Optical Eng. 34/10, pp. 2976–2980
27. G. Menges, E. Haberstroh, W. Michaeli, E. Schmachtenberg. (2002) Werkstoffkunde Kunststoffe, 5. Auflage, Kapitel. 12, Carl Hanser Verlag, München Wien
28. C.J. Nonhof. (1994) Laser welding of polymers, Polymer Eng. SCI 34/20, pp. 1547–1549
29. W.W. Duley, R.E. Mueller. (1992) CO_2 laser welding of polymers, Polymer Engineering and Science, Mid-May 1992, Vol. 32, No. 9, pp. 582–585
30. M. Sieffert. (2003) Farbstoffe und Pigmente- von schwarz bis weiß zu kunterbunt. Aachener Laser Seminare: Aachen
31. U.A. Russek. (2006) Prozesstechnische Aspekte des Laserdurchstrahlschweißens von Thermoplasten. Dissertation RWTH Aachen University, Shaker Verlag, Aachen
32. A. Boglea, A. Olowinsky, A. Gillner. (2007) Fibre laser welding for packaging of disposable polymeric microfluidic-biochips, Journal of Applied Surface Science, Vol. 254, pp. 1174–1178
33. A. Boglea, A. Olowinsky, A.Gillner. (2007) TWIST – a new method for the micro-welding of polymers with fibre lasers, in the proceedings of the ICALEO, October 29–November 1, 2007, OR, USA, pp. 136–142
34. A. Roesner, P. Abels, A. Olowinsky, N. Matsuo, A. Hino. (2008) Absorber-free Laser Beam Welding of Transparent Thermoplastics, ICALEO 2008, Temecula, California, USA, Paper ID 303
35. Kern W. (1993) Handbook of Semiconductor Wafer Cleaning Technology: Conventional RCA-Type Hydrogen Peroxide Mixtures, p. 19, Noyes Publications, New Jersey
36. Lasky J.B. (1986) Wafer bonding for silicon-on-insulator technologies, Applied Physics Letters, Vol. 48, pp. 78–80
37. M. Shimbo, K. Fukukawa, K. Fukuda, K. Tanzawa. (1986) Silicon-to-silicon direct bonding method, Journal of Applied Physics, Vol. 60, pp. 2987–2989
38. M.A. Schmidt. (1998) Wafer-to-Wafer Bonding for Microstructure Formation, Proceedings of the IEEE, Vol. 86, No. 8, pp. 1575–1585
39. M. Wiegand. (2001) issertation, Auswirkungen einer Plasmabehandlung auf die Eigenschaften des Niedertemperatur-Waferbondens monokristalliner Siliziumoberflächen, Mathematisch-Naturwissenschaftlich-Technischen Fakultät der Martin-Luther-Universität Halle-Wittenberg, Der Andere Verlag
40. V. Dragoi, M. Alexe, M. Reiche, U. Gösele. (1999) Low temperature direct wafer bonding of silicon using a glass intermediate layer, Proceedings of the 22nd Annual Conference on Semiconductors (CAS'99) 2, pp. 443–446, IEEE Cat. No. 99TH8389, Sinaia, Romania
41. R.F. Wolffenbuttel. (1994) Low-temperature Silicon Wafer-to-wafer Bonding Using Gold at Eutectic Temperature, Sensors and Actuators A, Vol. 43, pp. 223–229
42. R. Knechtel. (2005) Dissertation, Halbleiterwaferbondverbindungen mittels strukturierter Glaszwischenschichten zur Verkapselung oberflächenmikromechanischer Sensoren auf Waferebene, TU Chemnitz, Fakultät für Elektrotechnik und Informationstechnik, Verlag Dr. Hut München
43. Q.-Y. Tong, U. Gösele. (1996) A model of low-temperature wafer bonding and its application, Journal of Electrochemical Society, vol. 143, p. 1773

44. M. J. Wild. (2002) Dissertation RWTH Aachen, Lokal selektives Bonden von Silizium und Glas mit Laser, Shaker Verlag, Aachen
45. F. Sari, A. Gillner, et al.: Advances in selective Laser Radiation Bonding of Silicon and Glass for Microsystems, in Proceedings of the Third International WLT-Conference on Lasers in Manufacturing 2005, München, AT-Fachverlag Stuttgart, 791–796
46. M.J. Wild, A. Gillner, R. Poprawe. (2001) Locally selective bonding of silicon and glass with laser. Sensors and Actuators A, Vol. 93, pp. 63–69
47. M.J. Wild, A. Gillner, R. Poprawe. (2001) Advances in Silicon to Glass Bonding with Laser. Proceedings of the SPIE 4407, pp. 135–141
48. F. Sari, M. Wiemer, M. Bernasch, J. Bagdahn. (2008) Laser Transmission Bonding of Silicon-to-Silicon and Silicon-to-Glass for Wafer-Level-Packaging and Microsystems, The Electrochemical Society ECS, Trans, Vol. 16, No. 8, p. 561
49. F. Sari, W.-M. Hoffmann, E. Haberstroh and R. Poprawe, Applications of laser transmission processes for the joining of plastics, silicon and glass micro parts, Microsystem Technologies, Vol. 14, Issue 12, 1879–1886, 2008
50. J.-S. Park, A.A. Tseng. (2004) Transmission laser bonding of glass with silicon wafer in Proceedings of 2004 Japan-USA Symposium on flexible Automation, Paper No. UL-073, American Society of Mechanical Engineers, New York
51. J-S. Park, A.A. Tseng. (2005) Development and characterization of transmission laser bonding technique in Proceedings of IMAPS Int. Conf. Exhibition Device Packaging, Paper No. TA15, Int. Microelectronics and Packaging Society
52. J-S. Park, A.A. Tseng. (2006) Line bonding of wafers using transmission laser bonding technique for microsystem packaging in ITherm 2006 Proceedings, IEEE, Thermal and Thermomechanical Phenomena in Electronics Systems, pp. 1358–1364
53. A.A. Tseng, J.-S. Park. (2006) Using Transmission Laser Bonding Technique for Line Bonding in Microsystem Packaging in IEEE Transactions on Electronics Packaging Manufacturing, Vol. 29, No 4, pp. 308–318
54. A.A. Tseng, J.-S. Park. (2006) Mechanical strength and interface characteristics of transmission laser bonding for wafer-level packaging, IEEE Transactions on Electronics Packaging Manufacturing, Vol. 29, No. 3, pp. 191–201
55. A.A. Tseng, J.-S. Park. (2006) Effects of surface roughness and oxide layer on wafer bonding strength using transmission laser bonding technique in ITherm 2006 Proceedings, IEEE, Thermal and Thermomechanical Phenomena in Electronics Systems, pp. 1349–1357
56. A.A. Tseng, J.-S. Park, G.-P. Vakanas, H. Wu, M. Raudensky, T.P. Chen. (2007) Influences of interface oxidation on transmission laser bonding of wafers for microsystem packaging in Microsystem Technologies, Vol. 13, No. 1, pp. 49–59
57. A.A. Tseng, J.-S. Park. (2006) Effects of surface roughness and contact pressure on wafer bonding strength using transmission laser bonding technique, Journal of Microlithography, Microfabrication, and Microsystems – October–December 2006 – Vol. 5, No. 4, 043013, 11pp
58. A.W.Y. Tan, F.E.H. Tay. (2005) Localized laser assisted eutectic bonding of quartz and silicon by Nd:YAG pulsed-laser in Sensors and Actuators A: Physical, Vol. 120, No. 2, 17 May 2005, pp. 550–561
59. A.W.Y. Tan, F.E.H. Tay, J. Zhang. (2006) Characterization of localized laser assisted eutectic bonds in Sensors and Actuators A: Physical, Vol. 125, No. 2, 10 January 2006, pp. 573–585
60. U.M. Mescheder, M. Alavi, K. Hiltmann, Ch. Lietza, Ch. Nachtigall, H. Sandmaier. (2002) Local laser bonding for low temperature budget in Sensors and Actuators A: Physical, Vols. 97–98, No. 1, April 2002, pp. 422–427
61. U. Mescheder, M. Alavi, K. Hiltmann, Ch. Lizeau, Ch. Nachtigall, H. Sandmaier. (2001) Local Laser Bonding for Low Temperature Budget, Digest of Technical Papers of the Transducers '01, Eurosensors XV, pp. 620–623, Munich, Germany, June 2001
62. S. Theppakuttai, D. Shao, S.C. Chen. (2004) Localized laser transmission bonding for microsystem fabrication and packaging, Journal of Manufacturing Processes, Vol. 6, No. 1, pp. 24–31

63. J. Bagdahn. (2001) Festigkeit und Lebensdauer direkt gebondeter Siliziumwafer unter mechanischer Belastung: Der Micro-Chevron-Test (MC-Test), p. 58, Dissertation, VDI Fortschrittsberichte, VDI Reihe 9, Nr. 334, VDI Verlag Düsseldorf
64. DIN8505. (1979) "Löten; Allgemeines, Begriffe", Teil 1, Normausschuss Schweißtechnik (NAS) im DIN Deutsches Institut für Normung e.V.
65. DIN8505. (1979) "Löten – Einteilung der Verfahren, Begriffe", Teil 2, Normausschuss Schweißtechnik (NAS) im DIN Deutsches Institut für Normung e.V.
66. R.J. Klein Wassink. (1991) "Weichlöten in der Elektronik", 2. Auflage, Eugen G. Leuze Verlag, (ISBN: 3874800660)
67. L. Bosse, A. Schildecker, A. Gillner, R. Poprawe. (2002) High quality laser beam soldering, Journal of Microsystem Technologies, Vol. 7
68. L. Bosse, A. Koglin, A. Olowinsky, V. Kolauch, M. Nover. (2003) "Laser Beam Soldering – An Attractive Alternative to Conventional Soldering Technologies", Laser and Applications in Science and Technology, Proceedings of SPIE, San Jose, USA
69. U. Dilthey. (2000) Laserstrahlschweißen: Prozesse, Werkstoffe, Fertigung und Prüfung; DVS-Verlag Düsseldorf
70. L. Dorn. (1992) Schweißen und Löten mit Festkörperlasern; Springer Verlag
71. M. Glasmacher. (1998) Mikroschweißen mit Laserstrahlung, Meisenbach Verlag Bamberg
72. M. Beck. (1996) Modellierung des Lasertiefschweißens; B.G. Teubnerverlag, Stuttgart

Chapter 15
Ablation

Arnold Gillner and Alexander Horn

15.1 Micro- and Nanostructuring

15.1.1 Introduction

The significance of microtechnology has increased seriously within the last years. Recent studies forecast a market volume of 25 bn$ in 2009.[1] Microproducts open up new applications in several industrial and consumer areas. Mobile phones, PDAs, laptops, digital cameras would not have been possible with a significant decrease of the size of the components and also higher integration. Other fields of miniaturized products can be found in sensor industry, medical devices, and various products from automotive industry. Production and machining of micro parts were in the past mainly made by technologies developed from the electronic industry, which are particularly based on silicon etching technologies for the production of, e.g., sensor elements. Due to the increasing demand for microproducts in other production areas like the medical, automotive, optical, or chemical-industry, suitable processes for machining parts from non-silicon materials become more and more important. Due to their advantages, laser processes have been established within this growing field of microtechnology. The availability of new processes and beam sources qualifies the laser as a universal tool in this area.

Laser technology has been qualified for microtechnology because of its high lateral resolution by minimized focusability down to a few micrometers, low heat input, and high flexibility. Some examples for laser processes are micro welding, soldering, selective bonding of silicon and glass, micro structuring, and laser-assisted forming. Especially for micro machining, laser processes have been qualified for a wide range of materials starting from semiconductors in the field of microelectronics, hard materials like tungsten carbide for tool technology to very weak and soft materials like polymers for medical products. Even ceramics, glass,

---

A. Gillner (✉)
Mikrotechnik, Fraunhofer-Institut for Laser Technology, 52074 Aachen, Germany
e-mail: arnold.gillner@ilt.fraunhofer.de

[1] NEXUS Market Analysis for MEMS and Microsystems III, 2005–2009.

and diamond can be processed with laser technologies with accuracies less than 10 μm. In comparison to the classical technologies laser processes are generally used for small and medium lot sizes but with strongly increased material and geometry variability.

Micro ablation by means of laser-based vaporization of materials has been used in a large variety of applications to produce micromolds, functional components in electronics, and micro drilling applications. Most of these processes were focused on small parts, where the laser has its unique position in selectivity and low thermal input without changing properties of the entire part. Recent developments in solar cell manufacturing, display manufacturing, and the manufacturing of products based on electrical conducting polymers such as OLEDs have opened a new field on laser micro manufacturing with laser radiation, where the specific advantages and outstanding properties of laser radiation are highly required. Laser micro ablation and surface functionalization with

- micrometer and submicrometer scale accuracy in ablation depth
- nanometer scale in depth heat input
- micrometer scale lateral accuracy
- submicrometer scale in lateral dimensions with special process techniques

on almost all types of materials allow new manufacturing processes, which cannot be met by conventional processes [1]. In this way laser ablation and surface functionalization have already been proven as non-convertible tools in solar cell and display manufacturing.

Micro- and nanostructuring can be subdivided into four working regimes given by the precision Δ:

- conventional structuring ($\Delta > 8$ mm);
- precision structuring ($0.8\,\mu$m $< \Delta < 8$ mm);
- ultra-precision structuring ($0.1\,\mu$m $< \Delta < 8$ mm);
- nanostructuring ($\Delta < 0.1\,\mu$m).

Structuring needs a working tool, e.g., laser radiation or a threading die, and a positioning stage to move the working tool relative to the substrate. Micro- and nanostructures are generated by interaction of laser radiation with matter through modification or ablation of the matter.

15.1.2 General Aspects for Laser Ablation

The precision of a structure is described by the contour fidelity and is reflected by the overall precision of micro- and nanostructuring given by the sum of the precisions of

- the working tool (geometry of the laser beam);
- the process itself (laser–material interaction); and
- the positioning system for part and laser beam.

The precision of positioning stages is independent of the working tool and the involved processes during structuring and can reach today 1 nm at very small positioning velocities. In order to assure large productivity, large velocities up to 1 m/s are requested and limits today the precision of positioning to 100–300 nm.

Laser radiation is used as a working tool to structure matter by ablation defining the lateral precision. For this, laser radiation is projected by objectives onto the surface, thereby being formed to a specific spatial intensity distribution. Depending on the intensity, the absorption conditions of the materials and the interaction time, several aspects have to be considered, which define the ablation geometry and the precision of the ablation process.

15.1.2.1 Absorption Effects on Ablation Precision and Ablation Geometry

The precision of the resulting structures depends not only on the focus diameter and the positioning stage but also on the *precision of the process* resulting from the reactions of the matter during and after irradiation. So, the processing diameter depends not only on the beam radius w_0 but also on the physical properties of the material, like the reflectivity, absorptivity, melting, and evaporation enthalpy. Depending on the optical properties of the material and on the applied radiation, the precision of the process diameter depends on the absorption of the radiation, which can be *linear* or *non-linear*. Additionally, the ablation depends strongly on the *pulse duration* and on the *wavelength*.

After absorption of the optical energy the ablation starts. The laser-induced ablation itself can be distinguished between photo-induced and thermal ablation. The precision is larger for photo-induced ablation being comparable to the precision of the working tool, here resulting in the processing diameter. In case of thermal ablation, additional processes, like heat diffusion, decrease the precision and give rise to a larger processing diameter than the focus diameter [2].

Linear Absorption

Materials absorb laser radiation linearly, when an absorption band at the laser wavelength (color centers) or due free or quasi-free electrons (like metals or graphite) are given. This can be described by Lambert–Beer's law (see Chap. 3)

$$I(x) = (1 - R)I_0 e^{-\alpha x} \tag{15.1}$$

with R the reflectivity, α the absorption coefficient [cm^{-1}], and I_0 the threshold intensity for ablation [W/cm^2]. Based on this a logarithmic dependence of the ablation depth h [cm] on the intensity can be found

$$h = \frac{1}{\alpha} \ln\left(\frac{I}{I_0}\right) \tag{15.2}$$

The ablation depth defines the precision of the structures in depth.

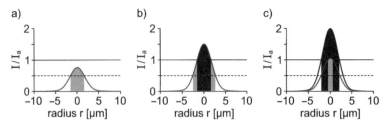

Fig. 15.1 Extension of the melted (*red*) and ablated (*blue*) region for intensities smaller (**a**) and larger (**b**) then I_a. For $I \approx I_a$ and $I > I_a$ demonstrates differences in the extension of the ablated region (**c**)

Applying Gaussian radiation with a focus diameter $2w_0$ the processing diameter can be controlled by the intensity of the radiation depending on the ratio of the applied intensity to the threshold intensity for ablation. Threshold intensities are given for homogeneous material by the threshold intensity for melting I_m and for evaporation I_v (Fig. 15.1). Solid materials feature a melt enthalpy being always smaller than the evaporation enthalpy, resulting in larger processing diameters for melting than for evaporation (Fig. 15.1b for $I_m = I_v/2$).

The processing diameter of focused Gaussian radiation can be reduced steadily by decreasing the intensity close to the threshold intensity for ablation I_a. Processing diameters much smaller than the focal diameter $2w_0$ are achievable (Fig. 15.1c).

Thermal and/or photo-physical processes govern ablation. Thermal processes are represented by heating of matter without change in the chemical constitution, whereas photo-physical processes imply changes in the chemical constitution or irradiated matter.

Depending on the absorptivity of matter, volume or surface absorption is distinguished. For example, BK7 glass exhibits for the fundamental radiation of Nd:YAG lasers ($\lambda = 1{,}064$ nm: photon energy $E_\gamma \approx 1$ eV) an absorptivity $< 5\%$. The laser radiation is absorbed in a volume (see Sect. 4.3.4) and induces in the example of glass to an uncontrolled ablation. Converting the IR radiation into the UV regime, e.g., $\lambda = 266$ nm exhibits an absorptivity α for BK7 glass $> 80\%$. Laser radiation is now absorbed in a very small depth compared to the processing diameter (see Sect. 4.3.3). Precise structuring of glass is possible.

Multi-photon Absorption

Matter featuring absorption bands in the deep-UV (< 250 nm, $E_{band} > 3$ eV) do not absorb laser radiation at moderate intensities $< 10^{10}$ W/cm^2 in the VIS and IR wavelength regime. The probability for one-photon absorption of this radiation can be neglected. Increasing the photon density per unit time and volume, equivalent to increasing intensity, makes multi-photon absorption probable. In case of multi-photon absorption two or more photons of the VIS–IR wavelength regime are

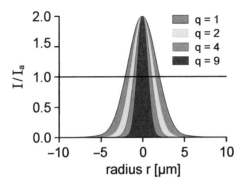

Fig. 15.2 Spatial intensity distribution of the radiation for different multi-photon coefficients q

absorbed contemporary, representing a one-photon absorption of a deep-UV photon with the same energy.

The precision of the structuring can be increased by using Gaussian radiation applying a q-photon absorption, q being the multi-photon coefficient, reducing the processing diameter by $q^{0.5}$ to w_{mp} (Fig. 15.2):

$$w_{mp} = \frac{w_0}{\sqrt{q}} \qquad (15.3)$$

15.1.2.2 Thermal Effects on Ablation Precision and Ablation Geometry

Today for precise laser ablation single nanosecond pulses with time spacings between the pulses in the range of 10 μs are used. Ablation rates between 0.1 and 1 mm³/min are achieved with depth accuracies of 1–2 μm. In the nanosecond range the thermal influence of the laser irradiation results in typical melting depths of several micrometers and a surface quality, which is controlled and defined by the melt resolidification and the surface tension of the melt. With this technology, ablation accuracies of typically 1 μm can be achieved and a surface roughness of 0.6 μm can be obtained. Recent investigations using ps lasers showed even higher accuracy, which is due to a totally different laser interaction regime with significant differing thermal influence and resulting quality. Generally the interaction of photons and matter is based on the absorption of the electromagnetic energy at the free and bound electrons of a material. The electrons are heated and transfer their energy after a characteristic time to the lattice of the material. The temperature increase and thermal behavior are then described by a two-temperature model which takes into account that laser–material interaction takes place generally between photons and electrons with an overheated electron gas and an energy transfer to the lattice by a material specific coupling coefficient (Fig. 15.3).

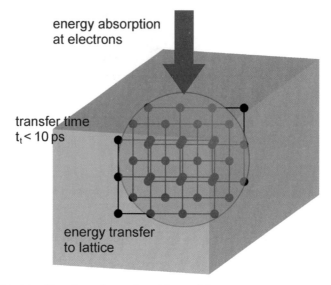

Fig. 15.3 Principle of laser beam interaction with materials

The effect of the absorbed optical energy on thermal ablation of matter is driven by the pulse duration and is subdivided into four regimes for linear absorbing matter:

- absorption of the optical energy by quasi-free electrons ($t_{\gamma e} < 10\,\text{fs}$);
- thermalization of the electrons called electron system ($t_{ee} < 100\,\text{fs}$);
- interaction between the electron and the phonon system ($t_{ep} < 10\,\text{ps}$);
- thermalization of the phonon system ($t_{pp} < 100\,\text{ps}$).

The durations of the single processes within the brackets are exemplary given for copper. Two limiting cases can be distinguished.

(I) Pulse duration larger than thermalization of phonons $\tau > t_{pp}$: For $\tau > t_{pp}$ the times for absorption of the photons, thermalization of electron, and phonon system are much smaller than the pulse duration resulting in the description of heating by one temperature for the electron and the phonon system. Ablation concerning the subsystems is an instantaneous process. The threshold fluence for ablation (fluence = energy per area) scales with the square root of the pulse duration and a thermal penetration depth can be defined as

$$\delta_{\text{therm}} = 2\sqrt{\frac{\kappa t_p}{c_p \rho}} \qquad (15.4)$$

depending on the thermal property of matter with κ the thermal conductivity, c_p the heat capacity, ρ the density, and t_p the pulse duration of the applied radiation. The thermal penetration depth δ defines the region beyond the focus diameter (tool diameter), which can be thermally modified, like amorphization of crystalline substrate.

This regime is called heat-affected zone (HAZ). The processing diameter is given by $2(w_0 + \delta)$.

(II) Pulse duration smaller electron–phonon relaxation time $\tau < t_{ep}$: For $\tau < t_{ep}$ the processes within the electron and the phonon system are decoupled. The temperature development for these systems are described by a two-temperature model representing two coupled differential equations:

$$C_e \frac{dT_e}{dt} = \frac{\partial}{\partial z}\left(\kappa_e \frac{\partial T_e}{\partial z}\right) + S - \mu\left(T_e - T_p\right)$$
$$C_p \frac{dT_p}{dt} = \mu\left(T_e - T_p\right) \tag{15.5}$$

C_e and C_p represent the heat capacities of the electron and the phonon system, κ the heat conductivity of the electron system, S the applied optical energy, μ the electron–phonon coupling constant, and T_e and T_p the temperatures of both systems. Ablation of materials is characterized by negligible melt and small mechanical load of the irradiated region. The processing diameter is comparable to the focus diameter.

In nanosecond laser processing this energy transfer happens during the duration of the laser pulse, whereas in picosecond laser processing the energy transfer happens after a certain interaction time. For metals, this transfer time is generally in the range of some picoseconds. In this case, the material is heated up after the end of the laser pulse, so that there is no interaction of the photons with melt and evaporated material. The result is a much more accurate ablation because the ablation is mainly due to vaporization of the material and not by melt expulsion. The second reason for the use of picosecond pulse durations in laser ablation is the very high intensity of the pulses. With peak intensities of more than 10^{10} W/cm² all materials are vaporized rather than melted. As a consequence of these high intensities, the ablation rate per pulse is quite low, because for vaporization an exceeding amount of energy is needed. In Fig. 15.4 a direct comparison of laser ablated steel with nanosecond pulses and picosecond pulses is shown. It clearly can be seen that in nanosecond ablation the residual melt after ablation is much stronger and the geometry is affected by melt resolidification. In picosecond ablation almost no melt can

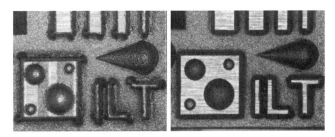

Fig. 15.4 Comparison of nanosecond laser ablation (*left*) and picosecond laser ablation (*right*) of steel

be found in the ablation area and clean surfaces with surface roughness $< 0.5\,\mu m$ can be produced by laser ablation.

15.1.3 Process Principles for Laser Ablation

Micro- and nanostructuring with laser radiation adopts one of the following techniques:

1. mask technique;
2. interferometric technique; and
3. scribing technique.

Mask and interferometric technique are appropriate for $2^1/_2$ dimensional structuring, for example, applied in lithography and generation of micro fluidic systems and nanoscaled optical devices like gratings. The scribing technique is adopted for full three-dimensional structuring and is comparable to mechanical milling.

15.1.3.1 Mask Technique

Laser radiation is formed and homogenized by a telescope and projected onto a mask representing the desired intensity distribution (e.g., a circular top-hat distribution, Fig. 15.5) [3, 4]. The mask is imaged by an objective to the final intensity distribution. The transversal miniaturization m is given by

$$|m| = \frac{f}{g-f} \tag{15.6}$$

a relation known from the geometrical optical, with the focal length of the lens f and the distance of the mask to the lens g (Fig. 15.5).

The precision of the working tool is described by Abbe's law, which gives the theoretical resolution limit Δx of an objective for non-coherent radiation

$$\Delta x = 1.22 \frac{\lambda f}{D_L} = 0.61 \frac{\lambda}{NA} \tag{15.7}$$

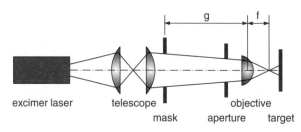

Fig. 15.5 Principle of mask projection imaging. Laser radiation is collimated and homogenized by a telescope and a homogenizer to illuminate the mask. The mask itself is imaged on the target

and depends on the objective diameter D_L and on the numerical aperture of the objective NA.

15.1.3.2 Interferometric Techniques

Due to the diffraction limit of focused laser beams direct nanostructuring of materials requires special setups or special processing approaches. Lithographic methods using special masks with dielectric phase shifting masks and non-centric illumination today allow structure sizes of 50–70 nm. Since those technologies require large machines and expensive equipments, lithographic-based nanostructuring is not suitable for the patterning of implants and consumer products with sophisticated surface functionalities. Nanostructuring by focused laser beams is only possible with non-linear absorption effects or interferometric techniques. Several optical approaches have been used to produce nanostructures in the size of less than 100 nm. Since this a serial technique with a scanned laser beam, the processing speed is quite low with scribing rates of only up to 0.01 mm^2/s at a laser repetition rate of 1 MHz.

For larger processing areas and higher processing speeds a new method for nanostructuring is required, which can be used also on cheap products and which is applicable in rough industrial surroundings. A new approach using interference-based laser structuring provides the ability for manufacturing sub-100 nm structure sizes [5]. This technology is based on splitting the laser beam into several single beams and combining them on the surface of the product. Here interference patterns are generated, which provide structure sizes from 50 to 300 nm. In Fig. 15.6 the principle setup of an interference ablation system is shown. For the optical setup, different laser sources can be used, depending on the required structuring geometry. Using a frequency tripled Nd:Vanadate laser at 355 nm allows the generation of

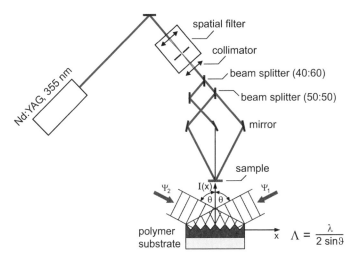

Fig. 15.6 Principle of interference patterning. Laser radiation is divided into several beams and when combined on the substrate surface form interferometric pattern

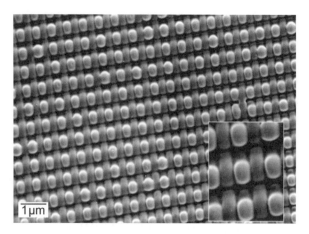

Fig. 15.7 Two-dimensional nanopatterning with multi-exposure

structures in the range of 100 nm at areas of several 100 μm at one laser pulse. For refining the laser beam a special filter is used and the laser is split into two separate beams by dielectric mirrors.

Nanostructures with an extreme surface to volume ratio show a particularly high potential, since over the manufacturing processes both chemical and structural characteristics can be combined. Thus super-hydrophobic surface properties can be adjusted in the sense of a synthetic lotus effect by a locally selective structuring of this surface in the submicrometer or nanometer range, e.g., on certain polymers. On medical-dosing equipment the adhesion of fluids can be avoided and thus an accurate dosage can be guaranteed with medical dosing assistance. In biological chips for medical analytics the flow characteristics and capillary effects can be functionally influenced.

With this setup nanoscaled structures with dimensions between 100 nm and 1 μm can be produced. Since with only two beams linear scales are made, more complex structures must be generated by moving the part in a definitive manner. Using a cross-wise ablation single spots with dimensions of 300 nm can be generated. In Fig. 15.7 an example for this two-dimensional approach is shown.

Using nanostructured surfaces specifically adapted functionalities, for example, for optical functions can be produced. In Fig. 15.8 an example for nanostructured polymer for the generation of multi-color rings only by diffraction methods with white light irradiation is shown.

Scribing Technique

Scribing is a common technique for structuring with laser radiation. Similar to milling, the working tool "laser radiation" is moved relatively to the substrate and removes matter in the focal regime by melting, vaporization, or photo-ablation.

The spatial intensity distribution of the laser radiation depends among others on the applied laser resonator. Of special importance for microstructuring concerning

15 Ablation

Fig. 15.8 Multi-color diffraction patterning by nanostructured surfaces

focusability is the Gaussian radiation, exhibiting a spatial Gaussian intensity distribution. Micro-structuring applying this radiation benefits of the property of Gaussian radiation to be focused to the smallest possible value, called diffraction-limited focus, resulting in a beam radius defined by equation 15.8.

$$w_0 \approx \frac{2\lambda f}{D} \approx \frac{\lambda}{\text{NA}} \qquad (15.8)$$

with λ the wavelength of the applied radiation, f the focal length of the objective, and D the beam diameter of the radiation close to the objective. The working tool diameter is given by $2w_0$ and defines the working tool precision. The numerical aperture is given by $\text{NA} = D/f$ for objectives with large focal length ($f \gg 15\,\text{mm}$) and by $\text{NA} = n \sin\theta$ for microscope objectives ($f < 15\,\text{mm}$), with the refractive index of the medium between the objective and the substrate and θ the aperture angle of the objective.

Microstructures, like a cavity, are generated by ablation with laser radiation displacing the laser radiation relative to the workpiece and carrying out a meander trajectory removing the material in layers (Fig. 15.9). In general, the laser displacement of the laser beam is realized by high-speed galvanometer mirrors. Focusing of the laser beam is made either by moving lenses before the scanning mirrors or by telecentric lenses.

The ablation rate per length depends on the ratio between the applied intensity to the threshold intensity for ablation and on the overlap o

$$o = 1 - \frac{v}{2 f_p w_p} \qquad (15.9)$$

where w_p represents the processing diameter, v the relative velocity, and f_p the pulse repetition rate. The processing diameter for mechanical ablation (e.g., milling) is given by the tool diameter, whereas the processing diameter for laser ablation is

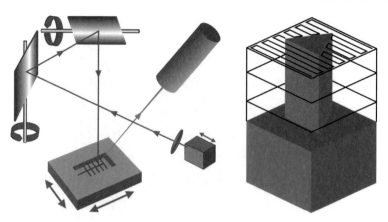

Fig. 15.9 Principle of three-dimensional laser machining by laser scribing technology

given by the focus diameter (tool diameter) and the process precision, described below. Applying laser radiation for micro- and nanostructuring the appropriate tool diameter is obtained by reducing the wavelength and increasing the numerical aperture. The last one is obtained by decreasing the focal length of the objective or by increasing the beam diameter close to the objective (Example: $\lambda = 355$ nm; NA $= 0.5 \Rightarrow w_0 \approx 0.7$ μm).

Microscope objectives are often used for focusing laser radiation to small focus diameters < 10 μm. The focus diameter is varied during processing by changing the objective or by changing the beam diameter in front of the objective or by positioning the laser radiation extra focally. Different to mechanical milling, laser radiation is contact free and massless resulting in reduced delay times for positioning of the working tool. Mechanical micro-milling with tool diameter < 100 μm exhibits positioning times in the range of minutes, whereas laser radiation is positioned within fraction of seconds [6, 7].

15.1.4 Examples

In the production of micro-scaled products and products with micro- and nanoscaled surface functionalities, laser ablation becomes a more and more important tool which is able to generate structure sizes in the range of 10–100 μm and with new machining strategies even in range smaller than one micrometer. Using ultrashort-pulsed lasers with pulse durations of 10 ps in pulse bursts of several pulses with a time spacing of 20 ns each and adapted pulse energies, the surface quality of metal micro ablation has been increased significantly. With a combination of ultrashort pulses and high-resolution interference methods, structures with dimensions of less than 300 nm can be generated either in polymer parts directly or in steel replication tools. For mass replication of those structures to achieve improvements in wetting capabilities or optical properties of surfaces, a new laser-supported embossing technology has been developed.

The availability of these new process variants and improved beam sources qualifies the laser as a universal tool especially in the area of micromold processing. The wide range of materials processed by lasers spread out from hard materials like tungsten carbide for tool technology and even ceramics, glass, and diamond with accuracies less than 10 μm, which make the technology useful for the processing of innovative tools, also for high temperature mass replication processes.

With the newly developed laser ablation technology using ultrashort-pulsed lasers in the picosecond range, accuracies of less than 5 μm and surface roughnesses less than 0.5 μm can be provided for the manufacturing of micromolding tools (Fig. 15.10).

This new laser ablation technology opens a new field of microprocessing of tools, which allows the manufacturing of micro- and even nanoscaled functional structures on polymer, metal, and glass components. With laser ablation using picosecond lasers a machining technology with accuracies < 1 μm is available which can be applied to all kinds of materials. In Fig. 15.11 the surface of a replication tool is shown, whereby laser ablation in tungsten carbide micro pits with sizes between 1 and 5 μm have been produced [8, 9]. Due to the process characteristics of laser ablation submicrometer-scaled substructures are produced, which further increase the surface area. Tools like this have been successfully tested for the replication of polymers to achieve functional surfaces (Fig. 15.11). In Fig. 15.11 the result of a wetting test shows that by microstructuring a super-hydrophobic effect can be produced [10, 11].

By mask technique ablation with UV-Excimer lasers especially polymer parts can be produced in a very flexible way. Among others micro fluidic systems require very small channels and reservoirs in the geometry range of several tens of micrometers. With mask-based excimer ablation a tool for rapid prototyping is available for manufacturing of single parts up to small lot sizes (Fig. 15.12) [12].

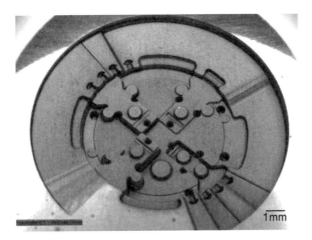

Fig. 15.10 Laser-manufactured micro injection molding tool insert

Fig. 15.11 Microstructured replication tool for self-cleaning surfaces (*left*). Super lotus effect on microstructured polymer surface (*right*)

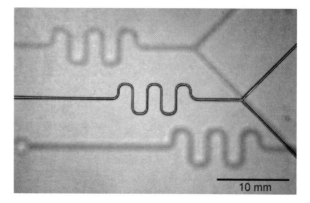

Fig. 15.12 Micro fluidic channels ablated with excimer laser radiation

15.2 Cleaning

Carsten Johnigk

In the last years besides conventional chemical, mechanical, or thermal cleaning techniques, laser cleaning has become increasingly important. The reasons are advantages in process engineering (e.g., contactless, careful and damage free, high precision, low-heat influence of the base material) as well as tightening legislation concerning hygiene observance and waste and/or pollutant avoidance [13, 14].

Laser cleaning offers a new range of opportunities and processing techniques for coating and surface layer removal that can often overcome current processing

---

C. Johnigk (✉)
Oberflächentechnik, Fraunhofer-Institut für Lasertechnik, 5207 Aachen, Germany
e-mail: carsten.johnigk@ilt.fraunhofer.de

limitations. This chapter discusses the basics of laser cleaning, e.g., laser–material interaction, laser characteristics, and processing strategies for laser cleaning. A wide field of application of laser cleaning from microtechnology up to the cleaning of large buildings and constructions as well as new developments like mobile laser cleaning systems and specific applications are presented [15, 16].

15.2.1 Basics of Cleaning with Laser Radiation

Cleaning/removal is based on locally confined, contactless interaction of pulsed or continuous laser radiation with the surface layer. Depending on the surface layer composition, thickness, and the processing parameters, different cleaning mechanisms can be distinguished [17, 18].

The most important mechanisms are the following:

- ablation by evaporation or decomposition of the surface layer (sublimation) and
- removal through thermally induced stress or by laser beam-induced shock waves.

On technical surface layers, several mechanisms often appear at the same time, whereby the dominant mechanism depends on the material properties and the processing parameters, in particular wavelength, power density, and interaction time.

If the surface layer absorbs the laser radiation of the selected wavelength well and if the underlying base material has a small absorption $A_D > A_G$ (cf. Fig. 15.13), the cleaning process is called "self-limiting." At the beginning of the process a large amount of the incident laser radiation is absorbed and converted into thermal energy. Due to thermal conduction the base material heats up with rising interaction times ($t_2 > t_1$). To avoid damage to the base material interaction time must be short. If the threshold power density for reaching the evaporation temperature of the surface layer is exceeded ($I_L > I_D$), this leads to the evaporation of the surface layer. If the surface layer is completely removed, laser radiation interacts with the base material and is reflected to a large amount, so that the cleaning process terminates and no damage to the base material occurs (Fig. 15.13).

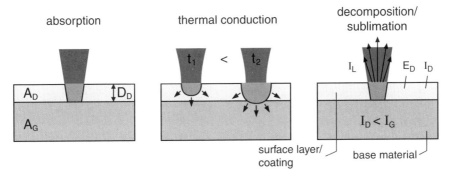

Fig. 15.13 Scheme of the ablation/cleaning process

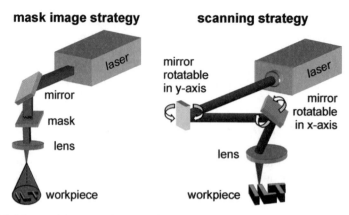

Fig. 15.14 Scheme of the processing strategies

If the wavelength of the laser radiation cannot be selected in such a way that the process is self-limiting, then either the processing parameters have to be adapted accurately or process control and/or regulation measures are necessary, in order to avoid damage to the base material.

Two different processing strategies are used to clean larger areas (Fig. 15.14). When the mask image strategy is used, a mask with the desired pattern is projected onto the surface. With this strategy special patterns can be produced or when using a square-formed mask, large-scale cleaning can be performed. Care must be taken when designing the projection ratio to ensure the power density is exceeded in the desired cleaning area, which may limit the pattern size.

When the alternative scanning strategy is used, the laser beam is deflected with rotatable mirrors (galvanometers) and afterward the beam is focused with a special designed lens (e.g., F-Theta objective) onto the surface. Typical spot diameters are in the range of 0.1–1.0 mm. Typically two perpendicular-oriented galvanometers are integrated into a scan-head; this makes it possible to move the beam in an X-Y plane. Advantages of galvanometers are high-speed movement of the laser beam with speeds up to several meters per second while having a high position accuracy and position repeatability. The dimensions of the X-Y plane (scan field) depend on the focal length of the F-Theta objective. With rising focal length the dimensions of the scan field will increase, but also the laser spot diameter, which leads to a decrease in power density. Therefore, F-Theta objectives with very long focal length which allow scan field dimensions in the range of meters can only be used when less power density is needed, or in combination with laser sources with high-peak output power or excellent beam quality.

15.2.2 Example Applications for Laser Cleaning

Microtechnical applications: With increasing miniaturization and integration density of many microtechnical products the cleaning methods must fulfill higher requirements. Tendencies are a reduction of the surface areas to be cleaned into the μm^2 range, the increasing use of sensitive functional surfaces, the rising requirements to the cleaning quality (residual layers and/or particles with smaller dimensions and/or smaller concentrations must be removed), and quality control measures. During production, post-treatment, and/or final assembly of these parts and assembly groups, a large number of cleaning and activation processes are necessary. Solder resist, contamination, residues of plastics, lacquers, and oxide coatings must be removed. In particular for subsequent junction processes (welding, soldering, adhesive bonding) and coating processes (e.g., galvanizing), the appropriate part surfaces must fulfill the necessary purity requirements.

For the cleaning of microtechnical parts mainly short-pulse lasers with wavelengths of $\lambda = 10.6\,\mu m$ (CO_2), $\lambda = 1.06\,\mu m$ (Nd:YAG), $\lambda = 532, 355$ (Nd:YAG frequency-converted) $\lambda = 248, 193\,nm$ (excimer) are used with pulse durations between 5 and 250 ns. Due to the high-power densities and the short interaction times the coatings are removed completely without damaging the base material.

An example from microtechnology is the cleaning of molds, which are used in the microsystem engineering as tools for the production of structured microparts. The molds are manufactured by galvanically filling up the female molds, which are made out of polymers. After the galvanic structure of the mold has been formed the female mold must be removed. This is achieved in a two-step laser cleaning process. In the first step more than 99% of the polymers are removed in a rough cleaning process with a TEA–CO_2- or CW CO_2 laser (cf. Fig. 15.15) at high speed. After this cleaning process, a residual layer with a thickness of less than 1 μm remains on the surface. This layer is removed in a fine cleaning process with an excimer laser. After this two-step cleaning process the polymers are removed completely without damaging the very sensitive surface with its structures.

The alternative cleaning process for removing the polymer by chemical etching leads to the destruction of the microstructures, which are present as single structure

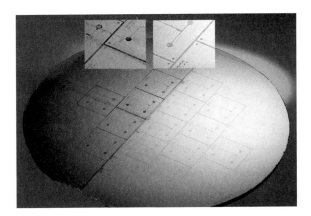

Fig. 15.15 Laser-cleaned mold for microsystem engineering. *Left side* after rough cleaning with a CO_2 laser. *Right side* after fine cleaning with excimer laser

with aspect ratio (height to width) of more than 50 and lateral dimensions of less than 20 μm.

In-line cleaning: A further interesting application for laser cleaning which is getting more and more important is the integration of the process into production lines for mass production. An important industry is the automotive and/or automotive supply industry, with the following potential applications:

- Local pre-treatment of pre-coated materials. The layers consist of KTL, water-based lacquers, powder coatings, or other plastic coatings. Laser cleaning can also be used for three-dimensional parts.
- Removal of grease and oil residues.
- Preparation of adhesive surfaces by removal of oxides and grease.
- Partial lacquer removal from galvanized steel sheets for contacting the electrical mass without damaging the zinc layer.

Other applications are the in-line cleaning of cylinder rolls or belt conveyors. In these applications laser cleaning is carried out simultaneously with the production process. As an example Fig. 15.16 shows the cleaning of a rotogravure cylinder, the residual "old" paint in the engraving must be removed.

The cleaning with excimer laser radiation (wavelength $\lambda = 248$ nm) is suitable due to the high absorption of ultraviolet laser radiation for the removal of thin paint layers in the range of a few micrometers. The chromium layer and the engraving of the rotogravure cylinder are not damaged if suitable laser beam parameters are used.

The beam delivery and shaping system consists of a cylinder lens telescope for beam shaping, reflective and/or transmissive beam homogenizer for the generation of a homogeneous power density distribution as well as zoom optics for adjustment of the square-shaped beam size. The laser beam is directed parallel to the cylinder axis.

Within a linear moving unit a deflection module causes the laser beam to be directed perpendicularly onto the cylinder surface. The linear moving unit can be moved over the complete length of the rotogravure cylinder ("flying optics"). As the rotogravure cylinder rotates the whole surface can be treated.

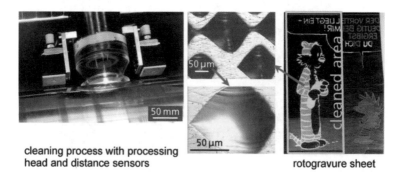

Fig. 15.16 Laser cleaning of a rotogravure sheet

Figure 15.16 shows on the left the zoom optic and exhaust system as well as the surface of the rotogravure cylinder which has to be cleaned, on the right it shows a partly cleaned area of the rotogravure cylinder. The residual "old" paint is completely removed without affecting the engraving.

Cleaning of large buildings and constructions: In addition to the spatially limited cleaning of parts, new applications where large surfaces have to be cleaned get more and more important for laser cleaning. The probably most well-known application is the cleaning of historic monuments. The tasks which have to be solved are very different and show the large potential of the application [19, 20].

Applications are the following

- Cleaning of wood surfaces of lime, gypsum, color, wallpaper, and dirt layers without damaging the original wooden base material.
- Removal of crusts and films from weather-worn stone surfaces, in order to remove stone-damaging deposits and to open the pores.
- Removal of corrosion crusts from bronze surfaces while conserving the natural patina.
- Removal of contamination and fungal attack, e.g., from paper.
- Opening and removal of paint designs, e.g., from wood.

But the previous work also shows that many fundamental investigations are still necessary for the determination of the parameter fields and for investigation of the optical and thermal material properties, in order to fulfill the high requirements when restoring and conserving monuments.

Removal of paint coatings plays a dominant role if large surfaces of technical constructions and buildings have to be cleaned. In many of these applications multilayer coatings with different compositions and layer thicknesses must be removed from sensitive substrates. One of the most examined applications is the paint removal from airplane components, which are submitted to regular maintenance.

When using chemical or blasting processes a large amount of secondary and sometimes hazardous waste results. Laser cleaning is used in particular for composite materials, whereby a selective paint removal down to the primer or a complete removal down to the base material is possible. Examples are the robot-supported paint removal of radomes and engine casings from the company SLCR [21, 22].

In the above applications the careful decoating without damaging the base material is of main interest. In contrast to this goal the paint removal from large steel constructions (e.g., ships, bridges, large tanks and high voltage transmission pylons) requires mobile laser systems, easy and flexible handling, robustness of the components, the complete collection of the waste products for environmental protection reasons, and high cleaning speeds.

In a joint project a mobile laser cleaning system was developed which fulfills all of the above requirements (Fig. 15.17). A Q-switch Nd:YAG laser is used providing pulses with pulse durations in the range of 100–150 ns. Best cleaning results are obtained with peak power densities in the range of 80–110 MW/cm^2. During the last decade of development, the output power of the laser source rose from 100 W to nearly 2000 W which leads to an increase of cleaning speed on the same factor

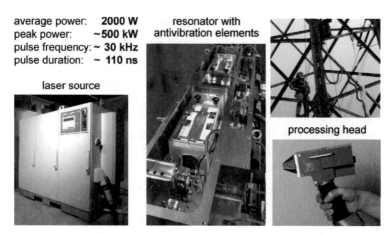

Fig. 15.17 Mobile laser cleaning system

(20). The laser beam is guided within a flexible fiber with a length of 50 m and a core diameter of 400 μm. The fiber connects the laser source with the hand-guided processing head. The processing head with integrated suction nozzle and integrated optical system (collimating optics, galvanometer scanner, and focusing lens) weighs about 2 kg. The galvanometer scanner deflects the laser beam in such a way that the focused beam moves on a line. Perpendicular to this direction the processing head is moved by the operator so that large-scale cleaning with a high cleaning speed can be realized.

This system opens a wide range of new applications especially when coating thicknesses and the dimensions of the surfaces rise. A system with modified optics was successfully tested for the cleaning of railway rails. It works by directing a laser beam at the rail, where it rapidly heats up and ablates any contamination, including leaves and their residues, which may cause low adhesion. The special designed optic aligns and focuses the laser to a 20 mm wide stripe transverse to the longitudinal direction of the rail. The whole device is mounted on a train. For each rail two laser systems with a respective output power of 1000 W are used, providing cleaning speeds of up to 60 km/h.

The future of laser cleaning will mainly depend on further development of the laser source (e.g., increasing output power, more and shorter wavelength, and shorter pulse durations) as well as further component development (e.g., processing heads in the multi-kW range with high processing speeds). This will offer new applications that will stimulate further development [23–25].

References

1. J.-C. Diels, Ultrashort Laser Pulse Phenomena, Academic Press, Boston (1996)
2. J. Jandeleit, A. Horn, R. Weichenhain, E.W. Kreutz, R. Poprawe, Fundamental investigations of micromachining by nano- and picosecond laser radiation, Applied Surface Science 127, 885–891 (1998)

3. G. J. Schmitz, C. Brücker, P. Jacobs, Manufacture of high-aspect-ratio-micro-hair sensor arrays, Journal of Micromechanics and Microengineering (15), Seiten 1904–1910 (2005)
4. M. Wehner, S. Beckemper, P. Jacobs, S. Schillinger, D. Schibur, A. Gillner, Processing of polycarbonate by high-repetition rate ArF excimer laser radiation, Proceedings of Lasers in Manufacturing 3, Seiten 557–561 (2005)
5. F. Korte et al., Sub-diffraction limited structuring of solid targets with femto-second laser pulses, Optics Express 7(2), 41–49 (2000)
6. R. Poprawe, A. Gillner, D. Hoffmann, J. Gottmann, W. Wawers, W. Schulz, High speed high precision ablation from ms to fs, Proc. SPIE 2005, 12 S., 200
7. A. Gillner, J. Holfkamp, C. Hartmann, A. Olowinsky, J. Gedicke, K. Klages, L. Bosse, A. Bayer, Laser applications in microtechnology, Journal of Materials Processing Technology 167, Seiten 494–498 (2005)
8. C. Hartmann, T. Fehr, M. Brajdic, A. Gillner, Investigation on Laser Micro Ablation of Steel Using Short and Ultrashort IR-Multipulses, LAMP 2006, Kyo-to, Japan (2006)
9. C. Hartmann, A. Gillner, U. Aydin, et al., Investigation on laser micro ablation of metals using ns-multi-pulses, Journal of Physics 59, 440–444 (2007)
10. E. Bremus-Köbberling, U. Meier-Mahlo, O. Henkenjohann, S. Beckemper, A. Gillner Laser structuring and modification of polymer surfaces for chemical and medical micro components, Proceedings of SPIE 5662, Seiten 274–279 (2004)
11. E. Bremus-Köbberling A. Gillner, Laser structuring and modification of surfaces for chemical and medical micro components, Proceedings of the 4th International Symposium on Laser Microfabrication LMP 4, Seiten 1–5 (2003)
12. M. Wehner, Excimer Laser Technology, Edt.: D. Basting, Lambda Physik AG Göttingen (2001)
13. R. Lotze, J. Stollenwerk, K. Wissenbach 'Selektives, präzises Abtragen von Deckschichten mit Laserstrahlung auf kompliziert geformten Werkstückgeometrien' in 'Laser in der Materialbearbeitung', Band 12 'Präzises Oberflächenreinigen von technischen und natürlichen Werkstoffen' 'Präzisionsabtragen mit Lasern', VDI-Herstellung und Druck (2000)
14. K. Wissenbach, C. Johnigk, E. Willenborg, Innovative Entwicklungen im Bereich des Laserstrahl-Reinigens – Ein Überblick Proceedings Aachener Kolloquium Lasertechnik, S. 498 (2002)
15. E. Büchter, W. Barkhausen, 'Reinigen und Entschichten mit Laserstrahlung' internet presentation Clean Lasersysteme GmbH, www.cleanlaser.com
16. N. N. 'Laser Cleaning and Surface Preparation' internet presentation, http://www.quantel-laser.com/industrial-scientific-lasers/media/produit/fichier/11_10pindusVA1106.pdf
17. J. Stollenwerk, H.-D. Hoffmann, J. Ortmann, G. Schmidt, K. Wissenbach, R. Poprawe, High Speed Removal and Structuring of Surfaces in the Automotive Industry with a New Diode Pumped Solid State Laser Source, Proceedings of the ICALEO 1999, San Diego, CA (1999)
18. R. Lotze, J. Birkel, K. Wissenbach, Entlacken mit Laserstrahlung – Neue industrielle Anwendungen – JOT, 8, S. 44 (1999)
19. H. Brünninghoff Reinigen von Hochspannungsmasten Konferenzbericht AKL (2002)
20. N. N. internet presentation Palfinger, www.palfinger.com
21. O. Schulz, 'SLCR Lasertechnik – umweltfreundliche Lösungen für die Oberflächenbearbeitung' Internetdarstellung SLCR Lasertechnik, www.slcr.de
22. H. Jetter, 'Cleaning at the Speed of Light' internet presentation Jet Lasersysteme GmbH, www.jetlaser.com
23. G. Wiedemann, 'Laserabtragen dünner Deckschichten – eine alternative Reinigungsmethode für die Restaurierung und Denkmalpflege – Möglichkeiten und Grenzen' Sonderdruck Arbeitsblätter für Restauratoren, Heft 2 (2001)
24. G. Wiedemann, 'Möglichkeiten und Grenzen des Einsatzes von Diodenlasern zum Abtragen von Oberflächenschichten' in 'Laser in der Materialbearbeitung', Band 12 'Präzisionsabtragen mit Lasern' VDI-Herstellung und Druck, S. 25 (2000)
25. G. Barbezat Laserunterstütze thermische Beschichtung Sonderdruck aus Sulzer Technical Review 3/2000

Chapter 16
Drilling

Kurt Walther, Mihael Brajdic, and Welf Wawers

16.1 Introduction

The non-contact and therefore wear-free drilling with laser radiation allows a high flexibility and the possibility for automatization. A great variety of technical relevant materials like metals, alloys, high-strength materials, ceramics, multi layer systems, semiconductors, carbon compounds, composites, diamond, or plastics can be drilled by laser radiation. Further advantages are the reproducibility, the drilling velocity, and the achievable aspect ratio. Laser drilling is therefore an alternative for drilling techniques like mechanical drilling, electro discharge machining, electrochemical machining, and electron beam drilling.

Depending on the application and the requirements regarding quality and efficiency, laser drilling is the only applicable technique, e.g., for drilling of holes with high inclination angles in turbine blades with ceramic thermal barrier coatings. Today, laser drilling is applied in industry due to the small achievable diameters (less than 100 μm), the easy integration in existing manufacturing chains as well as the flexibility (fast change of diameters or inclination angles, fast adaption to different materials). With short drilling times and an "on-the-fly" material machining small cycle times and machining costs are realized.

To cover a wide range of applications four different laser drilling techniques are used (Fig. 16.1), depending on the requirements concerning geometrical specifications (diameter, depth), quality (precision), and productivity (drilling time):

- single-pulse drilling;
- percussion drilling;
- trepanning;
- helical drilling.

Single-pulse drilling is used in applications that require a large number of holes with diameters ≤ 1 mm and depths ≤ 3 mm. The holes are produced with the laser radiation of a single pulse and pulse durations in the range of 100 μs to 20 ms.

K. Walther (✉)
Fraunhofer-Institut für Lasertechnik, 52072 Aachen, Germany
e-mail: kurt.walther@ilt.fraunhofer.de

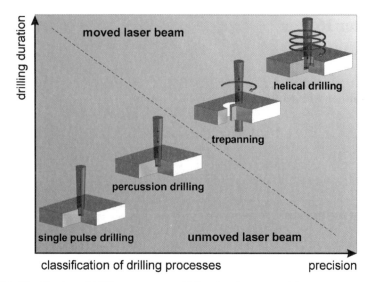

Fig. 16.1 Classification of drilling processes – drilling duration vs. precision

Percussion drilling is applied for holes with diameters ≤1 mm and depths up to 20 mm. Laser radiation with pulse durations from fs to ms is used. With the pulse duration, the intensity and the temporal pulse shape of the laser radiation the machined holes differ in quality regarding recast layer, aspect ratio, conicity, or cracks in the recast layer and the base material.

Trepanning is a combined drilling and cutting process typically applied with pulsed laser radiation. Machining a through hole with a single pulse or with percussion drilling is followed by a relative movement between laser radiation and workpiece. Providing the appropriate equipment (e.g., 5-axis positioning system) the machining of free-form holes with different shapes and contours on the hole entrance and the hole exit is possible. Pulse duration of the laser radiation used for trepanning is in the range of μs to ms.

For *Helical drilling* the laser radiation is rotated relative to the workpiece. Typically pulse duration in the nanosecond range are used. The drilling process is dominated by vaporization. This helps to avoid the formation of a large melt pool at the hole bottom. The helical drilled holes are very precise and exhibit a good microstructural quality.

A technical distinction is drawn between laser melt drilling and laser sublimation drilling (Fig. 16.2). In laser melt drilling, the irradiated intensity of the laser radiation is approx. $I \leq 10^8$ W/cm^2. A melt pool is formed and a part of the material is vaporized due to the increasing temperature. The recoil pressure of the vapor accelerates the melt and expels it along the hole wall. As a result of thermal conductivity and convective heat transfer, the melt is flowing away radially from the bottom of the hole and melts additional material, which widens the hole. The high drilling rates attainable with this method are achieved to the detriment of precision. For

16 Drilling

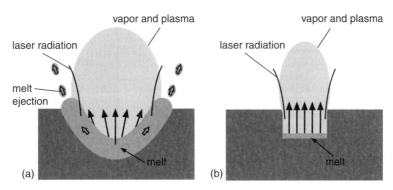

Fig. 16.2 Drilling with laser radiation: (**a**) melt drilling and (**b**) sublimation drilling

minimizing melt deposits the material is removed by means of laser sublimation drilling. The irradiated intensity has to be higher and the pulse duration is much smaller than in laser melt drilling (typically $< 10\,\text{ps}$). This results in higher precision and a smaller drilling rate.

16.2 Single-Pulse Drilling

16.2.1 Process Description

Single-pulse drilling with laser radiation is the most productive of all laser drilling methods and is used when a large number of holes have to be drilled, e.g., to produce filters and screens. The hole is drilled with the radiation of a single laser pulse. Nd:YAG laser systems are mainly used as the beam source for single-pulse drilling. Hole depths of up to 2 mm can be drilled with a single pulse using pulse durations in the microsecond to millisecond range and intensities $I > 1\,\text{MW}/\text{cm}^2$. Most of the material is expelled from the hole as melt. Hole diameters of approx. 10–500 μm are attainable. More than 100 holes per second can be produced with single-pulse drilling thanks to "on-the-fly" processing. This means that the workpiece is moved during processing at a speed that is adapted to the frequency of the laser radiation, eliminating positioning times for each individual hole.

High pulse-to-pulse stability of the laser radiation is required in single-pulse drilling, as fluctuations cannot be statistically evened out (in contrast to percussion drilling, trepanning, and helical drilling). A very high beam quality $M^2 \approx 1$ is needed, as the beam diameter determines the minimum possible diameter of the hole. The aspect ratio (ratio of hole depth to hole diameter) and the conicity of the hole are indicators of its geometrical quality. Maximum focusability, defined as the smallest diameter of the laser beam, is attained in basic mode operation (TEM_{00}). The laser beam is focused with lenses or mirrors.

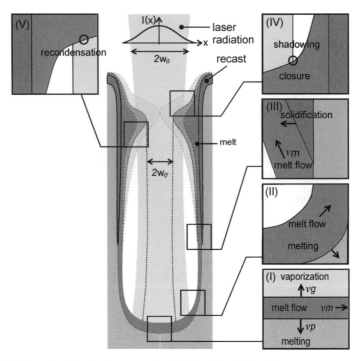

Fig. 16.3 Schematic of the four key aspects of single-pulse drilling [1]

The process can be divided into five key aspects (Fig. 16.3):

(I) Melting and Vaporization at the Bottom of the Hole

The material is heated to such an extent that a distinctive molten bath is formed. The melt front moves parallel to the laser beam axis at a speed of v_p. As soon as the vaporization temperature is reached, a metal vapor forms which flows away at a speed of v_g. In the boundary case, this expands at the velocity of sound. Owing to the recoil pressure of the vapor, the melt is continuously accelerated at the bottom of the hole and flows away at a speed of v_m.

(II) Widening of the Hole

The melt front moves radially outward perpendicular to the laser beam axis. By lateral convection of the melt, the hole is widened to the radius r_B.

(III) Solidification of the Melt at the Hole Wall

The melt is expelled upward on the hole wall. The melt cools down as it moves out of the hole and solidifies on the hole wall. The solidification front moves radially inward perpendicular to the laser beam axis.

(IV) Shadowing of the Laser Radiation and Closuring of the Hole

The thickness of the layer consisting of liquid and resolidified melt (recast) on the wall grows yielding to a closure of the hole by melt and therefore to a shadowing of the laser radiation.

(V) Recondensation

Recondensation of vapor and plasma at the hole wall can occur. This results in an additional energy transfer to the hole wall.

The hole depths and diameters that can be drilled, the drilling speed, and the speed of the melt front depend on numerous parameters relating to the laser radiation and the material being drilled. These include the spatial intensity distribution (caustic curve, divergence, Rayleigh length), the temporal pulse shape, pulse energy, pulse duration, focus position, type and pressure of the process gas used, and thermal and optical properties of the material.

16.2.2 Influence of Process Parameters

Tests on the influence of beam quality in the drilling of aluminum and titanium proved that the hole diameter and the attainable drilling depths are dependent on beam quality [2]. The dependence on beam quality is greater at lower pulse energies due to the reduction of scatter and interaction of the laser radiation with the laser-induced plasma. With otherwise identical test parameters, higher beam quality results in greater drilling depths and smaller hole diameters. The stability of the drilling process with regard to reproducibility of the hole geometries achieved is improved with a higher beam quality, resulting in smaller variations in geometry from hole to hole. Linear correlation analysis was used to examine the correlation of various laser beam parameters (spatial and temporal) and the dimensions of the holes drilled with individual pulses in titanium and aluminum [3]. The hole diameters varied by 5–10%. Compared with the other laser radiation parameters, the temporal pulse shape exerted the biggest influence on the resulting holes: With a large proportion of the pulse energy in spikes at the beginning of the pulse, deeper holes were drilled due to the increased expulsion of melt. Compared with the temporal pulse shape, the pulse energy displayed a lower correlation with the properties of the drilled hole. With lower pulse energies, temporal pulse shapes with shorter pulse durations and spikes of greater intensity were produced which supported expulsion of the melt, thus producing larger holes. The development of the hole geometry (depth and diameter) in single-pulse drilling as a function of the temporal pulse shape was examined in the drilling of stainless steel and CMSX-4 (Ni superalloy) [4]. Blockage of the hole can be observed on reaching a certain depth (Fig. 16.4). The longitudinal microsections depicted, which were produced by means of metallographic processing (grinding of the process sample along the axis of the hole), each represent the result of drilling at various stages of an individual pulse which

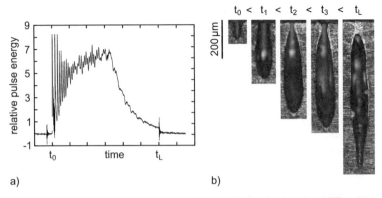

Fig. 16.4 Temporal pulse shape of the laser radiation used for single-pulse drilling (**a**) and the corresponding development of the hole geometry during a pulse (material: stainless steel) (**b**)

was switched off at certain points in time using a Pockels cell. This enables the hole geometry to be analyzed without the influence of the temporal pulse shape's negative slope. The hole tapers conically at the bottom because the laser radiation is shadowed by the blockage. Until complete perforation of the material, i.e., while the melt is still being expelled through the entrance to the hole, the spatial intensity distribution is the most important process parameter for the configuration of the hole geometry. The best results with regard to hole geometry and blockage formation are achieved with a Gaussian intensity distribution.

An approximately cylindrical hole geometry can be achieved by a suitable selection of the laser and process parameters [5]. Conical entry and melt deposits at the entrance to the hole can be very largely avoided through a temporally rectangular intensity distribution of the laser pulse. A constant value for the spatial intensity distribution in the laser pulse reduces conicity. As a result, the entry and exit diameters of the hole are the same. A small coaxial process gas pressure (≈ 0.6 MPa) at the start of drilling and large gas pressure shortly before complete perforation (>1.6 MPa) reduce the formation of melt drops at the exit to the hole. The focus of the beam parameters used here is set at 300 μm below the surface of the workpiece, regardless of the focus diameter and the laser output.

16.2.3 Application Examples

Single-pulse drilling is used in medical engineering, automobile manufacture, aviation, and toolmaking. In medical engineering, one use of single-pulse drilling is to produce suture holes on surgical needles [6] (Fig. 16.5). To reduce melt deposits on the edge of the drilled hole, the single pulse is temporally modulated at 20 kHz. The hole entries have diameters of 50–600 μm and are circular. Aspect ratios of 4:1–12:1 are achieved. Up to six needles per second are drilled.

In automobile manufacture, holes with a thickness of 0.95 mm are drilled in stainless-steel fuel filters at a rate of 120 holes per second (Fig. 16.6). The drilled

16 Drilling

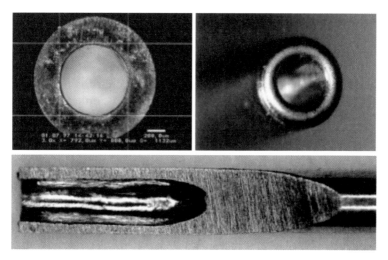

Fig. 16.5 Surgical needle [6]

Fig. 16.6 Microholes in fuel filter for the automotive industry [9]

holes have a diameter of 50–100 μm. On the fuel exit side, the holes have to be burr free. A high productivity rate of several hundred holes per second is achieved by means of "on-the-fly" processing [7, 8].

On commercial aircraft, holes are drilled in the outer skin of the tail fin to reduce air vortices and therefore air resistance [10]. This cuts fuel consumption by 5–7%. The density of the holes, which are approx. 1 mm deep, is around 4 million holes per square meter with diameters of 30–80 μm. Holes are drilled at a rate of 50 per second using laser radiation from a Q-switch Nd:YAG laser system with pulse energies of approx. 200 mJ [10]. In future, beam splitting will be used to increase productivity to ten times the number of holes drilled per second.

To produce aerostatic air bearings, 1–10 nozzles made of aluminum, stainless steel, brass, or plastic with diameters of 0.025–1 mm are produced in a single drilling operation using laser radiation. The number of holes drilled is between 100 and 1,000 for each air bearing. The workpiece thickness to be drilled through is typically 0.8 mm. Air bearings are used, for example, in semiconductors, measurement systems, and textile machinery [11].

16.3 Percussion Drilling

16.3.1 Process Description

Percussion drilling is a pulsed process, in which material is removed with consecutive pulses of laser radiation. This enables a higher aspect ratio to be achieved than with single-pulse drilling (approx. 50:1 compared with 20:1). The diameters of the drill holes produced typically range from 100 µm to 1 mm, and a drilling depth of 20 mm can be reached. Percussion drilling is performed using pulsed laser radiation with pulse durations in the femtosecond to millisecond range. The pulse duration, intensity distribution, and temporal pulse shape of the laser radiation have a direct effect on the quality of the drill hole in terms of the thickness of the melt film (recast), the aspect ratio, conical tapering, or cracks in the melt film and the substrate material. The depth of the drill hole increases with each pulse, as does the recoil pressure required to expel the melt out of the blind drill hole during fusion drilling. The recast on the drill-hole wall can accumulate until the melt forms a closure.

At the closure, the laser radiation is reflected, diffracted, and absorbed, and the solidified mass is re-melted. During percussion drilling, this process recurs at different points in the drill hole due to the consecutive pulses, and the material is partially removed from the hole. Single or multiple closures can also form during a pulse and can thus occur during single-pulse drilling. The absorption of laser radiation at a closure by solidified melt can result in local expansion of the drill hole (lateral convection) and thus a reduced reproducibility of the drill-hole geometry. Current investigations aim to identify the influence of process gases and beam parameters such as pulse duration, pulse energy, and intensity, in order to increase the drilling speed, reduce the amount of recast on the wall and at the entrance to the drill hole, and thus improve the productivity of the process and the quality of the hole.

16.3.2 Influence of Process Gases

During percussion drilling, most of the material is expelled from the drill hole in its molten state using pulsed laser radiation with pulse durations in the µs and ms range. Depending on the type of material involved, different process gases are used to influence, or support, the drilling process. The use of oxygen triggers chemical reactions that lead to an increase in temperature during the drilling process and cause the material on the drill-hole wall to oxidize. Nitrogen can be used to nitrate the material, while argon serves to block additional chemical reactions during the process. The driving forces generated by the gas stream can lead to either increased or reduced expulsion of the melt. If argon is applied with too much pressure, the melt is not fully expelled out of the hole (Fig. 16.7a). In the case of oxygen the pressure has no effect on expulsion (Fig. 16.7b) [12].

Studies on spatter (debris) formation when processing NIMONIC 263, and on the influence of process gas on the drilling process – [13, 14], and [15] – have shown that, with oxygen as a process gas, the occurring spatter is only 10–20% as thick as with compressed air, nitrogen, or argon. During experiments with oxygen as a process gas, the debris accumulating around the entrance to the drill hole was less adhesive than that which accumulated while using the other gases. This is true for many different types of metals. The exothermal reaction of oxygen with the molten material causes the melt to overheat, to be expelled in greater quantities in the form of droplets, and to turbulently solidify [15, 16]. When using the process gas argon, the melt has a higher cooling rate [14] and solidifies in multiple layers that can be metallographically detected by means of longitudinal sections of the drill hole [14, 17]. A numerical model of the drilling process, which factors in the influence of the process gas, shows that a greater recoil pressure results in a faster radial

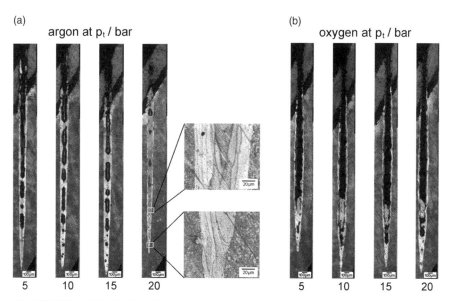

Fig. 16.7 Blind drill holes in multi-layer system, each produced with the same number of pulses. (**a**) Incomplete expulsion of the melt due to too much pressure while using argon as the process gas. (**b**) No influence by the pressure of the process gas on the expulsion of the melt while using oxygen [12]

displacement of the melt at the base of the drill hole [18]. This increases the rate at which the melt is expelled and causes more material to be melted.

In addition to the influence of process gases on melt expulsion, their influence on the average drilling speed was investigated [19]. The use of nitrogen results in a greater average drilling speed and a larger diameter at the hole exit than the use of oxygen. The gas stream does not enter the drilling channel before penetration. It is only after penetration that it helps to clear the melt out of the drilling channel [19]. Flow visualization is used to investigate the development of the gas stream after leaving the nozzle, as a function of the recoil pressure and the distance between the nozzle and the surface of the workpiece [20]. The gas stream slows and laterally diverts the melt expelled from the drill hole. Part of the vaporized material is locked in between the shock wave of the gas stream and the surface of the drill hole [21].

Further investigations have been carried out on the effects of oxidation during the drilling process using oxygen and argon as process gases [22–24]. The developing oxide layer influences the drilling result through changes to the absorption characteristics and the melting temperature, and through the additional reaction enthalpy. Single-pulse drilling ($I = 10^6$ W/cm^2, $t_P < 1$ ms) in steel and stainless steel under an oxygen and an argon atmosphere does not result in any significant changes to the absorption coefficients, and the melting temperatures of the oxides are comparable. In the case of aluminum, the absorption coefficient is doubled. At the same time, however, the process duration is increased due to the higher melting temperature of the oxide. At intensities $I > 10^9$ W/cm^2, no differences can be observed between the two process gases argon and oxygen. This can be attributed to the shorter heat-up period, which reduces oxidation. Regarding penetration times, measurements carried out on samples of aluminum (1.6 mm thick, process gas argon) and copper (0.8 mm thick, process gas oxygen) show that the penetration rate increases as the pressure of the process gas rises (from 2.4 to 6.5 bar) [23]. This is presumably because molten material is more efficiently removed from the laser impact zone.

16.3.3 Influence of Beam Parameters

Drilling materials for aerospace applications using a 10 kW Nd:YAG laser, it has been demonstrated that pulse shaping and the "back-off" technique (raising the focusing lens back up after drilling the hole) can help to control the taper of the hole, and thus to produce a variety of drill-hole geometries [25]. It has also been shown that "ramping" the pulses during percussion drilling helps to expel the melt more efficiently than when the pulses are not ramped (i.e., rectangular). Likewise it has been exhibited how modulating the temporal intensity curve of the laser beam in the μs range influences the taper of the drill hole [26].

Investigations on percussion laser drilling with pulse durations in the femtosecond to millisecond range show that the high ablation rates achieved in the millisecond range can equally be reached with laser radiation in the nanosecond range, resulting in a higher quality in terms of the thickness of the melt film [27, 28].

The plasma generated during the drilling process can either shield the surface of the material from the incident laser beam [29] or support the drilling process [30–32]. Studies on drilling in heat-resistant alloys demonstrate the applicability of modulated laser radiation with multiple ns pulses (bursts) as an alternative to established systems with pulse durations in the ms range for processing such objects as turbine components [33].

The ablation rate decreases with increasing drilling depth under otherwise constant process parameters [34]. The changing ablation rate can be explained by the developing conical geometry of the drill hole. The absorption of laser radiation at the base and on the wall of the drill hole and the heat conduction inside the material are largely determined by the given proportion of perpendicularly impinging intensity. If the radius of curvature at the base of the drill hole is less than the thermal penetration depth, the heat losses are greater and so is the minimum intensity required for ablation. If the absorbed intensity is lower, no more material is ablated.

The temporal pulse shape of the laser radiation influences the geometry of the drill hole. Tapering can be reduced by modulating the laser radiation [35, 26]. Various working groups have achieved an increase in the ablation rate by using double and multiple ps and ns pulses with pulse distances in the ns range [36, 31, 37]. When applying bursts of ns pulses with pulse distances in the µs range, the ablation rate increases with shorter pulse distances and a higher number of pulses per burst at a constant level of burst energy [38]. In the case of ns pulse bursts with pulse distances between 20 and 90 µs, further division of the pulses into a ps substructure leads to no further increase in ablation [39]. The superposition of pulsed (ms range) or cw laser radiation on laser radiation with a shorter pulse duration and a greater intensity results in increased expulsion of the melt and a faster drilling speed [40, 41].

During percussion drilling at low pulse repetition rates, the recast is composed of multiple layers (Fig. 16.8). These can fuse to form a single layer if the repetition rate is high enough to prevent the melt from re-solidifying between pulses.

At high repetition rates, the laser radiation is additionally coupled into the vapor, or plasma, generated by the preceding laser pulse. By temporally modulating the laser radiation, the plasma can be controlled, and it is possible to produce a variety of drill-hole geometries. The hole geometry can be influenced by temporally modulating the laser radiation at a constant pulse output and pulse energy (Fig. 16.9).

16.3.4 Applications

Percussion drilling is used in areas such as tool and die making, medical engineering, automotive engineering, aviation, and tool manufacturing. Applications include the drilling of cooling holes in the turbine blades of aircraft engines and stationary turbines; holes in filters, sieves, and nozzles for ink-jet printers; and feedthrough holes for solar cells [42]. In the aircraft industry, Nd:YAG laser systems are used for such applications as drilling holes in combustion-chamber plates or

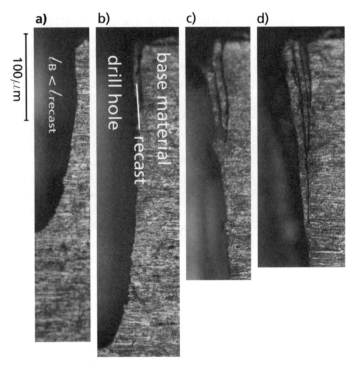

Fig. 16.8 Formation of multiple recast layers during percussion drilling: (**a**) one pulse, (**b**) two pulses, (**c**) three pulses, and (**d**) four pulses

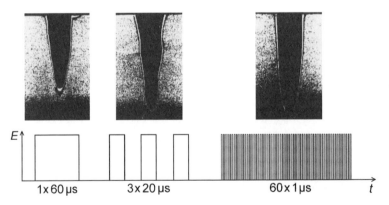

Fig. 16.9 Drill-hole geometries produced with differently modulated laser radiation and intensities above the plasma threshold [32]

the blades of aircraft turbines. During the manufacture of drawing dies, tiny holes with diameters down to between 10 and 50 μm are drilled in diamond [43, 44]. Laser drilling has also become the established method of producing micro-vias in printed circuit boards [43, 45]. Micro-vias are feed-through connections, i.e., drilled and subsequently plated holes that connect different signaling layers. Holes drilled in biomedical micro-filter membranes measure < 20 μm in diameter [46], while particularly fine filters are produced with even smaller holes measuring between 15 μm (Fig. 16.10) and 1μm across [47].

By drilling holes with diameters ranging between 10 and 30 μm spaced 60 μm apart in metal sheets, it was possible to produce "transparent" metal (Fig. 16.11).

Lubricating-oil holes in engine components have so far been drilled mechanically, or by erosion, due to the required drilling depths of between 8 and 20 mm, and the target diameters of 800 μm. It takes several minutes to drill each hole in the conventional way, whereas percussion drilling reduces this time to 9 s [48]. Holes in nozzles for the print heads of ink-jet printers are drilled by excimer laser (Fig. 16.12).

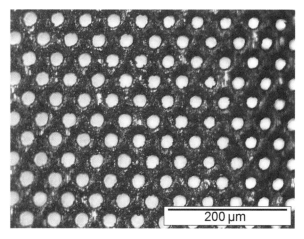

Fig. 16.10 Micro-filter: 2000 drill holes. Material: aluminum, 15 μm drilling depth [47]

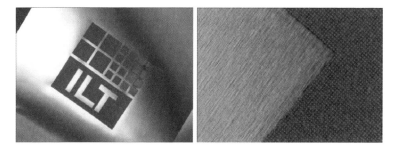

Fig. 16.11 "Transparent" metal: 6.5×10^5 drill holes with diameters between 10 and 30 μm spaced 60 μm apart (Source: ILT)

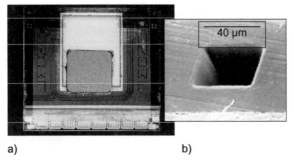

Fig. 16.12 *Left*: print heads for ink-jet printer, material: Vacrel (**a**). *Right*: print-head nozzle drilled by KrF laser, 248 nm (**b**) [49]

16.4 Trepanning

16.4.1 Process Description

Trepanning is a combined cutting and drilling process, typically performed using a pulsed laser. The duration of each pulse ranges from a few ns [39] to the ms range [50], depending on the material, the thickness of the workpiece, and the required quality of the drill hole (e.g., diameter, surface properties, melt film thickness, heat-affected zone).

In trepanning, a through hole is first pierced by percussion drilling (Sect. 16.3) and then, in a second step, the through hole is widened to its final diameter in a circular cutting motion (relative movement of the workpiece with respect to the laser beam). The required diameter can be achieved either directly in a single positioning step (Fig. 16.13) or in several consecutive positioning steps with the beam describing concentric circles or a spiral. The relative movement of the laser beam

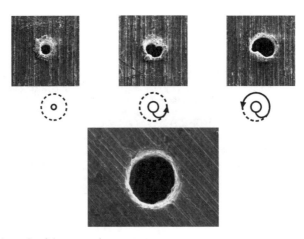

Fig. 16.13 Schematic of the trepanning process

with respect to the workpiece can be achieved using a positioning table (linear or rotational axes), scanning optics, or special-purpose trepanning optics.

The taper, roundness, and roughness of the drill hole can be improved by performing repeat cycles of the circular path, including in the reverse direction. This also reduces polarization effects on the drill-hole geometry.

The maximum thickness of workpieces that can be processed using this method is similar to that for percussion drilling (Sect. 16.3). Typical drill-hole diameters are 0.15–1 mm. The melt formed during the cutting movement is expelled through the exit area of the joint or drill hole, with the assistance of a process gas flowing coaxially to the laser beam [17, 50–52].

16.4.2 Process Gas and Gas Pressure

During trepanning, the thickness of the melt depends on the process gas pressure and the thickness of the alloy. The melt formed during trepanning closes the holes again at low process gas pressures ($p \leq 10$ bar). In order to blow out the melt from the bottom, the process gas pressure has to be larger than 14 bar [51, 53]. The deeper the trepanning kerf, the higher the pressure has to be to remove the melt from the hole.

The recast thickness during trepanning of Inconel 718 has been investigated and a process gas pressure window has been identified ($p \sim 15$ bar) where the recast layer thickness is reduced by the proper removal of molten material [50]. At larger pressure values (~ 25 bar) the recast layer thickness is increased and cracking can be detected in the recast layer and the base material. This can be ascribed to the forced convection cooling caused by the process gas [50, 54].

Inclining the surface of the alloy with respect to the laser beam means an increase of the effective length of the hole and of the irradiation time required for percussion drilling. For trepanning with argon as process gas, a pressure $p > 16$ bar is necessary in order to maintain a sufficient melt expulsion rate at the hole exit [51]. When drilling inclined holes, a large amount of gas is deflected from the surface, thus not penetrating the hole (Fig. 16.14a). The stagnation point is located beside the hole, leading to a gas flow crossing the entrance of the hole. The gas flow is partly reflected into the hole entrance, resulting in maxima and minima of the gas pressure distribution at the hole surface. Accumulations of resolidified melt can be detected by longitudinal sections showing qualitative correlations with the locations of the calculated minima in the pressure distribution within the hole (Fig. 16.14a).

A simulation tool for solving Euler equations by the finite volume method is adopted to analyze the gas jet of the conical nozzle (diameter 1 mm) placed above a 45° inclined sheet surface [55].

The arrangement of a centered nozzle (Fig. 16.14a) is compared to a nozzle laterally displaced (Fig. 16.14b) in the direction of the gas flow crossing the entrance of the hole in order to position the stagnation point directly above the hole. The simulation predicts an increase of the pressure along the hole surface and avoidance of oscillations in the pressure distribution for a nozzle displaced relatively to the hole.

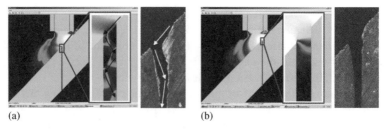

Fig. 16.14 Longitudinal section of 45° inclined trepanning kerfs (process gas argon) in CMSX-4 and simulation of pressure with (**a**) centered nozzle and (**b**) displaced nozzle [56]

The centered and the laterally displaced arrangement of the nozzle is used for the trepanning of 60° inclined holes in CMSX-4. When using the centered nozzle, an accumulation of melt in the entrance region of the hole is revealed by a cross-section of the trepanned hole (Fig. 16.14a). The thickness of the resolidified melt at the hole entrance can be reduced by using the laterally displaced nozzle (Fig. 16.14b) in order to obtain more cylindrical holes. The entrance diameter is widened mainly during initial percussion drilling. The melt expelled through the entrance of the hole erodes (by heat transfer) the hole entrance unidirectionally by flowing down the inclined upper surface.

Due to the avoidance of oscillations of the gas pressure at the hole surface, the melt expulsion is homogenous and the thickness of the resolidified melt at the hole surface is reduced [17, 55, 56].

16.4.3 Type of Process Gas

Trepanning of 2 mm thick CMSX-4 samples by microsecond pulsed laser radiation is investigated adopting different process gases (Fig. 16.15). The mean thickness of the resolidified melt in the holes trepanned with nitrogen is $30\,\mu m$. Chemical composition measurements of the resolidified melt show a nitration of the superalloy forming a melt with high viscosity that cannot be removed properly by the process gas flow. When using oxygen, the resolidified melt has a thickness of $20\,\mu m$ and exhibits bubbles and cracks. The oxygen is more enriched in the resolidified melt than in the bulk material. Compared to nitration, oxidation forms a melt with a lower viscosity. On the other hand, melt is not removed completely by the process gas flow. Argon is found to be the most effective process gas in reducing the thickness of the recast layer ($10\,\mu m$) for CMSX-4. The resolidified melt at the hole walls reveals an orientated microstructure [56].

Depending on the Rayleigh length of the focused laser beam employed, the laser intensity may not be sufficient to melt the material and expel it from the drill hole beyond a certain drill-hole depth. The melt remains in the drill hole as a closure or recast layer on the hole wall and cannot be entirely removed even after several

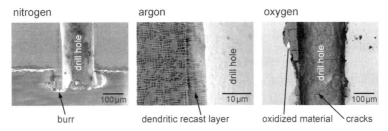

Fig. 16.15 Process gas influence during trepanning in CMSX-4 with a thickness of 2 mm [56]

trepanning cycles. The use of argon as process gas helps to cool the melt and supports the solidification process [56, 50, 54, 57, 58].

The use of oxygen as process gas generates higher temperatures and pressures in the hole during drilling as a result of the additional reaction enthalpy [22]. The melt is highly oxidized and has less tendency to adhere to the hole entrance and exit area than in holes drilled using argon as the process gas. The greater reaction enthalpy permits holes to be drilled in thicker materials, on condition that the application in question will tolerate the previously mentioned compromises with respect to the quality of the drill hole.

16.4.4 5-Axis Trepanning

The minimum diameter of a trepanned hole and the maximum workpiece thickness are predefined by the initial percussion-drilling operation (Sect. 16.3). Typical diameters range between 0.2 and 1 mm, and workpiece thicknesses between 0.5 and 3 mm. It is possible to represent different drilling geometries (conical, freeform) as a function of the positioning system (Fig. 16.16a). For instance, it is possible to increase the cylindrical profile of trepanned holes by describing the contours of a conical shape ($d_{ENTRANCE} < d_{EXIT}$) [59]. Simultaneous positioning on five axes permits almost any freeform surface to be produced, in addition to round drill holes.

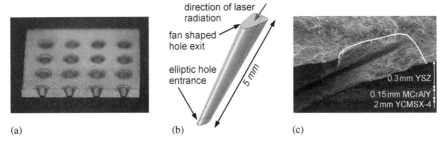

Fig. 16.16 (a) Longitudinal section through a trepanned conic drill hole in 1 mm stainless steel 1.4301, (b) schematic of the required geometry of a fan shaped drill hole, and (c) longitudinal section of a fan shaped drill hole

The example illustrated in Fig. 16.16b and c shows the geometry of a fan shaped cooling-air drill hole in a gas turbine component [60, 61].

Because trepanning is a combined process consisting of separate drilling and cutting steps, processing times are longer than in the case of single-pulse or percussion drilling. Typical times are 2–10 s per hole ($d = 2$ mm, $\varnothing = 0.7$ mm). Holes with low diameter tolerances can be produced by displacing the laser beam relative to the workpiece using a positioning system of the appropriate precision (deviation at entrance approx. $\pm 5\,\mu$m for $\varnothing = 0.35$ mm and $d = 2$ mm). As a result of the cyclic movement, asymmetries in the intensity distribution of the caustic curve have a negligible effect on the result, unlike in single-pulse and percussion drilling. The pressure of the process gas drives the melt toward the drill-hole exit as soon as the hole is drilled through. This limits the thickness of the melt film to a value in the region of $10\,\mu$m ($t = 2$ mm, μs pulse duration range) [51].

16.4.5 Applications

Typical areas of application for trepanning are power-generation and turbine engineering, the automotive industry, and toolmaking.

During the manufacture of drawing dies, tiny holes with diameters down to between 10 and $50\,\mu$m are drilled in diamond [43].

With vent holes the cavity of a mold can be evacuated to avoid trapping of air in injection molded parts (Fig. 16.17). The holes must not exceed a certain diameter (depending on the viscosity of the used plasticmaterial) to avoid a footprint of the hole on the surface of the injection-molded part.

Fig. 16.17 (**a**) Vent holes in a mold for CDs (Source: Krallmann GmbH) and (**b**) longitudinal section of a vent hole

16.5 Helical Drilling

16.5.1 Process Description

In helical drilling, a relative movement takes place between the workpiece and the laser beam. Two superposed rotation movements can be distinguished.

The revolution of the laser beam on the helical path (Fig. 16.18 left) and a rotation of the laser beam in itself (proper rotation) synchronized with the helical path [62].

Proper rotation is necessary on the one hand to compensate for a non-circular beam profile. On the other hand, proper rotation makes it possible to produce small drill holes with a diameter in the range of the beam cross-section, the helical path radius approaching zero. At the left of Fig. 16.18 is a 3-D helical drilling diagram showing the drill penetration into the material. The superposed rotational movements of the helical path and the proper rotation of the laser beam are illustrated on the right by branding marks at low laser output on the surface of the workpiece.

Full perforation only occurs after multiple revolutions. Since the use of pulsed laser radiation in the nano- to picosecond range means that only a very small volume of material is ablated by each pulse, helical drilling exhibits similar ablation mechanisms to laser ablation. If the geometry of the helical path is not circular, the technique may also be referred to as laser erosion [63].

Because of the helical beam path, helical drilling offers the possibility of altering the geometry of the drill hole by changing the angle of incidence between the rotating laser beam and the workpiece and the point of impact on the workpiece (Fig. 16.19). The angle of incidence determines the conical taper φ of the drill hole, while the distance a from the optical axis of the focusing lens determines the drill-hole diameter.

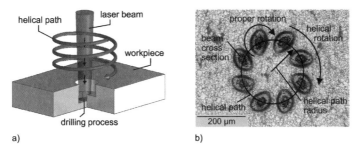

Fig. 16.18 Diagram of helical drilling (**a**). Branding marks at low laser output on the surface of the workpiece illustrates the rotating motion (**b**)

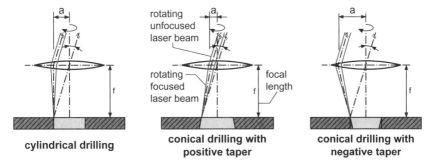

Fig. 16.19 Types of drill holes that can be produced by helical drilling

16.5.2 Characterization of the Process

Helical drilling offers the greatest precision of all currently known laser drilling methods. Its precision can be seen on the one hand in the roundness of the drill hole and the sharpness of the edges at the entrance and exit of the drill hole, and on the other hand in the diminished surface roughness and the thinner recast layer of solidified melt when drilling materials with a molten phase. The wall of the drill hole is accordingly more homogeneous.

This enhanced precision is due to the fact that the larger part of the material is ablated in the form of vapor. Helical drilling ablation models described in the relevant literature attribute the volatilization of the material to the use of laser radiation with a short pulse duration in the nanosecond range together with a high-peak pulse output and an extremely focused intensity profile whose diameter in the area of relevance to ablation is appreciably lower than the diameter of the drill hole [9, 64, 65].

These ablation models relate to helical drilling without proper rotation, in which a minimum helical diameter is always necessary. Investigations of helical drilling with proper rotation have shown that the material ablation is mainly vaporous, even when the helical diameter is approaching zero and the intensity distribution corresponds to that observed in percussion drilling [66], assuming that the rotational movement of the molten material is synchronous with the rotation of the laser beam. The melt flow is driven by the density gradient of the plasma in the area of the focus cross-section. The rotational movement causes the melt formed in the drilling channel to be deposited on the wall of the drill hole. At the beginning of the drilling process, some of the molten material flows out of the drilling channel and forms a crest of molten material at the entrance to the drill hole. At a penetration depth of a few $100\,\mu m$ a balance forms between the newly formed melt and the rotating melt peripheral to the wall of the drill hole. As the drilling process continues, the rotating melt is volatilized, thereby causing the initially formed crest to break off. Figure 16.20 shows the temporal progress of helical drilling after n pulses.

In this rotating melt model, it is the rotational movement of the laser beam that causes the enhanced precision in terms of the recast thickness and the melt deposits at the entrance to the drill hole. The pulse duration and thus also the pulse output is selected so as to achieve the desired intensity of vaporization in conjunction with the focus cross-section. Comparative tests show that re-solidified melt causes less impairment to the drill-hole geometry in helical drilling with short pulses in the ns range than in percussion drilling with ultra-short pulses in the fs range [67].

As the helical drill hole grows deeper, it takes a conical shape due to the laser beam scattered and reflected on the wall of the drill hole, in a similar way to percussion drilling. Figure 16.21 shows a diagram of the helical drilling process, divided into the starting and deep drilling phases.

The polarization of the laser beam affects the exit of the drill hole during helical drilling. A common feature of all presently known helical drilling processes such as rotating wedge plates, off-center rotating lenses, or scanner systems is that they only rotate the laser radiation and its spatial intensity distribution, but not the direction of polarization. In helical drilling with ultra-short pulses, the exit is elongated

16 Drilling

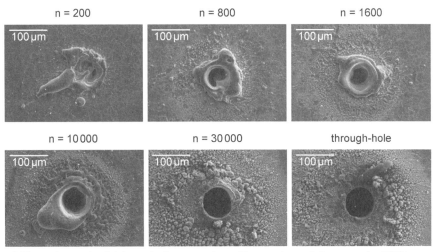

drilling entrance in stainless steel, n = number of laser pulse, rotation speed = 5000 rpm

Fig. 16.20 Temporal progress of helical drilling, drill-hole entrance after *n* pulses (source: ILT)

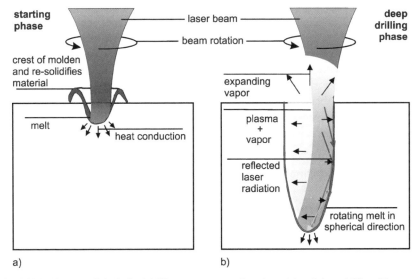

Fig. 16.21 Diagram of the helical drilling process: starting phase (**a**) and deep drilling (**b**)

perpendicularly to the direction of polarization, and rills are formed on the wall of the drill hole [68, 69]. A polarization-dependent elongation of the drill-hole exit is also observed when using laser radiation with a short pulse duration. The elongation of the drill hole caused by polarization can be reduced by using a laser beam with circular polarization or by synchronizing the path of polarization with the rotational movement of the laser beam (Fig. 16.22).

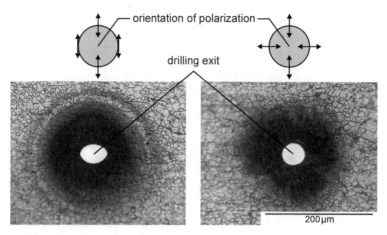

Fig. 16.22 Effects of synchronized polarization on the drill-hole exit during helical drilling

In contrast to other laser drilling methods in helical drilling, the removal of material does not come to a halt after drilling right through. Due to the progressive conical geometry of the drill hole, full penetration of the material begins at the center of the drill hole. Depending on the selected angle of conical taper, a certain volume of material to be ablated still remains on the periphery of the drill hole (Fig. 16.23).

In helical drilling, material removal is completed when the incident laser beam is reflected on the drill-hole wall along the entire length of the drill hole without ablating any further material. This depends on the selected angle of conical taper φ. The time taken to ablate the remaining volume (duration of widening) may be many times as long as the penetration time. Certain process strategies aim to lengthen the penetration time and thus shorten the duration of widening. This can be achieved,

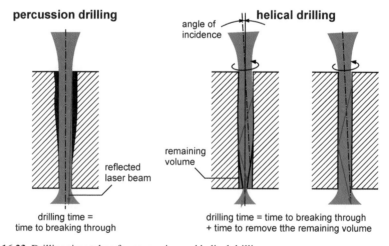

Fig. 16.23 Drilling time taken for percussion and helical drilling

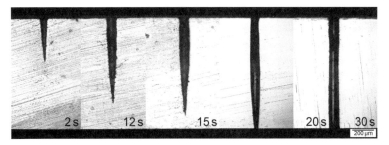

Fig. 16.24 Formation of the drilling channel up to penetration and after the drilling time

for example, by reducing the laser output up to the point at which the material is penetrated. This creates an oblique angle to form in the drilling channel, which results in a relatively large cross-section of penetration and a small residual volume. Figure 16.24 shows the formation of the drilling channel during helical drilling.

16.5.3 Helical Drilling Optic

The characteristic feature of helical drilling is the rotating movement of the laser beam relative to the workpiece. In the simplest case, the workpiece rotates and the laser beam remains stationary. This works for small, largely rotation-symmetrical components, which are mounted on a spindle or a motor-driven X–Y traversing unit. An example of such a setup is given in [70]. Industrial applications include drilling holes in jewel bearings and wire drawing dies.

A more versatile and therefore more widespread variant is that in which the laser beam rotates and the workpiece remains stationary. The most common principles for rotating the laser beam are based on eccentrically rotating lenses, scanner systems, rotating wedge prisms, and image rotator systems (Fig. 16.25).

Systems with eccentrically rotating lenses are mainly used for trepanning. There are no known industrial applications of such systems for helical drilling [71]. Scanner systems with two independently movable mirrors have the advantage that they can guide the laser beam over the workpiece on any desired path, which makes it possible to drill contoured holes. An example of a scanner system tailored to helical drilling applications is the Precession Elephant built by Arges. Helical drilling optics based on rotating wedge prisms have been technically refined over the last few years. While the early systems comprised two wedge prisms, the possibilities for setting the drilling parameters have now been improved by adding a third wedge prism [72, 73].

What the rotation principles described above have in common is that the laser beam is only moved relative to the workpiece, whereas its orientation toward the rotational axis remains constant. This causes any out-of-roundness imperfections in the beam profile to be reproduced in a less pronounced form in the geometry of the hole [65]. It is therefore necessary to achieve as round a beam profile as possible,

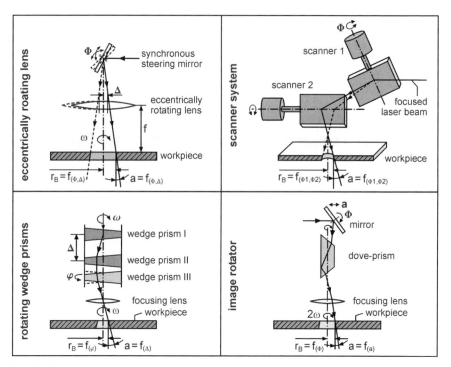

Fig. 16.25 Principles of laser beam rotation

particularly when drilling small holes with a diameter close to that of the focus cross-section. If the helix diameter approaches zero, the laser beam stops moving. For this reason, it is always necessary to set a minimum helix diameter as an additional allowance to the diameter of the producible hole.

Image rotation helical drilling optics, on the other hand, differ in that they are the only ones to offer a superimposed rotational movement of the laser beam [9, 62]. This means that, synchronously to the rotational movement relative to the workpiece, the laser beam also rotates around its own axis, the proper rotation, see Fig. 16.26 and also Fig. 16.18 right.

Figure 16.27 shows an industrially used image rotation helical drilling optics system based on a rotating Dove prism.

16.5.4 Applications

Helical drilling is used in cases where high-precision micro-drill holes of high surface quality with small melt deposits on the wall of the drill hole, a circular drill-hole cross-section, and the option to produce drill holes with a positive or negative conical taper are required. Examples include injection nozzles for the automotive industry, spinning nozzles in textile engineering, dosage nozzles for hydraulic systems,

16 Drilling

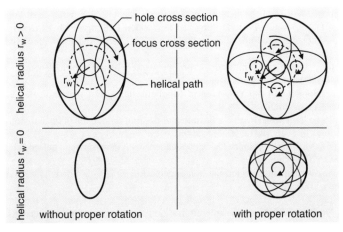

Fig. 16.26 Comparison of helical movement in the plane of incidence with and without proper rotation of the laser beam

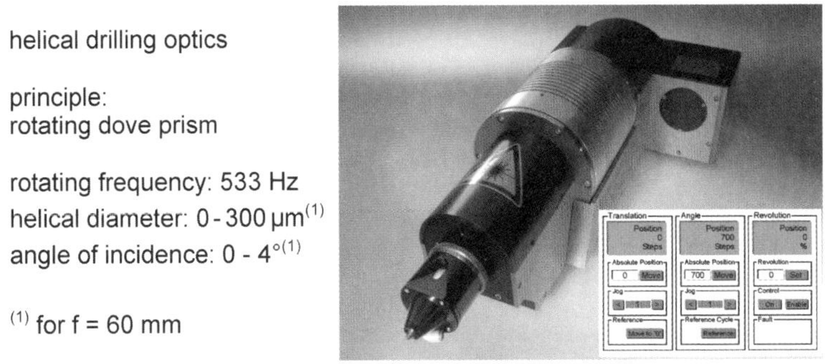

helical drilling optics

principle:
rotating dove prism

rotating frequency: 533 Hz
helical diameter: 0 - 300 µm$^{(1)}$
angle of incidence: 0 - 4°$^{(1)}$

$^{(1)}$ for f = 60 mm

Fig. 16.27 Helical drilling optics based on an image rotator (Dove prism) (Source: [ILT07])

drill holes in air bearings, drawing dies made of diamond for wire manufacturing, and starting holes for wire cut EDM.

References

1. W. Schulz & U. Eppelt, Basic Concepts of Laser Drilling. In: The Theory of Laser Materials Processing, Springer Series in Materials Science, Volume 119. ISBN 9781402093395 Springer Netherlands, 2009, p. 129
2. S. Kudesia, W. Rodden, D. Hand & J. Jones, Effect of beam quality on single pulse laser drilling, ICALEO, 2000
3. W. Rodden, S. Kudesia, D. Hand & J. Jones, Correlations between hole properties and pulse parameters in single pulse Nd:YAG laser drilling of aluminium and titanium, ICALEO, CD Version, 2001

4. L. Trippe, J. Willach, E. W. Kreutz, W. Schulz, J. Petereit, S. Kaierle, & R. Poprawe, Melt Ejection During Single-Pulse Drilling and Percussion Drilling of Micro Holes in Stainless Steel and Nickel-Based Superalloy by Pulsed Nd:YAG Laser Radiation, Proceedings of SPIE **5662**, 609–615, 2004
5. H. Rohde, Qualitätsbestimmende Prozeßparameter beim Einzelpulsbohren mit einem Nd:YAG-Slablaser Universität Stuttgart, 1999
6. U. Dürr, Bohren mit Nd:YAG-Lasern: Möglichkeiten und Grenzen in der Bohrqualität, Bayrische Laserseminare, 2003
7. J. Rapp, The laser with Bosch – a flexible tool in serial production of an automotive supplier, Proceedings of SPIE **4831**, 390–396, 2003
8. R. Holtz & M. Jokiel, Optimized laser applications with lamp-pumped pulsed Nd:YAG lasers, ICALEO, CD Version, 2003
9. T. Wawra, Verfahrensstrategien für Bohrungen hoher Präzision mittels Laserstrahlung, Utz, München, 2005
10. J. Bahnmüller, T. Herzinger, F. Dausinger, M. Beck & A. Giering, Characterization of pulsed Nd:YAG-laser beams as an optimization tool in laser drilling applications, ICALEO, E 82–91, 2000
11. C. Brecher, A. Mayer, R. Schug, G. Slotta & W. Thaler, Aerospin – aerostatisch, synchron und hochgenau, wt Werkstattstechnik online **6**, 459–464, 2005
12. K. Walther, M. Brajdic, J. Dietrich, M. Hermans, M. Witty, A. Horn, I. Kelbassa & R. Poprawe, Manufacturing of shaped holes in multi-layer plates by Laser-drilling, PICALO, 789–794, 2008
13. D. K. Y. Low, L. Li & A. G. Corfe, Effects of assist gas on the physical characteristics of spatter during laser percussion drilling of NIMONIC 263 alloy, Applied Surface Science **154–155**, 689–695, 1999
14. D. Low, L. Li & A. G. Corfe, The influence of assist gas on the mechanism of material ejection and removal during laser percussion drilling, Proceedings of the Institution of Mechanical Engineers, Part B: Journal of Engineering Manufacture **214**(7), 521–527, 2000
15. D. Low, L. Li & P. Byrd, The influence of temporal pulse train modulation during laser percussion drilling, Optics and Lasers in Engineering **35**(3), 149–164, 2001
16. B. S. Yilbas & M. Sami, Liquid ejection and possible nucleate boiling mechanisms in relation to the Liquid ejection and possible nucleate boiling mechanisms in relation to the laser drilling process, Journal of Physics D: Applied Physics **30**, 1996–2005, 1997
17. J. Willach, Herstellung von konturierten Bohrungen in Mehrschichtsystemen mit Nd:YAG-Laserstrahlung, Shaker Verlag, Aachen, 2005
18. D. K. Y. Low, L. Li & P. Byrd, Hydrodynamic physical modeling of laser drilling, Journal of Manufacturing Science and Engineering **124**(4), 852–862, 2002
19. M. Schneider, R. Fabbro, L. Berthe, L. Landais, M. Nivard & P. Laurens, Parametric study of drilling with new innovative laser source: application to percussion regime, ICALEO, 2004
20. M. Schneider, R. Fabbro, L. Berthe, M. Muller & M. Nivard, Gas investigation on laser drilling, ICALEO, 1094–1099, 2005
21. M. Schneider, L. Berthe, R. Fabbro, M. Muller & M. Nivard, Gas investigation for laser drilling, Journal of Laser Applications **19**(3), 165–169, 2007
22. R. S. Patel & M. Q. Brewster, Effect of oxidation and plume formation on low power Nd-Yag laser metal interaction, Journal of Heat Transfer (Transactions of the ASME, Series C) **112**(1), 170–177, 1990
23. R. S. Patel & M. Q. Brewster, Gas-assisted laser-metal drilling: Experimental results, Journal of Thermophysics and Heat Transfer **5**(1), 26–31, 1991
24. R. S. Patel & M. Q. Brewster, Gas-assisted laser-metal drilling: Theoretical model, Journal of Thermophysics and Heat Transfer **5**(1), 32–39, 1991
25. P. French, M. Naeem & K. Watkins, Laser percussion drilling of aerospace material using a 10 kW peak power laser using a 400 μm optical fibre delivery system, ICALEO, CD Version, 2003

26. D. Low & L. Li, Comparison of intra- and interpulse modulation in laser percussion drilling, Proceedings of IMechE **216**(B), 167–171, 2002
27. X. Chen, Short pulse high intensity laser machining, High Temperature Material Processes **4**, 151–160, 2000
28. D. Karnakis, G. Rutterford & M. Knowles, High power DPSS laser micromachining of silicon and stainless steel, WLT-Conf on Lasers in Manufacturing, München, 2005
29. A. A. Bugayev, M. C. Gupta & M. El-Bandrawy, Dynamics of laser hole drilling with nanosecond periodically pulsed laser, Optics and Lasers in Engineering **44**(8), 797–802, 2005
30. S. Paul, S. I. Kudryashov, K. Lyon, S. D. & Allen, Nanosecond-laser plasma-assisted ultra-deep microdrilling of optically opaque and transparent solids, Journal of Applied Physics **101**, 043106, 2007
31. A. Forsman, P. S. Banks, M. Perry, E. Campbell, A. Dodell & M. Armas, Double-pulse machining as a technique for the enhancement of material removal rates in laser machining of metals, Journal of Applied Physics **98**, 033302, 2005
32. H. Treusch, Geometrie und Reproduzierbarkeit Einer Plasmaunterstützten Materialabtragung Durch Laserstrahlung, Physik Technische Hochschule Darmstadt, 1985
33. V. Serebryakov, M. Volkov & X. Zhang, Optimization of laser microdrilling in heat-resisting alloys, Proceedings of SPIE, 5777, 858–863, 2005
34. T. Kononenko, S. Klimentov, S. Garnov, V. Konov, D. Breitling, C. Föhl, A. Ruf, J. Radtke & F. Dausinger, Hole formation process in laser deep drilling with short and ultrashort pulses, SPIE, 4426, 2002
35. D. Low & L. Li, Effects of Inter-Pulse and Intra-Pulse shaping during laser percussion drilling, Proceedings of SPIE **4426**, 191–194, 2002
36. C. Hartmann, A. Gillner, Ü. Aydin, T. Fehr, C. Gehlen & R. Poprawe, Increase of laser micro ablation rate of metals using ns-multi-pulses, COLA, 2005
37. M. Lapczyna, K. P. Chen, P. R. Herman, H. W. Tan & R. S. Marjoribanks, Ultra high repetition rate (133 MHz) laser ablation of aluminum with 1.2-ps pulses, Applied Physics A Materials Science & Processing **69**, 883–886, 1999
38. A. Binder, T. Metzger, D. Ashkenasi, G. Müller, T. Riesbeck & H. Eichler, High aspect-ratio laser-drilling of micro-holes with a Nd:YAG Master-Oscillator Power-Amplifier (MOPA) system, ICALEO, CD Version, 2002
39. M. Ostermeyer, P. Kappe, R. Menzel, S. Sommer & F. Dausinger, Laser Drilling in Thin Materials with Bursts of Ns-Pulses Generated by Stimulated Brillouin Scattering (SBS), Applied Physics A **81**(5), 923–927, 2005
40. C. Lehane & H. Kwok, Enhanced drilling using a dual-pulse Nd:YAG laser, Applied Physics A **73**(1), 45–48, 2001
41. K. Walther, M. Brajdic & E. W. Kreutz, Enhanced processing speed in laser drilling of stainless steel by spatially and temporally superposed pulsed Nd:YAG laser radiation, International Journal of Advanced Manufacturing Technology **35**(9–10), 895–899, 2008
42. D. Karnakis, Applications of laser microdrilling, Laser Micromachining: Developments & Applications, AILU Laser Micromachining: Developments & Applications, UK, 2005
43. J. Meijer, K. Du, A. Gillner, D. Hoffmann, V. S. Kovalenko, T. Masuzawa, A. Ostendorf, R. Poprawe & W. Schulz, Laser Machining by short and ultrashort pulses, state of the art and new opportunities in the age of the photons, CIRP Annals – Manufacturing Technology **51**(2), 531–550, 2002
44. H. R. Niederhauser, Laser drilling of wire drawing dies, Wire Industry **53**(10), 709–711, 1986
45. W. Lei & J. Davignon, Micro-via Drilling Applications with Solid-State Harmonic UV laser Systems, ICALEO, 2000
46. M. Baumeister, K. Dickmann & A. P. Hoult, Combining high-speed laser perforation and cold roll forming for the production of biomedical microfiltration membranes, Journal of Laser Applications **19**(1), 41–45, 2007
47. A. Gillner, Micro processing with laser radiation – Trends and perspectives, Laser Technik Journal **4**(1), 21–25, 2007

48. C. Lehner, K. Mann & E. Kaiser, Bohren von Kühl- und Schmierlöchern, Stuttgarter Lasertage (SLT), Stuttgart, 103–106, 2003
49. D. Basting & G. Marowsky, Excimer Laser Technology, Springer, 2005
50. W. Chien & S. Hou, Investigating the recast layer formed during the laser trepan drilling of Inconel 718 using the Taguchi method, The International Journal of Advanced Manufacturing Technology 33(3–4), 308–316, 2006
51. A. Horn, R. Weichenhain, S. Albrecht, E. W. Kreutz, Michel, J. Michel, M. Nießen, V. Kostrykin, W. Schulz, A. Etzkorn, K. Bobzin, E. Lugscheider & R. Poprawe, Microholes in zirconia coated Ni-superalloys for transpiration cooling of turbine blades, Proceedings of SPIE **4065**, 218–226, 2000
52. J. Willach, J. Michel, A. Horn, W. Schulz, E. W. Kreutz & R. Poprawe, Approximate model for laser trepanning with microsecond Nd:YAG laser radiation, Applied Physics A: Materials Science & Processing **79**(4–6), 1157–1159, 2004
53. J. Willach, A. Horn, E. W. Kreutz, Drilling of cooling holes and shaping of blow-out facilities in turbine blades by laser radiation, 7th Liège Conference on "Materials for advanced power engineering" 21, 743–750, 2002
54. S. C. Tam, C. Y. Yeo, M. W. S. Lau, L. E. N. Lim, L. J. Yang & Y. M. Noor, Optimization of laser deep-hole drilling of Inconel 718 using the Taguchi method, Journal of Materials Processing Technology 37 (1–4), 741–757, 1993
55. J. Willach, E. W. Kreutz, J. Michel, M. Nießen, W. Schulz & R. Poprawe, Melt expulsion by a coaxial gas jet in trepanning of CMSX-4 with microsecond Nd:YAG laser radiation, 4th International Symposium on Laser Precision Microfabrication, 5063, 435–440, 2003
56. E. Lugscheider, K. Bobzin, M. Maes, K. Lackner, R. Poprawe, E. W. Kreutz & J. Willach, Laser Drilled Microholes in Zirconia Coated Surfaces using Two Variants to Implement the Effusion Cooling of First Stage Turbine Blades, Advanced Engineering Materials **7**(3), 145–152, 2005
57. R. Poprawe, I. Kelbassa, K. Walther, M. Witty, D. Bohn & R. Krewinkel, Optimising and manufacturing a Laser-drilled cooling hole geometry for effusion-cooled multi-layer plates, ISROMAC-12, USA, 2008
58. E. W. Kreutz, L. Trippe, K. Walther & R. Poprawe, Process development and control of laser drilled and shaped holes in turbine components, Journal of Laser Micro/Nanoengineering **2**(2), 123–127, 2007
59. R. S. Patel & J. M. Bovatsek, Method for laser drilling a counter-tapered through-hole in a material, US, US6642477B1, 2003
60. I. Kelbassa, K. Walther, L. Trippe, W. Meiners & C. Over, Potentials of manufacture and repair of nickel base turbine components used in aero engines and power plants by laser metal deposition and laser drilling, International Symposium on Jet Propulsion and Power Engineering, **1**, 207–215, 2006
61. E. W. Kreutz, L. Trippe, K. Walther & R. Poprawe, Process development and control of laser drilled and shaped holes in turbine components, International Congress on Laser Advanced Materials Processing, Kyoto, Japan, 2006
62. A. Gillner & W. Wawers, Drilling holes, Industrial Laser Solutions, 29–31, 2006
63. F. Dausinger, T Abeln, D. Breitling, J. Radtke, V. Konov, S. Garnov, S. Klimentov, T. Kononenko & O. Tsarkova, Bohren keramischer Werkstoffe mit Kurzpuls-Festkörperlaser, LaserOpto **78**(31), 1999
64. M. Honer, Prozesssicherungsmaßnahmen beim Bohren metallischer Werkstoffe mittels Laserstrahlung, Utz, München, 2004
65. K. Jasper, Neue Konzepte der Laserstrahlformung und –führung für die Mikrotechnik, Utz, München, 2002
66. W. Wawers, Präzisions-Wendelbohren mit Laserstrahlung, Shaker, Aachen, 2008
67. J. Radtke, C. Föhl, K. Jasper & F. Dausinger, Helical drilling of high quality micro holes in steel and ceramics with short and ultrashort pulsed lasers, Laser in manufacturing, WLT-Conference on Lasers in Manufacturing, 331–340, 2001

68. M. Panzner, Bohren von Silizium durch den Einsatz diodengepumpter Festkörperlaser mit Frequenzvervielfachung. Forum Photonics BW, Stuttgart, 2002
69. U. Emmerichs, Präzisionsabtragen durch Ultrakurzpulslaser. Bayerisches Laserseminar Laserstrahlbohren, Bayerisches Laser Zentrum, Erlangen, 2003
70. J. Kaspar, A. Luft, S. Nolte, M. Will & E. Beyer, Laser helical drilling of silicon wafers with ns to fs pulses: Scanning electron microscopy and transmission electron microscopy characterization of drilled through-holes, Journal of Laser Applications **18**, 85–92, 2006
71. W. A. Dage & W. G. Frederick Jr., Optical Drilling Head for Lasers, 3799657, 1974
72. R. L. Leigthon, Laser hold drilling system with lens and two wedge prisms including axial displacement of at least one prism, 4822974, 1989
73. T. Graf & H. Hügel, 20 Jahre ganzheitliche Laserforschung am IFSW, Laser Technik Journal LTJ **5**, 8–14, 2006

Chapter 17
Cutting

17.1 Laser Oxygen Cutting

B. Seme and Frank Schneider

17.1.1 Introduction

Laser oxygen cutting uses oxygen as the process gas. In comparison with inert gases used for laser fusion cutting, such as nitrogen, oxygen provides not only for the ejection of the molten material out of the cut kerf but also for the burning of the material. In an exothermic reaction iron oxide is basically generated with a reaction enthalpy of approx. 4,800 kJ per 1 kg iron. When the oxygen reacts with the iron exothermically, an energy supply in addition to the laser radiation is provided. At the same laser power, cutting speeds for oxygen cutting are, in general, higher than for fusion cutting. According to the portion of the reaction energy to the energy by laser radiation, the difference in speed is significant for thick materials cut with low laser power and nearly negligible for thin materials and higher laser power. In contrast to shining cut surfaces of cuts by the fusion cutting process, the cut surfaces of the oxygen cuts have a thin oxide layer, which has to be considered for potential subsequent processes, e.g., painting or welding of the cuts. Laser oxygen cutting is used for sheet thicknesses up to approx. 25 mm. With higher sheet thicknesses, the processing speed is only marginally higher at lower quality compared with cost-saving conventional flame cutting. By a combination of laser oxygen cutting and flame cutting, the so-called burning stabilized laser oxygen cutting [1], or by the laser-assisted oxygen cutting, Lasox [2], significantly higher sheet thicknesses of more than 25 mm can also be cut.

B. Seme and F. Schneider (✉)
Trenn- und Fügeverfahren, Fraunhofer-Institut für Lasertechnik,
5207 Aachen, Germany
e-mail: frank.schneider@ilt.fraunhofer.de

17.1.2 Power Balance for Laser Oxygen Cutting

The total power P_c required for the cutting process consists of the power for heating P_w, for melting P_m, and of thermal leakage power (dissipation loss) P_{HL}. The heating power P_w is required for heating up the material that is ejected out of the cut kerf to the process temperature T_p between melting and evaporating temperature of the material.

The total power P_c has to be balanced by the power from the combustion of the material and the absorbed laser power. P_w and P_m can be calculated as

$$P_w + P_m = b_c d v_c \rho \cdot (c_p \cdot (T_p - T_\infty) + H_m) \tag{17.1}$$

where b_c is the cut kerf width, d the sheet thickness, v_c the cutting speed, ρ the density of the material, c_p the specific heat capacity, T_p the process temperature, T_∞ the ambient temperature, and H_m the specific enthalpy of fusion of the material. With the Peclet number $Pe = v_c b_c / 2\kappa$

$$P_w + P_m = 2Kd \cdot (T_p - T_\infty + \frac{H_m}{c_p}) \cdot Pe \tag{17.2}$$

where κ is the thermal diffusivity and K the thermal conductivity with

$$\kappa = \frac{K}{c_p \rho} \tag{17.3}$$

Without the derivation of the dissipation losses, P_{HL} is according to [3]

$$P_{HL} = 4Kd \cdot (T_m - T_\infty) \cdot \left(\frac{Pe}{2}\right)^{0.36} \tag{17.4}$$

Here T_m is the melting temperature of the material. Altogether P_c is given by

$$P_c = 4Kd \cdot \left[\left(T_p - T_\infty + \frac{H_m}{c_p}\right) \cdot \frac{Pe}{2} + (T_m - T_\infty) \cdot \left(\frac{Pe}{2}\right)^{0.36} \right] \tag{17.5}$$

The power arising from the iron combustion is

$$P_R = \rho b_c d v_c H_{FeO} X_{abbr} = 2Pe \cdot \frac{K}{c_p} \cdot d H_{FeO} X_{abbr} \tag{17.6}$$

where H_{FeO} is the reaction enthalpy (4,800 kJ/kg) for the formation of FeO from iron and oxygen and X_{abbr} is the burn-off, i.e., the mass ratio of iron reacting in the kerf. For a complete combustion X_{abbr} is 1. For the exothermal reaction of iron, a geometry-specific threshold temperature, the ignition temperature, must be exceeded. For temperatures higher than the ignition temperature, the heat of the reaction is higher than the dissipative heat losses and so the reaction is self-sustaining. The ignition temperature for mild steel is approx. 1,000 °C.

The second power source is the absorbed laser power P_{abs}. This is the laser power P_L at the workpiece multiplied by the absorption coefficient A:

$$P_{abs} = A \cdot P_L \qquad (17.7)$$

A comprehensive description of the process including the thermodynamics of the oxidation reaction can be found in [4].

17.1.3 Autogenous Cutting

In autogenous cutting processes the cutting power P_c is provided by the combustion power P_R. Usually a flame surrounding the oxygen gas jet is used for heating the material to the ignition temperature (Fig. 17.2).

The minimal burn-off of an autogenous cutting process is with $P_c = P_R$:

$$X_{abbr} = \frac{c_p}{H_{FeO}} \left[T_p - T_\infty + \frac{H_m}{c_p} + (T_m - T_\infty) \cdot \left(\frac{Pe}{2}\right)^{-0.64} \right] \qquad (17.8)$$

Since the maximum burn-off is 100% ($X_{abbr} = 1$), there is a minimal Peclet number and thus a minimal cutting speed that is necessary for a self-sustaining cutting process:

$$Pe_{min} = 2 \left[\frac{H_{FeO} - H_m}{(T_m - T_\infty)c_p} - 1 \right]^{-1.5625} \Rightarrow v_{c\ min\ autogen} = \frac{2\kappa}{b_c} Pe_{min} \qquad (17.9)$$

Here the approximation $T_p = T_m$ is used. The calculation shows that the necessary burn-off is independent of the sheet thickness. That means that in autogenous cutting there is no upper limit for the maximal sheet thickness. In fact even more than 3 m thick steel was cut.

Figure 17.1 exemplifies the total power P_c and the combustion power P_R for mild steel with a thickness of 10 mm, a cutting kerf width of 1 mm, and a 100% burn-off. Here the minimal Peclet number is $Pe_{min} = 0.253$. This results in a minimal cutting speed of $v_{c\ min\ autogen} = 0.24$ m/min for a self-sustaining process. For a cutting speed of 1 m/min a burn-off of 3.56 kW/6.28 kW = 57% would be sufficient to provide the total power P_c.

However, in practice, a 100% burn-off cannot be reached over the total speed range. For cutting speeds higher than v_{diff}, in this example approx. 0.5 m/min, the diffusion velocity of iron and oxygen through the molten iron oxide and the gas boundary layer limits the combustion of iron. This leads to a curve, indicated

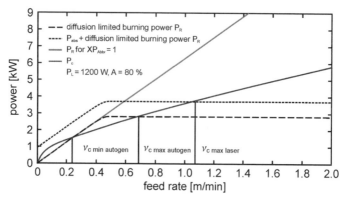

Fig. 17.1 Power balance for autogenous flame cutting of 10 mm mild steel S355J2G3

$K = 40$ W/mK; $T_m = 1{,}780$ K; $H_m = 205$ kJ/kg; $H_{FeO} = 4{,}800$ kJ/kg; $c_p = 650$ J/kgK; $\kappa = 8 \cdot 10^{-6}$ m^2/s; $b_c = 1$ mm; $T_\infty = 293$ K; $T_P = 2{,}000$ K; $\rho = 7{,}850$ kg/m^3; $d = 10$ mm

as diffusion-limited combustion power P_R in Fig. 17.1, that is flat for speeds higher than v_{diff}. Therefore, the maximum cutting speed in autogenous cutting is $v_{c\ \text{max autogen}} < 1$ m/min for a sheet thickness $d \geq 10$ mm. A detailed description of diffusion processes in autogenous cutting can be found in [5]. The combustion power can provide the total cutting power only in the range of cutting speeds between $v_{c\ \text{min autogen}}$ and $v_{c\ \text{max autogen}}$. Also in the range of this cutting speed a flame is necessary for cutting. If this additional heat source is turned off after ignition of the combustion, the combustion stops because the surface of the material is cooled by the oxygen gas jet. This lets the temperature at the upper edge of the cutting front fall below the ignition temperature. Additionally, oxide at the surface can cover the iron and inhibit the combustion.

The flame prevents the formation of scale and can remove present scale (which originates outside the flame). Through the permanent heating of the cut front upper surface to the ignition temperature, the ignition conditions for the iron combustion are hence maintained. The burner flame does not, however, heat the cutting front in the kerf and, therefore, does not contribute to the cutting power P_c. With autogenous cutting, sheet thicknesses of up to approx. 100 mm can be cut with roughness depths of under 50 μm with only very low kilowatt burner power. Due to the diffusion limitation of the iron combustion, the maximum cutting speed is limited, however, to $v_{c\ \text{max autogen}}$. An increase of speed is possible by coupling laser radiation in the cut kerf. This case is described in the following:

17.1.4 Procedural Principle

To the left, Fig. 17.2 shows the principle of laser oxygen cutting, which is identical to laser fusion cutting except for the use of the cutting gas, oxygen. The laser beam

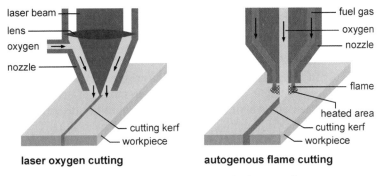

Fig. 17.2 Laser oxygen cutting and autogenous flame cutting in comparison

heats the workpiece to ignition temperature and injects additional energy into the cut kerf. In Fig. 17.1 the curve "P_{abs}+ diffusion-limited combustion capacity P_R" shows the available power for laser oxygen cutting in the case of a laser output P_L of 1,200 W on a workpiece with a sheet thickness of 10 mm and a kerf width of 1 mm. As an absorption degree A, 80% was assumed, since the oxide layer on the cut front caused by the combustion leads to an increase of absorption in comparison to fusion cutting. In the above example, the maximum cutting speed $v_{c\ max\ laser}$ amounts, thus, to approx. 1.1 m/min.

Figure 17.3 shows maximum cutting speeds for laser oxygen cutting of mild steel at 1,200 W and 2,300 W CO_2 laser power on the workpiece.

For $P_L = 1,200$ W and 10 mm sheet thickness, the maximum cutting speed amounts to 1.2 m/min. This agrees clearly with the above calculation (Fig. 17.1). In general the laser power needed for cutting using laser oxygen cutting has to increase with increasing sheet thickness. Figure 17.3 demonstrates that using 1,200 W laser power a sheet thickness of 25 mm can no longer be cut, even at very low cutting

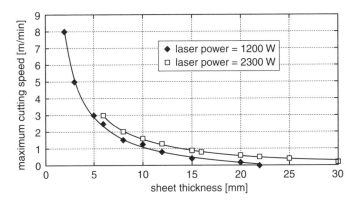

Fig. 17.3 Maximum cutting speed for laser oxygen cutting at 1,200 W and 2,300 W CO_2 laser beam power on the workpiece [5]

speeds. At 2,300 W laser power, the maximum cutting speed amounts to only approx. 0.5 m/min. This fact would be understandable only in view of the high reaction power, if cutting speeds above $v_{c\ max\ autogen}$ were to be achieved. In fact, for every laser power of laser oxygen cutting, there is a maximum sheet thickness that can be cut. The reason for this is potential instabilities of the iron combustion during laser oxygen cutting in contrast to autogenous cutting. That is, less reaction power is released than would normally be possible. To illustrate this, Fig. 17.2 shows a comparison of laser oxygen cutting with autogenous flame cutting. The essential difference between both processes consists in that the flame during autogenous cutting heats up a large surface of the outer edge of the cutting front outside the cutting nozzle and kerf, while during laser oxygen cutting the laser beam runs coaxial to the cutting nozzle and has a smaller diameter than the oxygen stream at the workpiece surface. For laser oxygen cutting with lower speeds (< 1 m/min) an ignition of iron combustion can occur within the oxygen stream but outside the laser beam, since the material around the laser beam is heated up to combustion temperature through heat conduction. Thus, the cutting front runs out of the laser beam. This "precombustion," first described by Arata [6], is, however, transient since the combustion reaction extinguishes without contacting the laser beam and is then ignited anew, when the laser beam catches up to the cutting front through the feed process. Such period interruptions of the burn-off appear as coarse striations ($R_z > 0.1$ mm) on the cut surfaces. Through the ignition of the iron combustion outside the laser beam, cut kerfs arise at lower feed speeds, which have the width of the oxygen beam, whereas at appropriate process cutting speeds, the kerf width approximately corresponds to the diameter of the laser beam on the workpiece surface.

Figure 17.4 shows a cut kerf of laser-cut mild steel sheet of 10 mm thickness. As a comparison, cut kerfs made using flame cutting and plasma cutting are also illustrated. With a laser power of 1,500 W, higher cutting speeds can be obtained than using flame and plasma cutting. The cut kerfs are significantly narrower using laser oxygen cutting than using the other processes and are formed nearly at right angles to the sheet surface. The width of the heat-affected zone for laser oxygen cutting amounts to a few tenths of a millimeter and for plasma cutting approx. 0.5 mm. Autogenous flame cuts are particularly thermally loaded due to the preheat flame, so that the width of the heat-affected zone increases to over 2 mm.

In order to reduce the formation of coarse striations in the sheet thickness range above 10 mm and in the speed range of under 1 m/min when using laser oxygen cutting, two measures are taken to reduce the above-mentioned "pre-heating" through dampening the iron combustion:

- Use of low oxygen-cutting gas pressures (less than 1 bar). As a comparison, in the sheet thickness range of 1–10 mm, pressures of approx. 1–5 bar are otherwise applied.
- Pulsing the laser radiation.

17 Cutting

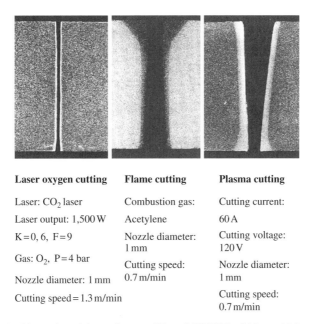

Laser oxygen cutting	Flame cutting	Plasma cutting
Laser: CO_2 laser	Combustion gas: Acetylene	Cutting current: 60 A
Laser output: 1,500 W	Nozzle diameter: 1 mm	Cutting voltage: 120 V
$K = 0, 6, F = 9$	Cutting speed: 0.7 m/min	Nozzle diameter: 1 mm
Gas: O_2, $P = 4$ bar		
Nozzle diameter: 1 mm		Cutting speed: 0.7 m/min
Cutting speed = 1.3 m/min		

Fig. 17.4 Cut kerf formation of thermally cut mild steel (S235JR) of 10 mm thickness [7]

In particular the modulation of the laser beam is applied to avoid burnouts at low cutting speeds and in those parts of the workpiece, which are thermally loaded, because the geometry leads to a heat accumulation, e.g., in sharp-cornered contours.

17.1.5 Burning-Stabilized Laser Oxygen Cutting

Burning-stabilized oxygen cutting can be applied for high sheet thicknesses above the range of the standard laser oxygen cutting process of approx. 25 mm. When using burning-stabilized laser oxygen cutting, in addition to the laser radiation, which is, as usual, placed coaxially to the cutting nozzle, a second heating source (e.g., a burner flame, an inductive heating, or a second laser beam) is used, with which the cutting front upper edge is heated to combustion temperature (Fig. 17.5). In contrast to laser oxygen cutting, this method requires a heated spot to be produced, whose diameter is larger than the diameter of the oxygen jet. The heated spot is found in the cutting direction in front of the cutting nozzle or it is placed around the nozzle. In both cases, it should overlap, at least partially, the oxygen stream. Using such an auxiliary heating source, the periodic extinguishing of the iron combustion at lower cutting speeds can be prevented (see Section 17.1). As a temporal average, in comparison to laser oxygen cutting, an increased combustion power is released. The iron combustion is, thus, intensified here and not, as in the procedure

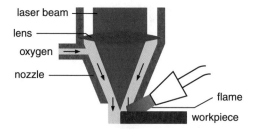

burning-stabilized laser oxygen cutting

Fig. 17.5 Sketch of burning-stabilized laser oxygen cutting

described in the last section, dampened. In this way, quicker cutting speeds are possible than when using laser oxygen cutting. If only the stabilizing preheating power of the burning-stabilized process is used for cutting, then the cutting process passes over into autogenous flame cutting. This means that, theoretically, as with autogenous flame cutting, any sheet thickness can be cut using the burning-stabilized laser oxygen cutting process. Figure 17.6 shows the maximum cutting speeds when using burning-stabilized laser oxygen cutting at 4.9 and 2.4 kW CO_2 laser beam power. The heated spot was produced with an acetylene burner (see Fig. 17.5). As a comparison, the maximum cutting speeds with laser oxygen cutting at 2.4 kW laser power and with autogenous flame cutting are also plotted. The illustration demonstrates that with the burning-stabilized laser oxygen cutting process, due to the additional laser power injected into the cut kerf, higher cutting speeds can be reached than with autogenous flame cutting techniques. Figure 17.6 also shows that with laser oxygen cutting at a laser power of 2.4 kW, the cutting speed can be increased by up to 30% through burning stabilization.

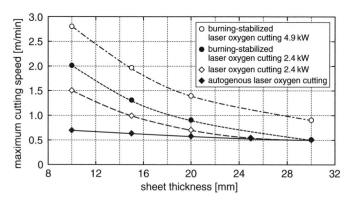

Fig. 17.6 Cutting speeds with burning-stabilized laser oxygen cutting and autogenous flame cutting [1]

17.2 Fusion Cutting

Frank Schneider

17.2.1 Introduction

For laser fusion cutting, an inert gas is applied to expel material. The power needed to warm, melt, and partially vaporize the cut kerf material is covered solely by the absorbed laser radiation.

Fusion cutting is conducted with significantly higher cutting gas pressures than is the case with laser oxygen cutting, from approx. 0.5 MPa up to over 2 MPa, depending on the material thickness. Thus, the term "high pressure cutting" is widely used for fusion cutting.

The field of application of this process is to cut material for which oxidation of the cut edges needs to be prevented, thus, above all, for the processing of stainless steels. With CO_2 lasers, stainless steel up to a material thickness of 40 mm can be industrially cut; typically, however, material thicknesses below 20 mm are processed. Figure 17.7 shows a typical cut edge in stainless steel with 10 mm thickness. As cutting gas, nitrogen is used. For a few applications, other inert gases are utilized, such as argon to cut titanium. For both fusion cutting and oxygen cutting equally, the largest field of application is the 2D cutting of flat material, preferably using CO_2 lasers. For 3D contour cutting, solid-state lasers are primarily used in order to take advantage of the beam guidance via fiber optic cables. Due to their good beam quality, the types used are particularly fiber lasers and disc lasers. The special advantages of the fusion cutting process are

- oxide-free cut edges and
- nearly perpendicular cut edges with high contour accuracy, especially in the corners.

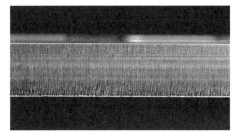

Fig. 17.7 Cut edge of stainless steel 1.4301, sheet thickness 10 mm, cutting speed 1.5 m/min

---

F. Schneider (✉)
Trenn- und Fügeverfahren, Fraunhofer-Institut für Lasertechnik, 5207 Aachen, Germany
e-mail: frank.schneider@ilt.fraunhofer.de

In comparison with laser oxygen cutting, laser fusion cutting has a higher gas consumption and larger laser power demands. With a given laser beam power, the maximum workable sheet thicknesses for fusion cutting are smaller than for oxygen cutting, or rather with the same material thickness, higher cutting speeds can be obtained with oxygen cutting in the thick sheet range with low laser power.

17.2.2 Process Parameters

17.2.2.1 Cutting Speed

For fusion cutting, the cutting speed can be limited by the insufficient ejection of the melt out of the cut kerf, or by a power demand not covered by the absorbed laser beam power. With an appropriate choice of the cutting gas parameters, the latter – the failure to fulfill the power balance – is the limiting sub-process. The power balance for fusion cutting can be set up analogous to that for laser oxygen cutting, while leaving out the exothermic combustion power P_R. Corresponding to the share of the warming and fusion power of the power balance, the cutting speed attainable is inversely proportional to the material thickness (Fig. 17.8). Deviations result at lower cutting speeds through a share of lateral thermal dissipation losses that cannot be ignored. Extensive description of the power balance for fusion cutting can be found in [8] and [9].

Application-oriented cutting speeds lie in the area of 80% of the maximum speed. The power reserve serves to catch process instabilities, for example, through thermal fluctuations of the focus position, of the nozzle distance, or of the laser power. With a

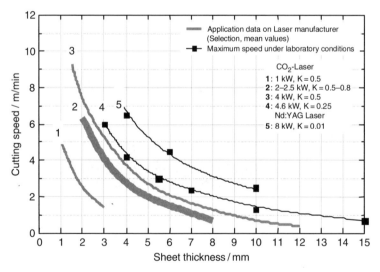

Fig. 17.8 Cutting speeds for laser fusion cutting of stainless steel with CO_2 lasers ($M^2 \approx 2$), selection of application data, average values

large excess of power for reasons of low cutting speed, a steep cutting front is formed and a large part of the laser radiation is transmitted through the kerf "unused." The temperature of the melt film surface and the melt film thickness goes down with decreasing cutting speed. With a low speed of the melt and higher viscosity, the removal of the melt at the lower cut edge is impeded, so that burr formation can occur at low cutting speeds. If the speed has to be reduced, for example, to machine contours, burr-free quality can be attained by pulsing the laser radiation.

17.2.2.2 Focusing

To set up an appropriate focusing for a cutting task, an appropriate combination of a small focus diameter and sufficiently long Rayleigh length has to be chosen, depending on the material thickness and the beam quality, usually with the goal of producing a narrow cut kerf to minimize the power demands and to have a sufficient tolerance in the focal position. The use of lasers with high beam quality supports both criteria. A Rayleigh length is commonly used in the range of half to double sheet thickness. With the given beam quality of the laser, the range of the usable F-numbers is thus given.

For fusion cutting, the focal position typically lies between the top and the bottom side of the material. A low position to the bottom side leads to an improved coupling of the cutting gas jet in the kerf and can be reasonable when burr formation otherwise occurs. The maximum speed – depending on the focal position – is reached at higher focal positions in the area of the middle of the material to the material surface.

17.2.2.3 Cutting Gas Flow

The parameters of the cutting gas flow aim at attaining as good a coupling of the gas stream as possible into the cut kerf in order to expel the melted material. High pressure, a small nozzle distance, and a large nozzle orifice have a positive effect, but there are undesirable negative influences that limit an appropriate operational range.

Cutting gas pressure that is too low produces an insufficient expulsion of melt and leads to burr formation as well as to reduced cutting speed. Pressure that is too high hardly has a positive effect on cutting speed or cutting quality, rather it creates high gas consumption and can be destabilizing by promoting plasma formation. The optimal cutting gas pressure p_0, measured in the pressure chamber of the nozzle, increases with the sheet thickness from approx. $p_0 = 0.5$ MPa (pressure difference to the ambient pressure p_u) at material thicknesses in the area of a millimeter to over 2 MPa for sheet thicknesses of 15 mm and more.

Commonly, conic nozzles with simple cylindrical or conical outlets are utilized. Since the cutting gas pressure during fusion cutting lies over the critical pressure of 1.89 p_u, the cutting gas streams to the nozzle opening at the speed of sound in an under-expanded jet. After exiting, the jet expands at supersonic speed, and above the workpiece a vertical shock wave forms. A part of the gas jet is coupled

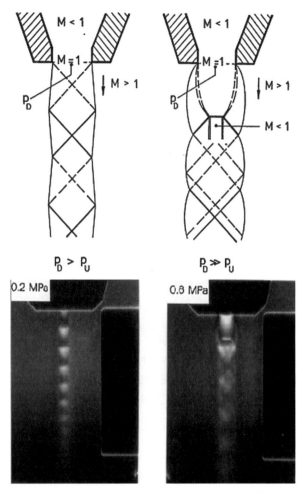

Fig. 17.9 Schematic and Schlieren optical representation of the free jet under a cutting nozzle

in the cut kerf and forms periodic compression and expansion zones. The effects of the supersonic phenomena as well as the investigation of the boundary layer of the jet can be observed with Schlieren diagnostics. Density gradients of the gas are visualized in this observation method as bright-dark gradients. Figure 17.9 shows a Schlieren optical picture of a free jet of a cutting nozzle. In the free jet, a vertical shock wave forms with a strongly under-expanded jet.

In order to calculate the gas consumption of a cutting head through a nozzle, the mass flow rate is given according to [10] for an isentropic expansion by

$$\dot{m} = \rho_0 c_0 A M a \left(1 + \frac{\kappa - 1}{2} M a^2 \right)^{\frac{\kappa+1}{2(\kappa-1)}} \quad (17.10)$$

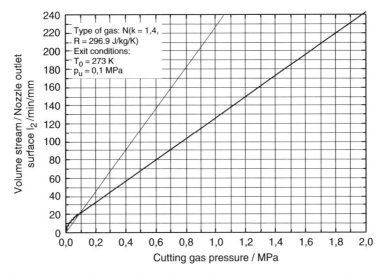

Fig. 17.10 Gas consumption dependent on the cutting gas pressure; volume flow in norm liters, $p_n = 0.10113$ MPa, $T_n = 273.15$ K (0 °C)

When cutting gas pressures are above critical pressure, the Mach number at the nozzle exit for conical nozzles is limited to Ma $= 1$ and the mass flow rate depends linearly on the exit surface A and the cutting gas pressure p_0. Figure 17.10 shows the gas consumption in the common specification of the volume flow under normal conditions, instead of the mass flow rate.

Typical nozzle diameters lie between 1 and 3 mm.

For a standard fusion cutting process, nozzle distances between 0.5 and 2.0 mm have proven useful, and typical values lie around 1 mm. As when pressure is too high, a nozzle distance to the workpiece that is too small can promote the formation of plasma. Technically more relevant, however, is a lower limit of the nozzle distance, which results from the process stability. When distances are too small and the material is uneven, a collision of the tip of the nozzle and the material surface cannot be excluded with certainty despite using a capacitive distance control as is standard in laser cutting.

17.2.3 Fusion Cutting with Mirror Optics and Autonomous Nozzles

The use of cutting lenses for CO_2 lasers is limited for laser beam power up to approx. 6–8 kW, because intensities are attained in this output range that lead to strong thermal loads on the lenses and corresponding effects, such as a shift of the focal position, until the thermally stable operating point is reached. At this load the optics are very sensitive to fouling and the lifetime of the lenses is reduced. Focusing mirror optics can be used, as they are for welding, since they are more robust and can be employed without limitation regarding laser beam power.

With common laser cutting heads, the space between the cutting nozzle and the focusing lens forms the pressure chamber to build up the cutting gas pressure. The lens' task is, thus, not only to focus the laser beam but also to tightly seal the pressure chamber to the laser side of the head.

When mirror optics are used to focus the beam, the space above the nozzle is open. To build up the cutting gas pressure, double-walled nozzles (autonomous nozzles) can be used in this case, by which the cutting gas pressure is built in a ring-shaped structure. These nozzles as well produce a cutting gas stream that runs coaxially to the laser beam and have characteristics corresponding to those produced with conical standard nozzles.

Figure 17.11 shows the principle setup of the autonomous nozzle. The cross section in the streaming channel has to be dimensioned in such a way that neither air is sucked in the opening through which the laser passes, thus contaminating the cutting gas, nor a leakage stream that is too large exits through this opening.

These nozzles can be used for laser oxygen cutting as well, but the range of parameters, in particular for fusion cutting of large material thicknesses, is covered in combination with mirror optics and high laser beam power.

Figure 17.12 shows the cut edges and a cross section of cuts in stainless steel.

17.2.4 Sample Applications

In the 2D field of application, flat bed machines with CO_2 lasers and a laser beam power between 2 and 6 kW are commonly used; this is seen, for example, in "laser job shops" by service companies for sheet metal cutting. These kinds of companies take advantage of the high flexibility and productivity of the machines, e.g., for producing small and medium batch sizes.

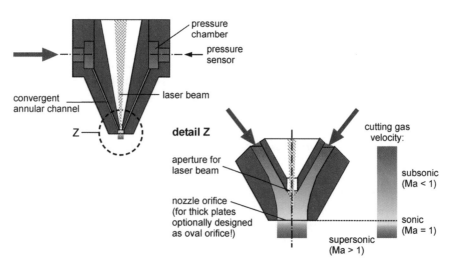

Fig. 17.11 Schematic of an autonomous cutting nozzle

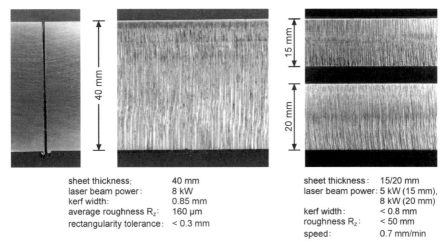

sheet thickness: 40 mm
laser beam power: 8 kW
kerf width: 0.85 mm
average roughness R_z: 160 µm
rectangularity tolerance: < 0.3 mm

sheet thickness: 15/20 mm
laser beam power: 5 kW (15 mm), 8 kW (20 mm)
kerf width: < 0.8 mm
roughness R_z: < 50 mm
speed: 0.7 mm/min

Fig. 17.12 Cut edges from fusion cuts in thick sheet (stainless steel) with autonomous nozzles

Applications in the 3D field lie, for example, in the processing of pipes. For this, special laser pipe cutting machines are available on the market, with which complete lengths of pipe can be automatically loaded and processed. In addition to a turning axis and the axial feed of the pipe, the machines have at least two further axes to position the cutting head vertical to the pipe axis. This way, due to extensive automation, even more complex contours can be cut, making innovative pipe connections possible.

A further example is the 3D cutting of deep-drawn parts, e.g., the trimming and cutting of holes and cut-outs in components from vehicle construction. Since high-strength working materials are being used increasingly, laser processing is becoming more and more important compared to mechanical cutting, with which such materials can only be processed with great difficulty.

17.3 High-Speed Cutting

Frank Schneider

17.3.1 Introduction

In standard laser oxygen and laser fusion cutting, the laser beam is absorbed at the cutting front, and the molten material is ejected by the cutting gas along the front.

F. Schneider (✉)
Trenn- und Fügeverfahren, Fraunhofer-Institut für Lasertechnik, 5207 Aachen, Germany
e-mail: frank.schneider@ilt.fraunhofer.de

Forced by shear stress and a pressure gradient from the cutting gas, the melt is accelerated from the top to the bottom side of the material.

The temperature of the melting layer is strongly dependent on the cutting speed. The reason is that with increasing cutting speed the heat flow through the melt to the liquid–solid interface has to increase, and also the mass flow of molten material and the thickness of the melting layer increases. To achieve a stationary balanced energy flow at the interphase under these conditions, the temperature on the melt surface has to increase, resulting in a higher portion of material that is heated up even beyond vaporization temperature [8].

The evaporated material leads to azimuthal pressure gradients induced by the vapor pressure on the cut front. These pressure gradients accelerate the melt to the edges of the emerging cut kerf. With increasing cutting speeds the melt flows more and more around the laser beam, prior to the ejection forced by the downward-oriented acceleration from the cutting gas. On the side of the interaction zone opposed to the cutting front, the melt is accumulated and a keyhole is formed. The ejection of the melt out of the cut kerf takes place some focal diameter behind the laser beam position and the keyhole. This cutting process with initially dominant azimuthal melt flow and a keyhole is called high-speed cutting [11, 12]. However, in many publications the term high-speed cutting is used also for a fast standard cutting process, regardless of the details of the laser–material interaction zone. Figure 17.13 shows sketches comparing the interaction zones in conventional and in high-speed cutting.

The dominantly azimuthal melt flow generates a convective energy transport to the edges of the kerf, leading to an enlarging of the cut kerf. Whereas in a standard cutting process the cut kerf width roughly equals the size of the laser beam focus diameter, in a high-speed cutting process the cut kerf width can be twice as wide as the laser beam diameter. Figure 17.14 shows the cut kerf width versus the maximum cutting speed. *Maximum* cutting speed implies that for each cutting speed in Fig. 17.14, just the minimal necessary laser power was applied.

Besides the melt flow, the absorption mechanism in high-speed cutting is also different compared to standard cutting. The laser radiation is absorbed at the front not

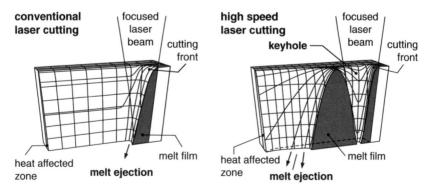

Fig. 17.13 Schematical comparison of the laser–material interaction zone for conventional cutting and high-speed cutting

17 Cutting

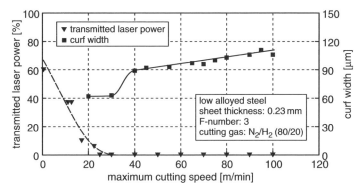

Fig. 17.14 Cut kerf width and transmitted laser power through the cut kerf at minimum laser power adapted to the cutting speed

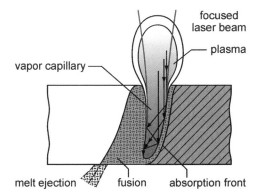

Fig. 17.15 Sketch of the keyhole illustrating increased absorption by multiple refection

only in direct absorption of the laser beam and absorption of its (multiple) reflected parts but also in absorption in the backside melt accumulation. At the backside reflected laser radiation can be absorbed at the front and thus contribute to good process efficiency (Fig. 17.15). The efficient absorption in high-speed cutting is evident by the amount of transmitted laser radiation through the cut kerf: employing the minimal necessary laser power per speed, at cutting speeds higher than 30 m/min, no transmitted laser power can be detected below the cut kerf (Fig. 17.14). When cutting with a considerable laser power reserve, the capillary is open at the bottom side. A closed capillary can be observed at speeds higher than 80% of the maximum speed.

17.3.2 Process Description

High cutting speeds, leading to a high ratio of vaporization and thus establishing the high-speed cutting process phenomena, are achieved in thin material thicknesses, with small focal diameter and high laser power. In sheet thicknesses in the range of 1 mm, typically below 1 mm, focusing numbers in the range of 3–5 can be employed. This leads to short but sufficient Rayleigh lengths suitable for the thin material thickness. The process demands excellent laser beam quality, provided by multi-kilowatt CO_2 lasers with $M^2 < 1.1$ (BBP < 3.8) and recent fiber and disk lasers with similar beam parameter products, providing fibers from 100 μm diameter down to single mode fibers with diameters smaller than 20 μm. For high-speed cutting these laser beams are focused to focus diameters below approx. 100 μm. This leads to cutting speeds over 100 m/min in 1 mm sheets.

Besides the laser beam parameters, the cutting gas flow has a significant impact on the process. The primary function of the cutting gas is the melt ejection. Moreover, the cutting gas shields the lens from sputter material, e.g., during piercing. Commonly the cutting gas flow is used in the cutting head to cool the lens. Using CO_2 lasers for high-speed cutting with intense plasma formation, the plasma can be influenced by the kind of gas and the gas pressure. Positive results are found by mixing gas with low atomic weight to the inert cutting gas, in particular by mixing hydrogen to nitrogen. Hydrogen enhances the recombination rate and "cools" the plasma. The result is a slight increase of the cutting speed and a stabilization of the process.

The use of oxygen is not advantageous in high-speed cutting. In standard cutting the exothermal reaction and the higher absorption increase the cutting speed in comparison with the use of inert gas in fusion cutting at the same laser power. With decreasing sheet thickness, this effect diminishes. In high-speed cutting of thin sheets in the thickness range of some tenths of a millimeter, the speed with oxygen is about 10% higher than with inert gas. However, the slight increase in speed is associated with a significant loss in quality, because the melt is not only oxidized at the absorption front but also uncontrolled in the melt pool behind the front.

In the high-speed range an excellent, burr-free cutting quality is obtained. However, at low speeds, e.g., during the acceleration of the machine, the melt is ejected in a standard cutting process driven by sheer stress and pressure gradients along the cutting front. With thin sheets, the distance where the gas can accelerate the melt is low, and in the low-speed range, the melt film is thin and the surface tension is high because of low temperatures. All of these constrain the ejection and lead to a drop-shaped burr. For a burr-free quality also in the low-speed range, the laser power must be modulated. High intensities during beam on times generate for short times high temperatures and vapor pressure that accelerate the melt and lead to an effective ejection.

The modulation parameter frequency and duty cycle should dynamically be adapted to the cutting speed. Important boundary conditions are sufficient average power and sufficient overlap of laser pulses.

With increasing sheet thickness, the change of the process from high-speed cutting with a keyhole formation to standard cutting is smooth. Employing high laser power, the process is also in the thickness range of 1 mm controlled by a high ratio of vaporization, but typical characteristics of the high-speed process as the enlarging of the kerf width compared to the beam diameter are not found. Using CO_2 lasers for this fast cutting process, which is not a high-speed process as defined above, a parameter optimization aims at reducing a shielding effect of the plasma, e.g., by using a moderate gas pressure and a high nozzle stand-off.

17.3.3 Application Examples

Figure 17.16 shows the maximum cutting speed for electrical steel (insulated, low-alloyed steel) dependent on the sheet thickness for a CO_2 laser at 2.5 kW laser power. The maximum speeds can be reached in 1D applications as slitting and trimming of strip steel [13–15]. For 2D applications, e.g., manufacturing of repair sheets for generators [16], the cutting speed usually is limited by the dynamics of the machines. However, at lower speeds the good process efficiency of the high-speed cutting process allows the use of lower laser power. Figure 17.17 describes the maximum cutting speeds for various non-ferrous metals.

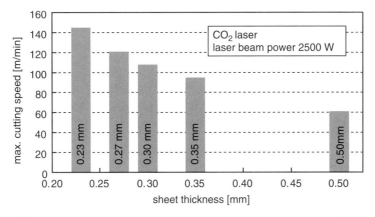

Fig. 17.16 Maximum cutting speeds for electrical steel (insulated, low-alloyed steel) [13]

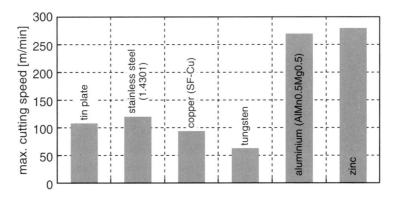

Fig. 17.17 Maximum cutting speeds for various materials [11]

17.4 Sublimation Cutting

B. Seme and Frank Schneider

17.4.1 Introduction

During laser beam sublimation cutting, the workpiece is converted from the solid to the gaseous aggregate state in the area of the kerf due to the influence of the laser radiation. According to the type of material to be cut, the transformation from the solid to gaseous aggregate state occurs through vaporization (e.g., Plexiglas), sublimation (e.g., graphite), or, in most cases, through chemical decomposition (e.g., polymers, wood). In the last case, complex molecules are decomposed into smaller molecules and volatile components, vaporize. Since the transition of the workpiece into the gaseous phase does not solely occur through sublimation, one also calls sublimation cutting "laser vaporization cutting." Because the working material vaporizes within the kerf, no process gas jet is needed to expel the material when using sublimation cutting. Commonly, however, an inert gas is utilized, coaxial to the laser beam, in order to protect the processing optics and to prevent the workpiece from oxidizing (see Fig. 17.18). Sublimation cutting occurs with nonmetals such as paper, wood, and several types of ceramics and plastics, all of which do not possess molten phases [9, 17].

Sublimation cutting is characterized by the following advantages in comparison to laser beam fusion cutting or oxygen cutting:

- Since very little molten material arises, smooth surface kerfs result without pronounced striation structures.

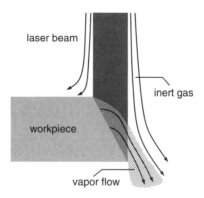

Fig. 17.18 Laser sublimation cutting

B. Seme and F. Schneider (✉)
Trenn- und Fügeverfahren, Fraunhofer-Institut für Lasertechnik, 5207 Aachen, Germany
e-mail: frank.schneider@ilt.fraunhofer.de

- Oxide-free cut kerfs arise as in laser fusion cutting. The workpiece can be processed further without post-treatment.
- The heat-affected zone at the margins of the cut edges and the total heat load of the workpiece are minimal.

In comparison to fusion cutting, sublimation cutting requires, however, significantly higher intensities and, hence, a laser of good beam quality.

17.4.2 Power Balance for Laser Sublimation Cutting

For the case that the entire material in the cut kerf vaporizes, the following power balance results for the cutting process:

$$AP_L = b_c d\, v_c \rho \left[c_p (T_V - T_\infty) + H_m + H_V \right] + P_{HL} \qquad (17.11)$$

The summands on the right side indicate the necessary output for the warming of ambient temperature to vaporization temperature, the power for the melting and vaporizing of the cut kerf volume, and the thermal dissipation losses. This power requirement has to be covered by the absorbed laser power. In the balance, A is the absorption degree of the workpiece, P_L the laser power, b_c the cut kerf width, d the workpiece thickness, v_c the cutting speed, ρ the work material density, c_p the specific heat conductivity of the workpiece, T_∞ the ambient temperature, T_V the vaporization temperature of the workpiece, H_m the specific fusion enthalpy of the workpiece, H_V the specific vaporization enthalpy, and P_{HL} the thermal dissipation losses from the cutting zone.

If the entire material is sublimated in the cut kerf, the sum of the fusion enthalpy and vaporization enthalpy has to be supplied as sublimation enthalpy during the transition from the solid to the gaseous state. That is, the specific sublimation enthalpy H_S is given by $H_S = H_m + H_V$, and the above-indicated power balance remains valid in the case of sublimation of the material in the cut kerf.

During chemical decomposition of the workpiece under the influence of the laser radiation, $H_m + H_V$ has to be replaced by the specific decomposition enthalpy H_Z and T_V by the decomposition temperature T_Z in the balance.

In Table 17.1, the necessary energies per volume for the heating and phase transformation of various metals or semi-metals are given. Here T_m indicates the fusion temperature. The total vaporization energy in each is a degree of magnitude higher than the necessary energy for the melting. For example, valid for iron is the following: $e_{ht}/(e_{hm} + e_m) = 8.5$. That is, in case different laser cutting processes can be chosen for a material, the sublimation cutting requires the highest power density. In order to reach the necessary high-power densities for sublimation cutting of the materials listed in Table 17.1, pulsed laser radiation is commonly used. The short effective duration of the intensive laser pulses limits, in addition, the heat transport into those materials with good heat conductivity. Hence the thermal dissipation losses are small and cuts are possible which have a very small influence

Table 17.1 Necessary energies (per volume) for heating and phase transformation of various materials [9]

	Heat energy (melt) $e_{hm} = \rho c_p / T_m - T_\omega$ (J/mm^3)	Latent heat of fusion $e_m = \rho H_m$ (J/mm^3)	Heat energy (vaporize) $e_{hv} = \rho c_p / T_m - T_\omega$ (J/mm^3)	Latent heat of vaporization $e_v = \rho H_v$ (J/mm^3)	Total heat of vaporization $e_{ht} = e_{hm} + e_m + e_{hv} + e_v$ (J/mm^3)	Melting energy $(e_{hm} + e_m)/e_{ht}$	Latent heat of vaporization e_v/e_{ht}
Si	2.27	4.28	1.54	27.37	35.46	18%	77%
Al	1.58	1.00	4.47	30.78	37.83	7%	81%
Fe	5.44	2.12	4.36	52.18	64.10	12%	81%
Ti	3.93	4.68	3.77	102.60	114.98	7%	89%
Cu	3.77	6.30	5.23	147.39	162.69	7%	91%

on the surrounding material. Sublimation cutting of the named materials is thus advantageous when complex contours should be cut out with very high precision from thin workpieces. In all of the other cases, fusion cutting is more advantageous when processing metals due to its lower power demand.

17.4.3 Sample Applications for Sublimation Cutting of Non-metals

For the sublimation cutting of non-metals such as ceramics, wood, paper, plastics, or leather, CO_2 lasers can be used efficiently, since the materials named exhibit a very high degree of absorption at a wavelength of 10.6 μm. Since, in addition, the heat conductivity of these materials is several degrees of magnitude lower than for metals (see Table 17.2), the thermal dissipation losses are much lower than when processing metals. The heat conductivity of the workpiece is, therefore, minimal. Figure 17.19 indicates cutting speeds for various non-metals at 500 W cw CO_2 laser power. Figure 17.20 shows a cut edge made in pine wood using CO_2 laser radiation.

With Nd:YAG lasers, many organic materials cannot be cut or only cut poorly, since they are transparent for radiation with a wavelength of 1,064 nm or only have a minimal degree of absorption. For some organic materials, the absorption can be increased using additives such as carbon dust so that here too the cutting with

Table 17.2 Comparison of heat conductivity of several metals and non-metals [18]

Material	Heat conductivity (W/mK)
Copper	4×10^2
Aluminum	2.4×10^2
Iron	8×10^1
Plexiglas	1.9×10^{-1}
Wood (Oak)	1.6×10^{-1}
Cotton	4×10^{-2}
Cork	3×10^{-2}

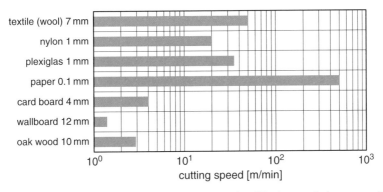

Fig. 17.19 Cutting speeds in several non-metals when using CO_2 laser radiation, power 500 W (cw) [18]

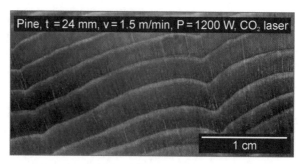

Fig. 17.20 Pine wood cut with laser radiation (Source: ILT)

Nd:YAG lasers is possible. The ability to cut strongly depends, however, on the type and amount of additive. Many non-organic non-metals can be cut well with pulsed Nd:YAG laser radiation. The Nd:YAG laser is mainly used here when precise cuts and complex contours are required.

17.5 Laser Fine Cutting

Arnold Gillner

17.5.1 Introduction and Application Areas

Laser fine cutting today is used for numerous applications in fine mechanics, medical industry, and electronics. The main fields of use can be found at thin materials with thicknesses < 0.5 mm and generally complex structures with feature sizes $< 200\,\mu$m. The main lasers used in this field are lamp-pumped solid-state lasers and fiber lasers for metals but also diode-pumped solid-state Nd:YAG and Nd:VO$_4$ lasers in Q-switch mode. The spectrum of materials and applications is summarized in Table 17.3.

Table 17.3 Application areas for laser fine cutting

Metals	Stents, stencils for solder paste printing, metallic prototypes, lead frame prototypes, canules for medical products, spinnerets
Ceramics	Printed circuit boards, spinnerets
Polymers	Printed circuit boards, labels
Semiconductors	Semiconductor components, solar cells

A. Gillner (✉)
Mikrotechnik, Fraunhofer-Institut for Laser Technology, 52074 Aachen, Germany
e-mail: arnold.gillner@ilt.fraunhofer.de

17 Cutting

Especially for the last two applications, where thermal influence has to be minimized and surface absorption plays a crucial role concerning the achievable quality, frequency-converted lasers at 532 and 355 nm are used. Due to the short wavelength the laser energy is absorbed at the surface of the material, rather than in volume, thus leading to clean ablation results and high cutting quality. Moreover, the achievable spot size is much smaller, leading to smaller cutting kerfs and less energy deposition.

17.5.2 Process Principle

Laser fine cutting of metals does not distinguish from the general cutting process applied for macro applications. The material, which has to be ablated, is heated and molten by the laser radiation. The melt is then driven out of the cutting kerf by a high-pressure gas jet. In contrast to macro-laser cutting, in fine cutting the laser radiation is applied in a pulse mode [19]. In this way, laser fine cutting is similar to single pulse drilling, with the difference that the single pulses are applied with an overlap of typically 50–90% leading to a continuous cutting kerf. In Fig. 17.21 the principle of laser fine cutting is shown. With this process approach, the necessary laser energy and laser intensity can be applied for melting the material, but the overall energy deposited into the material is kept low. Otherwise, the small geometries to be cut would be overheated and destroyed. The focus of the laser beam is set to the surface of the material to achieve the highest intensity and to reduce thermal impact.

By using pulsed laser radiation, the energy for cutting and ablation is applied only for short time intervals. Depending on material and material thickness the process parameters for different materials can be summarized with following values in Table 17.4:

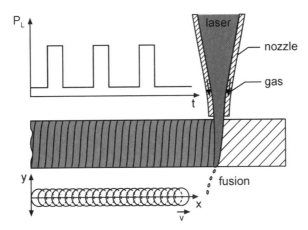

Fig. 17.21 Schematic presentation of laser fine cutting process

Table 17.4 Typical processing parameters for laser fine cutting

Material	Thickness (mm)	Pulse power	Speed (mm/min)	Pulse frequency
Metal	0.1–0.5	500 W–1 kW	50–200	1–5 kHz
Ceramics	0.3–0.5	1–5 kW	25–100	100 Hz–1 kHz
Polymers	0.05–0.2	> 10 kW	100–1,000	10–20 kHz

The advantage of the pulsed energy deposition is the moderate overall heating of the entire workpiece, whereas in laser cutting with continuous laser radiation overheating at small structures will cause increased molten areas and reduced accuracy. This holds especially at complex structures at thin materials, where the restrictions in acceleration of the machining system reduces the achievable cutting speed and where the machine has to reduce the speed nearly down to zero with subsequent overheating of the material. In this case the energy deposition has to be matched with the resulting speed by selective pulse repetition adaptation.

For the heating of the material without melting within a single laser pulse the following equation can be derived from the general heat conduction equation:

$$\begin{aligned} \Delta T &= 2(1-R) \frac{I_0}{K} v_{rep} \tau \sqrt{\kappa t} \cdot \text{ierfc}\left(\frac{z}{\sqrt{4\kappa t}}\right) \\ &= 2(1-R) \frac{I_0}{K} v_{rep} \tau \sqrt{\frac{\kappa t}{\pi}}, \; z = 0 \\ &= 2(1-R) \cdot \varepsilon \cdot v_{rep} \sqrt{\frac{t}{\rho \cdot c_p \cdot K \cdot \pi}} \end{aligned} \quad (17.12)$$

The overall temperature rise in the material follows the pulsing temperature rise at the surface of the workpiece slowly and reaches a steady-state value after a starting phase. The steady-state value is lower than the melting temperature of the material, so that the overheating of the workpiece is avoided. In Fig. 17.22 the temperature rise within each single laser pulse and the resulting overall temperature is shown schematically.

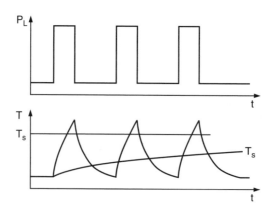

Fig. 17.22 Temperature rise during pulsed laser irradiation

17 Cutting

With this approach even filigree structures with structure sizes < 10 μm and cutting kerf widths of < 20 μm can be cut with high-quality laser beams. For these applications lasers with high beam quality M^2 < 1.5 and low beam parameter products are necessary to allow spot sizes of 10 μm and even smaller.

17.5.3 Laser Sources for Fine Cutting

Today, for laser fine cutting typically lamp-pumped solid-state lasers are used, which are modified for very high beam quality by special resonators or oscillator–amplifier configurations (see Fig. 17.23). Usually side-pumped rod- or slab lasers with special multipass resonators are used. Since fine cutting does not require very high pulse energies, but high peak intensities, the lamp-pumped rod solid-state lasers are used at pulse frequencies of up to 5 kHz and pulse durations of up to 100 μs.

For laser fine cutting of metal parts diode-pumped solid-state lasers are generally not the right choice for an economic cutting process. Since the optimum cutting conditions require pulse powers of up to 10 kW and pulse durations of several hundred microseconds, a diode-pumped laser system would require fast pulsing of diode-pumping source with powers up to several tens of kilowatts. On the one hand, fast switching of diodes results in short lifetime, and on the other hand, this approach would not be economical due to the large number of necessary diodes. Thus using diode lasers for long-pulse solid-state lasers is not an economical way.

Fig. 17.23 Lamp-pumped solid-state laser for fine cutting (Source: Lasag)

Nevertheless, diode-pumped solid-state lasers are increasingly used for cutting heat-sensitive materials and where the amount of melt has to be reduced to a minimum. In this case, diode-pumped Q-switch lasers are used with pulse durations in the range of 10–100 ns and pulse powers in the range of 10–100 kW. Due to the short pulse duration, the thermal influence to the material is minimized according to the thermal penetration depth.

$$d_w = \sqrt{4\kappa t} \quad (17.13)$$

Since the intensities using this lasers are generally higher than the vaporization threshold, the whole cutting process is based on sequential vaporization. The laser passes several times across the cutting contour and typically ablates several micrometers of material. With this approach heat accumulation can be minimized and laser cuts with very high quality can be produced. Moreover, diode-pumped Q-switch lasers provide the possibility of frequency conversion and resulting laser wavelengths in the visible light and UV range.

Starting from the base wavelength of the Nd:YAG laser, the following laser wavelengths can be set for industrial applications (Table 17.5):

Table 17.5 Laser wavelength by frequency conversion

Base wavelength	$\lambda = 1,064$ nm
Frequency doubling	$\lambda = 532$ nm
Frequency tripling	$\lambda = 355$ nm

17.5.4 Applications

17.5.4.1 Cutting of Stents

One of the most prominent applications of laser fine cutting is the manufacturing of cardiovascular stents. Stents are small expandable stainless steel tubes with a diameter of 1.6–2 mm, which are implanted in the cardiovascular vessels and expanded by a balloon catheter thus expanding and stabilizing the vessel walls. Since the structure of the stent has to be designed in a way that it has to provide a maximum of stability and at the same time a maximum of expansion by the catheter, the residual structures are as small as 50–100 µm. The only technology, which can be used for the production of those small features in miniaturized tubes, is laser fine cutting.

In Fig. 17.24 a laser-cut tube with the stent design is presented. Figure 17.25 shows the etched and expanded stent.

17.5.4.2 Cutting of Spinnerets

For the manufacture of polymer fibers for textiles, micro-structured spinning nozzles, so-called spinnerets are used. The fluid polymer mass is pressed through the nozzles which form the geometry of the fibers and which solidifies after the nozzle

17 Cutting

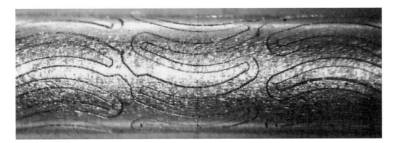

Fig. 17.24 Laser-cut stent (diameter 1.6 mm) after cutting

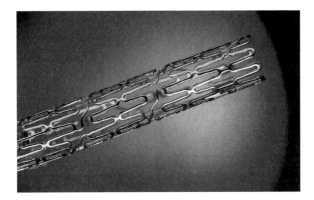

Fig. 17.25 Laser-cut stent (diameter 1.6 mm) after polishing

tool. For textiles with special features and functionalities, spinnerets with partly very complex shapes have to be generated. Sometimes feature geometries, like circles, stars, and triangles with feature size geometry of 100–500 µm and cutting geometries < 20 µm, are necessary. In Fig. 17.26 a threefold geometry is shown, which has been produced with a frequency-tripled short pulse laser.

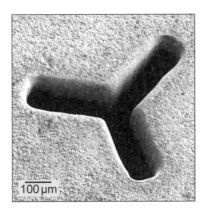

Fig. 17.26 Laser-cut spinneret (Source: IFSW)

17.5.4.3 Cutting of Flex Board with UV Laser Radiation

By frequency conversion the base wavelength of an infrared Nd:YAG laser can be transformed in the UV range. This leads to an adaptation of the absorption and reduction of the penetration depths to the surface of the material varying from several hundred nanometers to several microns depending on the material. With this reduction of the optical penetration depth the thermal effects for the entire workpiece can be further decreased. This holds especially for polymers, which are more or less transparent in the near-infrared and visible area, but absorbing very good in the UV. In Fig. 17.27 the absorption spectra for different materials are schematically shown depending on the wavelength. From this picture it is clear that for polymer cutting, where surface absorption is necessary, either far-infrared laser radiation with wavelength $> 3\,\mu m$ or UV radiation $< 400\,nm$ should be used.

According to the high absorption of polymers in the UV and long IR range, polymer cutting is typically performed with UV and CO_2 lasers [20, 21]. For micro-

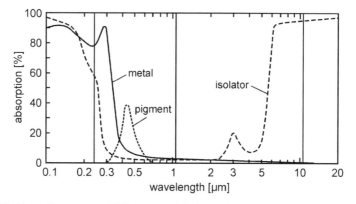

Fig. 17.27 Absorption spectra of different materials depending on the wavelength

Fig. 17.28 Laser-cut printed circuit board (Source: LaserMicronics)

processing of polymers excimer lasers as well as frequency-tripled solid-state lasers are used. Especially for cutting flexible printed circuit boards from polyimide, high repetition rate Nd:YAG and Nd:Vanadate lasers at 355 nm are used [22]. In comparison to excimer lasers with even shorter wavelengths, those lasers provide much higher repetition rates > 20 kHz, which allow higher processing speeds. High cutting speeds not only reduce the costs for manufacturing but also increase cutting quality since the interaction time for laser and materials is reduced and heat accumulation is minimized. In Fig. 17.28 as an example a laser-cut printed circuit board from FR4 and PI is shown.

References

1. J. Franke, W. Schulz, G. Herziger, Abbrandstabilisiertes Laserstrahlbrennschneiden – ein neues Verfahren, Schweißen und Schneiden 45 (1993), Heft 9, S. 490–493
2. W. O. Neill, J. T. Gabzdyl, New developments in laser-assisted oxygen cutting, Optics and Lasers in Engineering 34 (2000), 355–367
3. W. Schulz, D. Becker, J. Franke, G. Herziger, Heat conduction losses in laser cutting of metals, Journal of Physics D: Applied Physics 26 (1993), S. 1357–1363
4. J. Powell, D. Petring, R. V. Kumar, S. O. Al-Mashikhi, A. F. H. Kaplan, K. T. Voisey, Laser-oxygen cutting of mild steel: the thermodynamics of the oxidation reaction, Journal of Physics D: Applied Physics 42 (2009), 015504 (11pp)
5. J. W. Franke, Modellierung und Optimierung des Laserstrahlbrennschneidens niedriglegierter Stähle, Dissertation RWTH Aachen (1994), DSV-Berichte Band 161
6. Y. Arata et al., Dynamic behavior in laser gas cutting of mild steel, Transactions of JWRI, 8 (1979), S. 15–25
7. Schneiden mit CO_2-Lasern, Handbuchreihe Laser in der Materialbearbeitung, Band 1, VDI Technologiezentrum, VDI Verlag Düsseldorf (1993)
8. D. Petring, Anwendungsgestützte Modellierung des Laserstrahlschneidens zur rechnergestützten Prozeßoptimierung, Dissertation RWTH Aachen (1995)
9. A. F. H. Kaplan, Theoretical Analysis of Laser Beam Cutting, Shaker-Verlag, Aachen (2002), (Berichte aus der Fertigungstechnik)
10. E. Truckenbrodt, Fluidmechanik, Band 2, Springer-Verlag, Berlin Heidelberg New York (1980), S.30ff
11. K.-U. Preißig, Verfahrens- und Anlagenentwicklung zum Laserstrahl-Hochgeschwindigkeitsschneiden von metallischem Bandmaterial, Dissertation RWTH Aachen (1995)
12. D. Petring, K.-U. Preißig, H. Zefferer, E. Beyer, Plasma effects in laser beam cutting, 3. Internationale Konferenz Strahltechnik, Karlsruhe (1991) DVS-Berichte Band 135, S 251 ff
13. F. Schneider, D. Petring, R. Poprawe, Increasing Laser Beam Cutting Speeds, Proc. ICALEO 99, San Diego, USA, LIA Vol. 87, Section C, (2000), S. 132–141
14. D. Petring, F. Schneider, C. Thelen, R. Poprawe, Mit Sicherheit schnell: Neue Entwicklungen zum Laserstrahlschneiden von Fein- und Feinstblechen, LaserOpto 31 (1999) Nr.2, 1999, S. 70ff
15. F. Schneider, D. Bingener, H.-D. Riehn, Prozeßüberwachung und –regelung beim Laserstrahl-Hochgeschwindigkeitsscheiden, Laser und Optoelektronik 29(2), (1997), S. 59–65
16. L. Morgenthal, E. Pfeiffer, E. Beyer, Laser Konturschneiden von Elektroblechsegmenten, LaserOpto, Band 31 (1999) Heft 2, 1999, S. 66–69
17. G. Herziger, P. Loosen, Werkstoffbearbeitung mit Laserstrahlung, Grundlagen-Systeme-Verfahren, Carl Hanser Verlag München Wien (1993)

18. Laser Institute of America: Handbook of Laser Materials Processing, (2001), Editor in Chief: John F. Ready
19. M.C. Sharp, Cutting and drilling with average power pulsed Nd:YAG lasers including fibre optics delivery. Power Beam Technology, (September 1990), 239–247
20. F. Caiazzoa, F. Curcio, G. Daureliob, F. Memola Capece Minutolo, Laser cutting of different polymeric plastics (PE, PP and PC) by a CO_2 laser beam, Journal of Materials Processing Technology, 159(3), (10 February 2005), 279–285
21. S. Bednarczyk, R. Bechir, P. Baclet, Lasermicro-machining of small objects for high-energy laser experiments, Applied Physics A 69 [Suppl.], (1999), S495–S500
22. X. C. Wang, Z. L. Lia, T. Chena, B. K. Loka, D. K. Y. Lowa, 355 nm DPSS UV laser cutting of FR4 and BT/epoxy-based PCB substrates, Optics and Lasers in Engineering, 46(5), (2008), 404–409

Chapter 18
System Technology

Kerstin Kowalick

18.1 Process Monitoring

Kerstin Kowalick, Stefan Kaierle, and Boris Regaard

18.1.1 Introduction

Modern manufacturing demands the simultaneous optimization of cost, time to market, time to factory and, maybe most importantly, the quality. Advanced product design, high production accuracy as well as flexibility for handling design changes and small lot production are only few matters material processing stands to benefit from laser technology. New markets are continuously opened to laser-assisted manufacturing by enhanced laser performance and system improvement. Yet new production technologies and the increasing complexity of the equipment and processes make great demand on adherence to production parameter and raise sensitivity with regard to disturbance variables. In addition sources of error also result from heightening productivity up to physical limitations. Therefore, increasing demands on production quality as well as improved processing techniques lead to further requirements of quality assurance. It can also be observed in recent years that requirements for the all-embracing quality assurance have continuously increased. This is reflected in strict customer requirements of reliability and documentation of workpiece quality (e.g. process capability index, Cpk[1]), requirements by law (e.g. for safety relevant components), self-imposed quality standards of the manufacturer and not least the implementation of industrial standards such as DIN EN ISO 9000ff. Today's manufacturers are increasingly forced to keep record of their full production. To meet those demands frequently post-processing quality checks have to be conducted. However, these steps are usually adherent to significant efforts

---

K. Kowalick (✉)
Systemtechnik, Fraunhofer-Institut für Lasertechnik, 5207 Aachen, Germany
e-mail: kerstin.kowalick@ilt.fraunhofer.de

[1] A measurement of how well a process in chain production meets the specifications.

and are often restricted to surface checks. Destructive testing to check for inner defects can only be applied to few parts.

The appraisal of production quality during processing also reduces the exigency of subsequent destructive techniques for quality evaluation. Particularly long and expensive production cycles therefore call for online monitoring strategies identifying production anomalies immediately or even avoid them in advance in order to minimize scrap and costs.

18.1.2 Sensors

Modern laser systems, in addition to their basic components such as the beam source and the beam-guiding and handling optics, comprise a considerable number of sensors. These are responsible for monitoring, controlling or ideally regulating the state of the system and the laser process.

The values and states recorded by the sensors are usually converted to electrical signals and subsequently processed using evaluation algorithms. There are a variety of different types of sensor for measuring physical or chemical properties. In addition to the usual sensors for measuring such parameters as travelled distance, time, angle, acceleration or velocity, laser technology also relies in particular on sensor systems that measure electromagnetic radiation, mainly in the visible and near-infrared range (photodiodes, cameras, pyrometers) to monitor tool characteristics or process states during operation.

Acoustic sensor systems, on the other hand, are rarely used in industrial laser materials processing. This is because the analysis of acoustic emissions – such as those produced by changes in thermal gradients, the formation of holes and cracks, and rapid pressure fluctuations in the keyhole [1] – barely allows reliable conclusions to be drawn about the processing result. Supporting tasks, such as measuring the distance to or between objects or monitoring machine operating limits (e.g. triggering an emergency shut-off if the threshold is mechanically exceeded), are performed by mechanical sensors, usually in the form of push-button or switch sensors or inductive or capacitive transducers.

The laser process is influenced by a variety of factors. The following list classifies the most important parameters.

- Laser beam sources
  - Power
  - Spatial beam quality
  - Polarization
- Gases
  - Flow rate
  - Properties
- Workpiece

- Material properties
- Material metallurgy
- Structure (e.g. wall thickness, seam width)
- Shape (e.g. dimensional stability, geometrical accuracy, dimensions)

• Machine

- Beam-shaping optics
- Beam-guiding optics
- Workpiece clamping
- Generation of relative movement

• Method

- Production sequence
- Pre-treatment

• Planning accuracy

- Path planning
- Process planning

• Filler materials
• Human factors

However, a full top-down analysis of the entire machining process reveals many more parameters that influence the processing result either directly or indirectly in a cause-and-effect relationship [2, 3]. The various signals being measured need to be recorded and processed in different ways. The volume of measurement data differs from case to case, as does the frequency with which measurements are carried out and adjustments made. While some information, such as the position of the workpiece or the laser beam parameters, is calculated off-line, other data such as the feed rate or process emissions are analysed during the machining process at repetition rates up to the kilohertz range. In this way, any necessary adjustments, such as regulating the laser output as a function of the feed rate, can be made on the spot. However, even today's most sophisticated control circuits are unable to factor in all the parameters that influence the processing result. The parameters therefore have to be reduced to keep the volume of measurement data to a manageable scale. The cause-and-effect relationships of the observed parameters must be of a nature that can be described in models or determined in tests. Pure monitoring of the process and the machine, however, is not subject to this type of limitation.

Most of the mechanical and electrical sensors employed to monitor parameters are part of the system's standard configuration and work independently of the laser tool. They are, therefore, not relevant to the following evaluation. Given that it is impossible to record all the existing parameters and sources of interference, it is very important to monitor and analyse the machining zone during the laser process by means of optoelectronic sensors to ensure the best possible quality and to keep post-processing steps to a minimum.

The following descriptions and examples mainly refer to the monitoring of laser-processed metal parts. Emphasis is also given to the welding process since in industrial production it has been most widely established in comparison to other laser material processing methods.

18.1.3 Electromagnetic Radiation Sensors

The detection of electromagnetic process and laser radiation is normally restricted to the visible and near-infrared range. The various different detectors available are distinguished in particular by their spectral sensitivity, temporal resolution and the spatial resolution they are able to achieve in combination with the right lenses. The most important types include one-dimensional temporal-resolution sensors such as photodiodes, pyrodetectors and thermopiles, and spatial resolution measuring and imaging sensors (arrays of light-sensitive cells), such as CCD or CMOS cameras. Integrally measuring detectors only register the intensity of the monitored radiation and not its spatial distribution. Photodiodes are mainly used to monitor highly dynamic processes due to their high temporal resolution and are easy to handle, thanks to their comparatively moderate demands on data acquisition software and hardware. Pyrodetectors differ from photodetectors primarily in terms of their spectral sensitivity and generally lower temporal resolution. Imaging sensors or cameras are made up of arrays of light-sensitive cells, or pixels, uniformly arranged in rows and columns. They monitor in spatial resolution, which means they not only deliver information on the brightness of an object or beam source but also on its shape, size and position. However, the data stream delivered by a camera is higher by the number of its pixels than that of a simple photodetector, and the time required to read out all the image data results in a lower overall temporal resolution than with single light-sensitive cells. Consequently, imaging systems place higher demands on hardware and cost more to buy, but are nevertheless suitable for industrial use. In principle, sensors with spatial resolution can be built up from all integrally measuring detectors by suitably combining various elements. However, the most practicable solutions for process monitoring are camera systems based on CCD or CMOS sensors. These are increasingly being used to supplement or replace photodiodes both in research and in industry and are opening the door to a large number of new applications. Inline process monitoring, which places high demands on the frame rate (temporal resolution), is generally performed using CMOS technology where, unlike with CCD sensors, the electron charge is transformed into a measurable voltage at chip level. Each pixel contains a light-sensitive diode and its own evaluation electronics and can be addressed individually and read out at high frame rates. CMOS cameras are also smaller in size than CCD cameras. However, their comparatively large proportion of interconnecting electronics and transistors per pixel has a negative effect on their light sensitivity, and their design has an unfavourable impact on their dynamics and signal-to-noise ratio. But today's new CMOS sensors are becoming more and more like their CCD counterparts in these respects. In laser technology, CCD sensors are used in such areas as pulsed-laser-beam diagnosis.

18.1.4 Measurement Techniques Based on Optoelectronic Sensor Systems

In the measurement of characteristic variables from laser machining operations, a distinction is made according to whether direct or indirect measurement techniques are used. Whereas direct measurement methods enable the desired parameter to be measured straight from the source, indirect measurement techniques require a detour through one or several auxiliary variables from whose measurement quantitative or qualitative conclusions can be drawn about the value sought. With reference to the example of laser beam welding, the multitude of quality-determining parameters (possible defects) can be represented in Table 18.1 [3]. Only very few of the parameters listed can be directly measured. Characteristics such as tarnishing or humping are difficult to register by means of sensors. Variables such as cracks and lack of side wall fusion can only be determined when the welding process has been completed. Quality assurance is optimized by a balanced combination of techniques for monitoring the condition of the tool (laser), the handling system, beam guidance, the workpiece (e.g. shape, position), consumables (gases, filler materials) and for direct as well as subsequent observation of the processing zone.

Table 18.1 Quality-determining criteria for laser beam welding

• Splatters	• Root undercut
• Penetration notches	• Penetration depth
• Scratches	• Lack of fusion
• Humping	• Seam width
• Seam swelling	• Undercut
• Seam patterning	• Surface voids
• Cracks/surface cracks	• Pores
• Annealing colours	• Distortion
• Root width	• Root inspection

18.1.4.1 Position and Time of Measurement

It has become an established practice to break down the monitoring procedure into pre-, in- and post-process phases (Fig. 18.1) according to the time and position of the measurement in the laser processing operation. Pre-process monitoring takes place at a time and/or position before the laser processing operation. The aim is to avoid process defects by off-line and therefore time-uncritical assurance of constant input parameters (e.g. measurement of the beam's caustic curve) and online control of parameters near to the processing zone in terms of time and position (e.g. seam tracking). In-process monitoring includes the measurement of primary[2] and secondary process radiation which is emitted or reflected directly from the

---

[2] The term primary radiation denotes the direct or reflected radiation of the processing and illuminating laser, respectively, whereas secondary radiation refers to the radiation which is induced by the process itself.

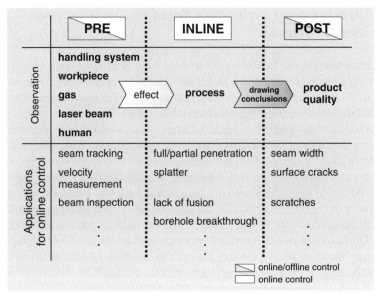

Fig. 18.1 Phases of process monitoring

processing zone. It enables conclusions to be drawn about the subsequent quality of the product. These techniques allow quality-relevant findings to be made which go beyond simple surface properties. Time-consuming manual or even destructive testing methods can be minimized. Post-process monitoring includes methods for assessing the already solidified processing zone and, in the case of online control using optoelectronic sensors, permits speedy automated evaluation of surface properties and rapid feedback to manipulated variables.

In the course of this chapter, various applications featuring inline monitoring of the processing zone as well as monitoring methods before and after the process operation will be explained more closely with reference to examples. In order to provide a better understanding, however, fundamental conditions and relationships will first be described.

18.1.4.2 Detector Arrangement

In practice, the preferred arrangement is to position the detectors above the workpiece because this permits a direct view into the zone of interaction between the laser and the workpiece. Measurements from below the component are not possible in many cases simply because the underside of the workpiece is not accessible.

The alignment of the detectors is stated relative to the axis of the laser beam and the direction of processing. Arrangements where the angle of the laser beam axis is around 0° are referred to as being coaxial and as quasi-coaxial at an angle of up to approx. 10°. At greater angles the arrangement is described as being lateral.

Regarding inline monitoring, tests of welding operations using CO_2 lasers reveal that the viewing angle of the sensors has a significant influence on the measured results from plasma monitoring with photodetectors. Depending on the viewing angle relative to the laser beam axis, the detectors cover different proportions of surface and capillary plasma. Sensors arranged quasi-coaxially to the laser beam mainly detect the plasma from the vapour capillary, whereas sensors arranged parallel to the top of the workpiece observe the surface plasma.

The coaxial arrangement is suitable in cases where the detectors can be integrated in the processing head, the beam path or the laser itself. The advantage of such installations is that the sensors are protected against contamination and other undesirable effects of the processing operation while at the same time permitting maximum visual access to the interaction zone. A potential problem with the lateral arrangement is that, when used for inline monitoring, the detector might become contaminated during processing, but this can either be avoided by appropriate design measures (e.g. inert gas nozzles) or by means of regular maintenance. In compact systems, lateral detectors can be fitted to the processing head or can be integrated in the processing gas nozzle.

Lateral detector arrangements (e.g. light-section sensor) enable a correlation to be established between the measurement signals and geometrical defects which, when for example welding lap and butt joints, can be used for pre- or post-process monitoring to identify sagging, edge misalignment, gaps and holes

18.1.4.3 Optical Components

The majority of the high-power lasers used in laser material processing emit wavelengths in the range of $1\,\mu m$ (e.g. Nd:YAG lasers, fibre lasers) or $10\,\mu m$ (CO_2 lasers). This makes it necessary to realize the focusing and deflection of the laser beam with various optical devices. In coaxial process monitoring, the laser beam and observation share the same beam path and so the process-monitoring system has to be adapted to the different substrate materials of the optical devices (i.e. lenses and mirrors) needed for guiding and focusing the laser beam (*see* Fig. 18.2 comparison of CO_2 and Nd:YAG).

As a rule, lasers with beam sources in the $1\,\mu m$ wavelength range use glass or, at higher outputs, fused silica as the substrate material for mirrors and lenses. Zinc selenide is a frequently used mirror and lens material for CO_2 lasers. Glass, silica glass and zinc selenide are transmittive for electromagnetic waves in the visible and near-infrared range. The properties of the mirror are determined by a specific coating. So-called dichroitic mirrors reflect the laser light and transmit the secondary radiation to be measured, or vice versa, depending on the coating. In beam splitters, the coating causes a fixed proportion of the laser light at the same wavelength to be reflected or transmitted according to the division ratio. Focusing is realized by

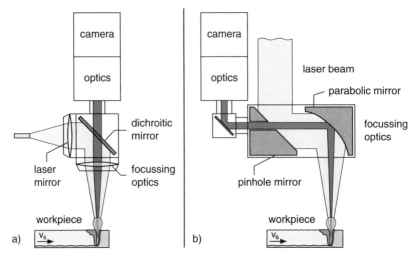

Fig. 18.2 Process-monitoring setup. (**a**) For Nd:YAG laser and (**b**) for CO_2 laser

means of lenses. At very high outputs, such as those required to weld thick metal plates, or in applications demanding a high degree of mechanical stability, metallic substrates are preferably used for CO_2 laser beam sources. At wavelengths in the 10 μm range, high thermal stress can lead to imaging defects or even cause the destruction of the optical components on transmissive optical devices. Metallic optical devices are easier to clean and less sensitive to contamination. The beam is deflected by reflection off a plane mirror. Parabolic mirrors are used to focus the beam. If metallic mirrors are used for guiding and focusing the laser beam, coaxial observation along the laser beam axis is more difficult than for lasers in the 1 μm wavelength range. Here the secondary process radiation can be observed through a pinhole (diameter typically 1–2 mm) drilled in the centre of one of the last mirrors in the beam path. Mirrors with such a drilled hole are designated as pinhole mirrors. The loss of laser power caused by the drilled hole can be regarded as negligible in actual practice as only a small fraction of the output is lost, and beam quality is not affected in any way relevant to the process. The losses occasioned by these mirrors are of the same magnitude as those caused by mirrors in the beam path, i.e. approx. 1% per mirror, which has to be compensated by means of cooling. A mirror with a large hole in its centre is described as a scraper mirror. The diameter of the hole is selected so as to allow the laser light to pass through the mirror unimpeded. Scraper mirrors filter out secondary radiation emitted at a narrow angle to the laser beam axis via the mirrored edge around the hole, for forwarding to the detector. Such mirrors can be used to obtain integrated measurements of light emitted by the process approaching the quality of coaxial measurements.

Another important element in process monitoring is optical filters damping signals or filtering out the spectral range of interest. The latter is of great relevance particularly with regard to detecting primary radiation reflecting from the processing zone.

18.1.4.4 External Secondary Light Sources

Certain aspects of relevance to the quality of the production process cannot be monitored by recording the primary radiation of the processing laser or secondary process radiation from the processing zone. This is true especially for pre- and post-process monitoring. Examples include the position of the laser beam in relation to the desired working position (seam tracking) or the profile of the solidified seam in laser beam welding. To detect such features, external light sources (mostly LEDs or diode laser) are used to illuminate the workpiece. For some measurement applications in pre- and post-process monitoring special light projections are formed (e.g. lines or circles) and recorded by the light-section method (Fig. 18.3).

Figure 18.4 illustrates additional illumination strategies which can be used to identify various contours of objects and surface properties.

At present there is no single optimum illumination strategy that is the right choice for the multiplicity of applications and criteria/failure to be observed. Furthermore, the type of illumination is in many cases restricted by the mechanical or optical conditions of the process or handling system.

One method, in which the processing zone is illuminated coaxially (bright field) to the processing laser beam by an additional laser of lower output, has opened up new areas of application for process monitoring for all process zones. Using a partially transmissive mirror or a pinhole mirror it is possible to couple in illumination radiation coaxially (Fig. 18.5). The reflected radiation of the illumination laser is likewise observed coaxially to the processing laser beam with a spatially resolving sensor. Among other things, this method enables the liquid/solid phase boundary of the material processed to be measured as well as the relative move-

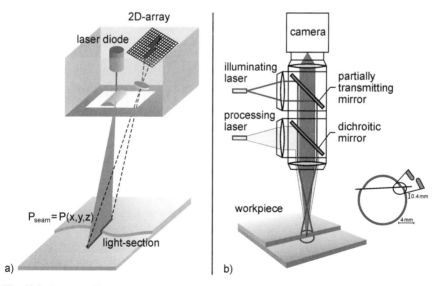

Fig. 18.3 Seam-tracking principle (**a**) based on a scanning triangulation sensor and (**b**) light section with coaxial ring projection

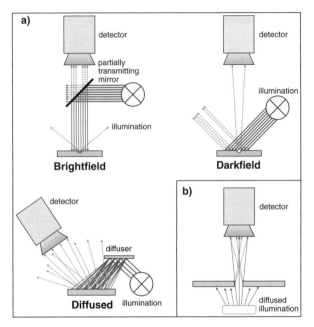

Fig. 18.4 Illumination strategies. (**a**) Incident light; (**b**) backlight

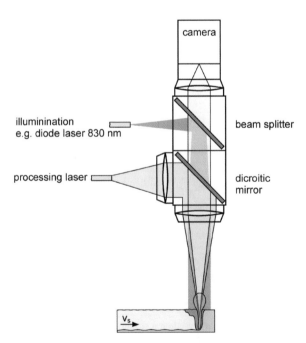

Fig. 18.5 Process-monitoring setup with coaxial illumination

ment between the processing head and workpiece. Both applications are described in more detail below.

18.1.4.5 Detected Radiation

In laser material processing, the light emitted by the laser beam is the primary source of the electromagnetic radiation. The most important lasers in industrial production are CO_2 lasers ($\lambda = 10,640$ nm), Nd:YAG lasers ($\lambda = 1,064$ nm), diode lasers (mainly used at $\lambda \approx 808$–980 nm) and fibre lasers (also available at various wavelengths but predominantly used at $\lambda \approx 1,030$–1,080 nm).

Online measurement of laser output and beam distribution involves decoupling a small percentage of the laser light (primary radiation). Process-relevant information can also be derived from the detection of back-reflected primary radiation from the laser processing zone. The use of an additional light source enables specific data to be recorded not only from the processing zone but also from the area around it. This can be useful when measuring the position of the laser beam in relation to the desired working position. The secondary sources of electromagnetic radiation are laser-induced plasma, metal vapour emission and emissions from the liquid material. These radiate from the zone of interaction between the laser beam and the workpiece or from direct surroundings (Fig. 18.6). It is well known that the secondary radiation contains information about the dynamics of the laser process and about the quality of the processed workpieces.

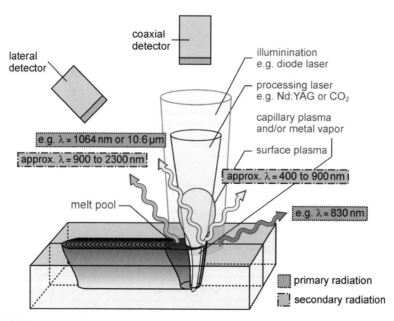

Fig. 18.6 Sources of electromagnetic radiation in laser material processing

18.1.4.6 Signal Evaluation

The techniques for signal evaluation are strongly related to the type of detector and the method used for process monitoring. Photodiodes, for example, just provide information about the temporal distribution of the measured signal intensity. The signal processing is therefore limited to an evaluation of the signal amplitude and frequency. A common approach used in most inline monitoring systems using spatially integrating sensors is the comparison of the signal amplitude (intensity profile) during a production process with predefined tolerance bands. Figure 18.7 shows a reference signal calculated from one or more signals recorded during reference processes (upper diagram). For the supervision of a production step a tolerance band is generated around the reference signal (lower diagram). With this setup a production process is rated as "good" if its detector signal lies always inside the predefined tolerance band.

Imaging sensors can also be used as "smart photo diodes" by programmatically selecting single pixels or groups of pixel as the source of a signal comparable to one of a photo diode. At present most spatially resolving inline process-monitoring systems, detecting process radiation, are using algorithms comparing process images with reference films (Fig. 18.8).

One particular area of interest in current research is the detection of specific process conditions on the basis of physical effects without having to compare image data with reference data. This will enable the development of widely applicable,

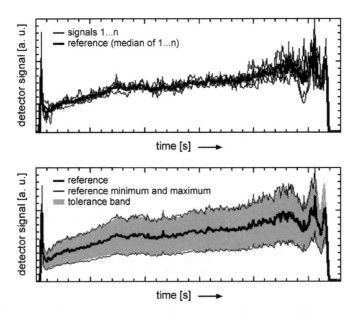

Fig. 18.7 Quality evaluation by comparison of the signal intensity monitored during welding with reference process (*top*), reference signals (*down*) tolerance bands

18 System Technology

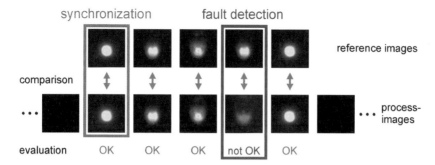

Fig. 18.8 Quality evaluation by comparison of the camera images monitored during welding with reference film

robust process-monitoring methods to detect defined classes of fault that can be applied without the need for special training.

With regard to pre- and post-process monitoring, especially applications using greyscale image analysis require complex imaging techniques in order to guaranty a stable recognition of the sought-after attribute. The extracting procedure depends on the surface properties. Figure 18.9 shows an example where edge detection filters were applied to a camera image of a welded seam in order to separate the seam from the unprocessed material. The extraction of the seam is useful to eliminate the influence of different surface conditions on the evaluation of the seam properties, especially when detecting surface voids [4]. In further steps the appearance of seam can be evaluated in terms of irregularities by using, e.g. the method of conjugate gradients or the threshold method. By extracting the seam in a first step the region of interest is reduced and the evaluation of the seam is thus faster than exploring the entire picture.

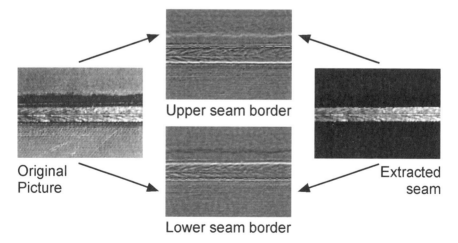

Fig. 18.9 Separation of the seam from camera images

18.1.5 Examples – Laserwelding of Metal

The welding of metals using laser beams is one of the first processes in which process-monitoring systems were used on an industrial scale. Most systems use optical sensors that, for example, determine weld seam width, joining errors, drop formation or splatter by coaxially or laterally observing primary and secondary radiation.

18.1.5.1 Inline Monitoring

The majority of the inline monitoring systems in industrial use today still operate on the principle of spatially integrated photo detectors with a high time resolution (5–20 kHz). Intensity is evaluated by means of frequency and amplitude signal analyses. If the focal position or the weld depth are altered during welding, this causes changes to the secondary process radiation. Process disorders can frequently be detected in this way, but it is difficult to discover precisely what has happened and why. Meanwhile, imaging sensors that register the emitted process radiation in spatial resolution on the basis of the plasma, metal vapour and the molten mass are becoming increasingly popular. The spatial resolution measurement enables errors to be detected with greater reliability.

For instance, on welding of tailored blanks or other sheet metal with a thickness below 3 mm, characteristic changes in the distribution of the secondary radiation can be measured when the process changes from full to partial penetration. These changes do not necessarily come along with a change in the overall amplitude of the measured radiation. Hence, changes might not be recognized by a photo detector-based system. Figure 18.10 shows a comparison of the camera images recorded during laser welding of a tailored blank and an approximation of photo detector signals (calculated from the camera images by spatial intensity integration). The camera images clearly indicate different process states while the spatial integrated (photo detector) signal only shows minor changes of the amplitude.

Most monitoring systems in use distinguish good welds from faulty welds by comparing signals acquired during production with reference pictures taken before. If certain attributes of a signal deviate too much from the reference, the weld is identified as "not good". Systems using this method can be put into practice without knowing specific features about the welding process. Yet the signals are strongly dependent on boundary conditions like the material lot, environmental conditions and the optical elements. Changing conditions might cause signal variations denoted as "NIO" ("not in order") even if the weld shows no failure (called "pseudo errors"). Specific expert knowledge is necessary to interpret the acquired signals correctly. Another drawback is that these systems have to be qualified and "learned" and that having to "re-learn" every time the process is modified costs additional material and time. It is therefore much more favourable to identify welding failures and certain process states by detecting determined signal attributes without comparing the signal to a reference. Corresponding algorithms which, for example, isolate areas

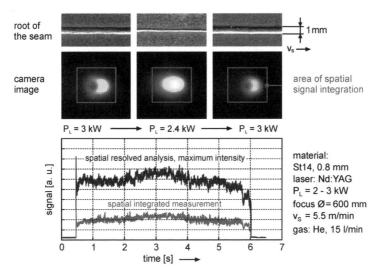

Fig. 18.10 Comparison of spatially resolved and spatially integrated measurement. *Top*: backside of a laser-welded tailored blank showing a broad seam (3 kW laser power) and a narrow seam (2.4 kW laser power). *Middle*: camera images of the keyhole area showing an intensity distribution with a local minimum (3 kW laser power) and an intensity distribution without a local minimum (2.4 kW laser power). *Bottom*: diagram showing a spatial resolved analysis of the recorded camera images (maximum intensity) and an approximation for the signal of a single photo detector (spatial integrated measurement)

of great intensity spatially against the workpiece can be used for splatter detection (Fig. 18.11).

In processes where a great deal of surface plasma is generated (e.g. the processing of aluminium with CO_2 radiation), it is difficult to assess the process quality on the basis of secondary process radiation due to the high radiation intensity and the absorption behaviour of the plasma. What is more, the interrelationship between the geometry of the molten mass and the secondary radiation emitted during the welding process is largely unexplained at the present time, which means that deviations in the shape and dimensions of the weld, such as irregularities in fillet welds, excessive material, edge misalignment, cannot be detected using this method.

A new concept for imaging monitoring systems has therefore been devised, based on the usually coaxial insertion of an additional beam source via a beam splitting mirror. The processing zone is illuminated in order to make the solid–liquid phase boundaries between the basic material and the molten mass visible and thus capable of being measured during processing. The light from the illumination source is reflected differently by the molten mass and the solid material, producing distinctly different textures. The smooth convex surface of the molten mass reflects the illumination light deterministically over a wide area, increasing its divergence. On the rough solid surface of the basic material, the light from the illumination source is reflected stochastically in numerous small patches. Working on the basis of Planck's law of radiation, the secondary process radiation reaches maximum intensities in

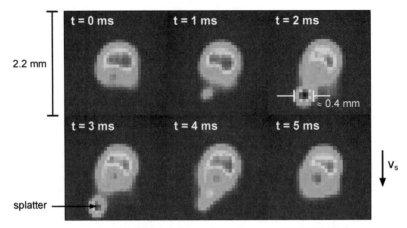

Fig. 18.11 Splatter detection

the wavelength range below 600 nm, particularly during the processing of metals. To enable the proportion of process radiation to be reduced relative to the reflected radiation from the external light source, diode lasers with wavelengths of around 800 nm have proven to be suitable as an illumination source. The spectral sensitivity of the camera is also great enough at these wavelengths. The use of narrow band pass filters enables radiation-intensive portions of the secondary radiation to be filtered out or weighted in comparison to the reflected radiation of the illumination source. Illumination sources with an output between 100 and 150 mW are sufficient for most applications.

The measurement principle utilizes the fact that technical surfaces of workpieces, e.g. used in laser welding applications possess a rough surface finish, thus causing alternating bright and dark areas on the camera image (Fig. 18.12).

The differentiation of solid and molten material can be done by texture analysis and is utilizable for visual inspection as well as automatic processing. However,

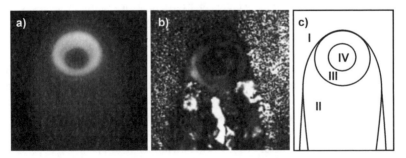

Fig. 18.12 Welding process radiation while welding (**a**) without external illumination, (**b**) illuminated with 150 mW diode laser, (**c**) phase boundaries. I: solid surface; II: melt pool geometry; III: thermal radiation surrounding the capillary; IV: capillary

texture analysis is sensitive to differing workpiece surfaces. A more robust approach uses the fact that the solid material is – essentially – stationary to the workpiece, while the molten material is unsettled.

The algorithm principle can be subdivided into four steps. First, the displacement of the solid surface between two consecutive images is determined. The solid surface structure of the workpiece is unique and stationary; therefore, a relative displacement between the sensor and the workpiece can be measured by finding the maximum cross correlation of a dedicated template area of one image in the consecutive image. By calculating the pixel-to-pixel difference of the two images, features stationary to the workpiece displacement are cleared, while non-stationary areas are enlightened. To eliminate small-scaled image irregularities, median filtering is used. The separation of stationary and non-stationary features is computed by binary threshold filtering. The reflection on molten areas is unsettled and fast altering; therefore, the pixel-to-pixel difference is unpredictable. However, by superposition of consecutive analysed images there is a good probability that sufficient brightness difference appears on the molten area. To determine the melt pool boundary, the computed image has to be cleared from noise, which is done by erosion- and dilatation filtering. The boundary then can be found by searching the first non-zero pixel in a row respectively column. The analysis of the melt pool contour can be utilized for robust gap- and lack of fusion detection.

Compared to state-of-the-art reference-based process-monitoring systems, this allows higher reliability and less pseudo errors, model-based error classification and the possibility of real-time process control.

A special challenge for the described image processing algorithm is surface contamination, e.g. by a grease film as well as coated workpieces.

18.1.5.2 Seam Tracking – Pre-process Monitoring

When materials are joined together, e.g. by laser beam welding, an accurate positioning of the laser beam relative to the joint is of paramount importance. This is true for all joints (butt joint, overlap joint, fillet joint, etc.) (Fig. 18.13). Additionally it has to be assured that no jumping in thickness of sheet metal parts that have to be joined occurs. Not least edge damages and geometrical data have to be identified in order to obtain high-quality welds at any time.

Predominant seam-tracking sensor concepts are based on the triangulation principle. First setups are using a deflecting mirror to scan the workpiece surface around the joint using point-shaped laser beams and line cameras [5]. The joint position is recognized as a discontinuity in the measured distance between sensor and workpiece surface (Fig. 18.13). The triangulation algorithm is fast and simple. However, the robustness is not optimal due to moveable parts and the time resolution is limited.

Nowadays, light-section sensors are most commonly used. They also utilize the triangulation principle, but stretch it to a second dimension. Instead of a point-shaped triangulation laser beam, a laser line is projected onto the workpiece surface. As all parts are static the setup is much more robust. The detector is two-dimensional

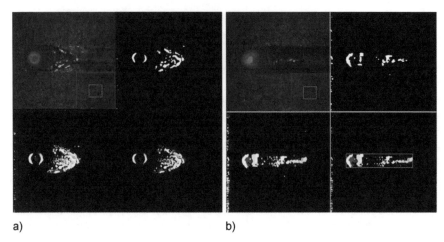

Fig. 18.13 Welding overlap joint. (**a**) Image processing for melt pool contour analysis (zero gap weld): 1 (*left top*): raw image; 2 (*right top*): pixel-to-pixel difference of adjusted consecutive images and binary threshold; 3 (*left bottom*): overlay of eight consecutive analysed images; 4 (*right bottom*): filtered image with calculated boundaries. (**b**) Melt pool of a lap weld with 0.15 mm gap

(CMOS or CCD camera) which allows much higher temporal resolution (dependent on the camera framerate and image processing algorithm). The measurement resolution in feed direction can be further increased by using multiple laser lines. The joint position is then measured at different positions in parallel and imaged on one camera, which is close to an increase of the camera framerate [5].

At least for now, from the industrial point of view the triangulation principle is used in most applications. The seam-tracking sensor is usually fixed to the welding head. The sensor moves with the TCP.[3] Since the TCP matches the current joint position, the sensor is continuously readjusted and therefore is able to cover a great deviation between joint and robot trajectory [6]. The tracking axis (Fig. 18.14) also may be abandoned and replaced by direct robot trajectory adjustment.

To reduce the sensor forerun and to reduce size, the sensor can be integrated in the welding head using a coaxial projection of a circular light section and observing the workpiece coaxially to the laser beam [7]. Further advantages are the possibility to recognize joints independently from the welding travel direction. However, the smaller triangulation angle caused by the setup principle requires a higher resolution of the camera.

Yet another concept using greyscale image analysis under incident illumination is existing (Fig. 18.15). This sensor type also uses a two-dimensional detector (camera) to observe the workpiece surface. The joint and workpiece are illuminated by diffuse vertical or slightly transversal light. The joint position is recognized by separating areas of different reflectivity (or brightness). Since no triangulation angle is needed small sensor designs are possible. The detection of very thin butt joints

---

[3] Tool Centre Point: the position where the focused laser beam hits the workpiece.

18 System Technology

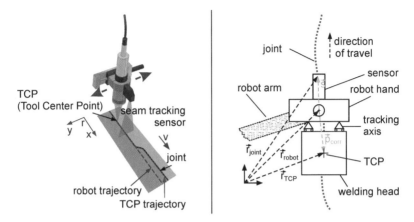

Fig. 18.14 Tracking axis

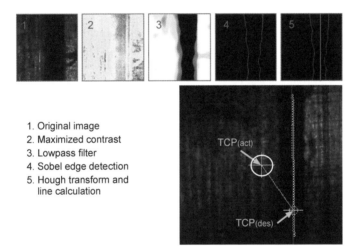

1. Original image
2. Maximized contrast
3. Lowpass filter
4. Sobel edge detection
5. Hough transform and line calculation

Fig. 18.15 Algorithm for seam tracking with greyscale image analysis

can be realized and the sensor adjustment in relation to the joint direction is not of relevance. Drawbacks are the limited observation angle which should not exceed $3°$–$7°$ to the surface normal and a limited illumination. Seam tracking with greyscale image analysis therefore has limits when applying to overlap joints with strong edge misalignment. Hence it is used in particular to measure gap width and edge damages.

Especially light-section sensors imply a constant linear robot movement, requiring

- a constant feed rate not assured at:
  - Deceleration on small curve radii or edges, where real and set speed differ from each other

- Inaccuracy of the robot
- Workpiece movement/distortion

• no transversal movement of the robot hand not assured at:

- Non-linear feed direction
- Workpiece movement
- Vibration

• no rotation of the welding head of robot hand not assured at:

- Inaccurate welding head installation
- Inaccurate workpiece clamping
- Distortion by robot movement

In numerous real applications, these requirements are not given, thus leading to positioning errors.

All error causes can be attributed to a lack of information. The seam-tracking sensor only measures the gap position relative to the current TCP position. The position of the TCP relative to the workpiece is not available.

The shown problems can be solved if the relative position of the welding head relating to the workpiece robot $\vec{r}_{robot}$ or alternatively the TCP position $\vec{r}_{TCP}$ is determined. Some robot systems maintain real-time output of the robot position in the required accuracy, but they are cost expensive and the interfacing of the sensor system to the robot control is laborious.

Therefore, a new concept has been developed also measuring the relative velocity between the workpiece and the sensor. It consists of a high-speed CMOS camera and a coaxially integrated laser diode illumination. The camera observes an area of approx. $6 \times 6\,mm^2$ carried out on the same image as the measurement of the gap position. The observed structure is unique and stationary on the workpiece. A relative displacement between the sensor and the workpiece therefore can be measured by finding the maximum cross correlation of areas in consecutive observed images (Fig. 18.16). The relative velocity corresponds to the displacement normalized to the framerate of the camera. The resolution can be enhanced by applying sub-pixel comparison.

If necessary, light-section measurement can also be integrated using an additional illumination. However, greyscale image analysis turned out to be very stable for different materials such as stainless steel, mild steel, copper and alloys with punched or laser cut edges. Knowing the relative velocity $\dot{\vec{r}}_{TCP}(t)$ and a reference position $\vec{r}_{const}$, the absolute TCP position related to the workpiece can be determined through integration.

$$\vec{r}_{TCP}(t) = \vec{r}_{const} + \int_0^t \dot{\vec{r}}_{TCP}(t) \cdot dt \qquad (18.1)$$

18 System Technology

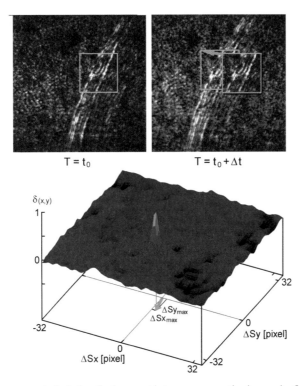

Fig. 18.16 Measurement of relative displacement between consecutive images by finding the maximum cross correlation of areas in consecutive observed images

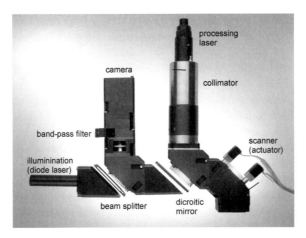

Fig. 18.17 Two-dimensional seam tracking (greyscale image analysis) with coaxial actuator

The absolute gap position can now be determined by

$$\vec{r}_{gap}(t) = \vec{r}_{TCP}(t) + \vec{s}(t) \tag{18.2}$$

For self-guided seam tracking, the relevant gap position related to the TCP position can be found by

$$\vec{r}_{target}(t) = \vec{r}_{gap}(i)\big|_{r_{gap,x}(i) = r_{TCP,x}(t)} \tag{18.3}$$

The correction vector equals the vector difference:

$$\vec{p}_{corr}(t) = \vec{r}_{target}(t) + \vec{p}_{axis}(t) - \vec{r}_{TCP}(t) \tag{18.4}$$

With this concept, the seam tracking is two-dimensional position-based instead of time-based. The previously described errors caused by a lack of position information are inexistent. Carrying out the laser beam correction with a robot-independent scanner system, seam tracking can actually be performed completely independent from the handling system.

In fact this reduces cost in robot systems due to lower accuracy and output requirements. It simplifies interfacing an installation and disengages from sensor calibration necessities.

18.1.5.3 Seam Inspection – Post-process Monitoring

Even after resolidification of the melt camera-based sensors are able to determine online the quality of the weld. To verify a portion of the ISO demands by use of post-process monitoring similar methods like known from pre-process monitoring are frequently used. Again the illumination of the seam surface turns out to be of great impact.

Figure 18.18 schematically demonstrates one possible setup for the detection of holes. The sensitivity of this method increases with protection against ambiance light and restriction of evaluation on areas of interest (e.g. the weld seam). The extraction of characteristics is done by component marking and the following classification by determination of size and shape (Fig. 18.19). However, in many cases the application of this method is in practice restricted by the missing accessibility of the backside of the workpieces.

The most important problems of tailored blank welding – undercut and welding interrupts – are detected by the light-section method. A diode laser with a line optics is utilized as light source. The angel between laser beam and camera is decisive for the projection of relief variation. Figure 18.20 displays the raw camera pictures with and without undercut and the result of automatic detection of undercut. In the case of welding interrupts the idea is to detect a stair between the different thick sheets surfaces, because there is a smooth passing between them in case with no interrupts [8].

18 System Technology

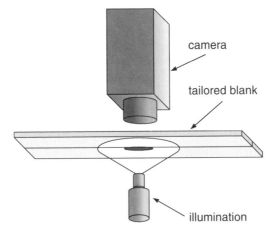

Fig. 18.18 Schematic setup for detection of holes

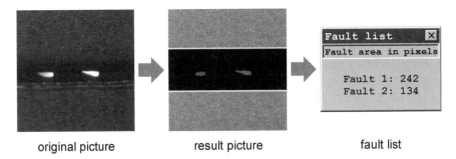

Fig. 18.19 Detection of holes and generated failure list

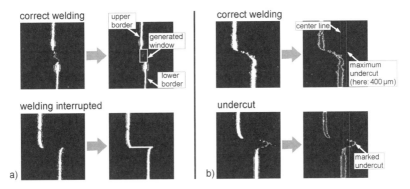

Fig. 18.20 Detection of (**a**) welding interrupt, (**b**) welding undercut

As mentioned in Sect. 18.1.4.6 Post-process-monitoring also makes use of grayscale image analysis. Complex imaging techniques enable locating different seam or surface irregularities. In the initial stage the extraction of areas is often useful to reduce influences of different surface conditions.

18.1.6 Conclusion

The majority of today's commercialized process-monitoring systems is used in macro processing of metal parts. Trends in laser material processing are set by development of innovative processes, new materials and strong miniaturization of workpieces. Currently established methods of process monitoring therefore have to be refined with regard to new requirements as illustrated exemplarily in Fig. 18.21 or new approaches have to be considered.

Nonetheless, also macro material processing is still faced to numerous challenges. Even with well-calibrated tools and carefully selected parameters it cannot be assumed that a process will run continuously free of perturbations. Unpredictable influences from the environment can have significant impact on the processing results thus leading to unacceptable errors. Partly this can be compensated by closed-loop controls. For the implementation of such closed-loop controls robust process parameters, both on the input and output side, are crucial. They have to be independent of fluctuations from batch to batch as, e.g. measurement of laser power, positioning at workpiece, melt pool and seam width/length. Using such parameters will be much more reliable compared to evaluating process or heat radiation which can only be measured indirectly. Former approaches to implement process monitoring and control in laser beam welding have been mainly based on the evaluation of process radiation thus predicting the quality or specific features of the process result. It appears to be evident that such indirect measurements of quality features can lead to considerable uncertainty since the indirect measurement method in this case bears a risk of false interpretation, especially if the signal is being measured only one-dimensional. It is also apparent that the measurement of distinct variables and parameters such as real position, speed, gap, laser power, melt pool or seam

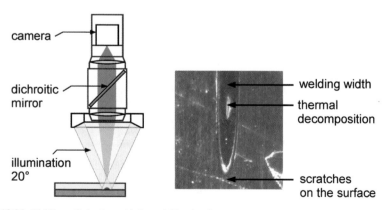

Fig. 18.21 Welding of plastics with lateral illumination

geometry will certainly lead to much improved analysis of the real state of the process and processing results. It is of great benefit to combine all the parameters (direct and indirect) in a summarized evaluation of process and machine state. However, today's monitoring systems are largely based on the evaluation of single indicators and thus mostly have a very limited reliability with regard to all monitored features [9]. The mechanical and optical boundary conditions of a laser plant or the process itself result in clear limits for the degree of possible integration of additional components. The simultaneous measurement of these parameters – most of them combined in one measurement device – will be a major step forward.

So far only limited research has been conducted to develop online monitoring systems facilitating the measurement of three-dimensional geometries of workpieces with high aspect ratios. Yet, the ongoing miniaturization calls for smaller tolerance limits and thus for more precise measurements.

Further development of such monitoring systems is of particular importance in context with the control of laser produced parts which are bound to crucial safety requirements.

A controllable process requires a good understanding of the effects of the influencing parameters. Therefore, the successful implementation of a closed-loop process control requires a good grasp of the process and its interdependencies. Practically this means, online process monitoring and closed-loop control offer high potential for rationalization of work flow if anomalies can be detected and evaluated reliably. Using currently available commercial optical-based monitoring systems a complete and automated determination of standardized quality features is not fully possible yet. In practice it can therefore not totally be avoided to include post-process non-destructive testing methods like radiological examination with X-rays or ultrasonic measurement or even imply destructive checks like strength tests or metallographic inspection. Even for well-established processing techniques the knowledge about sources for instabilities are not totally resolved. Process monitoring will therefore also in the near term be required for the enhancement of process understanding besides industrial quality assurance.

18.2 Numerically Controlled Tooling Machines for Laser Materials Processing

Oliver Steffens

18.2.1 Models of Tooling Machines

A system for processing with laser radiation is divided according to a DVS guideline into the following components [10, 11]:

O. Steffens (✉)
S&F Systemtechnik GmbH, 5207 Aachen, Germany
e-mail: oliver.steffens@ilt-extern.fraunhofer.de

- "Laser" with the components source of laser radiation, gas supply, cooling system
- "Beam manipulation" with the components beam guidance and beam forming
- "Handling of workpiece" with the components loading and unloading, positioning, movement as well as working gas and inert gas
- "Control" with the components laser control, numerical control and monitoring
- "Safety equipment" with the components safety tube, safety cabinet and beam shutter

This classification fitted on CO_2 systems is not generally valid. Hence, a more comprehensive classification is shown in Fig. 18.22 [12, 13]. This classification is independent of the used laser system and shows the flow of material, energy, control and information. "Supply" means the supply of additional materials, like welding wire and working gas. The supply takes place right at the region of processing.

The disposal encloses not only the disposal of material wastes but also dissipation of heat, respectively exhausting dust and smoke. Supply, disposal and handling – which enclose the workpiece admission up to the clamping devices, transport and generation of the relative movements between laser beam and workpiece – build the mechanical subsystem of the tooling machine. The optical system encloses beam generation and beam guidance from the source up to the processing area. Moreover, to take into account the sociological requirements of the system, the machine operator is integrated in the tooling machine. With his experience and creative competence he is able to set up the machine for the current task and to recognize and adjust failures of the system [14].

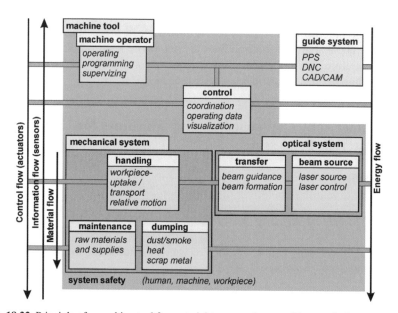

Fig. 18.22 Principle of a machine tool for material treatment by use of laser radiation

18 System Technology

One possibility of specifying the production process is the transfer of NC programs by management systems which do not belong directly to the tooling machine. Management systems are able to coordinate production processes beyond the considered system. In practice such systems have a data interface to the NC control and do release the machine operator from his program task. However, every single process is coordinated again by the NC control and is supervised by the machine operator.

The control is an interface between mechanical system, optical system and machine operator. Nowadays its functionality goes beyond its historical task of path interpolation. Besides the geometrical data the NC program contains instructions concerning the whole coordination of the process control. This actually means the control of the optical and mechanical system including supplement and disposal of waste.

The control flow moves from the control to the connected systems, which split up the data traffic to the actuators, transferring the flow of information into mechanical action. In addition to its control tasks the control takes up information of associated systems. This information is produced by sensors which record the actual state of a workpiece, mechanical or optical system and send the recorded data to the control. This information can be filtered and stored (recording operating data) or can be formatted for the machine operator (visualization). Survey of operating data in this sense means that recorded data are converted directly into a control action, e.g. by switching off the energy supply of the operating system, at least

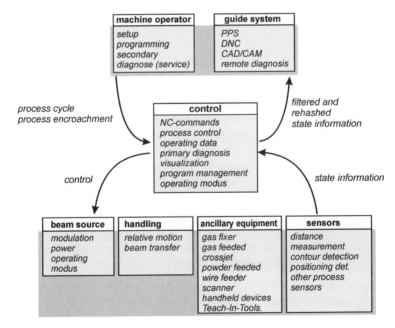

Fig. 18.23 Base model of an NC-controlled machine tool

when programmed sequences of motion cannot be achieved within a defined local tolerance limit.

The quantity of manufacturing processes shows the flexibility of lasers in production. Although the specifications of the different methods are different concerning their physical processes, there is a set of similarities, which in sum determine the requirements of the machine. To specify these requirements, another point of view is introduced as a basic model (Fig. 18.23). In this the numeric control is the main control instance. All data flow begins or ends with the control.

In the following, the functions and control-specific demands of the mended classes are being described, without mentioning the machine operator. His tasks can only be described in the full sense by looking at the whole productional surroundings. Further on he is no "technical" part of the machine. Moreover, the main systems represent that sort of systems that have a technical interface towards the control, but do not mean an enhancement of the functional range of application or production.

18.2.2 Components of the Basic Model

18.2.2.1 Source of Laser Radiation

The source of laser radiation includes laser control, which possesses a data interface to NC. Laser control determines modulation and power depending on the operation mode. The choice of source depends on the chosen processing method as well as to economic efficiency.

18.2.2.2 Handling

Handling is supposed to create a relative movement between working optics and workpiece in order to move the focus of the laser towards a defined trajectory within the working area. Handling also includes active beam transfer that can be realized by a mirror system or a fibre. In contrast, the passive beam transfer, which components are not integrated into the machine, cannot be regarded isolated (see additional equipment).

For the positioning of the working optics in relation to the workpiece surface robots are used. To cover a broad field of treatment tasks, special systems are used whose cinematics are chosen in a way that the focus of the laser beam can be positioned in a three-dimensional space. Besides, it should be guaranteed that the ray axis can be oriented by any angle to the workpiece surface. A choice of suitable cinematics of the robots is shown in Fig. 18.24.

Robots are distinguished by the following features:

- Number of axes
- Kind of movement of the main axes (cinematics)
- Working area
- Nominal load

18 System Technology

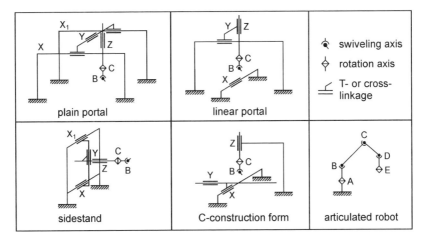

Fig. 18.24 Cinematics of robots for treatment of free-form surfaces [VDI 2861]

- Maximum speed of treatment
- Exactness of positioning, repeating accuracy and path accuracy

Because of their high geometrical precision, cartesian robots (constructed as a portal) are used in cutting and joining [15]. The axes of portal robots are arranged in the way of three linear axes that make up a right-handed, right-angled coordinate system [16]. Moreover, there can be up to two axes which enable the laser beam to tilt and turn relative to the workpiece. This happens either by moving the optics (using rotary axes) or by moving the workpiece itself (e.g. using a rotary table).

If the quality of the working results is not bounded to high geometrical precision – as, for example in hybrid processing or in de-focussed or scanned treatment of surfaces – low-priced articulated robots are used. These are characterized by three main axes. Their working area is spherical. The advantages of articulated robots are the low price – in comparison to portal robots – as well as the higher accessability of laser beam towards the workpiece (because of the spherically shaped working area). Their disadvantage lies in a lower exactness of positioning, because the faults of the single axes cumulate. Moreover, the machine operator cannot reconstruct the robot's movement as easily because having just one linear movement within the process area means that at least two axes are involved.

The relative movement between the TCP (for Tool Centre Point) and workpiece is generated by the coordinated proceeding of the robot axes within the process area. The movement of the TCP, which means the intersection of the laser beam and the workpiece, describes a trajectory with a defined speed. Additionally the angle of incidence is defined by the optical orientation towards the workpiece surface.

18.2.2.3 Additional Equipment

Additional equipment means all those tools which provide auxiliary and supplement material during the working process and therefore influence the result. The choice of equipment depends on the specific production process. Further equipment can be integrated, depending on the tasks of the machine operator (e.g. additional tools for teach-in program).

Following the functional analysis of the process, additional equipment can also be defined as tools, which control feeding, dose and combination of gas (working gas, protection gas, crossjet) [17] or which manipulate the beam.

While consistence as well as supply of wire, powders and gases are usually set up during the phase of working preparations – with the help of nozzles and other mechanical equipments [17], it can be meaningful to adjust the following parameters during the process:

- the pressure of gases (using pressure control devices)
- the volumetric flow rate of gases and pulver (using flow controller)
- the feed rate of wire (using wire feeder)

Purity and temperature of the materials, which are normally adjusted before beginning of the treatment, have additional impact on the process result. Very often, supply mechanisms are already integrated into the mechanical system of the machine. For example, a cutting head often possesses guidance for gases, cooling and powder [18]. The functions of further equipments vary depending on the specific demands and the technical development (e.g. Teach-In panel or scanner device).

18.2.2.4 Sensors

Sensors and actuators are taken as a collective term for all those components which either record and convert process data (sensors) or which influence the process as controlling element directly or indirectly (acutators). They serve to increase precision and tolerance of error as well as flexibility of the tool machine [3].

A certain combination given, sensors and actuators form a closed loop within the system, which means that actuators are controlled on the basis of sensor data. Besides it should be pointed out that laser source, handling and equipment have not only actuators but possess their own sensors as well. These sensors serve either the internal regulation of the desired values or the system state returning them to the controlling device or to another supervising unity. Sensors like 17.16 represent all additional sensors which measure certain sizes and give back this information directly to the control.

During the processing experiments in [3] a list was set up, containing all sensors and actuators applied during the process (online). On the base of Table 18.2 those sensors can be separated out which according to 17.16 already belong to handling, beam source or to additional equipment. For the class of the sensors remain:

- switches
- temperatures sensors
- capactive, inductive and tactile sensors
- camera systems (image processing)
- triangulation sensors

18 System Technology

Table 18.2 Sensors and actuators

(S)ensor/(A)ctuator	Belonging to	Comment
Switches and Buttons (S, A)	Handling	Digital information flow (on/off, open/close); by use of sensors, e.g. ok/error
	Additional equipment	
	Beam source	
	Sensors	
Way and angle sensor (S)	Handling	Integrated into the axis positioning cycle, weight balance
Force and torque sensor (S)	Handling	Axis positioning cycle
	Additional equipment	Other devices for the process
Acceleration sensors (S)	Handling	Axis positioning cycle
Temperature sensors (S)	Handling	Monitoring state of handling or measuring of process-relevant temperatures (surface treatment)
	Sensors	
Current-, voltage- and power consumption (S)	Handling	Monitoring of current-carrying system parts; by use of sensors: capacitive, inductive and tactile sensors (e.g. end switches, distance control)
	Beam source	
	Control	
	Sensors	
Camera-based Measurement Systems (S)	Sensors	Measuring the workpiece's position, properties and contours
Photodiodes and high speed CCD/CMOS cameras (S)	Sensors	Measurement of the emission of laser radiation for online quality check
Triangulation sensors (S)	Sensors	Measurement of distances
Sensors for laser power, beam guidance and intensity dispersion (S)	Beam source (Sensors)	Monitoring of the beam guidance and formation for quality assurance
Valves and engines (A)	Handling	Cooling, gas feed
	Additional equipment	

Sensors for monitoring of beam position, laser radiation power or intensitiy within the processing area cannot be used according to the present state of technology during the treatment process. These data are checked off-line before starting the treatment or in recurring service intervals (off-line used sensors for quality management).

Sensors, like additional equipment, can already be integrated into the mechanical system. So there exist commercially available cutting heads which already possess a capacitive distance sensor [19]. In Table 18.3 some examples for applications of sensors are listed.

Table 18.3 Application examples for sensors

Application example		
Sensor	Process	Description
Switches or buttons	All	Measurement of a three-dimensional surface and data feedback to a system for generation of geometric data (path planning) for the control [3]
Temperature sensor	Surface treatment	Measurement of the current temperature at the workpiece's surface near the treatment position and regulation of the laser power [20]
Capacitive approximation sensor	Cutting	Integration of the sensor in the cutting head; regulation of the working distance on a defined value by movement of the optics along the beam axis [21]
Tactile sensors	Hybrid-welding	Measurement of the root of a V joint in oil tanks and guidance of the laser beam focus [22]
Camera system (seam tracking)	Welding	Detection and measurement of the weld seam and automatical guidance of the laser beam focus (TCP) [23]
Camera system (image processing)	All	Detection of the workpiece's position and orientation within the working area before process treatment [24]
Triangulation sensors	Material removal	Material removal from surface layers (thickness of a few micrometers), Measurement of the surface and control of the laser power [25]
Caliper		Digitalization tools
Triangulation sensors		Measurement of the distance between optics and workpiece while Teach-In Phase

18.2.3 NC Control

18.2.3.1 Tasks of the NC Control of a Tooling Machine with Laser

Because of its universality regarding local, flexible and intelligent workstations, the laser beam presents itself as an ideal tool. The laser gains in importance not least by this quality in industrial manufacturing [26, 27]. The laser consists of the laser beam source in which the laser beam is produced and emitted by means of optical components, and of laser control, which is to control and supervise the beam source. Moreover, the laser is provided with a cooling system, security equipment as well as components and systems which supply the beam source, respectively the control with energy and auxiliary materials.

The transfer of laser beam to workpiece is achieved by beam guidance – consisting of mirrors or optical fibres according to construction of laser – and optical components for beam forming, e.g. focusing lenses. The task of handling of the tooling machine is the creation of a relative movement between laser beam and workpiece. For handling articulated robots or cartesian portal robots can be used. These robots consist of a number of axes whose movements are steered by motors, gears and drives. Normally the axes are set up in a fixed mechanical relation and thereby determine the cinematic of the robot.

The handling, which means industrial robots, beam transfer and laser beam, together with security systems and machine operator form the main and most important part of the tooling machine. Further on, other devices, sensors and actuators are applied to the system, which are used for the material treatment with laser radiation (e.g. supply of auxiliary materials, waste management, process sensors).

The numeric control is used for control and survey of the technical system components as well as for information exchange with the machine operator or any other connected systems. In that way the control forms the coordinating element within the system "tooling machine". According to the cinematic of the robot controls are either generally called NC control (NC for Numerical Control) or, having an articulated robot, a RC control (RC for Robot Control). In the following the term NC control is applied without distinction between NC and RC.

With rising degree of automation of manufacturing processes the task of the process control also increases by the NC in complexity [28, 29]. The treatment of a workpiece on a NC is described by a program (NC program or part program) [30, 31]. The data stored in the program are decoded by the control and is processed after geometrical and technological data separated.

In essence the tasks of the NC control can be divided into the control of the robot movements and into the control of the ray source and the additional equipments [32]. Moreover, it applies itself to take the data of NC of sensors, to process and to affect the process expiration on the basis of sensor data about the actuators. At least operating data can be visualized at the MMI (MMI for man–machine interface).

The tool owner can determine the process sequence by use of the MMI and affect and supervise with the help of the visualized operating data.

The NC controls the robot movements on the basis of interpolation points (geometry data in part program) and generates a trajectory by connecting the programmed points. The cinematic of the robot and the mechanical dimensions and limitations are passed in the machine date of the control and are considered by the interpolation algorithm of the trajectory. The control of additional equipments and laser device takes place on the basis of the technology data given in the NC program. Besides, an essential aspect by the treatment of the technology data is the temporal or local coordination of these issues with the robot movements. The assignment of geometrical data and technology data is given by the part program.

18.2.3.2 Internal Structure of an NC Control

The internal structure of a control was laid out on the different functions of the control. While the hardware of the NC adapted itself constantly to the technological innovations in the computer technology and information technology, the structural construction of commercially available NC controls stayed the same. The structure elements of the control are the MMI with a program interface, the NC kernel and the programmable logic control (PLC). Different procedures are used for the programming of a control. In laser technology three kinds of the NC programming predominate [33], the computer-aided indirect programming of the tooling machine, the direct programming by use of the editor of the NC and the direct programming by means of Teach-In. In every case the NC program about the MMI is filed into the memory of the control for the processing.

The NC kernel is responsible for the processing of part programs [34]. Movements of the robot and the processing of technology data during the process are coordinated by the NC kernel. The program decoder takes over the task to separate geometry and technology data and to convert in an internal structure to be treated for the control. The decoder reads for it in the first step the data from the NC program memory. These data are applied to the interpolator to calculate the trajectory in firmly given time intervals (interpolation time). Besides, cinematic and mechanical borders of the system are considered. If a technology function is called dependent on position, the Interpolator takes over the issues to the connected devices, either directly ("fast" periphery) or about the detour of the PLC. The PLC as a part of the NC control takes over the task from inputs and issues near connected devices. The necessity of the PLC is founded in the fact that it checks on the basis of logical connectives in the PLC program whether the machine tool is in a safe condition. In addition it is decoupled hardware- and software-sided by the NC kernel. All security equipments (e.g. protection doors, emergency stop, end switches) are to be connected to the entrances of the PLC [32]. Issues directly about the Interpolator take place when they are connected temporally closely with the steps of the process consequence (real time). The administration of these issues within the Interpolator needs additional time which gets lost for the continuous calculation of interpolation points along the programmed trajectory. The permissible number of the issues to be administered is limited on account of the capacity of the Interpolator and the given interpolation time. Certainly, a rise of the interpolation time is possible in modern controls. However, the increasing time of interpolation leads to a more inaccurate interpolation of the trajectory and to higher differences between desired and actual path. Therefore, direct controls of periphery from the Interpolator are limited to the necessary. However, with rising capacity and reliability of the processors tasks of the PLC can be shifted increasingly to the Interpolator. At present systems are already offered basing on PC technology, which distinguish logically between PLC and NC kernel, but processing takes place together on a single processor (PC-based controls). Process times are assigned with the time-slice procedure.

The sensors can supply additional information regarding the process. This information is processed either by the Interpolator or by the PLC program. On the basis of sensor data issues can be adapted in manipulation, beam source or additional components.

The control of the robot movement is realized by means of a closed-loop positioning cycle. Besides, the positioning cycle serves the physical control of the drives and the monitoring of the positions of the robot axes (Fig. 18.25). The functions of a two-stage position regulator are distributed to NC kernel and robot. Besides, the system borders are not unambiguous. It differs according to complexity of the positioning cycle, the used control and the applied drives.

18 System Technology

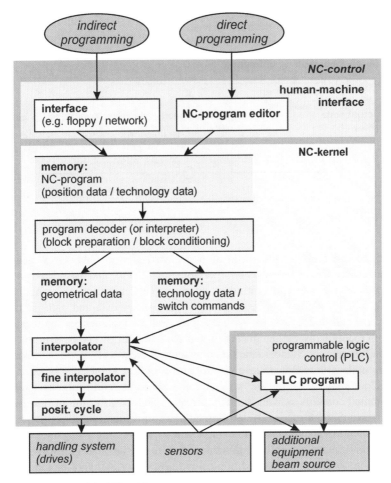

Fig. 18.25 Structure of the NC architecture

To acquire and visualize information from the connected devices data flow bottom-up into the NC kernel and from there to the MMI. Other tasks of the MMI are the filtering and putting into graphs of operating data as well as the administration of NC programs and other user's inputs and the supply of interfaces to other connected systems, like CAD/CAM systems, to shop-floor systems or DNC systems [35–37].

18.2.4 Functional Extensions of Numerically Controlled Machine Tools for the Material Treatment with Laser Radiation

18.2.4.1 The Programming System "Teasy"

The hand programming system "Teasy" was developed for the generation of NC programs by means of the Teach-In process and can also be used by the applica-

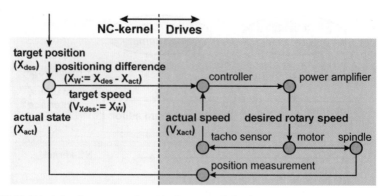

Fig. 18.26 Principle of a two-staged positioning control circuit

tion of test functions for the verification and correction of CAD/CAM generated programs. The "Teasy" consists of a portable input device and associated software which contains a diagrammatic user's interface with integrated NC program editor (Fig. 18.26). During the Teach-In phase *technology* data and path parameters can be inserted, led by menu. With complex order consequences the user is led based on dialogue by the parameterizing of the commands. The description of the German Institute for Standardization commands in clear text, as well as short help texts about the commands, relieve the programmer of the knowledge of all engines-specific commands. During the programming phase the editor allows starting of programmed contours as well as their correction and the insertion of new interpolation points. Supporting functions, like setting certain coordinate shifts or resetting are deposited as a macro and can be implemented by pressing a single button.

The representation of information on the screen was divided into windows. The information content is encased in logical entities. Other software modules can be implemented to realize assisting systems adapted to special machine tool functionalities. The functional character of these systems is integrated entirely into the "Teasy". Under the viewpoint of "open" systems the "Teasy" shows a flexible, adaptable, portable and scalable character [38].

On account of the open draft as well as the integrated functional character to the visualization, verification and correction of programs, the "Teasy" is suitable to the programmation in Teach-In procedure as well as to the correction of computer-generated (e.g. CAD/CAM) NC programs. In addition a part program can be loaded in the editor of the Teasy and can be verified block by block with the help of test functions. Geometrical differences caused by workpiece tolerances or inaccurate positioning of the workpiece as well as mechanical inaccuracies of the machine tool can be detected and corrected.

If the process parameters are ordered block likely in the part program, so that these can be assigned to the treatment sections, these parameters can also be adapted by the tool user according to his knowledge and experiences. An information feedback of the changes to the CAM module can be realized.

To use the CAD/CAM interface effectively, an internal regulation of the program structure is helpful, as well as other software modules in the planning level which recognizes changes of NC programs and leads them back into simulation modules and technology data banks.

18.2.4.2 Teach-In-Help

The Teach-In-Help corrects the distance between optics and workpiece surface by use of an integrated distance-measuring sensor automatically (Fig. 18.27). Moreover, on the basis of the programmed trajectory and by a measuring cycle the rotating axes are moved to position the optic towards the workpiece surface into a predefined angle.

In the measuring cycle the axes are moved to four points around the current TCP in a defined distance. After reaching the positions the distances between sensor and surface are measured by means of the appropriate sensor. By use of the measured data and regarding axis positions workpiece coordinates on the surface can be calculated.

In each measured point three coordinates define a level of the workpiece surface towards the optics. By using the four possible combinations {P1, P2, P3}, {P1, P2, P4}, {P1, P3, P4} and {P2, P3, P4} an average level is computed, and the normal vector to this level can be calculated. The round axes of the robot are moved so that the beam axis runs along the normal vector hitting the TCP. Within the "Teasy" the Teach-In-Help can be parameterized and started at the touch of a button during the programming phase. It relieves the tool user of the repeating activity to adjust distance and orientation of the optics relative to the workpiece surface. On account of the physical qualities of the distance sensor, the Teach-In-Help can be used if the surface of the workpiece shows suitable optical qualities (e.g. reflectivity) only. Particularly with bigger inclinations of the workpiece relative to the sensor head (3-D applications) and strongly reflecting surfaces the sensor reaches its physical borders.***

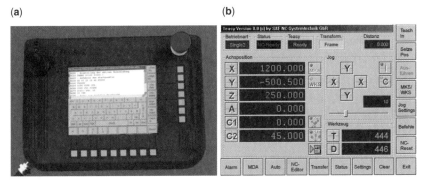

Fig. 18.27 Programming system "Teasy". (**a**) Handheld. (**b**) User interface

18.2.4.3 One-Dimensional/Three-Dimensional Distance Control

The correct focus position of the laser beam relatively to the workpiece surface is an essential factor of influence for a qualitatively good process result. A correction of the focus position can take place by use of sensors to measure the distance between optics and workpiece surface. The distance regulation falls in the area of responsibility "Path Control" and "Sensor Integration". Moreover, the quality of the process result can be supervised by the monitoring of the working distance during the process. A monitoring of the operating area of the sensors or the evaluation of a digital signal with touch of the workpiece raises the operational safety of the machine tool.

The distance control is differed in a one-dimensional distance regulation with which a separate axis is driven on account of the sensor data, and the three-dimensional the distance regulation with which the distance is corrected between optics and workpiece surface along the beam axis (Fig. 18.28). Measuring the distances takes place by distance-measuring sensor systems fixed appropriate to the optics.

Usually the one-dimensional distance regulation applies sensors which are not integrated into the optics. On account of the construction form and the measuring procedure it is not possible to measure directly at the point of treatment (TCP)

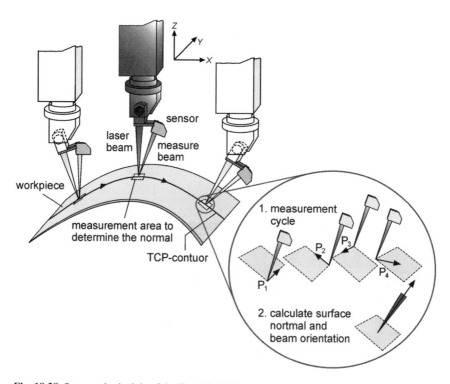

Fig. 18.28 Setup and principle of the Teach-In-Help

during the process. The measuring point lies around a defined distance to the TCP forerunning in working direction.

The three-dimensional distance regulation can be applied by us of a sensor which is integrated into a cutting head. In addition nozzles are available which lead the working gas or protection gas to the working point and dispose a capacitive sensor at the nozzle top also.

Inductive, capacitive, optical or tactile sensors are used. The sensor system generates an analogous value in dependence of the distance between sensor head and workpiece surface. To parameterize and control of the distance control a set of new NC commands are implemented [21].

To realize the distance control the NC calculated desired values for the axes movements are overwritten by desired values calculated by use of the sensor measurements in each interpolation cycle. The one-dimensional distance control requires the feed along one of the cartesian axis (leading axis without loss of generality X). The sensor measures along an orthogonal axis (following axis Y). The calculation of new desired values takes place in every interpolation cycle (Fig. 18.29). For that in a first step, the actual sensor measured distance value is used to calculate a new desired value for the TCP taking the forward run into account. These coordinates are stored in a ring buffer as second step.

Within the same interpolation cycle the ring buffer is used to read two values which coordinates X enclose the actual value X of the leading axis. By use of this value a new desired value is interpolated for the following axis Y. Taking the maximum velocities and acceleration into account the corrected desired value Y is written into the internal NC memory (as new interpolated position).

In contrast to the one-dimensional distance control-used sensors for three-dim -ensional distance control measure directly at the treatment area (TCP), so that to the algorithm requires the last entry in the ring buffer only. The integration of the

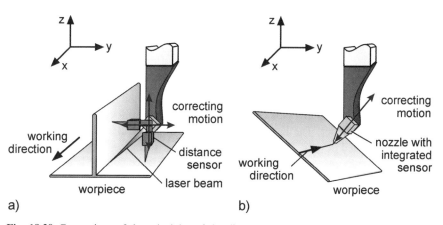

Fig. 18.29 Comparison of the principles of the distance controls (one-dimensional and three-dimensional). (**a**) one-dimensional distance control with two axes simultaneously (**b**) three-dimensional distance control along the beam axis

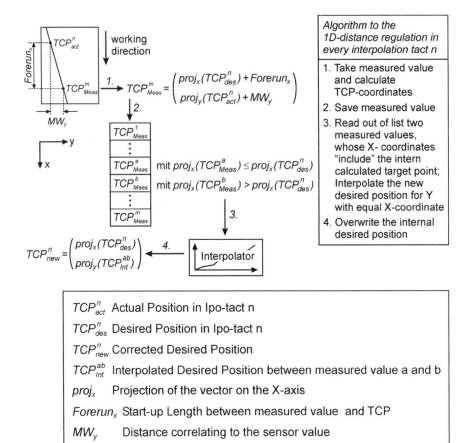

Fig. 18.30 Algorithm of one-dimensional distance control

sensor values within the NC internal interpolation is based upon the transformation of the measured values along the cinematic chain of the robot (Fig. 18.30).Therefore the actual position of the rotary axes are recorded (step 1). The transformed sensor values (step 2) are used to calculate an offset to the machine axes (step 3). Similar to the one-dimensional distance correction the maximum velocity and acceleration are taken into account. The last step is used to add the calculated offset to the NC internal calculated desired TCP position.

As example the usage of the three-dimensional distance control is illustrated in Fig. 18.31 which shows the operation of the "Rotocut" (Fig. 18.32). Within the cutting head a distance-measuring capacitive sensor is integrated, which generates a signal when touching the workpiece by the nozzle. This signal is evaluated by the PLC program and interrupts the treatment immediately when this signal rises.

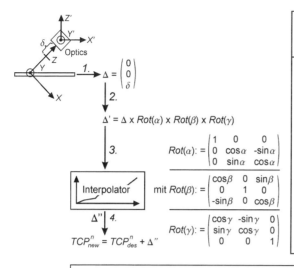

Fig. 18.31 Algorithm of three-dimensional distance control

18.2.4.4 Seam Tracking

Seam tracking is an application of high importance for the flexibility and automation. Seam tracking means the automatic control of robot axes in that way that the TCP is guided along a geometrical contour upon the workpiece's surface detected by the used sensor system [40]. For that optical sensor systems are used. These systems contain a laser diode which emitted light is projected upon the workpiece surface like a line by means of optics. This line is recorded by a CCD camera and is to be evaluated (light-section method, Fig. 18.33).

Within the camera image the detected contour is identifiable by the point of discontinuity. The width of the gap between the line segments can be used to calculate the gap width of the workpiece's seam. The measuring values are transferred to the control by a fieldbus protocol.

For seam tracking the sensor head is assembled at the optics in forward direction concerning the treatment feed to protect the sensor from process heat and the emitted

Fig. 18.32 Distance control by use of the RotoCut. Source of image: Thyssen Lasertechnik, Aachen

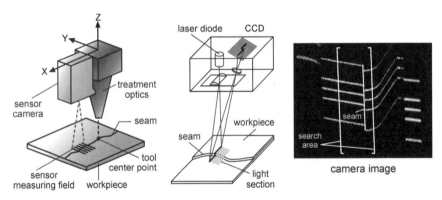

Fig. 18.33 Measurement principle of a seam-tracking sensor and the light-section method

light by the process plasma. On the other hand, the sensor should not influence the process itself. Further on it takes some time to record and evaluate a camera image as well as to transfer the data to the NC (dead time) which demands a certain forerun of the sensor.

If a safe recognition of the contour is possible and the geometric boundaries for using the seam tracking are fulfilled the seam tracking can be subdivided into a phase model to structure the sequences by their time order, Fig. 18.34). The param-

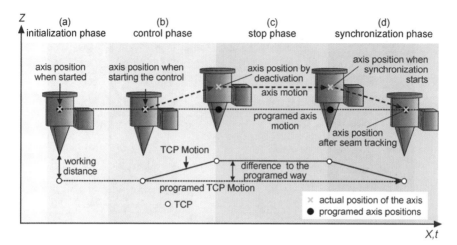

Fig. 18.34 Seam-tracking phases

eters of the single phases are set by NC instructions while the phase of initialization. This enables the seam tracking to be used without interruption of the robot's movement while processing [21, 41].

The tasks during the separated phases are:

- Phase of initialization: parameterization, move to start position, switch on sensor, regulate the forerun
- Phase of control: control axes by use of sensor values (if necessary control of the wire feeder or the laser power)
- Phase of stopping: switch-off sensor, freeze the distance between programmed and current path, leading the treatment optics in a area safe against collision
- Phase of synchronization: stop movements, synchronize the actual and desired axis positions, finish the seam tracking

The separate phases can be affected by a set of new NC commands and can be started or stopped. The advantage of this kind of realization is that existing programs can be adapted without great effort by adding the appropriate commands at the beginning and end of the processing cycle.

By use of the seam-tracking tolerances of the workpiece, or the workpiece's clamping position or tolerances caused by the process treatment can be compensated. Furthermore the afforded time to program the control is reduced. In practice the fields of application are limited because of the optical method to measure. In particular complex three-dimensional applications are very intricate because all robot axes have to be used to position the focus point of the laser beam upon the workpiece's surface and the rotary axes are used to adjust the appropriate angle of incidence. So the guidance of the measuring field of the sensor to detect the desired contour is difficult. In addition to that the orientation of the sensor head towards the workpiece is changing continuously. This results in a shift and tilt of the pro-

Fig. 18.35 Simultaneous usage of two seam-tracking systems to weld T-joints in shipbuilding. Source of image: Blohm und Voss, Hamburg

jected laser line within the CCD camera image and leads to geometrical errors when evaluating the two-dimensional camera image. To compensate this errors complex algorithms are needed.

An example of usage is the simultaneous operation of two sensor systems to weld T-joints in shipbuilding [41] (Fig. 18.35).

References

1. P. Abels, S. Kaierle, E.W. Kreutz, R. Poprawe: "Failure Recognition in Tailored Blank Welding by Image Processing", ICALEO 98, USA (1998)
2. John C. Ion: "Laser Processing of Engineering Materials: Principles, Procedure and Industrial Application". ISBN: 0750660791, Butterworth-Heinemann, Oxford (2005)
3. N.N.: Sonderforschungsbereich 368 "Autonome Produktionszellen", Arbeits- und Ergebnisbericht 1997/1998/1999. RWTH Aachen (1999)
4. S. Kaierle, M. Dahmen, A. Diekmann, E.W. Kreutz, R. Poprawe: "Failure Analysis for Laser Beam Welding". Proceedings of ICALEO 97, San Diego, California, USA (1997), 63–72
5. O. Rousselange: Entwicklung eines Sensors zur Detektion von Schweißfehlern beim "Tailored-Blank"-Schweißen mit Laserstrahlung, Diplomarbeit, Lehrstuhl für Lasersetechnik, RWTH Aachen (1998)
6. A. Reek: "Strategien zur Fokuspositionierung beim Laserstrahlschweißen", Dissertation, Forschungsberichte iwb Band 138, H. Utz Verlag, München (2000)
7. J.P. Boillot et al.: "Robot Feature Tracking Devices and Methods" Patent No. US 6,430,472 B1 (2002)
8. B. Regaard, S. Kaierle, W. Schulz, A. Moalem: "Advantages of External Illumination for Monitoring and Control of Laser Materials Processing" Proceeding of ICALEO 2005, Paper #2307, Miami (2005)

9. IFSW Institut für Strahlungswerkzeuge Stuttgart "Integration optischer Messmethoden zur Prozesskontrolle beim Laserstrahlschweißen", Laser Magazin (2003)
10. DVS 3203-1: Qualitätssicherung von CO2-Laserstrahl-Schweißarbeiten; Verfahren und Laserstrahlschweißanlagen; Ausgabe 1988–12, Beuth Verlag, Berlin (1988)
11. H.R. Schunk: Laser für die Oberflächenbehandlung: Konzeption geeigneter Systeme; VDI-Verlag, Düsseldorf (1992)
12. H. Oebels: Verfahrensentwicklung als integrierter Bestandteil moderner Qualitätssicherung am Beispiel des Laserstrahlschneidens, Dissertation, RWTH Aachen (1992)
13. O. Steffens: Adaption von numerischen Werkzeugmaschinensteuerungen an das Fertigungsumfeld bei der Materialbearbeitung mit Laserstrahlung, Dissertation RWTH Aachen, Shaker Verlag, Aachen (2003)
14. U. Wolf: Integration der Laserbearbeitung in ein Produktionssystem: Zur Vorgehensweise bei der Einführung innovativer Produktionstechnik Fügen mit CO_2-Laserstrahlung, Dissertation RWTH Aachen, Shaker Verlag, Aachen (1994)
15. C. Sander: Steuerungs- und Regelungstechnische Ansätze zur Nahtverfolgung beim Laserschweißen, Studienarbeit RWTH Aachen, Institut für Regelungstechnik, Aachen (1999)
16. DIN 66217: Koordinatenachsen und Bewegungseinrichtungen für numerisch gesteuerte Arbeitsmaschinen, Beuth Verlag, Berlin (1974)
17. Verein Dt. Ingenieure (Hrsg.): Materialbearbeitung mit dem Laserstrahl im Geräte- und Maschinenbau, VDI Verlag, Düsseldorf (1990)
18. A. Weisheit, K. Wissenbach: Optik für das Härten und Beschichten von Innenkonturen, in: Leistungen und Ergebnisse Jahresbericht 2000; Fraunhofer-Institut für Lasertechnik, Firmenschrift, Aachen (2000)
19. N.N.: http://www.precitec.com/PLMSysD.htm; Systemübersicht Lasermatic; Firma Precitec, Gaggenau-Bad Rotenfels (2000)
20. J. Sommer, S. Kaierle: Prozesskontrollierendes modulares Diodenlasersystem, in: Leistungen und Ergebnisse Jahresbericht 2000, Fraunhofer-Institut für Lasertechnik, Aachen (2000)
21. M. Fohn: Abstandsregelung V2.01, Bedienungsanleitung, Fraunhofer Institut für Lasertechnik, Aachen (1997)
22. M. Dahmen, S. Kaierle: Hybridschweißen von Öltanks, in: Leistungen und Ergebnisse Jahresbericht 2000; Fraunhofer-Institut für Lasertechnik, Firmenschrift, Aachen (2000)
23. O. Steffens: Anforderungsdefinition, Spezifikation und Implementierung eines Systems zur Bahnkorrektur bei der Werkstückbearbeitung mit numerischen Steuerungen in Echtzeit, Diplomarbeit, RWTH Aachen (1995)
24. M. Selders: Trends in der Messtechnik, in: >> inno << Innovative Technik – Anwendungen in Nordrhein-Westfalen, Nr. 14 Mai 2000, S. 16–17; IVAM Nordrheinwestfalen (Hrsg.), Dortmund (2000)
25. K. Wissenbach: Abtragen von Deckschichten mit Hochleistungslasern, in: Leistungen und Ergebnisse Jahresbericht 1998; Fraunhofer-Institut für Lasertechnik, Firmenschrift, Aachen (1998)
26. W. Trunzer, H. Lindl, H. Schwarz: Einsatz eines voll 3D-fähigen schnellen Sensorsystems zur Nahtverfolgung beim Laserstrahlschweißen; in: Laser in der Technik: Vorträge des 11. Internationalen Kongresses (Hrsg. Waidlich, W.), Seite 405–410, Springer Verlag, Berlin, Heidelberg, New York (1994)
27. D. Röhrlich: Hammer aus Photonen, Energiereiches Licht entpuppt sich als vielseitiges Werkzeug für die Materialbearbeitung; in: Wirtschaftswoche Nr. 16, Seite 190–192, Verlagsgruppe Handelsblatt, Düsseldorf (1999)
28. M. Weck: Werkzeugmaschinen Fertigungssysteme, Bd. 1 Maschinenarten, Bauformen und Anwendungsbereiche – 4. Auflage, VDI Verlag, Düsseldorf (1991)
29. N.N.: Offene Prozessschnittstellen für integrierte Steuerungsapplikationen, Resümee zum Forschungsbedarf im Rahmen der BMBF-Dringlichkeitsuntersuchung "Forschung für die Produktion von Morgen", Stuttgart (2000)
30. M. Weck: Werkzeugmaschinen, Bd. 3 Automatisierung und Steuerungstechnik – 3. Neubearbeitete und erweiterte Auflage, VDI Verlag, Düsseldorf (1989)

31. DIN 66025: Programmaufbau für numerisch gesteuerte Arbeitsmaschinen, Beuth Verlag, Berlin (1981)
32. B. Kief: NC/CNC Handbuch '92, Carl Hanser Verlag, München, Wien (1992)
33. P. Heekenjann, A. Köhler, C.K. van der Laan: Robotergestützte 3D-Lasermaterialbearbeitung; in: Laser, Ausgabe 2-1996, b-quadrat, Kaufering (1996)
34. H. Nitsch: Planung und Durchführung der Materialbearbeitung mit Laserstrahlung aus steuerungstechnischer Sicht, Dissertation RWTH Aachen, 1995, Aachen: Shaker (1996)
35. L. Stupp: Der Griff zur richtigen Tastatur; in: Werkstatt und Betrieb, Ausgabe 3/2000, Seite 24–31, Carl Hanser Verlag, München (2000)
36. W. Fembacher: Produktivitätssteigerung durch Steuerungsvernetzung; in: Werkstatt und Betrieb 3/2000, Seite 24–31, Carl Hanser Verlag, München (2000)
37. M. Leinmüller, N. Treicher: Entscheidungsunterstützung durch ein Produktionsinformationssystem; in: ZWF, 10/1999, Seite 597–600, Carl Hanser Verlag, München (2000)
38. K. Walter, Fertigungs- und Montagegerechte Konstruktion des Handprogrammiersystems "Teasy", Diplomarbeit, Fraunhofer-Institut für Lasertechnik, Aachen (2000)
39. J. Olomski: Bahnplanung und Bahnführung von Industrierobotern, Vieweg, Braunschweig, Wiesbaden (1989)
40. M. Fohn: Spezifizierung und Implementierung eines Profibus-Protokolls zwischen Sensorsystem "SCOUT" und der CNC "Sinumerik 840D" zur Nahtfolge, Diplomarbeit, RWTH Aachen (1999)
41. O. Steffens, S. Kaierle, R. Poprawe: Simultaneous Laser Beam Welding of T-Joints by Use of Seam Tracking Systems". Proceedings of the ICALEO 2000, Detroit, USA (2000)

Chapter 19
Laser Measurement Technology

19.1 Laser Triangulation

André Lamott and Reinhard Noll

19.1.1 Introduction

Most of the modern production processes have reached a high grade of automation. For reliable process guiding it is necessary to monitor and verify continuously the single process steps and the quality of the products manufactured. Non-contact online measurement systems ensure shortest time of reaction for an efficient process control. Laser measurement systems are predestinated for this application because of the following reasons:

- no mechanical contact to the measuring object,
- high measurement frequencies,
- independent of ambient light conditions and the temperature of the measuring object,
- fast adaptation to different surface properties of the specimens.

Laser light is the most versatile light source for optical measurement processes. Typical measurement distances range from a few millimeters to several meters. This allows for a flexible integration of laser measurement systems in process lines. Laser measurement techniques are independent of ambient light, because of the high spectral radiance of laser radiation (power per area, per solid angle, per wavelength interval). Furthermore, the laser light irradiance can be modulated as a function of time without inertia. This capability makes it possible to adapt the measurement system to varying surfaces and material properties. The average lifetime of current semiconductor lasers is more than 50,000 h, being significantly greater than the lifetime of conventional thermal light sources.

R. Noll (✉)
Lasermess- und Prüftechnik, Fraunhofer-Institut für Lasertechnik, 52074 Aachen, Germany
e-mail: reinhard.noll@ilt.fraunhofer.de

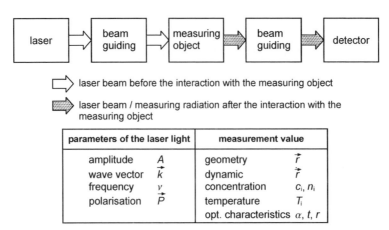

Fig. 19.1 Schematic of a laser measurement method. The interaction of the laser beam with a measuring object changes the laser parameters allowing the determination of different geometric, dynamic, chemical measuring quantities and optical properties of materials

All laser measurement techniques are based on the change of at least one parameter of the laser radiation after the interaction with an object, see Fig. 19.1 [1, 2]. These parameters are amplitude, wave vector (describes the direction of the laser beam and the wavelength), frequency, and polarization (describes the direction of the electrical field). Measurement quantities are mostly geometric dimensions, change in time of geometric measurements, concentration and particle densities, temperatures, and material characteristics. The following chapter describes measurement processes of geometric dimensions especially with the method of laser triangulation.

Laser systems are able to measure online geometric values relevant for the process and the product. The measuring object remains in the production line and is inspected under production conditions. There is neither a need to take out the specimen from the production line nor a need to transport the measuring object to a separate measuring room. Figure 19.2 shows schematically the integration of a laser measurement system in a production line and in an IT system. The data measured online are evaluated in real time and used for documentation, control of preceding process steps or sorting tasks, and for quality assurance purposes.

19.1.2 Measurement of Geometric Quantities

The principle of laser triangulation is shown in Fig. 19.3 [1]. There are different ways to measure the distance (1-D = one dimension), the contour (2-D = two dimensions), and the shape (3-D = three dimensions) of an object [3]. For a distance measurement a collimated laser beam is irradiated onto the object. The scattered laser light is imaged under a known angle on a position-sensitive detector such as, e.g., a CCD line array. When the laser beam hits the target at various distances, the

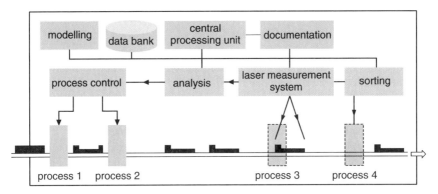

Fig. 19.2 Laser measurement system embedded in a production line and in an IT system

imaged spot on the detector is shifted. With the measured position of the imaged spot and the known direction of laser beam irradiation the distance between the measuring object and a reference plane can be determined. For a 2-D measurement of contours, the laser beam is formed to a laser line. This line is imaged under a known angle on a 2-D position-sensitive detector (for example, a CCD or CMOS array). This type of laser triangulation is called light section method.

Laser sources for triangulation sensors are usually continuous wave or pulsed semiconductor lasers. By modulation of the pulse width the available radiant energy for a measurement can be varied in a wide range to compensate variations of the intensity of the scattered laser light due to changing scattering properties of the surface of the specimen. Typical wavelengths used are in the red and near-infrared spectral range as, e.g., 660, 670, 685, and 780 nm. The radiant flux varies between 1 and 100 mW depending on the mean measurement distance, the measurement range, the wavelength, and the specific application. Laser line filters are often used in front of the detector to suppress interfering ambient light, while the laser wavelength is transmitted with only minor attenuation.

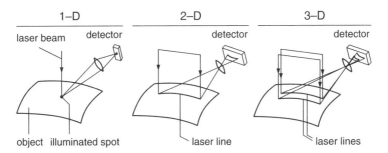

Fig. 19.3 Principle of laser triangulation methods. *Left*: distance measurement with a collimated laser beam, *center*: laser light section for the measurement of contours and profiles, *right*: projection of several laser lines for the measurement of shapes

19.1.3 Scheimpflug Condition and Characteristic Curve of Triangulation Sensors

Objective and detector of the triangulation sensor must be placed relative to the incident radiation in such a way that the illuminated spot is imaged sharply onto the detector plane for all distances of the measuring object within the measuring range. Therefore, the detector plane is tilted relative to the optical axis of the objective. Figure 19.4 illustrates this geometric arrangement in more detail.

To simplify the description a coordinate system (u, v) is chosen in a way that the v-axis runs parallel to the optical axis of the objective and the u-axis lies in the objective plane being oriented toward the z-axis, which is parallel to the incoming laser beam. The origin of the (u, v)-coordinate system is centered in the middle of the objective. The irradiated laser beam is tilted with respect to the v-axis. An illuminated point P_1 is imaged to a point P'_1 in the detector plane and P_2 to P'_2. Since P_2 is farther away from the objective than P_1, the image P'_2 is closer to the objective than P'_1. Obviously the detector must be tilted relatively to the v-axis to assure that the point P_2 is imaged sharply. By use of the thin lens equation $1/o + 1/i = 1/f$, with o object distance, i image distance, and f focal length, the curve can be calculated on which the images P' with the coordinates (u', v') are located.

The result is

$$u' = (m_\mathrm{L} - u_0/f)v' + u_0 \tag{19.1}$$

where m_L is the slope of the straight line describing the incident laser beam, u_0 the position of the laser beam at $v = 0$, and f the focal length of the objective.

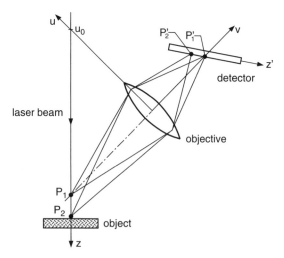

Fig. 19.4 Imaging geometry for laser triangulation

19 Laser Measurement Technology

Relation (19.1) describes a straight line. The detector must be oriented along this straight line to get a sharp image of the points P_1 and P_2 in Fig. 19.4. The same holds for all other points along the propagation direction of the incident laser beam. For $v' = 0$, the straight line crosses the u-axis at $u' = u_0$. At the same point it intersects the incident laser beam. The tilt of the detector plane must be chosen in a way that the laser beam axis, the u-axis in the objective plane, and the straight line where the detector plane is aligned to are crossing in one single point $(u_0, 0)$. This rule is called Scheimpflug condition, since the Austrian Captain Theodor Scheimpflug has formulated this rule for the first time in the beginning of the 20th century.

The characteristic curve of a triangulation sensor describes the relation between the position of the illuminated spot on the object and the position of the image in the detector plane. In the following, the coordinate axis z and z' as shown in Fig. 19.4 will be used, where the z-axis is parallel to the incident laser beam and the z'-axis is parallel to the straight line described by relation (19.1). The origins of these axes lie on the v-axis, cf. Fig. 19.4. The illuminated spot at the position z is imaged into the detector plane at the position z' given by

$$z' = z\sqrt{1 + \left(m_\mathrm{L} - u_0/f\right)^2} \, \frac{m_\mathrm{L} f}{z\left(m_\mathrm{L} - \frac{u_0}{f}\right) + \sqrt{1 + m_\mathrm{L}^2}\left(2u_0 - m_\mathrm{L} f - \frac{u_0^2}{m_\mathrm{L} f}\right)}$$

(19.2)

Figure 19.5 shows the characteristic curve according to Eq. (19.2) with exemplary parameters for u_0, m_L, and f. In this case, a measurement range of -120 to $+120$ mm in z-direction is imaged on a detector section having a length of about

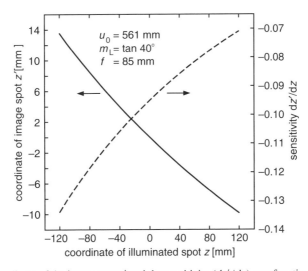

Fig. 19.5 Coordinate of the image spot z' and the sensitivity (dz'/dz) as a function of the illuminated spot position z

Fig. 19.6 Laser light section sensor for profile and contour measurements, *left*: object with projected laser line, *right*: sensor head

20 mm along the z'-direction. That is, for example, the typical extension of a CCD line array with 2,048 pixels.

Figure 19.6 shows a laser light section sensor for the measurement of contours and profiles. Typical technical data of triangulation sensors for distance (1-D) and contour measurement (2-D) are shown in Table 19.1 [2].

19.1.4 Application Examples

19.1.4.1 Thickness Measurement of Sheet Metal

The thickness measurement of an object is a 1-D measuring task (cf. Fig. 19.3, 1-D). State-of-the-art methods for thickness measurement are based on the measurement of the transmission of X-rays or gamma radiation. However, these methods require the knowledge of absorption coefficients, which depend on the chemical

Table 19.1 Technical data of triangulation sensors for distance and contour measurement

	Distance	Contour, profile
Laser source	Diode laser	Diode laser
Measurement range	5, 10, 20, ... up to 1,000 mm	50 mm (z) × 60 mm, 105 mm (z) × 120 mm
Linearity according to DIN 32 877 [3]	0.0015%	50 µm (z) × 120 µm, 105 µm (z) × 240 µm
Detector	CCD line sensor and CMOS array	CMOS array, 1,280 × 1,024 pixels
Measurement frequency	Up to 70 kHz	Up to 200 Hz
Surface properties of the measuring object	Automatic adaptation to varying scattering characteristics	Automatic adaptation to varying scattering characteristics

composition and the specific mass density of the object to be measured. The advantages of laser-based thickness measurement are high measurement frequency, high precision, no knowledge of material composition necessary, minimum maintenance effort, and simple safety measures (since no radioactive material or high voltages are needed) [4].

Figure 19.7 illustrates the setup for thickness measurement with two laser triangulation sensors measuring simultaneously the distance to the lower and upper side of a moving sheet metal [5]. With the two distance values the thickness of the sheet can be determined. The sensors are integrated in a C-frame, which is installed directly behind a roll stand. The determined thickness value is used to control the roll gap. The technical data for laser thickness measurement of sheet metal are listed in Table 19.2.

The accuracy in industrial routine operation for laser thickness measurement on moving cold sheet metal amounts to less than 3 µm. Figure 19.8 shows a view of the C-frame with the two laser sensors during the online measurement of the thickness directly behind a roll stand (upper right corner). Figure 19.9 shows the display in the control center, where the results of the thickness measurement as a function of time are visualized. In the presented example a sheet metal of varying thickness is rolled. The process of the so-called flexible rolling permits the manufacturing of sheets with defined thickness variations and smooth thickness transitions. Tailored

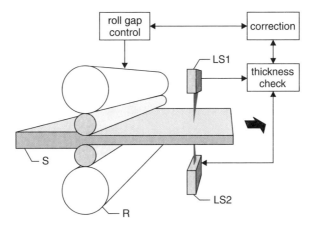

Fig. 19.7 Thickness measurement with two laser triangulation sensors. S = sheet metal, R = roll stand, LS1, LS2 = laser triangulation sensors

Table 19.2 Technical data of laser thickness measurement systems of sheet metal

Thickness range of sheet metal	10; 130 mm
Measurement frequency	Several tens of kilohertz
Sheet velocity	Up to 2,000 m/min
Diameter of the laser spot	45; 150 µm
Depth of C-frame	200–800; 400–2,000 mm
Accuracy of thickness measurement	3; 10 µm

Fig. 19.8 View of a C-frame with the two laser triangulation sensors for the online thickness measurement of sheet metal directly behind a roll stand

blank sheets are used for deep drawing of structural components for the automotive industry as, e.g., connection bearer, crossbar, boot platforms. Conventional thickness measurement systems cannot be used for online thickness measurement of Tailored Rolled Blanks, due to the high demands on dynamic and precision for this measurement task.

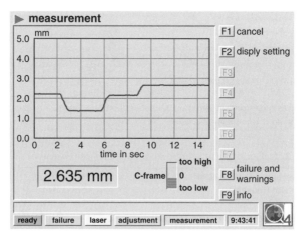

Fig. 19.9 Real-time display of the measured sheet metal thickness as a function of time

19.1.4.2 Straightness Measurement of Rails

The availability of compact diode lasers with high beam quality allows the realization of multi-point laser triangulation sensors, where up to 10 distances can be measured simultaneously with a single sensor [6]. An application is the straightness measurement of rails in an inspection center of a rail rolling mill [7].

The straightness of rails is an important quality feature for a great variety of rails such as vignol rails, grooved rails, full-web rails, crane rails, and tongue rails. Therefore, the rolled products run through a final inspection line, where the vertical and lateral straightness of the rail is measured. The rail to be inspected is transported on a roller table and measured during the passage with 14 multi-point laser sensors simultaneously. Figure 19.10 shows the arrangement of the sensors for the straightness measurement. Altogether 126 laser beam sources are used. All of these sensors detect a series of measurement points along the head of the rail. The sensors are designed in such a way that a lateral drift of the measuring object during the passage does not affect the measuring results. The sensors are integrated in a fully automatic inspection system. The data are processed electronically to determine the characteristics of the straightness online and to relate the measurement location to the respective length segment of the rail. The results are transmitted to the master computer and visualized for the operator. The inspection system is monitored on a regular basis by use of a reference artifact. This control of inspection, measuring, and test equipment monitoring is performed in compliance with ISO 9000, is supported by the software, and runs automatically. Furthermore, the system is able to adapt automatically to different rail profiles [8].

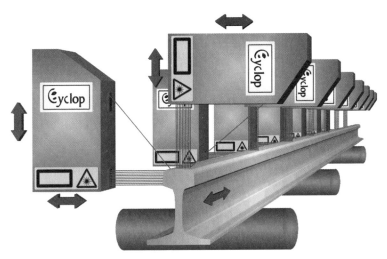

Fig. 19.10 Straightness measurement of rails with a multi-sensor arrangement. At the end of the manufacturing line the rails pass the inspection center on a roller table, where the vertical and side straightness is measured online with a set of triangulation sensors

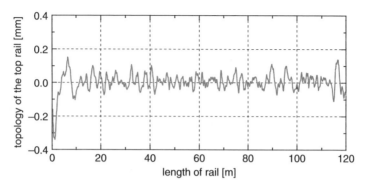

Fig. 19.11 Topology of the top of the rail as a function of the rail length measured on the moving rail passing the laser measurement system

The velocity of the rails during the inspection amounts up to 1.5 m/s. The measurement range of the sensors is 50 mm. During routine operation, the measurement systems achieve an uncertainty of better than 50 µm.

Figure 19.11 shows exemplarily the result of the inspection of a rail having a length of 120 m. The vertical axis shows the deviation between the top of the rail and an ideal straight line as a function of the rail length.

19.1.4.3 Inspection of Camshafts

Camshafts control the inlet and outlet valves of combustion engines. Assembled camshafts offer an attractive manufacturing alternative over conventional methods to meet the demand for high-performance camshafts at reduced costs. Near-net-shaped components are precision assembled onto a tube as, e.g., cams, sensor rings, axial bearings, drive gear. Figure 19.12 shows examples of different camshafts. Before final machining, the assembled camshafts must be inspected regarding different geometric characteristics, e.g., cam angles, cam profiles, axial positions, concentricities, diameters, cf. Fig. 19.13.

Conventionally the camshafts are inspected by tactile inspection machines. However, the disadvantages of such machines are the long setting up times and their susceptibility to failure.

In modern manufacturing, an inspection machine based on triangulation sensors has to fulfill the following requirements: (a) measurement objects with different surface characteristics from black to metallic shining, (b) 100% inspection, i.e., each produced part is measured, (c) inspection of all characteristics without special clamping devices, (d) automated loading and unloading of the inspection machine, (e) inspection time including handling times of the specimen < 20 s, (f) capability indices $c_g, c_{gk} > 1$ [10], (g) inspection place: factory floor, (h) typical production tolerances: cam angle ±0.5°, cam profile ±0.25 mm, diameter ±0.025 mm.

The developed inspection machine for that application is shown in Fig. 19.14. A robot loads and unloads the measurement system. The camshaft is clamped between

19 Laser Measurement Technology

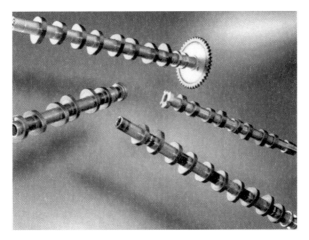

Fig. 19.12 Examples of assembled camshafts [9]

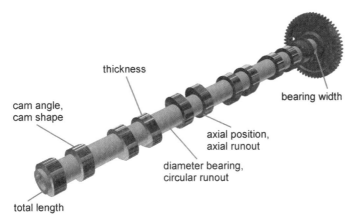

Fig. 19.13 Geometric features of camshafts to be inspected

two checking mandrels and is then rotated for the inspection. On a slide moving parallel to the camshaft, three sensors are mounted, whose measurement beams are directed onto the camshaft using different orientations. In this manner it is possible to measure also the axial positions of the cams and other parts of the shaft.

Figure 19.15 shows details of the camshaft and the sensors during a measurement.

The length of the camshaft varies between 300 and 700 mm. During inspection, the shaft rotates with a frequency of 3–6 Hz. The maximum axial velocity of the slide with the sensors is 1 m/s. One inspection machine checks about 1 million camshafts a year. During the 12 s of measurement time, overall 60 geometric characteristics are inspected. The measurement frequency amounts up to 30 kHz. Within a measurement range of 20 mm, a temporal repeatability of $2\,\mu m$ and a spatial

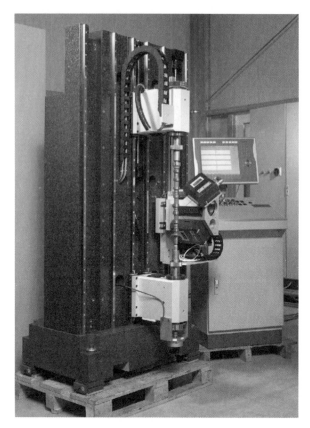

Fig. 19.14 Inspection machine for camshafts with laser sensors [11, 15]

repeatability of 5 μm are achieved [3]. As an example, Fig. 19.16 shows the display of the measurement results for cam profiles. The deviations from the ideal cam profile are displayed using an enlargement by a factor of 20.

The described machine inspects diameters with a standard deviation of 0.7 μm under repeatability conditions.

19.1.4.4 Rail Profile Measurement

Rail sections are typically measured with templates to find out if they satisfy the customer's demands. A number of characteristic and quality-relevant properties must be recorded, such as rail height, rail foot, rail web, rail head, and rail shoulder. The geometrical testing parameters are fixed in European and international standards [12, 13]. The measurement with a template on selected segments of the rail does not give a complete information of the geometry along the rail. Furthermore, the measurement is expensive because of different reasons, for example, the administrative and logistic effort which is necessary to provide all gauges and their regular

19 Laser Measurement Technology

Fig. 19.15 Detail of an inspection machine for camshafts showing parts of the camshaft and laser sensors [15]

recalibration. Therefore, there is a need for an automatic inspection along the entire length of the rail and to do so while the rail is passing on the roller table. The information gained online can be used for documentation purposes and the improvement of the manufacturing process.

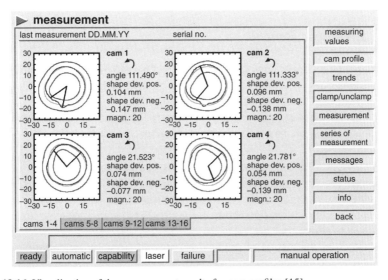

Fig. 19.16 Visualization of the measurement results for cam profiles [15]

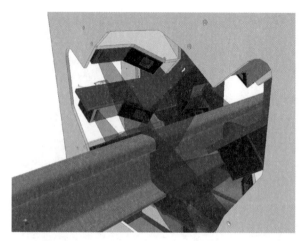

Fig. 19.17 Three dimensional CAD model of a measuring unit to measure the cross-sectional profile of rails. The measuring object shown here is the high-speed rail UIC 60 [15]

Figure 19.17 shows schematically the setup of the system [14]. Ten laser light section sensors are mounted on a frame where the rail to be inspected is passing through. The light section sensors generate a laser line over the entire circumference of the rail profile. The vertical positions of the light section sensors which acquire the rail head can automatically be set for different types of rail profiles. To avoid overlaps between the laser lines, different wavelengths are used and they are triggered with a time shift. The specifications of the sensors are summarized in Table 19.1, column "contour, profile." Over a period of 2 years, an availability of this measurement system of 99.8% is achieved.

19.1.5 Economic Benefit

Laser triangulation opens up new ways for a direct measurement of geometric quantities in industrial applications [16]. With the method of laser triangulation, geometric quantities – like distance, thickness, and profile – can be measured without contact, precise and fast in the production process. This enables a new automation potential not accessible so far. The prior-ranking benefits for the user are

- increase of process capability, higher process transparency;
- optimized process control, reduced reaction times, reduced downtimes of the production;
- nearly maintenance-free operation;
- reduction of set-up times;
- automated documentation of production data and automated monitoring of test equipment in compliance with ISO 9000.

19.2 Interferometry

Csaba Farkas and Reinhard Noll

Interferometric measurements are rooted in the wave nature of light that has been known since Young's double-slit experiment in 1802 [17, 18]. The principle of such measurements is based on the superposition of two or more light waves. Such a superposition of sufficiently coherent waves leads to a temporally constant intensity distribution, which is referred to as an interference pattern. Interferometric methods allow highly accurate measurements of geometric magnitudes, on the order of optical wavelengths. When measuring distances, the intensity of the interference pattern at some point is recorded as a function of time, while the position of a reference mirror is varied (see Sect. 19.2.1). For the measurement of surface topographies, the spatial interference pattern is analyzed. A detailed introduction to the fundamentals and the measuring principles of interferometry can be found in [19–22].

The interference of light waves can be described mathematically by a linear combination of solutions of the wave equation, which is itself a solution. This superposition principle is known as the Huygens–Fresnel principle in optics and is an important tool for the description of the propagation of waves for the phenomena of interference, diffraction, and holography. The superposition principle states that different waves of light do not influence each other and can be superposed. The resulting field $\vec{E}$ of the superposition of n waves of light at some point in space is the vector sum of the individual field vectors $\vec{E}_i$ at that point

$$\vec{E} = \sum_{i=1}^{n} \vec{E}_i \tag{19.3}$$

where the field strengths of the component waves $\left|\vec{E}_i\right|$ can be written as

$$E_i = \hat{E}_i \cos(\omega t - kz - \varphi_i) \tag{19.4}$$

For the sake of simplicity, it is assumed that the angular frequency ω and the wave vector k are the same for each wave and only the amplitudes $\hat{E}_i$ and the phases φ_i are different. Direct detection of the resulting electric field $\vec{E}$ is not possible, only the intensity $I \propto \left|\vec{E}\right|^2$ can be measured directly. In general, the component waves have different amplitudes, wavelengths, polarization directions, and phases. An observable, stationary interference pattern can only be produced if the relative phase of the component waves is constant. In other words, the phase relation of the interfering light waves must be constant over some period of time. Such waves are in general referred to as being coherent [19]. Thermal light sources, such as an incan-

---

C. Farkas (✉)
Messtechnik, Fraunhofer-Institut für Lasertechnik, 5207 Aachen, Germany
e-mail: csaba.farkas@ilt.fraunhofer.de

descent bulb, produce incoherent light, since the excited atoms emit spontaneously with completely irregular phases. The coherence of such sources can be increased by using a frequency filter and a slit, but this leads to a decrease of the intensity to as little as 10^{-10} of the original value, rendering such light sources generally unsuitable for interferometric techniques. Therefore, with the exception of white light interferometry (see Sect. 19.2.5), lasers are generally used as the coherent light source of an interferometer.

In general, a single laser is used as the source of the component waves in an interferometer, since two (or more) lasers are in practice not mutually coherent enough to produce a stable interference pattern for longer periods of time. Moreover, changes in the relative stability of multiple lasers can change the measurement signal of a metrological device. Therefore, in measurement applications, two or more component waves are always obtained by splitting a single wave. This can be done by splitting either the wave front or the amplitude (see [19]). The interferometers presented here all operate by splitting the amplitude of the source wave. Based on the number of component waves used to produce the interference pattern, devices are further categorized as two-beam or multiple-beam interferometers in the literature. The Michelson interferometer and its numerous variations are examples of two-beam interferometers. Examples of optical elements that are based on multiple-beam interference are diffraction gratings and Fabry–Perot etalons.

Figure 19.18 shows the schematic of a CNC milling machine with interferometric measurements of the x, y, and z dimensions. This allows a measuring accuracy up to ± 10 nm.

Most of the interferometers in use today are variants of the Michelson interferometer described in the following section. With this interferometer Michelson

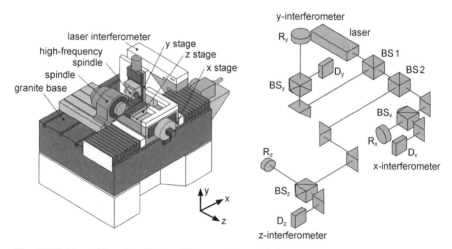

Fig. 19.18 The table of a CNC milling machine with interferometric length measurement (Source: Fraunhofer Institute for Production Technology, Aachen). BS_x, BS_y, BS_z: beam splitters; D_x, D_y, D_z: detectors; R_x, R_y, R_z: reference mirrors

performed interferometric length measurements in the year 1882 [23, 24]. Lacking a coherent light source, such as the lasers available today, a thermal light source was used which reduced the measuring range to under 1 mm [26]. Modern laser-based interferometers allow measurement ranges as high as 60 m [21, 27], with an accuracy of 1/20th to 1/100th of the wavelength of the light used when measuring path length differences [26, 27]. The Mach–Zehnder and the Fizeau interferometer, two variants of the Michelson interferometer that are used for measurements on transparent objects or in microscopes, are presented in Sects. 19.2.2 and 19.2.3.

Two methods for surface topography measurements are speckle interferometry (see Sect. 19.2.4) and white light interferometry (see Sect. 19.2.5). Both use a CCD camera or a photo plate to record the interference pattern, resulting in a picture of the investigated surface area of the sample. These two interferometers are in principle also variants of the Michelson interferometer.

19.2.1 Michelson Interferometer

The Michelson interferometer is based on two-beam interference and wave amplitude splitting. This interferometer is highly versatile and numerous variants of it are used. For example, it may be constructed as a fiber-optic device, significantly simplifying alignment and the isolation from vibrations. The main components of the basic setup of a Michelson interferometer, shown in Fig. 19.19, are a beam splitter and two mirrors. The beam splitter BS divides the light wave from the source L into two component waves. One wave is directed toward mirror M1 and is reflected back from it, while the other is reflected back by mirror M2. Mirror M2 is mounted such that it can be moved along the propagation direction of the light wave. The two component waves converge again at the beam splitter and are deflected toward the screen S, e.g., a ground glass, where the interference pattern can be observed. The compensation plate CP is needed to cancel the optical path length difference between the two component waves caused by the reflection on the back surface of

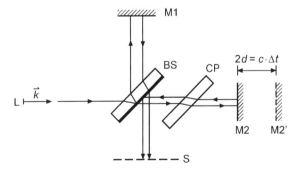

Fig. 19.19 Schematic setup of a Michelson interferometer. L: light source; M1, M2: mirrors; S: screen; CP: compensation plate; BS: beam splitter. The interference pattern can be observed on the screen

the beam splitter, by equating the number of passes through the plates for both arms. For this purpose, it is made of the same material as the beam splitter BS, has the same thickness, and is mounted with the same angle to the beam. With the proper alignment, such that the mirrors are perpendicular to each other, an interference pattern is produced on the screen S.

The phase difference Γ between the component waves for the case of interference of equal inclination that applies here is

$$\Gamma = 2d \cos \alpha \tag{19.5}$$

where d denotes the change in the position of the reference mirror such that $2d$ is the difference between the lengths of the paths of the component waves and α is the viewing angle with respect to the optical axis. If the difference in phase is

$$\Gamma = m\lambda \tag{19.6}$$

constructive interference results and an interference maximum can be seen. λ stands for the wavelength of the light, m is an integer that gives the order of the interference. For a given d, a wavelength λ, and an order of interference m, α is a constant, meaning that the phase difference and the resulting intensity is rotationally symmetric to the optical axis, so the interference pattern consists of concentric rings.

When light traveling in one medium (such as air) is reflected by an optically denser medium (such as glass), the phase of the wave is shifted by π, but no phase shift is produced if the media are switched. If the path length difference of the two component waves is $2d = 0$, the phase difference Γ is not 0, as would be given by Eq. (19.5), but π, since one component wave is reflected at the back of the beam splitter (by an optically denser medium), producing an intensity minimum on the screen. If d is increased by an amount equal to $\lambda/4$, a maximum will appear in the middle of the pattern, after another increase by $\lambda/4$ again a minimum, and so on. Hence, every displacement of the mirror M2 by $\lambda/2$ will produce a new ring.

The main field of application of the Michelson interferometer is the precise measurement of distances. The prototype meter bar in Paris was thus measured by Michelson [24]. The change in distance d is measured as a multiple of the wavelength λ of the laser used. This is done by slowly moving the reference mirror and measuring the intensity $I(t)$ of the moving interference rings at a point of reference on the screen. If m rings are counted, the corresponding change in distance is $d = m\lambda/2$. The width of the rings can be measured with an accuracy of approx. 1/100 [22]. The light source utilized is usually a helium–neon laser with a wavelength of $\lambda = 632.991399$ nm in vacuum. For measurements conducted in air, this value has to be divided by the index of refraction of air which depends on its composition, temperature, humidity, and pressure. At $(20 \pm 1)\,°\text{C}$, $(1013 \pm 10)\,\text{hPa}$, and a relative humidity of $(50 \pm 20)\%$ the average value is $\bar{\lambda} = (632.820 \pm 0.003)\,\text{nm}$. The accuracy of length measurements is therefore influenced by atmospheric conditions and can be improved by using a refractometer alongside the interferometric measurement to determine the index of refraction of the air. Keeping the difference between the lengths of the beam paths minimal further reduces the influence of the air's index of refraction. An example of an interferometer where this difference is zero is the Differenz–Planspiegel interferometer [28, 29].

The Michelson interferometer can also be used to measure the temporal coherence of a light source. A change in the position of the reference mirror introduces a delay of $\Delta t = 2d/c$ between the two component waves. The contrast of the interference pattern is maximum for $\Delta t = 0$ and decreases with an increase of Δt. The pattern is visible only as long as the delay time Δt is smaller or equal to the coherence time $T_c = L_c/c$ (L_c: coherence length) of the wave train, i.e., $\Delta t \leq T_c$. For all ordinary light sources, the coherence time of the wave train and its spectral width are related as $\Delta \omega T_c \approx 1$. Thus, the measurement of the spectral line width of a source is reduced to the simple measurement of length or duration.

Besides helium–neon lasers, small cost-effective laser diodes are also used in practice. Corner reflectors are often, utilized in place of plane mirrors, as these are less sensitive to turning or tilting. In order to avoid back reflections into the laser, which influences the frequency stabilization, polarizing beam splitters are employed.

A further increase of the resolution and an improvement of the insensitivity to external disturbances can be realized by using two coherent light waves with different frequencies, such as the output of a two-frequency helium–neon laser. The frequency difference in this case is in the range of a few hundred megahertz (heterodyne technique) [21].

19.2.2 Mach–Zehnder Interferometer

The Mach–Zehnder interferometer [25] is mainly used to determine variations of the index of refraction in a measurement volume. It consists of two mirrors and two beam splitters as shown in Fig. 19.20. The first beam splitter BS1 divides the incident light wave into two component waves. These are reflected at mirrors M1 and M2 and recombined by the second beam splitter BS2. The two optical paths can be far apart, allowing measurements of large objects. The interference pattern

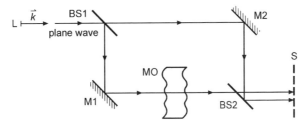

Fig. 19.20 Schematic setup of a Mach–Zehnder interferometer. L: light source, M1, M2: mirrors, S: screen, BS1, BS2: beam splitters, MO: measured object

generated depends on the optical path differences caused by the object being measured. These are a function of the thickness and index of refraction of the measured object in the measurement path, such that variations of these values become visible in the interference pattern. Figure 19.21 shows the experimental setup that was used to record the Z-Pinch plasma shown in Fig. 19.22 [30].

The original fields of application were the measurement of index of refraction distributions in transparent objects, such as lenses, as well as the investigation of

Fig. 19.21 Mach–Zehnder interferometer. The beam of a helium–neon laser (in the back on the right) is expanded and coupled to the interferometer with the help of two mirrors. The setup is consistent with the schematic shown in Fig. 19.20. The object to be measured is placed in the arm of the interferometer visible in the foreground. The resulting interference pattern can be seen on the screen (in the front on the right)

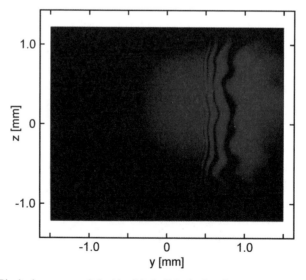

Fig. 19.22 Z-Pinch plasma, recorded with a Mach–Zehnder interferometer

flows, concentrations, or temperature distributions in transparent media. Another application is the inspection of the emitted wave fronts of laser diodes and other laser sources with regard to their phase and intensity distributions.

19.2.3 Fizeau Interferometer

Fizeau interferometers are often used in interference microscopes. Their advantage lies in a relatively simple setup, where only one reference surface has to be high-precision coated and polished. This surface alone defines the measurement and reference wave fronts. Figure 19.23 shows a schematic of the setup.

A part of the incident light wave is reflected by the reference surface and collinearly superimposed on the light wave reflected by the measuring object. The combined light waves are observed with a camera. Using this method, deviations of the flatness of a plane surface can be measured. Measuring curved surfaces is also possible by using appropriately matched reference surfaces. The planarity of the surface of transparent objects can also be inspected in a similar manner, see Fig. 19.23b. An improvement in the precision of the instrument is attainable by using a piezoelectric actuator to move the reference surface and thereby modulate the phase difference.

19.2.4 Speckle Interferometer

Speckle metrology includes various different methods that are used to measure geometric changes of objects with rough surfaces. "Rough" in this context means variations in height that are on the order of the wavelength of light or larger. The

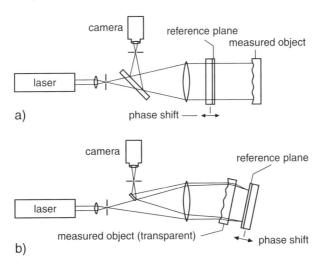

Fig. 19.23 Schematic setup of a Fizeau interferometer for the inspection of (**a**) opaque surfaces and (**b**) transparent objects

measurements make use of the speckle effect that occurs when coherent light falls on a rough, diffusely reflecting surface. One distinguishes between speckle photography (see [26], for example), which will not be discussed here, and speckle interferometry.

The emergence of speckles is illustrated in Fig. 19.24. The incident, coherent laser light is scattered in multiple directions by the rough surface of the object. Every point of the surface scatters light with a specific phase relative to the incident wave. Because of the microscopic structure of the surface, this phase fluctuates statistically for different surface points, such that the scattered waves form different interference patterns with different distances between interference fringes and different orientations depending on the position and angular orientation of the scattering points. These interference patterns are superimposed statistically in the observation plane, creating a speckle pattern that consists of light and dark regions randomly distributed in space. Imaging a coherently illuminated object with the help of a lens is shown schematically in Fig. 19.25. The so-called "subjective speckles" are formed in the image plane, whose average size Φ is given by the aperture of the imaging lens D, the laser wavelength λ, and the distance between the lens and the image plane b as described in relation (19.7). The average speckle size is in practice approx. 5–100 μm [20].

$$\Phi \cong 2.4 \frac{\lambda}{D} b \qquad (19.7)$$

Speckle interferometry analyzes interference patterns that are formed by the combination of a wave that is scattered by the object being measured and a reference wave such as a spherical wave or another speckle pattern. Figure 19.26 shows the schematic setup of a speckle interferometer that basically corresponds to a Michelson interferometer. A beam splitter divides a plane wave into two waves: one illuminates the diffusely scattering surface of the object being measured and the other that of the reference object. The two objects are imaged through the beam splitter by a lens onto the image plane, where the waves interfere and create a speckle

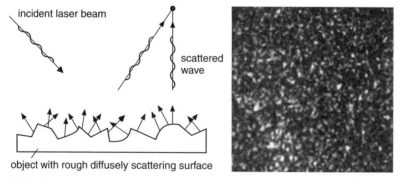

Fig. 19.24 Schematic of the scattering by a rough surface (*left*) and a resultant speckle pattern (*right*)

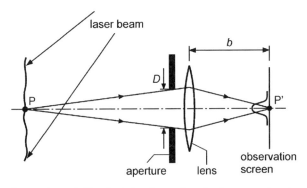

Fig. 19.25 Imaging a coherently illuminated surface onto a viewing screen by a lens. The resulting speckle size is given by Eq. (19.7). P: scattering center on the object; D: aperture of the lens; b: distance between the lens and the image plane; P′: image point on the viewing screen

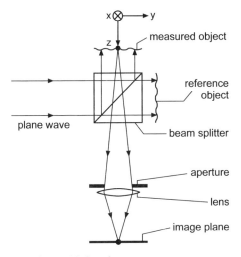

Fig. 19.26 Schematic setup of a speckle interferometer

pattern. Speckle patterns of different states of the object, such as a deformation, can be recorded and subsequently compared with each other (speckle correlation). Figure 19.27 shows such a correlation of two speckle patterns with rings of correlation fringes that were created by the bulging of a steel plate. The observed fringes correspond to locations where the speckle patterns are correlated. From the position and form of the fringes, the type and size of the spatial change of the object can be deduced. The deformation of the object between two neighboring dark fringes amounts to $\lambda/2 \cong 300$ nm.

In practice, a speckle interferometer has to be built as stable as a setup for holography. The advantages in comparison with holographic interference lie in the real-time capability of the system, since only speckles have to be resolved, making it

Fig. 19.27 *Left*: Speckle interferogram of a plate. *Right*: Inspection of freight vehicle tires using speckle interferometry

possible to use a camera with a resolution of approx. 30 line pairs per millimeter. For holography, up to 3000 line pairs per millimeter must be achieved. Furthermore, electronic speckle pattern interferometry (ESPI) makes it possible to perform quantitative analysis using digital image processing. As an example, the use of speckle interferometry for the detection of defects in the tires of freight vehicles is shown in Fig. 19.27 [31].

A speckle interferometer can also be used to detect surface vibrations [20, 26]. The resolution of oscillation amplitudes is in the range of approx. $0.12\,\mu$m when using a helium–neon laser. Modulating the reference wave can improve this up to $1/100$ nm (!).

19.2.5 White Light Interferometer

White light interferometry was already used as a method to measure lengths in 1892 by Michelson in measuring the standard meter [24]. The main area of application today lies in the measurement of surface profiles. In contrast to the interferometer types described in the preceding sections, a white light interferometer does not employ a laser as the light source, but uses a broadband "white" source instead. The coherence lengths of such sources amount to a few micrometers, whereas the coherence length of a laser is usually in the range of some centimeters to meters. Most interferometric measuring methods rely on a coherence length that is as large as possible, this being the limiting factor of the measuring range due to the principle of operation. The particular measuring method of white light interferometers allows the measuring range to be unlimited in principle, in spite of a short coherence length.

Figure 19.28 shows the schematic setup of a white light interferometer based on a Michelson configuration. One of the mirrors of an arm of the Michelson interferometer is replaced by the object to be measured. The light waves reflected from the

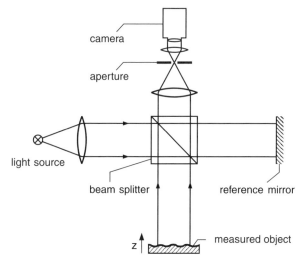

Fig. 19.28 Schematic setup of a white light interferometer

object and the reference mirror are superimposed and imaged onto a CCD camera. In order to produce stationary, i.e., observable, interference of the reflected waves, their relative phase has to be constant, i.e., the temporal and spatial coherence condition has to be met. Spatial coherence is fulfilled for all reflected light waves that are reflected by a region of the surface that is smaller than the region in which the incident wave (the light source) is spatially coherent. Stationary interference for this surface region arises, if temporal coherence is also given, i.e., the optical path difference between the two arms of the interferometer is less than the coherence length of the light source. In other words, the surface roughness of this region has to be smaller than the coherence length. In the case of speckle interferometry described in the previous section, this condition is true within a speckle.

The coherence length of white light is considerably smaller than that of a laser. Therefore, the range in which the length of the interferometer arms can be varied while retaining the observability of white light interference is also much smaller.

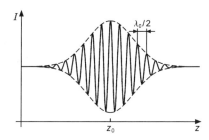

Fig. 19.29 Correlogram: light intensity at a point in the image plane as a function of the z-position of the measured object

Figure 19.29 shows a so-called correlogram that plots the light intensity at some point of the image plane as a function of the z-position of the corresponding point of the measured object. The width of the correlogram is proportional to the coherence length and inversely proportional to the spectral width of the light source. The interference maxima are most distinct if the two arms of the interferometer have exactly the same length (position z_0). As the object is moved along the z-direction, the maxima become smaller and eventually disappear completely.

If the object is moved in the z-direction, by a piezoelectric actuator for example, then the coherence condition is satisfied by different parts of its surface in sequence. A white light interferogram can be recorded for each pixel of the CCD camera during the movement of the object and the position of the maxima can be determined. This requires the width of the correlogram to contain approx. 5–100 periods of the intensity modulation. The surface topography can then be reconstructed from the z-position of the maximum and the lateral coordinates of each pixel.

Unlike classical interferometry, white light interferometry does not evaluate the phase in order to determine the surface position, but only the maxima of the intensity modulation. In doing so, it yields two important advantages for industrial applications: first, avoiding the ambiguity of classical interferometry and second, making it possible to measure rough surfaces, which lead to large speckle noise when using other methods. The longitudinal measuring range is only limited by the translating stage of the instrument. The longitudinal uncertainty does not depend on the optical setup, but instead to first order on the surface roughness of the object and is

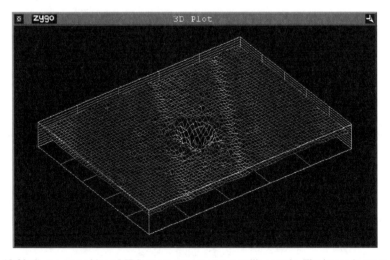

Fig. 19.30 Crater created by a LIBS measurement on a metallic sample. The image is not true to scale; the crater depth is approx. 15 μm and the dimensions of the depicted rectangle are 710 μm × 530 μm

typically in the range of 1 μm. The coaxial illumination and viewing of the object avoids shadowing effects. The entire surface profile in the lateral measurement range determined by the optical setup is recorded during the measurement process, so that, unlike with other methods, no lateral scanning is necessary. A single LED (for example, wavelength 875 nm, 6 μm coherence length) may suffice as the light source. The main drawbacks of this approach lie in the dynamic range, which is considerably limited by the CCD camera. For example, highly reflective metal surfaces can lead to overexposure. Also, because of the need to move the object, the measurement is relatively slow (several seconds).

Two examples of measurements with a white light interferometer are presented in Figs. 19.30 and 19.31. Figure 19.30 shows a crater left on the surface of a metallic sample surface after a LIBS measurement (see Sect. 19.6). Figure 19.31 is the result of the measurement of a contact array of a semiconductor on a ceramic substrate.

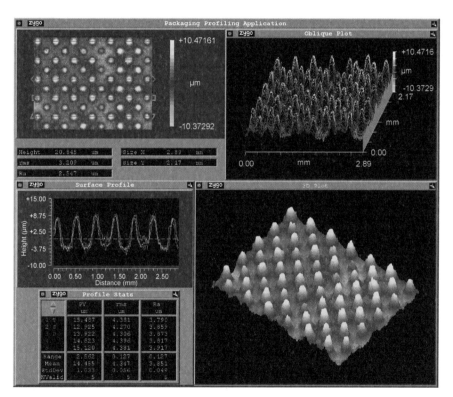

Fig. 19.31 Measurement of a ball grid array (BGA) on a ceramic substrate. Display of cross section profiles and statistical analysis [32] (used with kind permission of ZygoLOT Europe, Darmstadt, 27.10.2003)

19.3 Laser-Induced Fluorescence

Christoph Janzen and Reinhard Noll

19.3.1 Basics of Fluorescence

Fluorescence is a fast decaying emission of light from electronically excited states of atoms and molecules. If a reversal of the spin electronic wave function (triplet–singlet transition) is included, the phenomenon is called phosphorescence. Because such transitions are quantum mechanically prohibited, excited particles remain relatively long in a triplet state. Triplet state lifetimes and the effect of phosphorescence are typically in the range of milliseconds to seconds. Fluorescence is a much faster process, the lifetimes of excited singlet states typically range from picoseconds to nanoseconds.

Molecules can absorb energy not only in the form of electronic excitation but also as vibrational energy of the nuclei. The various levels of vibrational excitation of an electronically excited molecule are called vibronic levels. Electronic transitions (optical absorption or emission) proceed on a much shorter timescale than the vibrations of the atomic nuclei around their equilibrium positions. The duration of an optical transition is typically in the order of 10^{-15} s, during this time the nuclei do not almost change their relative positions with respect to each other. In a term scheme, in which potential energy curves of different electronic states are drawn together with their vibrational levels – so-called Jablonski term scheme – optical emissions or excitations are therefore marked as vertical transitions (no change in the coordinates of the nuclei).

The wave functions of the nuclei in ground and excited states have a large overlap integral for transitions of high probability. With the help of so-called Franck–Condon factors the transition probabilities between two vibronic states can be calculated.

At room temperature practically all molecules occupy their vibrational ground state. Therefore, all optical excitations start from this initial state, see Fig. 19.32 showing a term scheme for the fluorescence of molecules. The larger the potential curves of primary and excited state are shifted against each other, the more likely the vertical transition leads to an excited electronic state that also has a vibrational excitation. The vibrational energy of molecules in condensed phases is distributed very quickly to the surrounding molecules (thermal relaxation). Fluorescence, therefore, usually is emitted from the vibrational ground state of the electronically excited state. For the emission of light, the Franck–Condon principle of vertical transitions must also be fulfilled, hence the emission process leads to a vibrationally excited level of the electronic ground state. This leads to the fact that the fluorescence

---

C. Janzen (✉)
Messtechnik, Fraunhofer-Institut für Lasertechnik, 5207 Aachen, Germany
e-mail: christoph.janzen@ilt.fraunhofer.de

19 Laser Measurement Technology

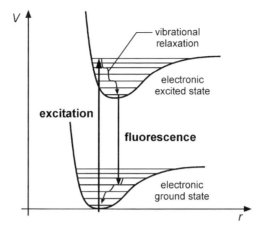

Fig. 19.32 Fluorescence in molecules: The potential curves of ground and excited states are shifted. In case of vertical transitions this leads to a simultaneous excitation of electronic and molecular vibrations. After the non-radiant relaxation to the vibrational ground state of the electronically excited state the fluorescent transition leads to a vibrationally excited state of the electronic ground state. The fluorescence is shifted to longer wavelengths, the so-called Stokes shift

experiences a red shift in relation to the excitation wavelength, which is called Stokes shift.

In solutions the interaction between fluorescent molecules and the solvent molecules leads to a broadening of the vibronical transitions. Often the individual transitions cannot be differentiated any longer and only broad peaks can be detected. In gas phase experiments the individual vibronic levels can usually be separated spectrally.

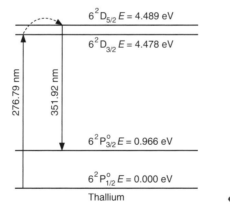

Fig. 19.33 Fluorescence in atoms: Since atoms possess neither vibrational nor rotational levels, their fluorescence spectra generally consist of individual, separated lines. The different electronic states have different spins, orbital angular momenta, and total angular momenta, which are described by the term symbols. An introduction to the denotation of atomic terms is given in [33]. The dashed arrow illustrates a transition between the closely neighboring energy levels that is induced by the particle temperature or by impacts with neutral particles

Unlike molecules, atoms do not possess vibrational or rotational energy levels, only electronic excitations are possible, see Fig. 19.33 with an example of an atomic term scheme. Therefore, excited atoms show a fluorescence emission that consists of individual, spectrally narrow lines. Only under increased pressure these transitions undergo a broadening caused by impacts between the particles (e.g., high-pressure mercury lamp).

19.3.2 Fluorescence Spectroscopy

Fluorescent particles can be excited in different ways. Atoms can be excited with X-rays or electron beams, the line emission of hot plasmas is based on the fluorescence of thermally excited atoms. Molecules are usually excited with UV or visible light.

In analytical chemistry fluorescence is applied for a sensitive and quantitative detection of organic molecules [34]. Compared to absorption techniques, fluorescence has the inherent advantage that instead of a signal decline (relative measurement) an absolute intensity is measured. For this reason the limits of detection are up to 3 orders of magnitude more precise than with absorption techniques. Typical detection limits lie in the ppb range ($1 : 10^9$).

Two principal approaches for analyzing fluorescence are possible. For the recording of a fluorescence emission spectrum a sample is illuminated with monochromatic light of a fixed wavelength and the emitted fluorescence is detected and spectrally resolved. Provided that the resolution is sufficient, the vibrational modes of the electronic ground state can be resolved. For samples in solution usually only a wrapped contour can be measured.

An alternative way to measure fluorescence consists of a setup with a tunable excitation source and fluorescence detection with a fixed spectral window with an edge filter or a monochromator to assure that no excitation radiation reaches the detector. While scanning the excitation source the red-shifted fluorescence is detected. The amount of detected fluorescence correlates with the absorption of the excitation light, so a fluorescence excitation spectrum is recorded that can reveal the vibrational structure of the electronically excited state.

Figure 19.34 shows the principles of fluorescence excitation and fluorescence emission spectroscopy, Fig. 19.35 shows the principle of absorption spectroscopy.

19.3.3 Laser-Induced Fluorescence Spectroscopy

The usage of a laser for exciting fluorescence features several advantages compared to conventional light sources [36]. The laser delivers narrow bandwidth, monochromatic light that has excellent collimation and focusing qualities. All fluorescence applications demand a separation of excitation wavelength and fluorescence emission, because otherwise stray light might disturb the measurements. If narrow bandwidth laser light is used for excitation, the fluorescence emission can be separated

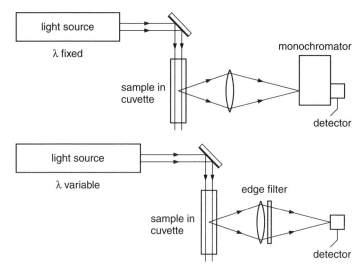

Fig. 19.34 Methods of fluorescence spectroscopy. In contrast to absorption spectroscopy the different methods of fluorescence spectroscopy (emission and excitation) are background-free techniques. The limits of detection are significantly lower. *Top*: If the sample is illuminated with monochromatic light, the fluorescence can be dispersed spectrally and a fluorescence emission spectrum results. *Bottom*: A fluorescence excitation spectrum is recorded by scanning the excitation light source and simultaneously detecting the fluorescence in a defined spectral window using filters or a monochromator (observation usually at the maximum of the fluorescence emission) [35]

efficiently from the laser wavelength with the aid of optical filters. If the electronic transitions of the sample have narrow bandwidths, the laser has additional advantages compared to broader light sources. Investigations of molecules in the gas phase or in so-called "molecular beam" experiments with molecules moving in the same direction without interacting with each other are usually carried out with tunable laser systems.

From high-resolved fluorescence spectra structural information like potential curves, vibrational frequencies, dissociation processes, the symmetries of electronic states, or rotational constants can be derived. Additionally, transient species like radicals, photo fragments, or intermediates can be investigated with pulsed laser

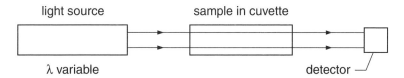

Fig. 19.35 Principle of absorption spectroscopy. A tunable light source generates a beam of light, which is directed through a sample onto the detector. The light intensity is measured as a function of the wavelength. The measuring result is the absorption spectrum of the sample. An alternative approach consists of a polychromatic light source in connection with a spectral resolving detection (e.g., a CCD spectrometer) [35]

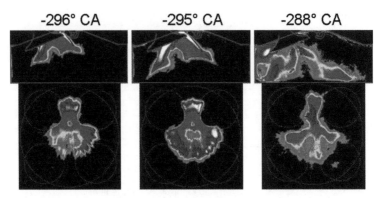

Fig. 19.36 A fluorescence analysis shows the combustion process in a motor. The distribution of fuel during the injection in the combustion chamber is visualized with PLIF (planar laser-induced fluorescence) at different positions of the crank angel (CA). The upper series of pictures shows a side view of the combustion chamber (inlet valve and spark plug are shown in *red*), the lower series of pictures show a top view. Fluorescence excitation was performed with a KrF excimer laser at 248 nm. The asymmetric distribution of fuel can be attributed to the influence of the open inlet valves (Pictures: D.H. Fuchs, Fuel distribution in gasoline engine monitored with PLIF, Lambda Highlights, Haft 62, pp 1–3 (2003), Lambda Physik)

radiation. Examples are the detection of concentration and distribution of nitrogen oxides, carbon monoxide, hydroxyl radicals, or hydrocarbons during a combustion process. These investigations are important for the optimization of motors. Fig. 19.36 shows an example [37].

19.3.4 Fluorescence Markers in the Life Sciences

Fluorescent dye molecules can be chemically coupled to biomolecules with specific binding capabilities, for example, antibodies. With this principle, specific structures in cells or in tissue can be labeled with fluorescent molecules. This makes the basic principle of a fluorescence marker. A selective fluorescent coloring can be visualized under a fluorescence microscope. During the last years the growing availability of various fluorescence techniques has led to a replacement of radioactive markers with fluorescence markers in many fields of application. Unlike with radioactive materials no special safety procedures are necessary and the disposal of used sample material is trouble free.

Fluorescent markers are used in bioanalytics and medical diagnostics in multiple ways. Flow cytometry is based on the detection and sorting of single, dye-labeled cells with laser-induced fluorescence. In the field of high-throughput analytics in micro-titer plates and in modern DNA- or protein-biochips fluorescence markers are widely used. In microscopy fluorescence markers are necessary for the selective labeling of cells or cell components. This enables a functional imaging [38]. In the field of medicine fluorescent markers can be used to selectively label tumors. Figure 19.37 shows an example.

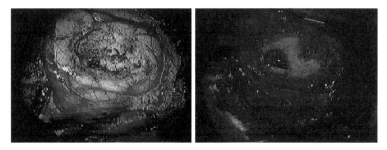

Fig. 19.37 Tumors can be specifically labeled with fluorescent dyes. The *left* picture shows a brain tumor under *white* light illumination. The tumor cannot be distinguished from the normal tissue. The *right* picture shows the tumor under illumination with *blue* light. The *red* fluorescence of the dye that specifically labels the tumor makes the tumor tissue clearly visible. (Pictures: Prof. W. Stummer, Laser Klinikum Universität München)

19.3.4.1 DNA Sequencing

The genetic information of all organisms is coded in the sequence of the four building blocks of the DNA: adenine, thymine, cytosine, and guanine. In the international human genome project the sequence of all 3×10^9 base pairs of the human DNA was decoded [39]. Fluorescence markers have played an important role in the investigation of the human genome.

A classical method for the DNA sequencing is the so-called chain-termination method. The unknown DNA section is put in a reaction vessel as a single strand together with the four nucleotides that contain the four DNA bases and some enzymes for the replication of the single strand to a double strand. In four different experiments a chemically modified nucleotide is added to each reaction vessel, respectively. This nucleotide can be built into a DNA instead of the native nucleotide, but it terminates the chain because no additional nucleotides can be added to it. If the single-strand DNA is complemented to the double strand with the help of a polymerase enzyme, the reaction stops when the first chain-terminating nucleotide is added to the strand. After a short time DNA fragments of different lengths will be present in every reaction vessel that all end with the same nucleotide, respectively. In every experiment the nature of the chain-terminating nucleotide determines the common last building block of the DNA fragments. The chain-terminating nucleotide is labeled with a fluorescence dye. The different DNA fragments have to be separated before they can be characterized. The DNA solutions are spotted on a polyacrylamide gel and with the help of an electric field, the charged DNA molecules move with different speeds according to their respective size through the gel. Smaller fragments are detained less by friction and move faster than larger fragments. After the electrophoretic separation the position of each DNA fragment can be visualized with a fluorescence excitation. As every DNA fragment from one reaction vessel ends with the same nucleotide, the electrophoretic separation of all four reaction vessels can be combined to yield the complete sequence of the analyzed DNA.

The usage of four different fluorescence dyes for the four chain-terminating nucleotides enables the sequencing in one reaction vessel, provided that the fluorescence detection system can discriminate the four dyes. In one experiment up to 20,000 base pairs can be analyzed. Figure 19.38 gives a schematic view of the sequencing technique.

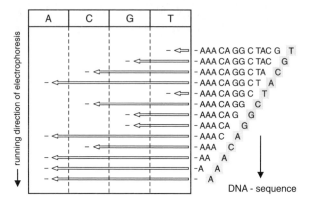

Fig. 19.38 DNA sequencing with the chain-termination method. In the four columns the DNA fragments are electrophoretically separated according to their size. Every column has a common chain-terminating nucleotide. This nucleotide is given in the gray field in the sequence on the right side. The DNA sequence can be read from the order of the chain lengths of the different fragments. (A=adenine, T=thymine, C=cytosine, G=guanine) [35]

19.3.4.2 Fluorescence Correlation Spectroscopy (FCS)

The fluorescence correlation spectroscopy is a method for analyzing the diffusional behavior of molecules. Since binding processes between molecules have an effect on the movement in solution, the interactions between molecules can be analyzed. Specific interactions between biomolecules determine nearly all biochemical processes in living organisms. Examples for specific binding processes ("lock–key principle") are the following:

- interactions between receptors and ligands. Control and regulation mechanisms of the cell are based on these interactions. Their examination is important for the search for new active pharmaceutical ingredients;
- protein–protein interactions. Enzymatic reactions or antibody–antigen interactions are examples;
- DNA–protein interactions;
- interactions between nucleic acids.

The FCS technique is based on the analysis of diffusion. A laser is focused to a diffraction-limited spot with a high numerical aperture microscope objective. Fluorescence from this extremely small focus volume is transferred with a confocal optics on a very sensitive detector, e.g., an avalanche photodiode. Optical filters are used to block stray light. If only a few fluorescent molecules are present in the focal volume, the Brownian motion of the molecules and the diffusion of the particles will lead to a strong fluctuation of the fluorescence intensity. The recorded fluorescence intensity is directly related to the movement of the molecules. The autocorrelation of the time-resolved intensity delivers the average time it takes for a molecule to cross the focal volume (diffusion time) and the particle number in the focal volume (concentration).

As smaller molecules diffuse faster than larger ones they show shorter diffusion times. If a protein that is labeled with a fluorescence dye is added to a binding partner, the change in diffusion time reveals the chemical-binding process and the reaction can be observed. This can be used to determine binding constants; for further literature, see [40]. Figure 19.39 shows a schematic approach of the FCS principles.

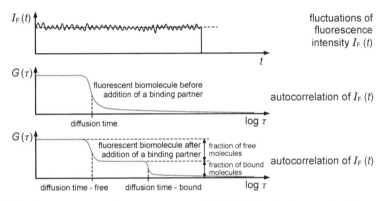

Fig. 19.39 Fluorescence correlation spectroscopy (FCS). $I_F(t)$ is the time-resolved fluorescence intensity. Below the autocorrelation $G(\tau)$ is shown (qualitatively). In the first autocorrelation curve only one fluorescent species is present. If a binding partner is added, there are free and bound fluorescent particles in solution that have different diffusion times. From the autocorrelation curve the relative fractions of free and bound molecules and the binding constant can be derived. Pictures are provided by ILT

19.3.4.3 Time-Resolved Fluorescence Analytics

Besides the spectral color there is another feature that characterizes the fluorescence of a molecule, its lifetime. For measuring the fluorescence lifetime, a pulsed light source can be used that has shorter pulses than the analyzed lifetime. Lasers that emit pico- or femtosecond pulses can be used to analyze even short lifetimes in the order of nanoseconds. The straightest way to record fluorescence lifetimes is to excite a large number of molecules with a short laser pulse and record the fluorescence decay curve. For simple cases, the fluorescence intensity decays with an exponential function, after τ seconds the intensity has dropped to the fraction $1/e$ of the initial fluorescence intensity with τ being the fluorescence lifetime. When working with tight focused laser beams and low sample concentrations, it is not possible to record a complete decay curve with only one excitation pulse, because the number of detected fluorescence photons is not sufficient.

One way of recording fluorescence decay curves in microscopy applications is the so-called time-correlated single photon counting (TCSPC). With this technique the analysis of single molecules is possible. It is also frequently used in the laser scanning microscopy for the fluorescence lifetime imaging (FLIM). Instead of a fluorescence intensity the fluorescence lifetime gives the contrast in the images.

A laser with a repetition rate of several megahertz emits short pulses (picoseconds to femtoseconds) and is focused with a microscope objective. The sample contains only a small concentration of fluorescent molecules. The excited fluorescence is collected with a confocal optical system and detected with a single photon counting detector. The laser power is so low that only a few percent of all laser pulses lead to the detection of a fluorescence photon, so it is very unlikely that more than one photon per laser pulse will be detected. Upon detection of a photon an electrical pulse is generated that is registered by a fast electronics. Additionally a synchronization signal is emitted from the laser with every single laser pulse. The electronics precisely measures the time difference between the laser synchronization pulse and the detector pulse. This is done with a "time to amplitude" converter, the laser synchronization pulse starts a voltage ramp that stops, when the detector pulse reaches the electronics. The ramp voltage is digitized with an AD converter and can be correlated with the detection time of the fluorescence photon. With this technique a time measurement is transformed to a voltage measurement. Time resolutions below a picosecond can be achieved. The arrival times of many photons build a histogram that reveals the decay curve. The more photons are recorded, the more precise the fluorescence lifetime can be estimated. Figure 19.40 illustrates this.

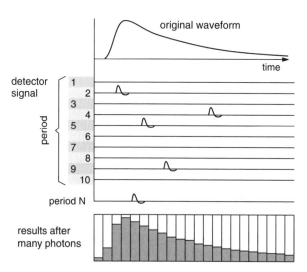

Fig. 19.40 Time-correlated single photon counting. The figure shows that many single photons whose arrival times are recorded add up to a histogram that reveals the decay curve. Pictures are provided by ILT

19.3.5 Economic Relevance of Laser-Induced Fluorescence

Laser-induced fluorescence plays an important role as an analytical measurement principle in many applications. The variable fields of application include routine analytics, fluorescence microscopy, clinical diagnostics, the control of chemical or pharmaceutical reactions, or basic research. Fluorescence has a role of special importance in the field of biochips and in vitro diagnostics (in vitro means "in glass," to be differentiated from "in vivo," which means in a living organism). The total worldwide market for medical diagnostics in 2005 was $37 billion. This includes applications like DNA sequencing, immunoassays, blood testing, nucleic acid diagnostics, and flow cytometers. The worldwide annual growth rate was between 5% (in vitro diagnostics) and 18% (DNA sequencing) [41].

Even bigger market potentials are estimated for protein biochips that are still in development.

19.4 Confocal Microscopy

Tilman Schwendt and Reinhard Noll

19.4.1 Motivation

Compared to conventional microscopes, a confocal laser scanning microscope (LSM) has higher contrast, better resolution in space, and an enhanced depth of focus. Due to these enhancements, it has been established in widespread fields of research in biology, medicine, and materials science [42–45].

In biology and medicine, confocal LSMs are applied for fluorescence microscopy to observe living cells or cell constituents, as shown in Fig. 19.41. Thereby, different elements of cells are marked with an appropriate fluorophore to visualize the diverse chemical and biological activities in the cell. Conventional microscopes can only generate images of inferior quality because of their objectives' low depth of focus, which is only a fractional amount of the cell thickness. For this reason, the conventional microscope image is a superimposition of diffuse parts from above and below the focal plane superimposed onto the well-focused part from the focal plane.

In case of a confocal LSM, the laser beam is focused on a small spot (ideally diffraction limited), and the sample and only the fluorescence occurring in the focus are measured. Images with microscopical resolution can be achieved by scanning the sample and digitizing the electronic pulses deriving from the photomultiplier tube (PMT). A focused laser beam leads to high intensities so that even small concentrations of a fluorescent dye can be detected with a sufficient signal-to-noise ratio. In principle, a confocal LSM can be operated with any wavelength; accordingly, an optimal wavelength can be chosen for all fluorescence markers.

T. Schwendt (✉)
Lasermess- und Prüftechnik, Fraunhofer-Institut für Lasertechnik, 52074 Aachen, Germany
e-mail: tilman.schwendt@ilt.fraunhofer.de

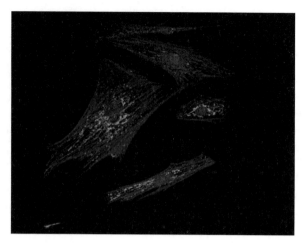

Fig. 19.41 Cell imaging: Nuclei (*red*), mitochondria (*green*), actin filaments (*blue*). (Source: Fraunhofer ILT)

In materials science a confocal LSM is mainly used for the analysis of structured and machined surfaces. Unlike with biological applications, the elastically backscattered light is measured instead of the laser-induced fluorescence [42].

19.4.2 Basic Principles

Figure 19.42 shows the basic configuration of a confocal laser scanning microscope. A collimated laser beam is directed via a dichroic beam splitter into a microscope objective with a high numerical aperture (NA) and is focused by the objective in a small focal volume (diffraction limited). Light emitted or reflected from this point of the sample is recollected by the same objective lens. Another objective (tube lens) focuses the light in a point of a plane, which is located at the conjugated (confocal) position of the focal plane.

A confocal pinhole is placed at this position and blocks all the light not deriving from the focal plane; thus it cannot be detected by the PMT. The area of interest is scanned sequentially pixel by pixel and line by line. The optical slice generated in this way is a high-contrast $x - y$ image of the sample with a high resolution along the principal z axis. Moving the focal plane stepwise permits one to achieve 3-D optical stacks from various single images which can be digitally edited afterward. The scanning process occurs either by misaligning the sample relative to the fixed optical setup, or, like in almost every commercially available LSM, by scanning the focused laser beam across the plane of interest. The diameter of the confocal pinhole determines the suppression of light deriving from object points outside of the focal plane. Therefore, the aperture of the confocal pinhole determines the thickness of optical slices whereas the numerical aperture of the microscope objective affects the size and length of the focal point and, accordingly, the lateral and axial resolution.

19 Laser Measurement Technology

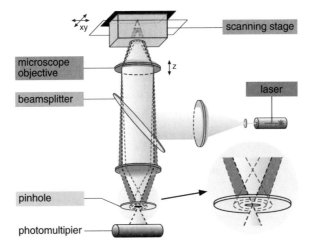

Fig. 19.42 Schematic setup of a confocal LSM

19.4.3 Resolution

When giving quantitative information concerning discrimination of depth and resolution of confocal LSM, one has to distinguish accurately, whether point-shaped or laminar, reflective, or fluorescent objects are considered since their image properties may differ drastically. Furthermore, the diameter of the image of the confocal aperture in the object plane (this is the pinhole diameter divided by the magnification of the objective) is of decisive relevance. In case of a diameter larger than an Airy disc generated by the microscope objective, the resolution is determined by the laws of geometrical optics. If the image of the confocal aperture in the object plane is smaller than 0.25 times the Airy disc, geometrical optics cannot be applied and the resolution of the microscope must be calculated considering the wave nature of light. In case of a fluorescent sample, both of these circumstances will be compared with conventional fluorescent microscopes in Table 19.3.

In the following, two different cases will be considered:

$$d_{AD} < d_{PH}$$

Axial and lateral resolution is enhanced in the confocal case by reason of replacing the longer emission wavelength by the shorter emission wavelength, whereby a factor $\lambda_{em}/\lambda_{exc}$ is obtained. Additionally, the quality of the image is improved due to an increase of contrast according to the suppression of scattered light by the confocal pinhole. The thickness of optical slices (discrimination of depth) is only given in a confocal setup and is determined primarily by the magnitude of the confocal pinhole (second term in the square). The first term is of wave optical form and is constant for any given emission wavelength.

Table 19.3 Comparison of resolution of a conventional and a confocal microscope considering fluorescent objects. n is the refraction index of air or immersion oil, λ_{em} the wavelength of emission, λ_{exc} the wavelength of excitation, $\bar{\lambda}$ the medial wavelength of emission and excitation (Eq. (19.8)), NA the numerical aperture of the objective, $f_{objective}$ the focal distance of the objective, f_{Tube} the focal distance of the tube lens, d_{ph} the diameter of the pinhole, and d_{AD} the diameter of the Airy disc, according to [42]

Conventional microscopy	Confocal microscopy $d_{AD} < d_{PH} < \infty$	Confocal microscopy $d_{PH} < 0.25 d_{AD}$
Axial resolution $\frac{n \cdot \lambda_{em}}{NA^2}$	Axial resolution $\frac{0.88 \cdot \lambda_{exc}}{n - \sqrt{n^2 - NA^2}}$	Axial resolution $\frac{0.64 \cdot \bar{\lambda}}{n - \sqrt{n^2 - NA^2}}$
Lateral resolution $\frac{0.51 \cdot \lambda_{em}}{NA^2}$	Lateral resolution $\frac{0.51 \cdot \lambda_{exc}}{NA^2}$	Lateral resolution $\frac{0.37 \cdot \bar{\lambda}}{NA^2}$
Thickness of optical slices Not defined	Thickness of optical slices $\sqrt{\left(\frac{0.88 \cdot \lambda_{exc}}{n - \sqrt{n^2 - NA^2}}\right)^2 + \left(\frac{\sqrt{2} \cdot n \cdot \frac{f_{objective}}{f_{tube}} d_{PH}}{NA}\right)^2}$	Thickness of optical slices $\frac{0.64 \cdot \bar{\lambda}}{n - \sqrt{n^2 - NA^2}}$

The choice of an appropriate pinhole diameter is critical. On the one hand, a smaller diameter improves the discrimination of depth, while on the other hand, the detectable intensity is reduced. Usually a pinhole diameter $d_{PH} \approx d_{AD}$ is chosen [46].

$$d_{PH} < 0.25 d_{AD}$$

When using a pinhole diameter less than 0.25 times the Airy disc, the appearance of diffraction at the pinhole aperture has to be considered as well. Furthermore, the emission wavelength λ_{em} contributes to the result, so a medial wavelength $\bar{\lambda}$ which is defined as

$$\bar{\lambda} = \sqrt{2} \frac{\lambda_{em} \cdot \lambda_{exc}}{\sqrt{\lambda_{exc}^2 + \lambda_{em}^2}} \qquad (19.8)$$

has to be inserted.

Thus, resolution is enhanced. A further reduction of the diameter leads to a decrease of intensity but not to further improvement of resolution. In case of a non-fluorescent object, the backscattered light is measured and, therefore, the excitation wavelength λ_{exc} has to be replaced everywhere in Table 19.3.

19.4.4 Typical Applications

19.4.4.1 Biology

The advantage of confocal microscopy is demonstrated in Fig. 19.43. This figure presents the image of a cell labeled with two different fluorophores. The cell is in the meta-/anaphase of cleavage. The plasma membrane (outer outshining area) and the spindle apparatus (inner outshining area) of this cell are marked with fluorescent

19 Laser Measurement Technology

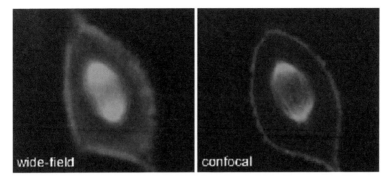

Fig. 19.43 Fluorescence microscopical image of cleavage in meta-/anaphase [47]. (**a**) Wide-field image and (**b**) confocal image. Pictures are provided by Zeiss

anaphylactics. Figure 19.43a shows an image generated with a conventional microscope. The membrane coloring at the cell border can be seen at the cell border, covering nearly the whole picture. The spindle fibers (middle of the cell) are blurred, but of notable intensity in the centromere region. Figure 19.42b shows the same picture taken with a confocal microscope. The outer plasma membrane is defined (optical slice by suppressing the light outside the focal area), and the fibers of the spindle apparatus are identifiable.

19.4.4.2 Materials Science

Another example to underline the power of confocal microscopy is shown in Fig. 19.44. This image shows a form electrode cut in sintered wolfram–copper. Remarkable is the measured depth of 1 mm, which is achieved by the digital combination of single optical slices to an overall image.

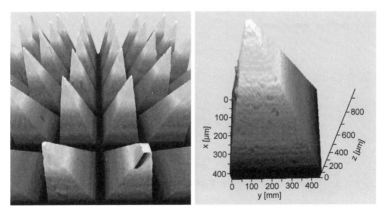

Fig. 19.44 Form electrode cut in sintered wolfram–copper. Sample: Probe: Prof. Uhlmann, Technische Universität Berlin, Institut für Werkzeugmaschinen und Fabrikbetrieb, Germany. Pictures are provided by Zeiss

19.4.4.3 Two-Photon Microscopy

The confocal two-photon microscopy (2PM) is an improvement of confocal fluorescence microscopy, which was developed to analyze biological problems in the last decade [47–49]. The 2PM is realized by using the following effect: the excitation of a fluorophore to an excited state can be realized not only by the absorption of a single photon but also by the absorption of two photons of approximately twice the wavelength during an extreme short time period (10^{-15} s) [50]. This means that the dye fluorescein with an absorption maximum at 500 nm can be excited by laser light at a wavelength of 1,000 nm. The two-photon excitation likelihood increases quadratically with the excitation intensity. Outside the focal area the intensity decreases quadratically with the distance from the focal point – as it does at confocal one-photon microscopy. Figure 19.45 shows the differences between confocal and two-photon microscopy. In 2PM fluorescence excitation only occurs in the laser focus and not in the entire beam cone as it takes place in case of a confocal microscope. The 2PM setup has two advantages. First, bleaching of the fluorophores and generation of phototoxicity is limited to the focal spot or rather to the focal plane when images are recorded. On the other hand, the overall fluorescence collected by the objective can be used for image generation. Hence, it is not necessary to integrate a pinhole which blocks all the light not deriving from the focal plane. Therefore, the fluorescence does not need to be propagated over the scanning device again and can be detected in a "non-descanned" position which is located closer to the sample. This allows the detection of some parts of fluorescence scattered in the sample and, thus, the measured fluorescence intensity increases.

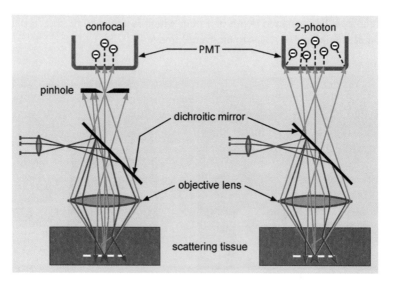

Fig. 19.45 Comparison of confocal and two-photon excitation. When using a two-photon setup no pinhole is required and scattered fluorescence photons can be detected as well

Therefore, the two-photon excitation likelihood decreases with the fourth power outside the focal area, and the excitation of the fluorophore is limited to the focal volume. The advantages of the 2PM in comparison with the 1PM are

- better localization of the fluorescence excitation;
- pinholes are not required;
- bleaching and generation of toxic co-products only in the focal area (maximal observation period is longer);
- wavelengths of excitation and emission do not overlap;
- entire fluorescence can be detected;
- low signal-to-noise ratio and high fluorescence signal;
- UV fluorophores can be excited by light at a wavelength of 680–850 nm;
- phototoxical damage of living preparation is reduced noticeably; and
- high depth of penetration in the most biological compounds (700–1,100 nm).

The major disadvantage of this method is that the two-photon excitation likelihood is much lower than the excitation probability of the one-photon excitation. Therefore, high intensities delivered by ultrashort pulsed laser sources and high intensities are required for the two-photon excitation.

19.5 Reading of Optical Storage Media

Reinhard Noll

19.5.1 Motivation

The compact disc (CD) was developed as a permanent and inexpensive mass storage medium by Philips and Sony and was introduced in the market in the year 1982. The information of the CD is stored physically in microstructures, the so-called pits, which are indentations aligned in a spiral track running from the center of the CD to the outside [51–53]. Due to the wavelength of the semiconductor laser ($\lambda = 780$ nm) used to read the CD, the storage capacity of a conventional CD is limited to about 0.7 GB corresponding to a length of the spiral track for about 5.5 km. Since about 1996 a further development of the CD, the digital versatile disc (DVD), was introduced having a storage capacity of about 4.7 GB. This was achieved by use of a shorter laser wavelength (650 nm) which allowed to further reduce the pit structure and hence to increase the storage capacity. In this case the length of the spiral track was extended to 12 km. Additionally some DVDs have two storage layers on top of each other – thereof one is semi-transparent – allowing to double the capacity (double-layer DVD). In 2006, a new format called blue ray disc (BD), designed by Sony, Philips, and Panasonic, was released as the successor to DVD. Because of

---

R. Noll (✉)
Lasermess- und Prüftechnik, Fraunhofer-Institut für Lasertechnik, 52074 Aachen, Germany
e-mail: reinhard.noll@ilt.fraunhofer.de

its shorter wavelength of 405 nm, the capacity was further increased to 25 GB for a single layer and the track length is 27 km. A dual-layer blue ray disc can store 50 GB. The reading of the information is based on the same physical principle for all systems: laser focusing and interference.

The economic relevance of optical storage media is indicated by the production numbers. Until 2001 more than 100 billion CDs were produced. In 2002 alone, more than 25 billion optical discs were produced throughout the world [54]. The estimated market volume for optical storage media for the year 2010 is US $31 billion with an average growth rate of about 20% [55].

19.5.2 Fundamentals

The reading unit of a CD or DVD player, the pick-up, is able to follow the track. It comprises laser beam source, optics, and photo detector. The laser beam is focused onto the protective layer having a thickness of 1.2 mm (CD) and 0.6 mm (DVD), respectively, yielding a light spot on the surface of the transparent layer of about 0.8 mm (CD) and 0.25 mm (DVD). The actual focusing down to the pit size (1 μm for CD and 0.6 μm for DVD) in the plane of the reflective layer is caused by the refractive index of the polymer of the transparent layer, see Fig. 19.46 (the thickness of the transparent layer is shown not to scale). Hence, damages at the surface of the transparent layer with dimensions in the order of 0.5 mm do not cause a reading error.

However, the pit structure itself would be unusable if much smaller damages occur. The height difference between pits and lands amounts to $\lambda/4$ causing an

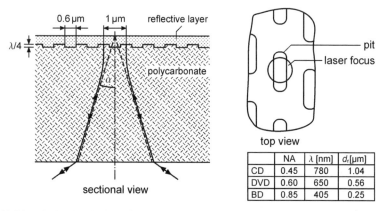

Fig. 19.46 *Left*: Sectional view of an optical storage medium showing the beam propagation of the laser beam reading the pit and land structure. The focusing to the size of a pit is due to the refractive index of the protective layer consisting of polycarbonate (the thickness of this layer is shown not to scale), α aperture angle; *right*: Top view of an optical storage medium showing a detail with a pit and the laser focus. *Bottom right*: Table with data of CD, DVD, and BD in terms of numerical aperture (NA) = $\sin\alpha$, wavelength λ, and focal diameter d_f [59]

optical retardation of $\lambda/2$ between the light scattered at the pit and the land while propagating to and from the reflective layer. So the backscattered light partially interferes destructively. If the laser beam irradiates only the land, then the beam is reflected nearly completely. By this a high signal contrast is achieved between land and pit.

19.5.3 Technical Realization of the Pick-Up System

During playback of a CD (DVD, BD) vibrations and axial run-outs occur, which have to be compensated due to the tiny focusing of the laser beam. Hence, the pick-up system comprises a precise focus and track control, see Fig. 19.47 [56]. The laser diode is positioned at the focal distance of a collimating lens, which forms the divergent light rays to a parallel beam. For track control the laser beam is split into three beams. They are generated by passing the laser beam through a diffraction grating leading to several diffracted beams. This group of beams is focused again obtaining a central laser focal spot and a series of weaker foci on both sides. This diffraction pattern irradiates the surface of the CD. The data are read with the main beam, whereas the secondary beams are used to follow the track. The beam splitter passes the light from the laser diode and reflects the back-propagating light to the photodiode. The beam splitter consists of two prisms with a common 45° base acting as a polarizing beam splitter. In a defined orientation linearly polarized light passes the prisms straightforward, whereas light with a perpendicular polarization is completely reflected at the interface of the two prisms. The laser light propagates through the collimator lens and the $\lambda/4$-plate, which is an anisotropic crystal

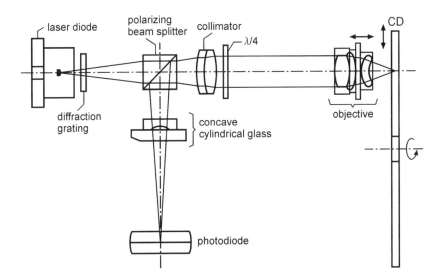

Fig. 19.47 Schematic setup of the reading unit of a CD player

generating a circularly polarized beam which allows to separate the forward propagating beam from the reflected beam. Finally the three light beams are focused by an objective onto the CD. The objective is mounted on a two-axes element, where the axis perpendicular to the CD enables a control of the focusing and the axis running parallel to the CD serves for track control. After reflection at the CD the circularly polarized light changes its sense of rotation and after passage of the objective and the $\lambda/4$-plate the polarization is linear again but now turned by 90° against the initial orientation of the polarization. Hence each of the three back-propagating beams is reflected by the polarizing beam splitter, passes a cylindrical lens, and irradiates the photodiode, whose photocurrent contains the information of the stored data. The detector consists of a matrix of four quadratic photodiodes arranged rhombically (quadrant detector). A non-uniform illumination of this detector yields the information to control the servos of the CD player for tracking and focusing. The data of the CD are retrieved via the sum of all four diode currents to maximize the signal level.

For the autofocus control the features of astigmatism are used. The focusing is monitored with the help of the cylindrical (astigmatic) lens in front of the photodetector, see Fig. 19.48 left, [55]. If the distance between objective and CD is correct, the cylindrical lens focuses a circular spot on the quadrant detector. A different distance causes a focus error, which results in a different beam diameter. A beam diameter which is not at the set point of the beam diameter is imaged by the cylindrical lens as an elliptic spot on the photodetector. Depending on the type of focus error this ellipse changes its horizontal and vertical orientation. A differential amplifier generates a signal corresponding to the focus error, which is transmitted to the focus servomechanism to readjust the objective lens thus compensating varying distances to the CD.

19.5.4 Further Development of DVD

The storage capacity of a DVD can be further increased by use of a two-layer disc, where the upper layer is semi-transparent. By this approach about 8 GB can be achieved for a single side of a disc. Bonding two dual-layer discs together results in the double-sided DVD with a storage capacity of 16 GB, see Fig. 19.49.

The development of blue laser diodes opens the chance to further increase the storage capacity of a DVD, since the better focusing capability allows to read even smaller pit structures [57]. In the beginning of the year 2008 many manufacturers and retailers were focusing on the blue ray disc. The blue ray disc has a protective coating of 0.1 mm and stores 25 GB on a single side, see Fig. 19.50. The distances between the information pits have to be smaller. Whereas for a CD the track pitch amounts to 1.6 μm, this value shrinks for a DVD down to 0.74 μm and for a blue ray disc to 0.32 μm.

In the digital future so-called surface recording discs will be developed, where the data are stored directly at the surface and are read by a near-field optic [59]. This disc shall store 100 GB.

19 Laser Measurement Technology

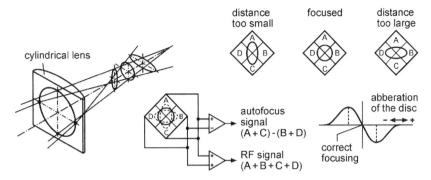

Fig. 19.48 Principle of the autofocus control of a CD and DVD player

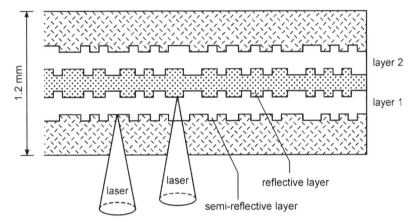

Fig. 19.49 Cross section of a DVD to quadruple the storage capacity by use of the two sides of the DVD and two layers, where one of these is semi-transparent

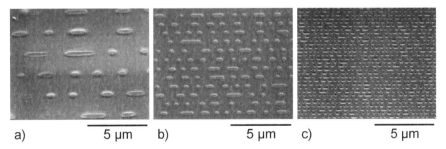

Fig. 19.50 Images of the pit structure of (**a**) CD, (**b**) DVD, and (**c**) blue ray disc taken by a scanning electron microscope [58]

19.6 Laser-Induced Breakdown Spectroscopy

Tobias Kuhlen, Ümit Aydin, and Reinhard Noll

19.6.1 Motivation

The chemical composition of a substance is an essential property for many tasks in process controlling and quality assurance in industrial production and environmental engineering. Conventional methods for chemical analysis normally comprise the following steps:

1. taking samples of the substance or removal of a workpiece from the production line;
2. transportation to analyzing system or laboratory;
3. preparation for chemical analysis;
4. separation of species (optional);
5. determination of composition.

The first two steps are necessary for conventional analysis because the analyzing systems are normally not on-site. The third step turns the sample into a physical–chemical state required for the particular method. Step 4 is a separation process like dissolving in an acid to obtain the insolvable fraction. Laser-based methods – such as laser-induced breakdown spectroscopy [60] – are predestined for chemical online analysis and allow simplifying or superseding steps 1 to 4.

19.6.1.1 Application and Market

Automated spectral analysis allows the determination of more than 20 elements in the fraction of a second. Limit of detections down to ppm (parts per million) can be reached while measuring a few micrograms of sample material. The measurements can be conducted under different atmospheric conditions or even in closed containers with the help of an optical window. Both the laser beam and the radiation to be measured can be led through the same optical window. Over the past years, various areas of application for laser-induced breakdown spectroscopy were studied and initial industrial applications were realized [61]:

- material analysis and mix-up examination [62–66];
- slag analysis [67, 68, 80];
- molten steel analysis [69–71];
- high-speed identification of polymers and metals for material-specific recycling [61, 72–75];
- classification of technical glasses [61];

T. Kuhlen (✉)
Messtechnik, Fraunhofer-Institut für Lasertechnik, 5207 Aachen, Germany
e-mail: tobias.kuhlen@ilt.fraunhofer.de

- identification of components in nuclear power plants [61, 75];
- multi-element analysis of liquids, aerosols, and particles [76–79];
- microanalysis of inclusions in metallic matrices [64, 81–84];
- monitoring of depth profiles [85, 86];
- online analysis of top gas [87];
- stand-off detection of explosives [88].

19.6.2 Differentiation to Conventional Methods

The optical emission spectroscopy (OES) in principle allows the identification and quantitative analysis of all elements of the periodic table. Conventional excitation sources for OES, like spark discharge (SD), show disadvantages with respect to the capability for automatic online inspection. Normally SD-OES requires a preparation or a finishing treatment of the workpiece. With laser-induced breakdown spectroscopy on the other hand, the laser itself can carry out the sample preparation. For example, the laser is able to ablate oxide or scale layers before the analysis takes place. Another advantage is the variable measurement distance between laser and sample, which can reach from a few centimeters up to several meters. The laser allows analysis of conductive as well as of insulating materials.

19.6.3 Basics

19.6.3.1 Evaporation and Plasma Ignition

For laser-induced breakdown spectroscopy, material is vaporized and excited to emit radiation. To get effective emission, the sample is locally heated up to its boiling or decomposition temperature by a focused pulsed laser beam. A part of the irradiated pulse energy is absorbed by the sample. Thereby, the heat conduction of the sample influences the level of heating. It dissipates the absorbed energy out of the radiated zone [1].

The parameters influencing the interaction between laser beam and sample are the following:

- beam diameter on the surface of the sample $2w$;
- optical penetration depth δ_{opt};
- heat penetration depth δ_h.

The optical penetration depth equals the distance inside the sample, within which the intensity of the laser beams drops down to approx. 37% of the value at the sample surface. It depends on the absorption characteristics of the material. For metals and laser wavelengths in the visible and near-infrared δ_{opt} typically amounts to $10^{-6} - 10^{-5}$ cm. The heat penetration depth describes how far the heat penetrates into the sample after the start of the irradiation. As a function of time, the heat penetration depth is given by

$$\delta_h = 2\sqrt{\kappa t} \tag{19.9}$$

where κ is the thermal diffusivity, $\kappa = 1\,\mathrm{m^2/s}$.

The heat equation describes the sample heating. Analytical solutions can be given for specific cases. We consider the following cases:

(a) Surface absorption ($\delta_{\mathrm{opt}} \ll \delta_h$)

The optical penetration depth is significantly smaller than the heat penetration depth. This applies, e.g., for steel at radiation times $t > 10^{-9}\,\mathrm{s}$ ($\kappa_{\mathrm{steel}} = 0.04\,\mathrm{cm^2/s}$). In the heat equation, the source term is described as surface source. In this case, the temperature as a function of time assuming a Gaussian intensity profile on the sample surface at the center of the beam is

$$T(t) = \frac{A I_0 w}{\sqrt{2\pi \kappa \rho c}} \arctan \frac{\sqrt{8\kappa t}}{r_b} \tag{19.10}$$

A absorption coefficient;
I_0 irradiance of the laser beam at its center at $r = 0$, $I_0 = 1\,\mathrm{W/m^2}$;
T temperature on the sample surface at the center of the laser beam, $T = 1\,\mathrm{K}$;
κ thermal diffusivity, $\kappa = 1\,\mathrm{m^2/s}$;
ρ density, $\rho = 1\,\mathrm{kg/m^3}$;
c specific heat, $c = 1\,\mathrm{J/(kg\,K)}$;
t time, $t = 1\,\mathrm{s}$;
w beam radius, $w = 1\,\mathrm{m}$.

As an example, it is possible with the help of Eq. (19.10) to determine the time, after which the temperature on the sample surface reaches the boiling point and therefore enhanced vaporization takes place for a given absorbed laser irradiance. The absorbed irradiance required for vaporization decreases with increasing time of irradiation. The laser irradiance on the sample surface has to be chosen higher according to the absorption coefficient to reach the vaporization threshold. Typical absorption coefficients for steel and aluminum are 0.03 to 0.1. For Q-switched lasers with pulse length between 10 and 100 ns intensities over $10^7\,\mathrm{W/cm^2}$ are necessary.

(b) Volume absorption ($\delta_{\mathrm{opt}} \gg \delta_h$)

The optical penetration depth is significantly greater than the heat penetration depth. This case holds, e.g., for plastic material. The optical penetration depth is approx. 3.5 mm for polymers like polyamide at the wavelength of Nd:YAG lasers. With a thermal diffusivity of $\kappa_{\mathrm{Polyamide}} = 1.3 \cdot 10^{-3}\,\mathrm{cm^2/s}$, the optical penetration depth is greater than the heat penetration depth calculated with Eq. (19.9) for times $t < 20\,\mathrm{s}$. If heat conduction can be ignored, the increase of temperature can be calculated directly from the locally absorbed laser irradiance. The decrease of the laser irradiance inside the absorbing medium is given by

$$I(z) = (1 - R)I_0 \exp(-z/\delta_{opt}) \qquad (19.11)$$

R reflectivity;
I_0 laser irradiance at the sample surface;
z coordinate in propagation direction of the laser beam, $z = 0$ mm is the sample surface.

The factor $(1 - R)$ in Eq. (19.11) describes the part of the incident irradiance that enters the sample. Contrary to case (a) the surface absorption cannot be taken as absorption A because the absorption is not limited to the sample surface. The absorbed energy per unit volume after the time t is

$$-(dI/dz)t = (1 - R)I_0/\delta_{opt} \exp(-z/\delta_{opt})t \qquad (19.12)$$

The absorbed energy causes the temperature of the unit volume to rise by the temperature difference ΔT, described by

$$\Delta q = \rho c \Delta T \qquad (19.13)$$

Δq heat energy per unit volume, $\Delta q = 1$ J/m^3;
ρ density, $\rho = 1$ kg/m^3;
c specific heat capacity, $c = 1$ J/(kg K);
ΔT temperature rise, $T = 1$ K.

Based on $-(dI/dz)t = \Delta q$, the following equation results from (19.12) for the sample surface ($z = 0$ mm):

$$(1 - R)I_0 = \rho c \Delta T \delta_{opt}/t \qquad (19.14)$$

If ΔT equals the temperature difference from initial to evaporation or decomposition temperature of the plastic, the necessary irradiance can be calculated as a function of time of irradiation using Eq. (19.14).

19.6.3.2 Evaporated Mass of Material

To estimate the amount of material vaporized by a laser pulse we use a simplified energy balance. We assume that the absorbed laser energy is transferred completely into the evaporation process:

$$AW_{\text{Laser}} = \rho V(\varepsilon_V + c\Delta T) \qquad (19.15)$$

A absorption coefficient;
W_{Laser} incident energy of a laser pulse;
ρ density;
V ablated volume;

ε_V evaporization enthalpy;
c specific heat capacity;
ΔT temperature difference between room and boiling temperature.

Equation (19.15) does not consider losses caused by heat conduction and radiation, melt expulsion, and re-condensation of vaporized material. Additionally for further simplification, we have ignored that the parameters A, ρ, c are not constant during heating and vaporization. The parameters are taken as constant at their initial values. For example, the approximated amount of material removed from a steel sample by a laser pulse with $AW_{\text{Laser}} = 10$ mJ calculated with (19.15) is $\rho V = 1.6 \cdot 10^{-6}$ g ($\varepsilon_V = 6 \cdot 10^3$ J/g, $c = 0.51$ J/(gK), $\Delta T = 2,600$ K).

The density of the resulting vapor can be estimated with help of the CLAUSIUS–CLAPEYRON relation. For orientation only, typical vapor densities are between 10^{18} and 10^{20} cm^{-3} for absorbed irradiances between 10^6 and 10^7 W/cm^2. Because of the high temperature of the vapor, free electrons exist, which can take energy out of the radiation field and transfer it to the vapor's atoms by collision processes. Collisions between electrons and atoms are the reason why electrons take energy out of the oscillating electrical field of the laser beam. Without the collisions electrons would absorb and emit energy periodically and therefore would not gain energy on average. The electrons reach high average energy levels, which are sufficient to ionize a fraction of the vapor's atoms. This generates a plasma – a physical state determined by the collective effect of ions and electrons. Inside the plasma, atoms and ions are excited and emit radiation.

With the help of balance equations for particle densities, momentums, and energies, the state of the plasma can be described. Model calculations show that within a few nanoseconds after beginning of the laser irradiation the electrons reach energies between 0.25 and 1 eV. The electron density rises to values between 10^{16} and 10^{18} cm^{-3} because of impact ionization processes. The plasma temperature typically is between 5,000 and 20,000 K.

19.6.4 Method Description

Laser-induced breakdown spectroscopy is a method for elemental analysis, which is based on the detection of spectral lines emitted by transitions of atoms and ions between outer electron orbits. For this, a high-energy laser beam is focused on a solid, liquid, or gaseous object. In doing so, a fast localized energy transfer is realized. The material is atomized locally and the atoms are excited. The relaxation of the excited atoms causes characteristic emission of these elements. Spectrally resolved detection of the emitted line radiation allows to determine the quantitative composition of the sample. The signal intensity is a measure for the concentration of the particular element.

Figure 19.51 illustrates this method. The laser beam is focused on the sample (Fig. 19.51 (1)). The material heats up locally, because of its natural absorption,

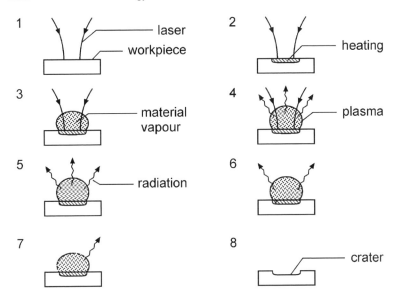

Fig. 19.51 Illustration of the LIBS method

and with sufficient laser irradiance, a fraction of the sample is vaporized (Fig. 19.51 (2–3)). The vaporized material partly absorbs the laser beam. The vapor reaches a sufficiently high temperature to excite a part of the atoms to higher energy states. Partly, the energy is even high enough to ionize some atoms. The system of neutral particles, ions, and electrons is called a plasma (Fig. 19.51 (4)).

The excited atoms or ions inside the laser-induced plasma transfer their excitation energy – among others – by the process of spontaneous emission. The frequency spectrum of this emission is characteristic for the composition of the analyzed material (Fig. 19.51 (5–7)). The radiation is detected and evaluated.

Later on the plasma decays and – because of the vaporized material – a crater is formed in a solid sample (Fig. 19.51 (8)). The diameter of the crater typically measures between 10 and 300 µm.

The diagram on the left side of Fig. 19.52 shows schematically an emission spectrum of a laser-induced plasma. In the spectrum are several emission lines. These lines are characteristic for the different elements comprised in the analyzed sample. Each line can be associated to a specific element. The line's height is a measure of the concentration of the element inside the sample. The line intensity, however, is also affected by a set of other factors. Normally the intensity of the line of the element j to be determined (the so-called analyte line) is set in a ratio to a reference line r. For example, a dominating element inside the analyzed material (normally an element of the matrix) is taken as reference line.

To gain quantitative results, a calibration of the method is necessary using samples with known composition. Figure 19.52 (right) shows a schematic of a calibration curve. It shows the ratio of element and reference line intensities as a function of the known elemental concentration of the calibration samples. Usually,

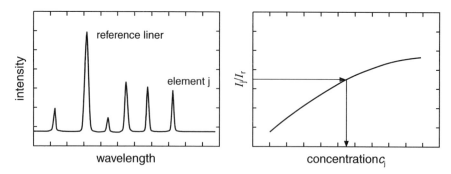

Fig. 19.52 Schematic of an emission spectrum of a laser-induced plasma (*left*) and a calibration curve (*right*)

the curves are nonlinear. For an unknown sample, the concentration of an element is determined from the line ratio with the help of the calibration curve (see arrows in Fig. 19.52 (right)).

A typical setup for LIBS is shown in Fig. 19.53. A Q-switched solid-state laser can be used as a light source. Pulses generated by those lasers have a duration between 5 and 100 ns with an energy between 100 µJ and 2 J. With this pulse energy, it is possible to reach an irradiance of more than 10^6 W/cm^2 in the focus of lens L, which normally is necessary for vaporization of a material. The plasma radiation is measured with a spectrometer via a fiber optic cable. For recording of the spectrum, basically two different methods are used. One alternative is to place detectors on specific positions inside the spectrometer, where the emission lines of the relevant elements are. For this, photomultipliers are used predominantly. The advantage of this method is that only information necessary for the analysis is recorded and processed.

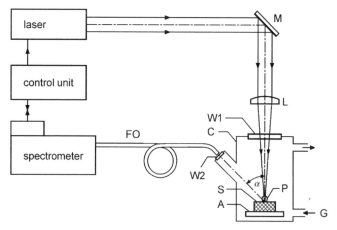

Fig. 19.53 Typical setup for LIBS. M, mirror; L, focusing lens; W1, entrance window; W2, exit window; C, measurement chamber; P, plasma; S, sample; A, positioning system; G, gas stream; α, viewing angle; FO, fiber optic cable; D, detector

The second alternative images the whole spectral range on a diode array. The spectrum shown in Fig. 19.52 (left) was recorded that way. For both methods, the registered signals are digitized and evaluated with a PC.

19.6.5 Time-Resolved Spectroscopy

The laser-induced plasma exists only for a short time. Hence the excited atoms and ions are emitting transiently. To estimate the order of magnitude of the plasma lifetime, the geometry of the plasma is assumed to be spherical. Inside the sphere is a hot gas, whose particles move with an average thermal velocity v_{th}. Because of this nondirectional movement the sphere begins to disintegrate. A measure for that time is calculated with

$$\tau = d_{Plasma}/v_{th} \tag{19.16}$$

where τ, lifetime of the plasma; d_{Plasma}, diameter of the plasma; and v_{th}, average thermal velocity of atoms and ions inside the plasma.

For a plasma of iron atoms and ions at a temperature of 9,000 K and a diameter of $d_{Plasma} = 2$ mm, the lifetime estimated by (19.16) is $\tau = 1.1\,\mu$s. In practice, one can observe line emission of the plasma in a time interval from approx. 200 ns to 10 μs after irradiation of the laser pulse onto a measuring object.

Figure 19.54 schematically shows emission spectra at three different times after the irradiance of the laser pulse. At time t_1, the plasma shows a mostly continuous spectrum caused by free–free transitions of electrons. Only a small number of atoms and ions are excited for emission. Accordingly, line intensities are low. At time t_2,

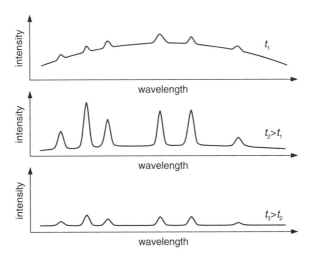

Fig. 19.54 Schematic illustration of the emission spectra of a laser-induced plasma at different times after irradiation of the laser pulse onto the sample

the plasma has cooled down and the line emissions of atoms and ions gain intensity. At time t_3, the plasma temperature is even lower and the line emission intensities are decreasing.

Time-resolved spectroscopy observes the emission of radiation within a certain time interval. At time t_2 in Fig. 19.54, there are, e.g., beneficial conditions to record emission lines with high intensities relative to the background. In the setup of Fig. 19.53, a signal electronic (part of the control unit) enables that the photomultipliers or diode arrays only record the line emissions in a specific time interval after plasma ignition.

19.6.6 Data Evaluation

Line intensities of elements of interest are determined from the measured spectra, which normally are gained in a digital form. The observed intensity at a given emission wavelength is in general a superposition of different parts, as seen in Fig. 19.55. In the following we will assume that the spectral intensity $S_\lambda(\lambda)$ at the wavelength λ is a combination of only two parts – namely the line radiation I_λ and the background radiation u_λ.

$$S_\lambda(\lambda) = I_\lambda(\lambda) + u_\lambda(\lambda) \tag{19.17}$$

$S_\lambda(\lambda)$ spectral intensity, power per surface unit and wavelength interval, $S_\lambda = 1\,\text{W/m}^3$;
u_λ spectral intensity of the background radiation, $u_\lambda = \text{W/m}^3$;
I_λ spectral intensity of the line radiation, $I_\lambda = \text{W/m}^3$;
λ wavelength of the emitted radiation.

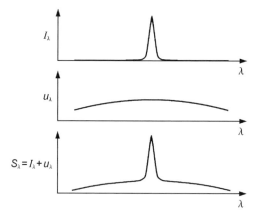

Fig. 19.55 Superposition of line emission and background emission of the laser-induced plasma

For evaluation, the so-called net line intensity is calculated as

$$J = \int_{\lambda_1}^{\lambda_2} (S_\lambda(\lambda) - u_\lambda(\lambda))d\lambda = \int_{\lambda_1}^{\lambda_2} I_\lambda(\lambda)d\lambda \tag{19.18}$$

J net line intensity, $J = \text{W/m}^3$;
λ_1, λ_2 integration limits, within the wavelength interval (λ_1, λ_2) the spectral intensity I_λ has non-zero values, $\lambda_i = \text{m}$.

The values of the function u_λ in the region of the emission line are interpolated from u_λ in the line-free neighborhood of that line.

The observed emission lines are assigned to the different atomic and ionic lines $i = 1, \ldots, m$ with the help of tables of spectral lines [89–91]. For a chosen number of lines, Eq. (19.18) is used to calculate the net line intensity J_i.

The intensities J_i are each a measure of the number of particles N_i of the respective element i. Furthermore, the plasma is assumed to be optically thin. In this case the emitted radiation is not significantly absorbed inside the plasma. The line emission intensity of the whole plasma volume is therefore directly proportional to the number of emitting particles N_i inside the plasma. Generalized, the factors influencing J_i can be described as

$$J_i = N_i \cdot F_i(I_0, E_{\text{Laser}}, r_b, t_{\text{delay}}, t_{\text{int}}, p_a, c_1, \ldots, c_m) \tag{19.19}$$

The function F_i describes the dependence of the line intensity J_i on the laser irradiance I_0, the laser pulse energy E_{Laser}, the beam radius w, the recording time t_{delay}, the integration time t_{int}, the ambient gas pressure p_a as well as the concentration c_j of all elements in the measuring object. The latter dependence results from the absorption and vaporization process, which is not independent from the composition of the sample. Also, the emission of the plasma depends on the collisions, which involve all species. In analytics, these effects are called matrix effects. In general, the function F_i is not known and has to be found empirically with calibration measurements. Practically, line ratios, like $J_1/J_2 = N_1/N_2$, often are used to compensate at least partly the influence of the factors mentioned.

19.6.7 Measurement Range

Typical parameters and data for LIBS are summarized in Table 19.4. Because the vaporization of the material is achieved in a purely optical way for LIBS, it does not matter whether the sample is an electric conductor or an isolator. The free distance between optics and sample is given by the focal length of the lens L in Fig. 19.53. Increasing the focal length for a given aperture of the lens increases the diameter

Table 19.4 Parameters and data for LIBS

Parameter	Data
Material	Metallic, non-metallic
Free distance between optics and workpiece	1 cm–1 m
Lateral resolution	1–300 μm
Detection sensitivity	100 ppb–1,000 ppm
Measurement rate	0.1–1,000 Hz

of the focus. In that case, higher laser pulse energies are necessary to reach the threshold irradiance for vaporization.

The lateral resolution is limited by the diameter of the laser beam on the surface of the sample. This in turn depends on the focal length, the illuminated aperture of the focusing lens, and the beam quality of the laser.

The values stated can be achieved with commercial Nd:YAG lasers. The detection sensitivities are element and matrix dependent. Typical values are listed in Table 19.4, which are determined using calibration curves [1]. The measurement rate is in general limited by the repetition rate of the Q-switched laser. To obtain higher accuracy usually several plasmas are ignited and their spectra are evaluated to obtain average values of the net line intensities of interest.

19.6.8 Examples of Applications

19.6.8.1 Fast, Spatially Resolved Material Analysis

For the production of high-quality steel products, like thin wires for energy-saving steel belt tires or thin foil for lightweight packaging, high-quality steel is needed. Inclusions lower the quality of the respective steel grade, so that, e.g., the diameter of wires in steel belt tires or the thickness of sheets for beverage cans cannot go below a certain limit.

Therefore, for the development of high-quality steels a method of analysis is required that can detect inclusions, like Al_2O_3, AlN, TiC, SiC, CaO, ZrO_2, in steel in a fast, reliable, and cost-effective way.

The steel cleanness analysis by LIBS is a method to solve that problem [81, 84]. To do so, samples are scanned with a precisely focused, pulsed laser beam, and the induced plasmas are evaluated spectrally. Each location is measured only with a single laser pulse. The frequency of measurement is up to 1,000 Hz.

The use of a laser offers the following advantages: in contrast to spark OES, a contamination with electrode material is not possible. The zone of interaction between laser radiation and sample material can be determined in a very accurate way, in contrast to spark OES. Concerning wet-chemical processes and the SEM-EDX method (scanning electron microscope with energy dispersive X-ray fluorescence detection), complex sample preparation and long measuring times can be avoided.

19 Laser Measurement Technology

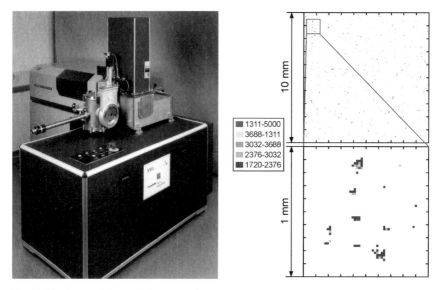

Fig. 19.56 Picture of the analysis system for steel cleanness inspection SML 1 (*left*), manganese enrichment of a steel sample from ThyssenKrupp Stahl (*right*) (*top*: whole scanned area with 10 × 10 mm² and 500 × 500 measurement points; bottom: 1 × 1 mm² detail of the area marked above)

For the task mentioned above, Fraunhofer ILT has developed an analysis system to inspect steel cleanness. Figure 19.56 (left) shows the SML 1 system (scanning microanalysis with laser spectrometry). In Fig. 19.56 (right), the raw signals of the photomultiplier for Mn-263 nm are shown in a 2D map for a measurement field of 10 × 10 mm. Only those signals, which exceed the sum of the average plus five times the standard deviation of the overall intensity distribution, are shown color coded. Local inclusions of manganese are clearly visible. A simultaneously measured elemental map of sulfur shows a local correlation between sulfur intensity peaks and manganese intensity peaks.

This strong local correlation indicates inclusions consisting of manganese sulfide MnS. With high-speed LIBS, the steel cleanness analysis can be accelerated significantly. The high-speed identification of other inclusions, like oxides or nitrides, is also possible with this method. The sensibility is even high enough to analyze segregation zones.

Numerous disadvantages can be avoided in comparison to conventional methods.

Advantageous in particular are the high savings of time and costs for measurement and sample preparation. Table 19.5 shows technical details of SML 1.

19.6.8.2 Automated Test for Mix-Ups in the Production Line of a Manufacturer of Pipe Fittings

The oil and gas industry, the chemical industry, as well as wastewater treatment plants use pipe fittings, e.g., 90° bows, T-pieces, reducers, and end caps. Because

Table 19.5 Technical data of SML 1

Parameter	Data
Measurement frequency	Up to 1 kHz
Analyzed elements	24 elements, including O, N, C, P, and S
Maximum number of elemental channels	48
Maximum scanning range	$110 \times 45\,\text{mm}^2$
Sample types	Samples with surface flatness $< 10\,\mu\text{m}$
Diameter of laser focus	$< 10\,\mu\text{m}$
Lifetime of the laser pump source	$> 10^9$ pulses
Maximum irradiance in the laser focus	$\approx 10^{12}\,\text{W/cm}^2$

of the different demands with respect to their corrosion resistance, pressure, and mechanical stability, various steel grades are used. The spectrum ranges from high-alloy steel to nickel-based alloys. Rising quality requirements and legal regulations for environment protection demand tests of each produced single workpiece. Based on LIBS, a test equipment has been designed that makes it possible to identify more than 35 different steel alloys. For an inspection, the amount of material ablated is only a few micrograms. An influence on the corrosion resistance of the measured workpieces by the laser radiation was tested according to DIN 50914 and DIN 50921. These tests for inter-crystalline and pitting corrosion showed no influence on the measured pipe fittings [92, 93].

Figure 19.57 shows a picture of the testing equipment LIFT (laser identification of fittings and tubes) with workpiece handling. A user inserts a specimen to the rotary table. The table makes a 90° rotation and positions the workpiece in front of the measurement window of LIFT. The test itself takes 2 s only. Within this time, 100 spectra are recorded and evaluated. If the measured material grade matches the expectation, an inkjet printer will mark the workpiece within the next cycle. Then a grabber, which transports the workpiece to the packaging line, will finally place it onto a belt conveyor.

Fig. 19.57 Testing equipment LIFT inside a production line (*left*), laser-induced plasma on the surface of a pipe fitting (*right*)

If LIFT recognizes a mix-up, the product is not marked and is discharged by reversing the direction of movement of the belt conveyor. One LIFT machine tests more than 20,000 parts per month. Since installation of the equipment, more than 2.5 million products have been tested automatically.

References

1. A. Donges, R. Noll, Lasermesstechnik – Grundlagen und Anwendungen, Hüthig Buch Verlag Heidelberg, ISBN 3-7785-2216-7, (1993), 318 S
2. R. Noll, Lasertriangulation, Handbuch Vision, Fraunhofer IRB Verlag, Stuttgart, 2007, S. 56–60
3. DIN 32 877: 2000–08, Optoelectronic measurement of form, profile and distance, Beuth Verlag GmbH, Berlin, Germany, August 2000
4. R. Noll, M. Krauhausen, Online-Lasermesstechnik für Walzprodukte, stahl und eisen 127, (2007), 99–105
5. R. Noll, M. Krauhausen, Online thickness measurement of flat products, Industrial Laser Solutions, 9–11, (July 2002)
6. Patent Triangulationsverfahren, DE 195 32 767
7. K. Becker, W. Gohe, H.-O. Katthöfer, A. Klein, L. Muders, A. Zimmermann, Herstellung und On-line-Prüfung von Schienen für den Hochgeschwindigkeitsverkehr, stahl und eisen119, (1999), 99–106
8. R. Noll, M. Krauhausen, Online laser measurement technology for rolled products, Ironmaking & Steelmaking 35, (2008), 221–227
9. Bildquelle: www.thyssenkrupppresta.com
10. Norm Ford EU 1880 (1997)
11. R. Noll, M. Krauhausen, Lasertriangulation für die Online-Messung geometrischer Größen in der Produktion, Handbuch zur Industriellen Bildverarbeitung, Hrsg. N. Bauer, Fraunhofer IRB Verlag, Stuttgart, (2007), 260–275
12. European Committee for Standardization, European Standard prEN 13674-1:1999, Railway Applications – Track – Rail – Part 1: Flat bottom symmetrical railway rails 46 kg/m and above. June 1999
13. American Railway Engineering Association, Manual for Railway Engineering. Chapter 4, Part 2, (2001)
14. C. Krobb, M. Krauhausen, R. Priem, H.-O. Katthöfer, M. Hülsmann, Measurement of rail profiles using 2D laser sensors, Int. Symposium on Photonics in Measurement, Aachen, 2002, VDI-Berichte 1694, P. 149–154
15. The Figures 19.14 to 19.17 were provided by the company NoKra Optische Prüftechnik und Automation GmbH, Max-Planck-Str. 12, 52499 Baesweiler, www.nokra.de
16. R. Noll, M. Krauhausen, Online Lasermesssysteme für die Stahl- und Automobilindustrie, VDI-Berichte Nr. 2011, (2008), 3–12
17. Th. Young "On the Theory of Light and Colours", Philos. Trans. Roy. Soc. London 92 (1802), 12–48
18. Th. Young "Experiments and calculations relative to Physical Optics", Philos. Trans. Roy. Soc. London 94 (1804), 1–16.
19. E. Hecht, "Optik", Oldenbourg, München, (2001)
20. R. Jones, C. Wykes, "Holographic and Speckle Interferometry", Second Edition, Cambridge Studies in Modern Optics: 6, Cambridge University Press, Cambridge, (1989)
21. P. Hariharan, "Optical Interferometry", Academic Press, Sydney, (1985)
22. J. R. Meyer-Arendt, "Introduction to Classical and Modern Optics", Prentice-Hall, New Jersey, (1995)

23. A. A. Michelson, "Interference Phenomena in a new form of Refractometer", Am. J. Sci. (3) 23 (1882), 395–400 und Philos. Mag. (5) 13, (1882), 236–242
24. A. A. Michelson, "Détermination expérimentale de la valeur du mètre en longueurs d'ondes lumineuses", Trav. Mem. Bur. Int. Poids Mes. 11, (1895), 1–42
25. L. Mach, "Modifikation und Anwendung des Jamin Interferenz-Refraktometers", Anz. Akad. Wiss. Wien math. Naturwiss.-Klasse 28, (1891), 223–224 und L. Zehnder, "Ein neuer Interferenzrefraktor", Z. Instrumentenkd. 11, (1891), 275–285
26. A. Donges, R. Noll, "Lasermesstechnik – Grundlagen und Anwendungen", Hüthig Verlag, Heidelberg, (1993)
27. D. Rosenberger, "Technische Anwendungen des Lasers", Springer, Berlin, (1975)
28. J. Müller, M. Chour, "Zweifrequenz-Laserwegmesssystem für extreme Verfahrgeschwindigkeiten und hohe Genauigkeit", Technisches Messen 58, (1991), 6, 253–256
29. M. Chour, M. Netzel, "Symmetrisches Trägerfrequenzinterferometer", Patentschrift DD 292 696 A5 vom 08.08.91
30. R. Noll, C. R. Haas, H. Kunze, "Development and diagnostics of a Z-pinch plasma target", J. de Physique C7, (1988), 177–184
31. R. Noll, "Speckle-Interferometrie", Praxis der Naturwissenschaften 43, (1994), 37–44.
32. www.zygolot.de
33. H. Haken, H. C. Wolf: Atom- und Quantenphysik - Eine Einführunq in die experimentellen und theoretischen Grundlagen, Springer Verlag, Berlin, (1980)
34. D. A. Skoog, J. J. Leary, Instrumentelle Analytik, Springer Verlag Berlin, (1996)
35. Römpp Chemie Lexikon, Online Version 2.0, Thieme Verlag, Web Seite: www.roempp.com
36. D. L. Andrews, Applied Laser Spectroscopy, VCH Publishers Inc., New York, (1992); W. Demtröder, Laserspektroskopie, Springer Verlag Berlin Heidelberg, (1997); J. R. Lakowicz, Princicples of Fluorescence Spectroscopy, Second Edition, Kluwer Academic/Plenum Publishers, New York, (1999)
37. Dr. H. Fuchs, Fuel distributions in engine is monitored with planar laser induced fluorescence, Lambda Highlights, 62, (2003), 1–3, Lambda Physik
38. J. M. Berg, L. Stryer, J. L. Tymoczjo, Lehrbücher der Biochemie, Biochemie, Spektrum Akademischer Verlag, (2003)
39. www.dhgp.de
40. R. Rigler, E. S. Elson, Fluorescence Correlation Spectroscopy, Springer Verlag Berlin, (2001)
41. G. Lee, K. Chu, L. Conroy, L. Fix, G. Lui, C. Truesdell, "A Study of Biophotonics: Market Segments, Size and Growth", Optik und Photonik, 2, (2007), WILEY-VCH Verlag GmbH & Co. KGaA, Weinheim
42. Die konfokale Laser Scanning Mikroskopie, http://www.zeiss.de
43. Ch. Zander, J. Enderlein, R. A. Keller, Single Molecule Detection in Solution, Wiley-VCH, Berlin, (2002)
44. J. B. Pawley, Handbook of Biological Confocal Microscopy, Plenum Press, New York, (1955)
45. T. Wilson, Confocal Microscopy, Academic Press, London, (1990)
46. R. H. Webb, Confocal optical microscopy, Rep. Prog. Phys. 59, (1996), 427–471
47. http://www.itg.uiuc.edu/technology/atlas/microscopy/
48. W. Lukosz, Optical systems with resolving powers exceeding the classical limit, J. Opt. Soc. Am. 56, (1966), 1463–4172
49. W. Denk, Two photon laser scanning microscopy, Science, 248, (1990), 73–76
50. Multiphoton Laser Scanning Mikroskopie, http://www.zeiss.de
51. C. Biaesch-Wiebke, CD-Player und R-DAT-Recorder: digitale Audiotechnik in der Unterhaltungselektronik, Vogel-Verlag, Würzburg, (1992)
52. K. Pohlmann, Compact-Disc-Handbuch: Grundlagen des digitalen Audio, technischer Aufbau von CD-Playern, CD-Rom, CD-I, Photo-CD, IWT-Verlag München, (1994)
53. G. Winkler, Tonaufzeichnung digital – Moderne Audiotechnik mit Computerhilfe, Elektor Verlag, Aachen, (1990)

54. www.bayer.com/annualreport2002/features/makrolon1.html
55. www.electronics.ca/prescenter/articles/92/1/globaldatastoragemarket
56. T. Kuhn, J. Asshoff, Optische Datenspeicher: Der CD-Player, Fachdidaktik-Seminar Physik, www.muenster.de/~asshoff/physik/cd/cdplayer.htm, (1997)
57. Blu-Ray-Disc, http://www.glossar.de/glossar/z_dvd.htm
58. Fraunhofer Institute for Laser Technology, Aachen, September 2008
59. F. Zijp, Near-field optical data storage, proefschrift, Technische Universiteit Delft, (2007)
60. L. J. Radziemski, D. A. Cremers, "Spectrochemical Analysis Using Laser Plasma Excitation", in Laser-induced plasmas and applications, Dekker Inc., New York, (1989), edited by L.J. Radziemski and D.A. Cremers, p. 295
61. R. Noll, V. Sturm, M. Stepputat, A. Whitehouse, J. Young, P. Evans, Industrial applications of LIBS, in Laser-induced breakdown spectroscopy, Ed. Miziolek, Palleschi, Schechter, Cambridge University Press, (2006), Chapter 11, 400–439
62. I. Mönch, R. Noll, R. Buchholz, J. Worringer, Laser identifies steel grades, Stainless Steel World, 12(4), (2000), 25–29
63. R. Noll, A. Brysch, F. Hilbk-Kortenbruck, M. Kraushaar, I. Mönch, L. Peter, V. Sturm, Laser-Emissionsspektrometrie – Anwendungen und Perspektiven für Prozesskontrolle und Qualitätssicherung, LaserOpto 6, (2000), 83–89
64. R. Noll, H. Bette, A. Brysch, M. Kraushaar, I. Mönch, L. Peter, V. Sturm, Laser-induced breakdown spectrometry – applications for production control and quality assurance in the steel industry, Spectrochim. Acta, Part B 56, (2001), 637–649
65. J. Vrenegor, V. Sturm, R. Noll, M. Hemmerlin, U. Thurmann, J. Flock, Preparation and analysis of production control samples by a two-step laser method, 7th Int. Workshop Progress in Analytica Chemistry in the Steel and Metal Industries, ed. J. Angeli, Glückauf GmbH, Essen, (2006), 81–86
66. M. Kraushaar, R. Noll, H.-U. Schmitz, Multi-elemental analysis of slag from steel production using laser induced breakdown spectroscopy, Int. Meet. on Chemical Engineering, Environmental Protection and Biotechnology, Achema 2000, Lecture Group: Laboratory and Analysis Accreditation, Certification and QM, (2000), 117–119
67. H.-U. Schmitz, R. Noll, M. Kraushaar, Laser-OES: A Universal Technique for Slag Analysis, Proc. 50th Chemists' Conference, Leamington Spa, (1999), 11–15
68. V. Sturm, L. Peter, R. Noll, J. Viirret, R. Hakala, L. Ernenputsch, K. Mavrommatis, H. W. Gudenau, P. Koke, B. Overkamp, Elemental analysis of liquid steel by means of laser technology, Int. Meet. On Chemical Engineering, Environmental Protection and Biotechnology, Achema 2000, Materials Technology and Testing, (2000), 9–11
69. V. Sturm, L. Peter, R. Noll, Steel analysis with Laser-Induced Breakdown Spectrometry in the Vacuum Ultraviolet, Appl. Spectrosc. 54, (2000), 1275–1278
70. L. Peter, V. Sturm, R. Noll, Liquid steel analysis with laser-induced breakdown spectrometry in the vacuum ultraviolet, Appl. Optics 42, (2003), 6199–6204
71. M. Stepputat, R. Noll, R. Miguel, High speed detection of additives in technical polymers with laser-induced breakdown spectrometry, VDI-Bericht, Anwendungen und Trends in der optischen Analysenmesstechnik, VDI Verlag, Bericht-Nr. 1667, (2002), 35–40
72. R. Sattmann, I. Mönch, H. Krause, R. Noll, S. Couris, A. Hatziapostolou, A. Mavromanolakis, C. Fotakis, E. Larrauri, R. Miguel, Laser-Induced Breakdown Spectroscopy for Polymer Identification, Appl. Spectrosc. 52, (1998), 456–461
73. M. Stepputat, R. Noll, On-line detection of heavy metals and brominated flame retardants in technical polymers with laser-induced breakdown spectrometry, Proc. Int. Symposium on Photonics in Measurement, 11./12. June 2002, Aachen, VDI-Verl., Düsseldorf, ISBN 3-18-091694-X, (2002)
74. Ü. Aydin, R. Noll, J. Makowe, Automatic sorting of aluminium alloys by fast LIBS identification, 7th Int. Workshop Progress in Analytica Chemistry in the Steel and Metal Industries, ed. J. Angeli, Glückauf GmbH, Essen, (2006), 309–314

75. A. I. Whitehouse, J. Young, I. M. Botheroyd, S. Lawson, C. P. Evans, J. Wright, Remote material analysis of nuclear power station steam generator tubes by laser-induced breakdown spectroscopy, Spectrochim. Acta, Part B 56, (2001), 821–830
76. R. E. Neuhauser, U. Panne, R. Niessner, G. Petrucci, P. Cavalli, N. Omenetto, On-line and in situ detection of lead in ultrafine aerosols by laser-excited atomic fluorescence spectroscopy, Sens. Actuat. B 39, (1997), 344–348
77. R. E. Neuhauser, U. Panne, R. Niessner, Laser-induced plasma spectroscopy (LIPS): a versatile tool for monitoring heavy metal aerosols, Anal. Chim. Acta 392, (1999), 47–54
78. J. Caranza, B. B. Fisher, G. Yoder, D. Hahn, On-line analysis of ambient air aerosols using laser-induced breakdown spectroscopy, Spectrochim. Acta Part B 56, (2001), 851–864
79. U. Panne, R. Neuhauser, M. Theisen, H. Fink, R. Niessner, Analysis of heavy metal aerosols on filters by laser-induced plasma spectroscopy, Spectrochim. Acta Part B 56, (2001), 839–850
80. M. Kraushaar, R. Noll, H.-U. Schmitz, Slag analysis with laser-induced breakdown spectrometry; Appl. Spectroscopy 57, (2003), 1282–1287
81. H. Bette, R. Noll, High-speed laser-induced breakdown spectrometry for scanning microanalysis, J. Phys. D: Appl. Phys. 37, (2004), 1281–1288
82. H. Bette, R. Noll, High-speed, high-resolution LIBS using diode-pumped solid-state lasers, in Laser-induced breakdown spectroscopy, Ed. Miziolek, Palleschi, Schechter, Cambridge University Press, Chapter 14, (2006), 490–515
83. H. Bette, R. Noll, H.-W. Jansen, H. Mittelstädt, G. Müller, Ç. Nazikkol, Schnelle ortsaufgelöste Materialanalyse mittels Laseremissionsspektrometrie (LIBS), LaserOpto 33(6), (2001), 60–64
84. H. Bette, R. Noll, G. Müller, H.-W. Jansen, Ç. Nazikkol, H. Mittelstädt, High-speed scanning laser-induced breakdown spectroscopy at 1000 Hz with single pulse evaluation for the detection of inclusions in steel, J. Laser Appl. 17, (2005), 183–190
85. H. Balzer, M. Höhne, V. Sturm, R. Noll, New approach to monitoring the Al depth profile of hot-dip galvanised sheet steel online using laser-induced breakdown spectroscopy, Anal. Bioanal. Chem. 385, (2006), 225–233
86. H. Balzer, M. Hoehne, S. Hoelters, V. Sturm, R. Noll, E. Leunis, S. Janssen, M. Raulf, P. Sanchez, M. Hemmerlin, Online depth profiling of zinc coated sheet steel by laser-induced breakdown spectroscopy, 7th Int. Workshop Progress in Analytica Chemistry in the Steel and Metal Industries, ed. J. Angeli, Glückauf GmbH, Essen, (2006), 237–242
87. V. Sturm, A. Brysch, R. Noll, H. Brinkmann, R. Schwalbe, K. Mülheims, P. Luoto, P. Mannila, K. Heinänen, D. Carrascal, L. Sancho, A. Opfermann, K. Mavrommatis, H.W. Gudenau, A. Hatziapostolou, S. Couris, Online multi-element analysis of the top gas of a blast furnace, 7th Int. Workshop Progress in Analytica Chemistry in the Steel and Metal Industries, ed. J. Angeli, Glückauf GmbH, Essen, (2006), 183–188
88. R. Noll, C. Fricke-Begemann, Stand-off detection of surface contaminations with explosives residues using laser-spectroscopic methods, in Stand-off Detection of Suicide Bombers and Mobile Subjects, ed. H. Schubert, Springer, (2006), 89–99
89. A. N. Zaidel', V. K. Prokov'ev, S. M. Raiskii, V. A. Slavnyi, E. Ya. Shreider, Tables of Spectral Lines, Ifi/Plenum, New York, (1970)
90. http://cfa-www.harvard.edu/amdata/ampdata/kurucz23/sekur.html
91. http://physics.nist.gov/cgi-bin/AtData/main_asd
92. R. Noll, I. Mönch, O. Klein, A. Lamott, Concept and performance of inspection machines for industrial use based on LIBS, Spectrochimica Acta B, 60, (2005), 1070–1075
93. R. Noll, L. Peter, I. Mönch, V. Sturm, Automatic laser-based identification and marking of high-grade steel qualities, Proc. 5th Int. Conf. on Progress in Analytical Chemistry in the Steel and Metals Industries, 12–14 May 1998, European Commission, CETAS, Ed. Tomellini, ISBN 92-828-6905-9, (1999), 345–351

Appendix A
Optics

A.1 Derivation of the FRESNEL Formulae

Reflection and refraction of plane electromagnetic waves at interfaces are described by the FRESNEL formulae. In the following a derivation of the FRESNEL formulae is presented.

Figure A.1 shows the coordinate system that is used. For the incident, reflected, and refracted waves, respectively, the plane wave ansatz is made [1]:

$$\vec{E}_i = \vec{E}_0 \, \exp(i\,\vec{k}_i \cdot \vec{r} - i\,\omega\,t) \tag{A.1}$$
$$\vec{E}_r = \vec{E}_{0_r} \, \exp(i\,\vec{k}_r \cdot \vec{r} - i\,\omega\,t) \tag{A.2}$$
$$\vec{E}_t = \vec{E}_{0_t} \, \exp(i\,\vec{k}_t \cdot \vec{r} - i\,\omega\,t) \tag{A.3}$$

The MAXWELL equations yield the following:

$$\vec{B}_i = \frac{\vec{k}_i \times \vec{E}_i}{\omega} \tag{A.4}$$

$$\vec{B}_r = \frac{\vec{k}_r \times \vec{E}_r}{\omega} \tag{A.5}$$

$$\vec{B}_t = \frac{\vec{k}_t \times \vec{E}_t}{\omega} \tag{A.6}$$

with

$$\vec{k}_i = k_i\,\vec{e}_i = k_i\,(\cos\alpha\,\vec{e}_z + \sin\alpha\,\vec{e}_x) \tag{A.7}$$
$$\vec{k}_r = k_i\,\vec{e}_r = k_r\,(-\cos\alpha'\,\vec{e}_z + \sin\alpha'\,\vec{e}_x) \tag{A.8}$$
$$\vec{k}_t = k_i\,\vec{e}_t = k_t\,(\cos\beta\,\vec{e}_z + \sin\beta\,\vec{e}_x). \tag{A.9}$$

The absolute values of the wave vectors are given by

$$k_i = k_r = \frac{\omega}{c}\,n_1 \tag{A.10}$$

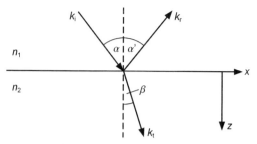

Fig. A.1 Coordinate system and notation

$$k_t = \frac{\omega}{c} n_2 \tag{A.11}$$

with the indices of refraction

$$n_1 = \sqrt{\mu_1 \varepsilon_1} \tag{A.12}$$
$$n_2 = \sqrt{\mu_2 \varepsilon_2} \tag{A.13}$$

If the radiation is absorbed in medium 2 the field strength decays exponentially in the direction of propagation. Absorbing materials can be described using a complex dielectric constant. This implies complex index of refraction and thus a complex wave number. In the expression for the plane waves it has to be considered that the absorption sets in at the interface $z = 0$. The planes of constant amplitude are parallel to the interface, whereas the planes of constant phase are perpendicular to the wave vectors. The wave in the absorbing medium no longer is a simple plane wave but an inhomogeneous wave. This can be accounted for if the components of the unit vectors $\vec{e}_t$ are allowed to be complex valued. Then β becomes complex too, which implies that β no longer has a simple physical interpretation as the angle of the refracted wave with respect to the surface normal. The boundary conditions at $z = 0$ have to be fulfilled for all values of x. This implies equality of the spatial variations of the three fields at $z = 0$ and thus the equality of the phase factors:

$$\vec{k}_i \cdot \vec{r}|_{z=0} = \vec{k}_r \cdot \vec{r}|_{z=0} = \vec{k}_t \cdot \vec{r}|_{z=0} \tag{A.14}$$
$$k_i \sin \alpha = k_r \sin \alpha' = k_t \sin \beta \tag{A.15}$$

Because $k_i = k_r$ it follows $\alpha = \alpha'$. With Eqs. (A.10) and (A.11) SNELL's follows:

$$n_1 \sin \alpha = n_2 \sin \beta \tag{A.16}$$

If n_2 is complex then β is complex too. Equation (A.16) comprises the kinematics of reflection and refraction.

In order to determine the amplitudes of the reflected and transmitted waves appropriate boundary conditions have to be set up:

Appendix A

- normal components of $\vec{D} = \varepsilon \, \varepsilon_0 \, \vec{E}$:

$$\left[\varepsilon_1 (\vec{E}_i + \vec{E}_r) - \varepsilon_2 \, \vec{E}_t\right] \cdot \vec{n} = 0 \tag{A.17}$$

- normal components of $\vec{B}$ (see Eqs. (A.4), (A.4), (A.4)):

$$\left[\vec{k}_i \times \vec{E}_i + \vec{k}_r \times \vec{E}_r - \vec{k}_t \times \vec{E}_t\right] \cdot \vec{n} = 0 \tag{A.18}$$

- the tangential components of $\vec{E}$ and $\vec{H} = \vec{H} = \dfrac{1}{\mu \, \mu_0} \vec{B}$ are continuous:

$$[\vec{E}_i + \vec{E}_r - \vec{E}_t] \times \vec{n} = 0 \tag{A.19}$$

$$[\dfrac{1}{\mu_1} (\vec{k}_i \times \vec{E}_i + \vec{k}_r \times \vec{E}_r) - \dfrac{1}{\mu_2} \vec{k}_t \times \vec{E}_t] \times \vec{n} = 0 \tag{A.20}$$

In case of arbitrary polarization the wave can be split into two waves, one with perpendicular polarization and the second one with parallel polarization of the electric field vector with respect to the plane of incidence. Both cases will be treated separately. In case of perpendicular polarization Eqs. (A.19) and (A.20) result in (Eq. (A.17) adds no further information, whereas Eq. (A.18) together with SNELL's law gives the same result as Eq. (A.20)):

$$E_i + E_r - E_t = 0 \tag{A.21}$$

$$\sqrt{\dfrac{\varepsilon_1}{\mu_1}} (E_i - E_r) \cos \alpha - \sqrt{\dfrac{\varepsilon_2}{\mu_2}} E_t \cos \beta = 0 \tag{A.22}$$

With this the FRESNEL formulae in case of perpendicular polarization follow:

$$\dfrac{E_r}{E_i} = r_s = \dfrac{n_1 \cos \alpha - \dfrac{\mu_1}{\mu_2} n_2 \cos \beta}{n_1 \cos \alpha + \dfrac{\mu_1}{\mu_2} n_2 \cos \beta} \tag{A.23}$$

$$\dfrac{E_t}{E_i} = t_s = \dfrac{2 n_1 \cos \alpha}{n_1 \cos \alpha + \dfrac{\mu_1}{\mu_2} n_2 \cos \beta} \tag{A.24}$$

If the electric field vector is parallel to the plane of incidence Eqs. (A.17) and (A.18) result in (Eq. (A.20) adds no further information, whereas Eq. (A.19) together with SNELL's law gives the same result as Eq. (A.18)):

$$(E_i - E_r) \cos \alpha - E_t \cos \beta = 0 \tag{A.25}$$

$$\sqrt{\frac{\varepsilon_1}{\mu_1}} (E_i + E_r) - \sqrt{\frac{\varepsilon_2}{\mu_2}} E_t = 0 \qquad (A.26)$$

With this the FRESNEL formulae in case of parallel polarization follow:

$$\frac{E_r}{E_i} = r_p = \frac{\frac{\mu_1}{\mu_2} n_2^2 \cos\alpha - n_1 n_2 \cos\beta}{\frac{\mu_1}{\mu_2} n_2^2 \cos\alpha + n_1 n_2 \cos\beta} \qquad (A.27)$$

$$\frac{E_t}{E_i} = t_p = \frac{2 n_1 n_2 \cos\alpha}{\frac{\mu_1}{\mu_2} n_2^2 \cos\alpha + n_1 n_2 \cos\beta} \qquad (A.28)$$

With the help of SNELL's law it follows for $\cos\beta$ that

$$n_2 \cos\beta = \sqrt{n_2^2 - n_1^2 \sin^2\alpha} \qquad (A.29)$$

With this the FRESNEL formulae can be written as a function of the angle of incidence alone. This also holds in case of complex n_2 in which β becomes complex too.

A.2 Dielectric Characteristics of Plasmas

A plasma is in general a mixture of free electrons, positive ions, and neutral atoms (or neutral molecules). A plasma is electrically neutral, charge neutrality is only violated in the vicinity of boundaries or within spatial dimensions that are comparable or smaller than the DEBYE radius. Due to their much smaller mass the electrons are at equal energy much faster than the ions. Thus in general the electron current density is much higher than the ion current density, which implies that the electrons mainly determine the electric characteristics of plasmas. The electron current density is

$$\vec{j} = -e \, n_e \, \vec{v}_e \qquad (A.30)$$

with

e – elementary charge
n_e – electron density
$\vec{v}_e$ – mean or drift velocity of the electrons.

The movement of electrons that are accelerated by an electric field is obstructed by collisions with other plasma particles. The impact of the collisions on the drift velocity can be modeled by a velocity proportional friction force. The equation of

Appendix A

motion that governs electron drift velocity reads

$$m_e \frac{d\vec{v}_e}{dt} + m_e \nu_m \vec{v}_e = -e \vec{E} \quad (A.31)$$

with

m_e – electron mass.

With the assumption that the momentum transfer frequency ν_m does not depend on the velocity it follows in case of time harmonic electric fields that

$$\vec{v}_e = -\frac{e}{m_e} \vec{E} \frac{1}{\nu_m - i\omega} \quad (A.32)$$

The use of complex numbers in mathematically describing real physical quantities is treated in Appendix A.3. Comparison with Eq. (A.30) results in OHM's law:

$$\vec{j} = \sigma \vec{E} \quad (A.33)$$

with the complex plasma conductivity

$$\sigma = \frac{e^2 n_e}{m_e \nu_m} \frac{\nu_m}{\nu_m - i\omega} \quad (A.34)$$

In case of time harmonic fields the MAXWELL equations read

$$\vec{\nabla} \times \vec{E} = i \omega \vec{B} \quad (A.35)$$
$$\vec{\nabla} \times \vec{H} = \mu_0 (\sigma \vec{E} - i \omega \varepsilon_0 \vec{E}) = i \omega \varepsilon_0 \mu_0 \varepsilon \vec{E} \quad (A.36)$$

In the second equation the particle and the displacement currents are combined. The complex dielectric constant of the plasma is given by

$$\varepsilon = 1 - \frac{\sigma}{i \omega \varepsilon_0} \quad (A.37)$$

$$\varepsilon = 1 - \frac{\omega_p^2}{\omega^2 + \nu_m^2} + i \frac{\nu_m}{\omega} \frac{\omega_p^2}{\omega^2 + \nu_m^2} \quad (A.38)$$

$$\omega_p^2 = \frac{e^2 n_e}{m_e \varepsilon_0} \quad (A.39)$$

With Eqs. (A.35) and (A.36) and the assumption of spatial homogeneity of the dielectric constant and that the plasma is electrically neutral the wave equation follows:

$$\Delta \vec{E} + \frac{\omega^2}{c^2} \varepsilon \vec{E} = 0 \tag{A.40}$$

This wave equation is solved by plane waves with complex wave vectors $\vec{k}$:

$$\vec{E} = \vec{E}_0 \exp[i(\vec{k} \cdot \vec{r} - \omega t)] \tag{A.41}$$

In case of propagation in the x-direction it follows that:

$$\vec{E} = \vec{E}_0 \exp[i(k_r x - \omega t)] \exp[-k_i x] \tag{A.42}$$

k_r is the real part and k_i the imaginary part of the complex wave vector:

$$|\vec{k}| = k_0 n_c \tag{A.43}$$
$$n_c = \sqrt{\varepsilon} \tag{A.44}$$
$$k_0 = \frac{\omega}{c} \tag{A.45}$$

with

k_0 – wave vector in vacuum
n_c – complex index of refraction of the plasma.

The time-averaged energy flux density is given by the time-averaged POYNTING vector[1]:

$$\vec{S} = \frac{1}{2} \mathrm{Re}\left[\vec{E}_0 \times \vec{H}_0^*\right] \tag{A.46}$$

The star indicates the complex conjugated value. In case of a plane wave it follows in scalar form:

$$\bar{S} = \frac{k_r}{k_0} \frac{|E_0|^2}{Z_0} \exp(-2 k_i x) \tag{A.47}$$

In case of zero momentum transfer frequency ν_m the plasma dielectric constant is given by

$$\varepsilon = 1 - \frac{\omega_p^2}{\omega^2} \tag{A.48}$$

and the index of refraction by

---

[1] See Appendix A.3.

Appendix A

$$n_c = \sqrt{1 - \frac{\omega_p^2}{\omega^2}} \tag{A.49}$$

If the electron plasma frequency ω_p exceeds the frequency ω of the electromagnetic wave the dielectric constant ε becomes negative. Then the index of refraction and the wave vector are imaginary and the intensity reflectivity R of the reflection at a plasma vacuum interface becomes 1,[2] i.e., the wave is reflected totally and thus the time-averaged POYNTING vector vanishes. $k_r = 0$ means that the wave cannot propagate within the plasmas and decays exponentially:

$$\vec{E} = \vec{E}_0 \exp(-i\omega t) \exp(-k_i x) \tag{A.50}$$

The electron density at which the plasma frequency equals the circular frequency of the electromagnetic wave is called critical density. The exponential decay of the field strength above the critical density in case of vanishing collision frequency is not due to absorption but solely due to reflection. In case of finite collision frequency the wave vector is no longer purely imaginary and the POYNTING vector has a finite value. The damping of the wave within the plasma is then due to both reflection and absorption.

A.3 Description of Electromagnetic Fields by Complex Quantities

Physical observables like the electric and magnetic field are real quantities. But in many cases of interest it is advantageous to use complex quantities to mathematically describe electromagnetic fields. In this case it has to be kept in mind that only real quantities are physically meaningful. The description by complex numbers is possible because the MAXWELL equations are linear in the field quantities and thus real and imaginary parts are independent of each other. In computing products of field quantities, e.g., the POYNTING vector, it has explicitly to be accounted for that fields are real quantities. The real field of a plane wave that propagates in z-direction can be written as

$$\vec{E} = \mathrm{Re}\left[\vec{E}_0 \exp[i(kz - \omega t)]\right] \tag{A.51}$$

$$= \frac{1}{2}\left(\vec{E}_0 \exp[i(kz - \omega t)] + \vec{E}_0^* \exp[-i(kz - \omega t)]\right) \tag{A.52}$$

$$\vec{H} = \mathrm{Re}\left[\vec{H}_0 \exp[i(kz - \omega t)]\right] \tag{A.53}$$

---

[2] See Eq. (2.17).

$$= \frac{1}{2} \left(\vec{H}_0 \, \exp\left[i(k\,z - \omega\,t)\right] + \vec{H}_0^* \, \exp\left[i(k\,z - \omega\,t)\right] \right) \tag{A.54}$$

As long as the fields only enter linearly the complex quantities can be used. This makes the computations often much easier because the exponential functions are in general easier to handle than the sin and cos functions. Inserting this into the MAXWELL equations results in an equation for the complex amplitudes $\vec{E}_0$ and $\vec{H}_0$. When determining physical quantities the expressions (A.51) and (A.51) have to be used:

$$\vec{E} = \vec{E}_{0_r} \cos(k\,z - \omega\,t) - \vec{E}_{0_i} \sin(k\,z - \omega\,t) \tag{A.55}$$

with

r – indicates the real part of $\vec{E}$
i – indicates the imaginary part of $\vec{E}/$

$$\vec{E} = \vec{e}_x \, [E_{0_{rx}} \cos(k\,z - \omega\,t) - E_{0_{ix}} \sin(k\,z - \omega\,t)] \tag{A.56}$$
$$+ \vec{e}_y \, [E_{0_{ry}} \cos(k\,z - \omega\,t) - E_{0_{iy}} \sin(k\,z - \omega\,t)]$$
$$+ \vec{e}_z \, [E_{0_{rz}} \cos(k\,z - \omega\,t) - E_{0_{iz}} \sin(k\,z - \omega\,t)]$$

This can be cast into the form

$$\vec{E} = \vec{e}_x \, [|E_{0_x}| \sin(k\,z - \omega\,t + \phi_x)] \tag{A.57}$$
$$+ \vec{e}_y \, [|E_{0_y}| \sin(k\,z - \omega\,t + \phi_y)]$$
$$+ \vec{e}_z \, [|E_{0_z}| \sin(k\,z - \omega\,t + \phi_z)]$$

with $\tan\phi_x = -\dfrac{E_{0_{ix}}}{E_{0_{rx}}}$, $\tan\phi_y = -\dfrac{E_{0_{iy}}}{E_{0_{ry}}}$, $\tan\phi_z = -\dfrac{E_{0_{iz}}}{E_{0_{rz}}}$.

While computing the POYNTING vector the real parts of the fields have to be taken:

$$\vec{S} = \text{Re}(\vec{E}_0 \, \exp[i(k\,z - \omega\,t)]) \times \text{Re}(\vec{H}_0 \, \exp[i(k\,z - \omega\,t)]) \tag{A.58}$$

The real parts can be written as the sum of the complex quantity and its complex conjugated value:

$$\vec{S} = \frac{1}{2} \left(\vec{E}_0 \, \exp[i(k\,z - \omega\,t)] + \vec{E}_0^* \, \exp[-i(k\,z - \omega\,t)] \right) \tag{A.59}$$
$$\times \frac{1}{2} \left(\vec{H}_0 \, \exp[i(k\,z - \omega\,t)] + \vec{H}_0^* \, \exp[-i(k\,z - \omega\,t)] \right)$$

This gives

$$\vec{S} = \frac{1}{4}\left(\vec{E}_0 \times \vec{H}_0^* + \vec{E}_0^* \times \vec{H}_0 \right. \\ + (\vec{E}_0 \times \vec{H}_0) \exp[2i(kz - \omega t)] \\ \left. + (\vec{E}_0^* \times \vec{H}_0^*) \exp[2i(kz - \omega t)]\right) \quad \text{(A.60)}$$

With this it follows that

$$\vec{S} = \frac{1}{2}\left(\text{Re}(\vec{E}_0 \times \vec{H}_0^*) + \text{Re}((\vec{E}_0 \times \vec{H}_0) \exp[2i(kz - \omega t)])\right) \quad \text{(A.61)}$$

The POYNTING vector consists of a constant part and an oscillating part. The oscillating part vanishes in taking the time average. The time-averaged POYNTING vector is the intensity of the wave and reads

$$\bar{\vec{S}} = \frac{1}{2}[\text{Re}(\vec{E}_0 \times \vec{H}_0^*)] \quad \text{(A.62)}$$

In vacuum the $\vec{E}$ and $\vec{H}$ vectors are perpendicular to each other and their absolute values are related by

$$|H_0| = \frac{|E_0|}{Z_0} \quad \text{(A.63)}$$

with

Z_0 – vacuum wave resistance.

With this it follows that

$$\bar{\vec{S}} = \frac{1}{2}\frac{|E_0|^2}{Z_0} \quad \text{(A.64)}$$

In the literature the fields are often defined omitting the factor 1/2 in Eqs. (A.51) and (A.53). Then Eq. (A.64) becomes

$$\bar{\vec{S}} = 2\frac{|E_0|^2}{Z_0} \quad \text{(A.65)}$$

Reference

1. J. D. Jackson, Classical Electrodynamics, John Wiley & Sons, New York, 1975

Appendix B
Continuum Mechanics

The following is mainly based on the excellent book [1].

B.1 Coordinate Systems and Deformation Gradient

In the field of continuum mechanics matter (solids, fluids, gases) is treated as a continuum in 3-dimensional space. The points that constitute the continuum are called **material points**. The **material points** can be identified uniquely by their coordinates at a reference time. At later times the positions of the **material points** will in general no longer coincide with the reference values. The change of position of the **material points** can be described using a time-dependent point transformation. The coordinates ξ_1, ξ_2, ξ_3 that the **material points** are assigned to at the reference time τ are called material coordinates. The coordinates x_1, x_2, x_3 of the 3-dimensional physical space are called space coordinates. It holds

$$\vec{\xi} = \vec{\xi}(\vec{x}, t; \tau) \tag{B.1}$$
$$\vec{\xi}(\vec{x}, \tau; \tau) = \vec{x} \tag{B.2}$$

as well as

$$\vec{x} = \vec{x}(\vec{\xi}, t; \tau) \tag{B.3}$$
$$\vec{x}(\vec{\xi}, \tau; \tau) = \vec{\xi} \tag{B.4}$$

The differentials of the spatial coordinates are given by

$$dx_i = \frac{\partial x_i(\vec{\xi}, t)}{\partial \xi_k} d\xi_k = F_{ik} d\xi_k \tag{B.5}$$

Double indices are summed over. The matrix

548   Appendix B

$$\hat{F} = (F_{ik}) = \frac{\partial(x_1, x_2, x_3)}{\partial(\xi_1, \xi_2, \xi_3)} = \begin{pmatrix} \frac{\partial x_1}{\partial \xi_1} & \frac{\partial x_1}{\partial \xi_2} & \frac{\partial x_1}{\partial \xi_3} \\ \frac{\partial x_2}{\partial \xi_1} & \frac{\partial x_2}{\partial \xi_2} & \frac{\partial x_2}{\partial \xi_3} \\ \frac{\partial x_3}{\partial \xi_1} & \frac{\partial x_3}{\partial \xi_2} & \frac{\partial x_3}{\partial \xi_3} \end{pmatrix} \quad (B.6)$$

is called deformation gradient. The determinant of this matrix is the JACOBI determinant Δ of the coordinate transformation $\left(\vec{\xi} \leftrightarrow \vec{x}\right)$:

$$\Delta = \det\left(\frac{\partial(x_1, x_2, x_3)}{\partial(\xi_1, \xi_2, \xi_3)}\right) \quad (B.7)$$

It always holds that $\Delta > 0$. Now it is assumed that the three material coordinate vectors $d\vec{\xi}_a$, $d\vec{\xi}_b$, $d\vec{\xi}_c$ with lengths a, b, and c, respectively, are perpendicular to each other (Fig. B.1). The volume element $dV_0 = (d\vec{\xi}_a \times d\vec{\xi}_b) \cdot d\vec{\xi}_c$ is then $(a\,b\,c)$. With this and Eq. (B.5) the coordinates of the vectors $d\vec{x}_i$ are given by

$$\begin{aligned} x_{1a} &= F_{11}\,a & x_{2a} &= F_{21}\,a & x_{3a} &= F_{31}\,a \\ x_{1b} &= F_{12}\,b & x_{2b} &= F_{22}\,b & x_{3b} &= F_{32}\,b \\ x_{1c} &= F_{13}\,c & x_{2c} &= F_{23}\,c & x_{3c} &= F_{33}\,c \end{aligned}$$

With this it follows that

$$d\vec{x}_a \times d\vec{x}_b = (F_{11}a, F_{21}a, F_{31}a) \times (F_{12}b, F_{22}b, F_{32}b)$$
$$= ab\,[(F_{21}F_{32} - F_{31}F_{22}), (F_{31}F_{12} - F_{11}F_{32}),$$
$$(F_{11}F_{22} - F_{21}F_{12})] \quad (B.8)$$
$$(d\vec{x}_a \times d\vec{x}_b) \cdot d\vec{x}_c = abc\,[(F_{21}F_{32} - F_{31}F_{22}), (F_{31}F_{12} - F_{11}F_{32}), \quad (B.9)$$
$$(F_{11}F_{22} - F_{21}F_{12})]$$
$$(F_{13}, F_{23}, F_{33})$$

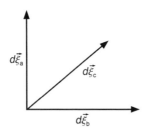

Fig. B.1 Material coordinates system

Appendix B

$$(d\vec{x}_a \times d\vec{x}_b) \cdot \vec{x}_c = dV$$
$$= dV_0[(F_{21}F_{32}F_{13} - F_{31}F_{22}F_{13}),$$
$$(F_{31}F_{12}F_{23} - F_{11}F_{32}F_{23}),$$
$$(F_{11}F_{22}F_{33} - F_{21}F_{12}F_{33})]$$

or

$$dV = dV_0 \, \Delta \tag{B.10}$$

Scalar as well as vector functions can be formulated in material as well as in spatial coordinates.

$$\phi(\vec{\xi}(\vec{x}, t; \tau)) = \tilde{\phi}(\vec{x}, t) \tag{B.11}$$
$$\vec{V}(\vec{\xi}(\vec{x}, t; \tau)) = \tilde{\vec{V}}(\vec{x}, t) \tag{B.12}$$

It holds

$$\frac{\partial \phi}{\partial x_i} = \frac{\partial \phi}{\partial \xi_j} \frac{\partial \xi_j}{\partial x_i} \tag{B.13}$$

B.2 Deformation

When considering three material points at a reference time $t = \tau$ and at a later time instance $t = t_0$ then in general the relative positions of these three points will have changed. If the relative positions did not change rigid body motion took place. If the points are infinitely close to each other then the vectors connecting the points are given by (see Fig. B.2)[1]

$$d\vec{x}^T \cdot \delta\vec{x} - d\vec{\xi}^T \cdot \delta\vec{\xi} = (\hat{F}d\vec{\xi})^T \cdot (\hat{F}\delta\xi) - d\xi^T \cdot \delta\vec{\xi} \tag{B.14}$$

with

$$(\hat{F}d\vec{\xi})^T = d\vec{\xi}^T \hat{F}^T. \tag{B.15}$$

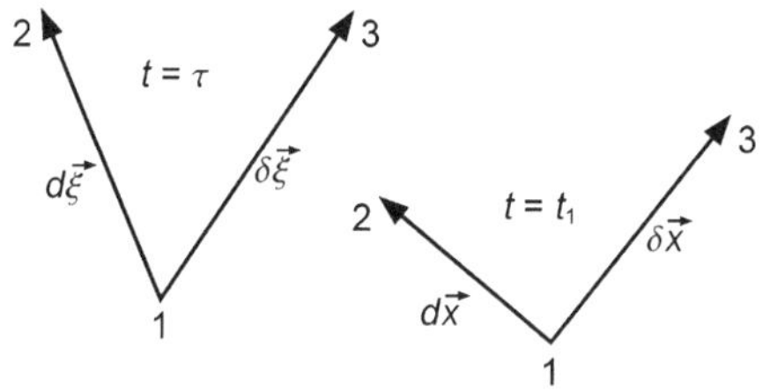

Fig. B.2 Deformation in material and spatial coordinates respectively

---

[1] T denotes the transposed matrix.

it follows that

$$d\vec{x}^T \cdot \delta\vec{x} - d\vec{\xi}^T \cdot \delta\vec{\xi} = d\vec{\xi}^T \hat{F}^T \hat{F} \delta\vec{\xi} - d\vec{\xi}^T \cdot \delta\vec{\xi}$$
$$= 2 d\vec{\xi}^T \left[\frac{1}{2}(\hat{F}^T \hat{F} - 1)\right] \delta\vec{\xi} \quad (B.16)$$

The expression in braces is GREEN's strain tensor:

$$\hat{G} = \frac{1}{2}\left(\hat{F}^T \hat{F} - \hat{1}\right) \quad (B.17)$$

GREEN's strain tensor is symmetrical. With the definition of the displacement vector $\vec{u}$

$$\vec{x} = \vec{\xi} + \vec{u}(\vec{\xi}) \quad (B.18)$$

the deformation tensor $\hat{F}$ is given by

$$F_{ik} = \frac{\partial x_i}{\partial \xi_k} = \delta_{ik} + \frac{\partial u_i}{\partial \xi_k} \quad (B.19)$$

or in short

$$\hat{F} = \hat{1} + \vec{\nabla}_\xi \vec{u} \quad (B.20)$$

Inserting this into Eq. (B.17) yields

$$\hat{G} = \frac{1}{2}\left[\left(1 + \vec{\nabla}_\xi \vec{u}\right)^T \left(1 + \vec{\nabla}_\xi \vec{u}\right) - 1\right]$$
$$\hat{G} = \frac{1}{2}\left[\left(1 + \vec{\nabla}_\xi^T \vec{u}\right)\left(1 + \vec{\nabla}_\xi \vec{u}\right) - 1\right]$$
$$\hat{G} = \frac{1}{2}\left[1 + \vec{\nabla}_\xi^T \vec{u} + \vec{\nabla}_\xi \vec{u} + \vec{\nabla}_\xi^T \vec{u} \vec{\nabla}_\xi \vec{u} - 1\right]$$
$$\hat{G} = \frac{1}{2}\left[\vec{\nabla}_\xi^T \vec{u} + \vec{\nabla}_\xi \vec{u} + \vec{\nabla}_\xi^T \vec{u} \vec{\nabla}_\xi \vec{u}\right] \quad (B.21)$$

In components GREEN's strain tensor reads

$$\gamma_{ik} = \frac{1}{2}\left[\frac{\partial u_k}{\partial \xi_i} + \frac{\partial u_i}{\partial \xi_k}\right] + \frac{1}{2}\sum_l \frac{\partial u_l}{\partial \xi_i}\frac{\partial u_l}{\partial \xi_k} \quad (B.22)$$

With the assumption that

$$\left|\frac{\partial u_i}{\partial \xi_k}\right| \ll 1 \quad (B.23)$$

Appendix B

quadratic terms in Eq. (B.22) can be neglected and the geometrically linearized GREEN's strain tensor follows

$$\gamma_{ik} = \frac{1}{2}\left[\frac{\partial u_k}{\partial \xi_i} + \frac{\partial u_i}{\partial \xi_k}\right] \quad (B.24)$$

With this Eq. (B.16) reads in components as

$$d\vec{x}\cdot\delta\vec{x} - d\vec{\xi}^T\cdot\delta\vec{\xi} = 2\sum_i\sum_k \gamma_{ik}\, d\xi_i\, \delta\xi_k \quad (B.25)$$

B.2.1 Physical Meaning of the Components of GREEN's Strain Tensor

With

$$d\vec{\xi} = \delta\vec{\xi} = (dl_0, 0, 0) \quad (B.26)$$

it follows from Eq. (B.25) (see Fig. B.3) that

$$dl^2 - dl_0^2 = 2\gamma_{11} dl_0^2$$
$$dl = \sqrt{1 + 2\gamma_{11}}\, dl_0 \quad (B.27)$$

With the assumption $|\gamma_{11}| \ll 1$ the square root can be expanded as

$$dl = (1 + \gamma_{11})\, dl_0$$

or

$$\gamma_{11} = \frac{dl - dl_0}{dl_0}$$

This implies that the diagonal elements of GREEN's strain tensor describe the elongation. With

$$d\vec{\xi} = (dl_0, 0, 0)$$
$$\delta\vec{\xi} = (0, dl_0, 0)$$

Fig. B.3 Physical meaning of the components of G

it follows from Eq. (B.25)

$$d\vec{x}^T \cdot \delta\vec{x} - d\vec{\xi}^T \cdot \delta\vec{\xi} = 2\gamma_{12}\, dl_0^2$$

and because $d\vec{\xi} \cdot \delta\vec{\xi} = 0$ it follows that

$$d\vec{x}^T \cdot \delta\vec{x} = 2\gamma_{12}\, dl_0^2$$

The scalar product on the right-hand side is

$$d\vec{x}^T \cdot \delta\vec{x} = |d\vec{x}^T||\delta\vec{x}|\cos\phi$$

With the vector length Eq. (B.27) it follows that

$$d\vec{x}^T \cdot \delta\vec{x} = dl_0\sqrt{1+2\gamma_{11}}\, dl_0\sqrt{1+2\gamma_{22}}\,\cos\phi$$

and thus

$$\cos\phi\sqrt{1+2\gamma_{11}}\sqrt{1+2\gamma_{22}} = 2\gamma_{12}$$

$$\gamma_{12} = \frac{1}{2}\frac{\cos\phi}{\sqrt{1+2\gamma_{11}}\sqrt{1+2\gamma_{12}}} = \frac{1}{2}\frac{\sin\alpha}{\sqrt{1+2\gamma_{11}}\sqrt{1+2\gamma_{12}}}$$

In case of $|\gamma_{ii}| \ll 1$ it approximately holds

$$\gamma_{12} \approx \frac{1}{2}\sin\alpha$$

and in case of small angles

$$\gamma_{12} \approx \frac{\alpha}{2} \qquad (B.28)$$

The non-diagonal elements of GREEN's strain tensor thus describe the changes of angles.

B.3 Derivation with Respect to Time

Time derivatives can be defined for material as well as spatial coordinates:

- material derivative

$$\frac{D\phi}{Dt} := \frac{\partial\phi(\vec{\xi},t)}{\partial t} \qquad (B.29)$$

Appendix B

- spatial derivative

$$\frac{\delta \phi}{\delta t} := \frac{\partial \phi(\vec{x}, t)}{\partial t} \tag{B.30}$$

The velocity and acceleration of a particle are defined as

$$\frac{D\vec{x}}{Dt} = \vec{v} \tag{B.31}$$

$$\frac{D\vec{v}}{Dt} = \vec{a} \tag{B.32}$$

If a scalar field function ϕ is given in spatial coordinates

$$\phi(x_1, x_2, x_3, t) = \phi(x_1(\xi_2, \xi_2, \xi_3, t), x_2(\xi_2, \xi_2, \xi_3, t),$$
$$x_3(\xi_2, \xi_2, \xi_3, t), t) \tag{B.33}$$

then it follows for the material time derivative

$$\frac{D\phi}{Dt} = \frac{\partial \phi(x_1, x_2, x_3, t)}{\partial t} + \frac{\partial \phi(x_1, x_2, x_3, t)}{\partial x_k} \frac{\partial x_k(\vec{\xi}, t)}{\partial t} \tag{B.34}$$

With Eqs. (B.29) and (B.31) it follows that

$$\frac{\partial x_k(\vec{\xi}, t)}{\partial t} = \frac{Dx_k}{Dt} = v_k \tag{B.35}$$

and with Eq. (B.30) one gets in components

$$\frac{D\phi}{Dt} = \frac{\delta \phi}{\delta t} + (\vec{\nabla} \phi) \cdot \vec{v} \tag{B.36}$$

The same applies to vector functions. With the definition of the gradient of a vector function

$$(\vec{\nabla} \vec{f}) = \left(\frac{\partial f_i}{\partial x_j}\right) \tag{B.37}$$

it follows that

$$\frac{D\vec{f}}{Dt} = \frac{\delta \vec{f}}{\delta t} + (\vec{\nabla} \vec{f})\vec{v} \tag{B.38}$$

and especially for $\vec{v}$

$$\frac{D\vec{v}}{Dt} = \vec{a} = \frac{\delta \vec{v}}{\delta t} + (\vec{\nabla} \vec{v})\vec{v} \tag{B.39}$$

B.4 REYNOLDS's Transport Theorem

While integrating a function of time and spatial coordinates $\phi(\vec{r}, t)$ over a moving material body B:

$$\psi(t) = \iiint_{B(t)} \phi(\vec{r}, t)\, dV \qquad (B.40)$$

the integration region will in general be time dependent. This has to be accounted for when performing the time derivative. When transforming the integral to material coordinates the integration region B_0 is time independent. With Eq. (B.10) it follows that

$$\psi(t) = \iiint_{B(t)} 0\, \phi(\vec{x}(\vec{\xi}, t), t)\, \Delta(\vec{\xi}, t)\, dV_0 \qquad (B.41)$$

The time derivative then is

$$\frac{D\psi}{Dt} = \iiint_{B_0} \left[\frac{D\phi}{Dt} \Delta + \phi \frac{D\Delta}{Dt} \right] dV_0 \qquad (B.42)$$

With [1]

$$\frac{D\Delta}{Dt} = \Delta\, \vec{\nabla} \cdot \vec{v} \qquad (B.43)$$

one gets

$$\frac{D\psi}{Dt} = \iiint_{B_0} \left[\frac{D\phi}{Dt} + \phi \vec{\nabla} \cdot \vec{v} \right] dV_0$$

$$= \iiint_{B(t)} \left[\frac{D\phi}{Dt} + \phi \vec{\nabla} \cdot \vec{v} \right] dV \qquad (B.44)$$

This is REYNOLDS's transport theorem. With Eq. (B.36) it follows that

$$\frac{D\psi}{Dt} = \iiint_{B(t)} \left[\frac{\delta\phi}{\delta t} + (\vec{\nabla}\phi) \cdot \vec{v} + \phi\, \vec{\nabla} \cdot \vec{v} \right] dV \qquad (B.45)$$

or

$$\frac{D\psi}{Dt} = \iiint_{B(t)} \left[\frac{\delta\phi}{\delta t} + \vec{\nabla} \cdot (\phi\vec{v}) \right] dV \qquad (B.46)$$

Using GAUSS's theorem

$$\iiint_{B(t)} \vec{\nabla} \cdot (\phi \vec{v}) = \iint_{\partial B(t)} (\phi \vec{v}) \cdot \vec{n} \, dA \qquad (B.47)$$

it follows that

$$\frac{D\psi}{Dt} = \iiint_{B(t)} \frac{\delta \phi}{\delta t} dV + \iint_{\partial B(t)} (\phi \vec{v}) \cdot \vec{n} \, dA \qquad (B.48)$$

$\partial B(t)$ — boundary of the body

B.5 Mass Conservation

The mass M of a body B is given be integrating its mass density over the volume of the body:

$$M = \iiint_{B(t)} \rho \, dV \qquad (B.49)$$

When the body B is moving the integration volume changes with time, whereas the mass of a body is conserved, which implies that its time derivative vanishes. With Eq. (B.44)

$$\frac{DM}{Dt} = 0 = \iiint_{B(t)} \left[\frac{D\rho}{Dt} + \rho \vec{\nabla} \cdot \vec{v} \right] dV \qquad (B.50)$$

B can be chosen arbitrarily so that the integrand (assuming sufficient continuity of the integrand) must be zero

$$\frac{D\rho}{Dt} + \rho \vec{\nabla} \cdot \vec{v} = 0 \qquad (B.51)$$

With Eq. (B.36) it follows that

$$\frac{\delta \rho}{\delta t} + \vec{\nabla} \cdot (\rho \vec{v}) = 0 \qquad (B.52)$$

This is the differential form of mass conservation. The time derivative of the integral of the product of an arbitrary field function ϕ with the mass density using Eq. (B.44) is given by

$$\frac{D}{Dt} \iiint_{B(t)} \rho\phi \, dV = \iiint_{B(t)} \left[\frac{D(\rho\phi)}{Dt} + \rho\phi \, \vec{\nabla} \cdot \vec{v} \right] dV$$

$$= \iiint_{B(t)} \left[\frac{D\rho}{Dt}\phi + \rho\frac{D\phi}{Dt} + \rho\phi \, \vec{\nabla} \cdot \vec{v} \right] dV \quad (B.53)$$

With Eq. (B.51) one gets

$$\frac{D}{Dt} \iiint_{B(t)} \rho\phi \, dV = \iiint_{B(t)} \rho \frac{D\phi}{Dt} \, dV \quad (B.54)$$

This relation was deduced for scalar functions but also applies in case of vector functions:

$$\frac{D}{Dt} \iiint_{B(t)} \rho\vec{a} \, dV = \iiint_{B(t)} \rho \frac{D\vec{a}}{Dt} \, dV \quad (B.55)$$

B.6 Momentum Conservation

The momentum of a body is defined as

$$\vec{P} := \iiint_{B(t)} \rho\vec{v} \, dV \quad (B.56)$$

Let $\vec{K}$ be the sum of all forces acting on the body B, then the time evolution of the momentum $\vec{P}$ of the body is given by

$$\frac{D\vec{P}}{Dt} = \vec{K} \quad (B.57)$$

A basic theorem of continuum mechanics states that the forces $\vec{K}$ that act on a body can be split into a contribution that act on the surface and a part that act on the volume (e.g., gravity):

$$\vec{K} = \iint_{\partial B(t)} \vec{t} \cdot \vec{A} + \iiint_{B(t)} \rho \vec{f} dV \quad (B.58)$$

with

$\partial B(t)$ – boundary of the body
$\vec{t}$ – surface forces.

Appendix B

The vector $\vec{t}$ is called stress vector and has the unit force per area. With Eq. (B.55) momentum conservation becomes

$$\iiint_{B(t)} \left[\rho \frac{D\vec{v}}{Dt} - \rho \vec{f} \right] dV = \iint_{\partial B(t)} \vec{t}\, dA \qquad (B.59)$$

$\vec{t}$ can be described using CAUCHY's stress tensor $\hat{T}$

$$\vec{t} = \hat{T} \cdot \vec{n} \qquad (B.60)$$

or in components

$$t_i = \tau_{ij} n_j \qquad (B.61)$$

$\vec{n}$ is the normal of the area element that $\vec{t}$ acts on. Thus the force that acts on an area element depends not only on the stress tensor, which is a function of space only, but also on the unit vector. This implies that $\vec{t}$ is not a vector field, whereas the stress tensor $\hat{T}$ is a tensor field. The quantities τ_{ii} are called normal stresses and the quantities $\tau_{ij}, i \neq j$ are called shear stresses. From angular momentum conservation it follows that CAUCHY's stress tensor is symmetric [1]:

$$\tau_{ij} = \tau_{ji} \qquad (B.62)$$

The surface integral in the momentum equation (B.59) can be cast to a volume integral:

$$\iint_{\partial B(t)} \hat{T} \vec{n}\, dA = \iiint_{B(t)} \vec{\nabla} \cdot \hat{T}\, dV \qquad (B.63)$$

The divergence of a tensor as defined in Eq. (B.63) is a vector:

$$\vec{\nabla} \cdot \hat{T} = \frac{\partial \tau_{ij}}{\partial x_j} \qquad (B.64)$$

Again double indices are summed over. Momentum conservation thus reads

$$\iiint_{B(t)} \left(\rho \frac{D\vec{v}}{Dt} - \vec{\nabla} \cdot \hat{T} - \rho \vec{f} \right) dV = 0 \qquad (B.65)$$

Because B is arbitrary the integrand must vanish (assuming continuity of the integrand). With Eqs. (B.36) and (B.30) it follows that

$$\rho \frac{D\vec{v}}{Dt} = \rho \frac{\partial \vec{v}}{\partial t} + \rho (\vec{\nabla} \vec{v}) \vec{v} = \vec{\nabla} \cdot \hat{T} + \rho \vec{f} \qquad (B.66)$$

This is the momentum conservation in differential form. In components it reads

$$\rho \frac{\partial V_i}{\partial t} + \rho \frac{\partial V_i}{\partial x_j} V_j = \frac{\partial \hat{T}_{ij}}{\partial x_j} + \rho f_i \tag{B.67}$$

Double indices are summed over. Multiplication of the mass conservation equation (B.52) by V_i:

$$V_i \frac{\partial \rho}{\partial t} + V_i V_j \frac{\partial \rho}{\partial x_j} + V_i \rho \frac{\partial V_j}{\partial x_j} = 0 \tag{B.68}$$

and adding Eq. (B.67) yields

$$\frac{\partial (\rho V_i)}{\partial t} + \frac{\partial (\rho V_i V_j)}{\partial x_j} = \frac{\partial \hat{t}_{ij}}{\partial x_j} + \rho f_i \tag{B.69}$$

This is the conservative form of the momentum conservation.

B.7 Material Equations

In case of pure mechanical material properties material equations relate stresses and strain. In case of solids the movement of the solid body is described by the strain tensor, whereas in case of fluids the strain velocity tensor applies.

B.7.1 Elastic Solids

Stress and strain of a solid are related by the general relation:

$$\hat{T}(\vec{\xi}, t) = f(\hat{G}(\vec{\xi}, t)) \tag{B.70}$$

In case of small strain

$$|\gamma_{ij}| \ll 1$$

i.e., in the geometric linear case it holds

$$\hat{T}_{ij} = f_{ij}(\gamma_{11}, \gamma_{12}, \gamma_{13}, \gamma_{22}, \gamma_{23}, \gamma_{33}) \tag{B.71}$$

If the functions f_{ij} are linear homogeneous in γ_{ij} (physical linear model), generalized HOOK's law follows:

$$\tau_{11} = c_1 \gamma_{11} + \cdots + c_6 \gamma_{33}$$
$$\tau_{12} = c_7 \gamma_{11} + \cdots + c_{12} \gamma_{33}$$

$$\tau_{33} = c_{31}\,\gamma_{11} + \cdots + c_{36}\,\gamma_{33}$$

The 36 constants $c_1, c_2, \ldots, c_{36}$ reduce to 21 due to symmetry considerations [1]. Assuming complete isotropy of the solid this 21 constants reduce to only 2 (there are crystals that make necessary to keep all 21 constants). In case of isotropic materials HOOK's law applies:

$$\hat{T} = 2\mu\hat{G} + \lambda(\operatorname{Tr}\hat{G})\hat{I} \tag{B.72}$$

$\operatorname{Tr}\hat{G}$ is the trace of the strain tensor and μ and λ are the LAMB constants. $\hat{I}$ is the identity tensor. In the field of continuum mechanics often alternative constants are used:

- YOUNG's modulus E
- POISSON's ratio ν
- compressibility κ

The two sets of constants are related by

$$\mu = \frac{E}{2(1+\nu)}$$
$$\lambda = \frac{E\nu}{(1+\nu)(1-2\nu)}$$
$$\kappa = \frac{3}{3\lambda + 2\mu}$$

and the inverse expressions

$$E = \frac{\mu(3\lambda + 2\mu)}{\lambda + \mu}$$
$$\nu = \frac{\lambda}{2(\lambda + \mu)}$$

λ must not be negative which implies that $0 \leq \nu < 0.5$. Using this constants Eq. (B.72) can be written as

$$\tau_{ij} = \frac{E}{1+\nu}\left(\gamma_{ij} + \frac{\nu}{1-2\nu}(\gamma_{11} + \gamma_{22} + \gamma_{33})\delta_{ij}\right) \tag{B.73}$$

The components of the strain tensor as a function of the stresses are

$$\gamma_{ij} = \frac{1+\nu}{E}\left(\tau_{ij} - \frac{\nu}{1+\nu}(\tau_{11} + \tau_{22} + \tau_{33})\delta_{ij}\right) \tag{B.74}$$

B.7.2 NEWTONian Fluids

In case of flows with friction the stress tensor is composed of a spherical symmetric and a friction contribution:

$$\hat{T} = -p\hat{I} + \hat{T}_r \tag{B.75}$$

with

p – hydrostatic pressure.

Because in fluids there is no distinguished configuration it can always be assumed that the deformation gradient is given by $\hat{I}$ which implies that GREEN's stress tensor vanishes. In fluids the deformation gradient is replaced by the deformation velocity:

$$\dot{\hat{F}} = \frac{\partial^2 x_i(\vec{\xi}, t)}{\partial \xi_k \partial t} = \frac{\partial}{\partial \xi_k} \frac{\partial x_i(\vec{\xi}, t)}{\partial t} \tag{B.76}$$

With Eqs. (B.30) and (B.31) it follows that

$$\dot{\hat{F}} = \frac{\partial V_i}{\partial \xi_k} = \frac{\partial V_i}{\partial x_l} \frac{\partial x_l}{\partial \xi_k} \tag{B.77}$$

Using short-hand notation this reads

$$\dot{\hat{F}} = \left(\vec{\nabla}\vec{v}\right)\hat{F} \tag{B.78}$$

Because in fluids there is no distinguished configuration every configuration can be taken as reference configuration. This implies that $\hat{F}$ can be set to be the unit tensor for all times. Thus it follows that

$$\dot{\hat{F}} = \vec{\nabla}\vec{v} = \frac{\partial V_i}{\partial x_j} \tag{B.79}$$

The matrix Eq. (B.79) can be split into a symmetric and an antisymmetric part:

$$\vec{\nabla}\vec{v} = \hat{D} + \hat{W} \tag{B.80}$$

$$D = \frac{1}{2}\left(\vec{\nabla}\vec{v} + \vec{\nabla}^T\vec{v}\right) = \frac{1}{2}\left(\frac{\partial V_i}{\partial x_j} + \frac{\partial V_j}{\partial x_i}\right) \tag{B.81}$$

$$W = \frac{1}{2}\left(\vec{\nabla}\vec{v} - \vec{\nabla}^T\vec{v}\right) = \frac{1}{2}\left(\frac{\partial V_i}{\partial x_j} - \frac{\partial V_j}{\partial x_i}\right) \tag{B.82}$$

Appendix B

where T indicates the transposed tensor. The antisymmetric part $\hat{W}$ describes a rigid body rotation [1], in case of fluids this contribution vanishes. The friction part of the stress tensor of fluids is postulated to be of the form

$$\hat{T}_r = f(\hat{D}) \ , \quad f(0) = 0 \tag{B.83}$$

In case of NEWTONian fluids the following relation applies [1]:

$$\hat{T}_r = 2\,\eta\,\hat{D} + \lambda(Sp\hat{D}) \tag{B.84}$$

with

η – shear viscosity
λ – volume viscosity.

Equation (B.69) then becomes

$$\frac{\partial(\rho\,V_i)}{\partial t} + \frac{\partial(\rho\,V_i\,V_j)}{\partial x_i} = -\frac{\partial p}{\partial x_i} + \lambda\,\frac{\partial}{\partial x_i}\frac{\partial V_j}{\partial x_j}$$
$$+ \eta\,\frac{\partial}{\partial x_j}\left(\frac{\partial V_i}{\partial x_j} + \frac{\partial V_j}{\partial x_i}\right) + \rho\,f_i \tag{B.85}$$

In case of incompressible fluids the mass conservation Eq. (B.52) yields

$$\frac{\partial V_j}{\partial x_j} = \vec{\nabla}\cdot\vec{v} = 0 \tag{B.86}$$

The divergence of $\hat{D}$ is thus given by

$$\vec{\nabla}\hat{D} = \frac{\partial d_{ij}}{\partial x_j} = \frac{1}{2}\frac{\partial^2 V_i}{\partial x_j^2} = \frac{1}{2}\Delta\vec{v} \tag{B.87}$$

The momentum conservation in case of incompressible NEWTONian fluids thus reads as follows:

$$\rho\,\frac{\partial V_i}{\partial t} + \frac{\partial(\rho\,V_i\,V_j)}{\partial x_i} = -\frac{\partial p}{\partial x_i} + \eta\,\frac{\partial^2 V_i}{\partial x_j^2} + \rho\,f_i \tag{B.88}$$

or in short

$$\rho\,\frac{\partial\vec{v}}{\partial t} + \rho\,(\vec{v}\,\vec{\nabla})\,\vec{v} = -\vec{\nabla}\,p + \eta\,\Delta\vec{v} + \rho\,\vec{f} \tag{B.89}$$

B.8 Energy Conservation

With the assumption that the energy of a body B can uniquely be split into a part that does not depend on the movement, namely its internal energy, and its kinetic energy, the total energy E of the body is given by

$$E = \iiint_{B(t)} \rho \, \varepsilon \, dV \tag{B.90}$$

with

$\varepsilon = e + \dfrac{1}{2} \vec{v}^2$ – specific total energy density
e – specific internal energy
v – velocity.

The integration runs over the material body B. If the body moves the integration region changes. According to the first law of thermodynamics the time change of the internal energy equals the power of the external forces that act on the body and the heat flux through the boundary of the body:

$$\underbrace{\frac{D}{Dt} \iiint_{B(t)} \rho \, \varepsilon \, dV}_{1} = \underbrace{\iint_{\partial B(t)} (\vec{t} \cdot \vec{v} - \vec{q} \cdot \vec{n}) \, dA}_{2,3}$$

$$+ \underbrace{\iiint_{B(t)} (\rho \, \vec{f} \cdot \vec{v} + w) \, dV}_{4,5} \tag{B.91}$$

with

1. total energy
2. power of the stresses acting on the boundary of the body
3. heat flux through the boundary ($\vec{n}$ is directed outward)
4. power of the volume forces
5. volume heat source.

With REYNOLDS's transport theorem Eq. (B.44) results in the energy conservation equation

Appendix B

$$\iiint_{B(t)} \left(\frac{\partial \rho \varepsilon}{\partial t} + \vec{\nabla} \cdot (\rho \, \varepsilon \, \vec{v}) \right) dV = \iint_{\partial B(t)} (\vec{t} \cdot \vec{v} - \vec{q} \cdot \vec{n}) \, dA$$
$$+ \iiint_{B(t)} (\rho \, \vec{f} \cdot \vec{v} + w) \, dV \quad \text{(B.92)}$$

With

$$V_i \, t_i = V_i \, (\tau_{ij} \, n_j) = (\tau_{ji}^T \, V_i) \, n_j \quad \text{(B.93)}$$

it follows that

$$\iint_{\partial B(t)} \vec{v} \cdot \vec{t} \, dA = \iint_{\partial B(t)} \vec{v} \cdot \hat{T} \, \vec{n} \, dA = \iint_{\partial B(t)} (\hat{T}^T \, \vec{v}) \cdot \vec{n} \, dA$$
$$= \iint_{\partial B(t)} \vec{\nabla} \cdot (\hat{T}^T \, \vec{v}) \, dV \quad \text{(B.94)}$$

and with

$$\iint_{\partial B(t)} \vec{q} \cdot \vec{n} \, dA = \iiint_{B(t)} \vec{\nabla} \cdot \vec{q} \, dV \quad \text{(B.95)}$$

Equation (B.92) yields

$$\iiint_{B(t)} \left(\left[\frac{\partial \rho \varepsilon}{\partial t} + \vec{\nabla} \cdot (\rho \, \varepsilon \, \vec{v}) \right] - \vec{\nabla} \cdot (\hat{T}^T \, \vec{v}) \right.$$
$$\left. - \rho \, \vec{f} \cdot \vec{v} + \vec{\nabla} \cdot \vec{q} - w \right) dV \quad \text{(B.96)}$$

Because B is arbitrary the integrand has to vanish (assuming continuity of the integrand):

$$\frac{\partial \rho \varepsilon}{\partial t} + \vec{\nabla} \cdot (\rho \varepsilon \, \vec{v}) = \vec{\nabla} \cdot (\hat{T}^T \, \vec{v}) + \rho \, \vec{v} \cdot \vec{f} - \vec{\nabla} \cdot \vec{q} + w \quad \text{(B.97)}$$

This is the differential form of the energy conservation of the total energy. In components this reads as

$$\frac{\partial \rho \varepsilon}{\partial t} + \frac{\partial \rho \varepsilon V_i}{\partial x_i} = \frac{\partial}{\partial x_j} (\tau_{ij} \, V_i) + \rho \, f_i \, V_i - \frac{\partial q_i}{\partial x_i} + w \quad \text{(B.98)}$$

In case of a friction-less fluid the first term on the right-hand side becomes using Eq. (B.75):

$$\frac{\partial}{\partial x_j}(\tau_{ij} V_i) = \frac{\partial}{\partial x_j}(-p \delta_{ij} V_i) = -\frac{\partial(p V_i)}{\partial x_i} \tag{B.99}$$

This is the compression work. Equation (B.98) applies to the total energy. Expanding the terms on the left-hand side yields the following:

$$\frac{\partial \rho e}{\partial t} + \frac{\partial \rho e V_i}{\partial x_i} + \rho V_i \frac{\partial V_i}{\partial t} + \frac{1}{2}(V_i V_i)\frac{\partial \rho}{\partial t} + \frac{1}{2}(V_i V_i)\frac{\partial \rho V_j}{\partial x_j} + \frac{1}{2}\rho(V_i V_j)\frac{\partial V_i}{\partial x_j} =$$
$$\frac{\partial}{\partial x_j}(\tau_{ij} V_i) + \rho f_i V_i - \frac{\partial q_i}{\partial x_i} + w \tag{B.100}$$

Double indices are again summed over. The fourth and fifth terms on the left-hand side equal the left-hand side of the mass conservation Eq. (B.52) multiplied by $\frac{1}{2}(V_i V_i)$ and thus are zero. The third and sixth terms equal the left-hand side of the momentum conservation Eq. (B.67) multiplied by V_i. With Eq. (B.67) and

$$\vec{\nabla} \cdot (\hat{T}^T \vec{v}) - \vec{v} \cdot (\vec{\nabla} \cdot \hat{T}) = \frac{\partial}{\partial x_j}(\tau_{ij} V_i) - V_i \frac{\partial \tau_{ij}}{\partial x_j}$$
$$= \tau_{ij}\frac{\partial V_i}{\partial x_j} = \mathrm{Tr}(\hat{T} \vec{\nabla} \vec{v}) \tag{B.101}$$

Equation (B.100) becomes

$$\frac{\partial \rho e}{\partial t} + \frac{\partial \rho e V_i}{\partial x_i} = \tau_{ij}\frac{\partial V_i}{\partial x_j} - \frac{\partial q_i}{\partial x_i} + w \tag{B.102}$$

In short

$$\frac{\partial \rho e}{\partial t} + \vec{\nabla} \cdot (\rho e \vec{v}) = \mathrm{Tr}(\hat{T} \vec{\nabla} \vec{v}) - \vec{\nabla} \cdot \vec{q} + w \tag{B.103}$$

Using the stress tensor of NEWTONian fluids Eqs. (B.75) and (B.84) the first term on the right-hand side becomes

$$\tau_{ij}\frac{\partial V_i}{\partial x_j} = -p \delta_{ij}\frac{\partial V_j}{\partial x_i} + 2\eta d_{ij}\frac{\partial V_i}{\partial x_j} + \lambda \frac{\partial V_j}{\partial x_j}\delta_{ij} \tag{B.104}$$

and with this

$$\tau_{ij}\frac{\partial V_i}{\partial x_j} = -(p - \lambda)\frac{\partial V_j}{\partial x_j} + \eta\left(\frac{\partial V_i}{\partial x_j} + \frac{\partial V_j}{\partial x_i}\right)\frac{\partial V_i}{\partial x_j} \tag{B.105}$$

Appendix B

The internal energy is a function of the temperature and the volume so that the differential of the internal energy can be written as

$$de = \left(\frac{\partial e}{\partial T}\right)_v dT + \left(\frac{\partial e}{\partial v}\right)_T dv \qquad (B.106)$$

with

$v = V/\rho$ – specific volume.

The expression in parentheses in the first term on the right-hand side is the specific heat capacity at constant volume:

$$\left(\frac{\partial e}{\partial T}\right)_v = c_v \qquad (B.107)$$

The second term in Eq. (B.106) describes the dependency of the internal energy on the volume. In case of ideal gases this term vanishes, in case of fluids and solids, respectively, this term can be neglected in comparison with the first one. The energy thus is

$$e = \int_{T_0}^{T} c_v(T') \, dT' + e_0 \simeq c_v (T - T_0) + e_0 \qquad (B.108)$$

If the second term on the right-hand side can be neglected the energy conservation reads as follows:

$$\frac{\partial \rho \, c_v \, T}{\partial t} + \vec{\nabla} \cdot (\rho \, c_v \, T \vec{v}) = -\vec{\nabla} \cdot \vec{q} + \tau_{ij} \frac{\partial v_i}{\partial x_j} + w(\vec{r}) \qquad (B.109)$$

In case of fluids and solids c_v and c_p, respectively, only differ marginally, so a unique value can be used. In case of constant velocity the second term on the right-hand side of Eq. (B.109) vanishes. But also in case of the velocity not being constant this term can be neglected when dealing with incompressible fluids. The heat flux is according to FOURIER's first law proportional to the negative gradient of the temperature:

$$\vec{q} = -K \, \vec{\nabla} \, T \qquad (B.110)$$

Equation (B.110) expresses that heat flows from warm to cold regions. The proportionality constant K is the heat conductivity.

B.9 Compilation of Mathematical Formulas Used in Energy Transport Computations

In Chap. 4.2 the method of GREEN's functions is employed to compute temperature distributions. The method of GREEN's functions makes necessary to solve integrals. The GREEN's function Eq. (4.13) reads as

$$G(\vec{r}, t|\vec{r}', t') = \frac{1}{[4\pi \kappa (t-t')]^{3/2}} \exp\left(-\frac{[\vec{r} - (\vec{r}' + \vec{v}(t-t'))]^2}{4\kappa(t-t')}\right) \quad \text{(B.111)}$$

or with $\vec{v} = v\,\vec{e}_x$

$$G = \frac{a^{3/2}}{\pi^{3/2}} \exp\left(-a(x - [x' + v(t-t')])^2\right) \exp\left(-a(y-y')^2\right) \exp\left(-a(z-z')^2\right)$$

$$a = \frac{1}{4\kappa(t-t')} \quad \text{(B.112)}$$

B.9.1 Integration Over Space

With a point source

$$w = \frac{2 P_L}{\rho c} \Theta(t')\,\delta(x')\,\delta(y')\,\delta(z') \quad \text{(B.113)}$$

the spatial integration in Eq. (4.14) over the space yields

$$T = \frac{2 P_L}{\rho c} \int_0^t \frac{a^{3/2}}{\pi^{3/2}} \exp\left[-a\left((x - v(t-t'))^2 + y^2 + z^2\right)\right] dt' \quad \text{(B.114)}$$

With a line source

$$w = \frac{w'}{\rho c} \Theta(t')\,\delta(x')\,\delta(y') \quad \text{(B.115)}$$

with

w' – absorbed power per length

and

Appendix B

$$\int_{-\infty}^{\infty} \exp(-a\,\xi^2)\,d\xi = \sqrt{\frac{\pi}{a}}. \tag{B.116}$$

it follows that

$$T = \frac{w'}{\rho\,c} \int_0^t \frac{a}{\pi} \exp\left(-a\,(x - v\,(t - t')])^2 - a y^2\right) dt' \tag{B.117}$$

And in case of a surface source

$$w = 2\,I_L(t)\,\delta(z) \tag{B.118}$$

and velocity $\vec{v} = v \vec{e}_z$

$$T = \frac{2\,I_L}{\rho\,c} \int_0^t \sqrt{\frac{a}{\pi}} \exp\left(-a\,(z - v\,(t - t'))^2\right) dt' \tag{B.119}$$

A GAUSSian source

$$w(x', y', z', t') = \frac{2\,P_L}{\rho\,c} \frac{2}{\pi\,w_0^2} \exp\left(-\frac{2\,(x'^2 + y'^2)}{w_0^2}\right) \delta(z')\,\Theta(t') \tag{B.120}$$

$$\int_{-\infty}^{\infty}\int_{-\infty}^{\infty} \frac{2\,P_L}{\rho\,c} \frac{2}{\pi\,w_0^2} \exp\left(-\frac{2\,(x'^2 + y'^2)}{w_0^2}\right) dx'\,dy' = \frac{2\,P_L}{\rho\,c}$$

leads while integrating over y' to

$$\int_{-\infty}^{\infty} \exp\left[-b\,y'^2 - a\,(y - y')^2\right] dy' =$$

$$\int_{-\infty}^{\infty} \exp\left[-(a+b)(y'^2 - 2 y' \frac{a}{a+b} y) - a y^2\right] dy' \tag{B.121}$$

$$a = \frac{1}{4\,\kappa\,(t - t')}$$

$$b = \frac{2}{w_0^2}$$

Quadratic supplement together with Eq. (B.116) yields:

$$\int_{-\infty}^{\infty} \exp\left(-(a+b)(y'^2 - 2y'\frac{a}{a+b}y + \left[\frac{a}{a+b}\right]^2 y^2) + \left[\frac{a}{a+b}\right]^2 y^2 - ay^2\right) dy' =$$

$$\int_{-\infty}^{\infty} \exp\left(-(a+b)(y' - \frac{a}{a+b}y)^2\right) \exp\left(-\frac{y^2}{\frac{1}{a} + \frac{1}{b}}\right) dy' =$$

$$\sqrt{\frac{\pi}{\frac{1}{4\kappa(t-t')} + \frac{2}{w_0^2}}} \exp\left(-\frac{y^2}{4\kappa(t-t') + w_0^2/2}\right) \quad \text{(B.122)}$$

Together with the integration over x' (quadratic supplement like above) and z' (simple integration over a DIRAC-δ function) it follows:

$$\frac{P_L}{\rho c} \exp\left(-\frac{(x - v(t-t'))^2}{4\kappa(t-t') + w_0^2/2}\right) \exp\left(-\frac{y^2}{4\kappa(t-t') + w_0^2/2}\right) \exp\left(-\frac{z^2}{4\kappa(t-t')}\right) \quad \text{(B.123)}$$

With this the temperature is

$$T(x, y, z, t) - T_\infty = \int_0^t \frac{2P_L}{\pi \rho c} \frac{1}{\sqrt{4\pi\kappa(t-t')}} \frac{1}{4\kappa(t-t') + w_0^2/2} \quad \text{(B.124)}$$

$$\cdot \exp\left(-\frac{(x - v(t-t'))^2 + y^2}{4\kappa(t-t') + w_0^2/2}\right) \exp\left(-\frac{z^2}{4\kappa(t-t')}\right) dt' \quad \text{(B.125)}$$

B.9.2 Integration Over Time

With $v = 0$ the integrals Eqs. (B.114), (B.117), and (B.119) have the form

$$\int_0^t \frac{1}{[4\pi\kappa(t-t')]^{n/2}} \exp\left(-\frac{r^2}{4\pi\kappa(t-t')}\right) dt' \quad \text{(B.126)}$$

with: $n = 1, 2, 3$
With the substitution

$$\frac{r^2}{4\pi\kappa(t-t')} = \xi^2 \quad \text{(B.127)}$$

it follows that

Appendix B

$$\frac{1}{\pi^{n/2}} \frac{1}{2\kappa} r^{(2-n)} \int_{\frac{r}{\sqrt{4\kappa t}}}^{\infty} \xi^{(n-3)} \exp(-\xi^2) \, d\xi \tag{B.128}$$

Case $n = 3$:
  With

$$\int_a^\infty \exp(-\xi^2) \, d\xi = \frac{\sqrt{\pi}}{2} \mathrm{erfc}(a) \tag{B.129}$$

it follows[2] that

$$\frac{1}{4\pi\kappa} \frac{1}{r} \mathrm{erfc}\left(\frac{r}{\sqrt{4\kappa t}}\right) \tag{B.130}$$

Case $n = 2$:
  With Eq. (B.128) it follows that

$$\frac{1}{\pi} \frac{1}{2\kappa} \int_{\frac{r}{\sqrt{4\kappa t}}}^{\infty} \exp(-\xi^2)/\xi \, d\xi \tag{B.131}$$

With

$$\int_a^\infty \exp(-\xi^2)/\xi \, d\xi = \frac{1}{2} \int_{a^2}^\infty \exp(-\xi)/\xi \, d\xi = \frac{1}{2} E_1(a^2) \tag{B.132}$$

one gets[3]

$$\frac{1}{4\pi\kappa} E_1\left(\frac{r^2}{4\pi\kappa}\right) \tag{B.133}$$

Case $n = 1$:
  With Eq. (B.128) it follows that

---

[2] For erfc see Sect. B.9.3, p. 570.
[3] For E_1 see Appendix B.9.4, p. 571.

$$\frac{1}{\pi^{1/2}} \frac{1}{2\kappa} r \int_{\frac{r}{\sqrt{4\kappa t}}}^{\infty} \exp(-\xi^2)/\xi^2 \, d\xi \qquad (B.134)$$

With

$$\int_{a}^{\infty} \exp(-\xi^2)/\xi^2 \, d\xi = \frac{1}{a} \exp(-a) - \sqrt{\pi} \, \text{erfc}(a) = \frac{a}{\sqrt{\pi}} \, \text{ierfc}(a) \qquad (B.135)$$

it follows [4] that

$$\sqrt{\frac{t}{\kappa}} \, \text{ierfc}\left(\frac{r}{\sqrt{4\kappa t}}\right) \qquad (B.136)$$

The solution of the integral in Eq. (B.125) with $v = 0$ and $x = 0, y = 0, z = 0$ yields

$$T(0,0,0,t) = \frac{2 P_L}{\rho c} \frac{1}{\sqrt{2\kappa} \, \pi^{3/2} \, w_0} \arctan\left(\sqrt{\frac{8\kappa t}{w_0^2}}\right) \qquad (B.137)$$

B.9.3 Error Functions

$$\text{erf}(x) = \frac{2}{\sqrt{\pi}} \int_{0}^{x} \exp(-\xi^2) \, d\xi \qquad (B.138)$$

$$\text{erfc}(x) = \frac{2}{\sqrt{\pi}} \int_{x}^{\infty} \exp(-\xi^2) \, d\xi \qquad (B.139)$$

$$\text{ierfc}(x) = \frac{1}{\sqrt{\pi}} \exp(-x^2) - x \, \text{erfc}(x) \qquad (B.140)$$

$$\text{erfce}(x) = \exp(x^2) \, \text{erfc}(x) \qquad (B.141)$$

---

[4] For ierfc see Appendix B.9.3, p. 570.

Appendix B

B.9.4 Exponential Integral

$$E_1(x) = \int_x^\infty \frac{e^{-\xi}}{\xi}\, d\xi \tag{B.142}$$

B.10 Diffusion in Metals

Equation (6.4) reads

$$\frac{\partial c}{\partial t} = D \frac{\partial^2 c}{\partial z^2} \tag{B.143}$$

A new dimensionless variable is introduced

$$\eta = \frac{z - z_0}{\sqrt{4Dt}} \tag{B.144}$$

With

$$\frac{\partial c}{\partial t} = \frac{\partial c}{\partial \eta} \frac{\partial \eta}{\partial t} \tag{B.145}$$

$$= -2D \frac{\eta}{4Dt} \frac{\partial c}{\partial \eta}$$

and

$$\frac{\partial c}{\partial z} = \frac{\partial c}{\partial \eta} \frac{\partial \eta}{\partial z}$$

$$\frac{\partial^2 c}{\partial z^2} = \frac{\partial}{\partial z}\left(\frac{\partial c}{\partial \eta} \frac{\partial \eta}{\partial z}\right)$$

$$= \left(\frac{\partial}{\partial z} \frac{\partial c}{\partial \eta}\right)\frac{\partial \eta}{\partial z} + \frac{\partial c}{\partial \eta} \frac{\partial^2 \eta}{\partial z^2}$$

$$= \left(\frac{\partial}{\partial z} \frac{\partial c}{\partial \eta}\right)\left(\frac{\partial \eta}{\partial z}\right)^2$$

$$= \frac{\partial^2 c}{\partial \eta^2} 4Dt \tag{B.147}$$

Equation (B.143) becomes

$$\frac{d^2 c}{d\eta^2} + 2\eta \frac{dc}{d\eta} = 0 \tag{B.148}$$

With

$$a = \frac{dc}{d\eta} \tag{B.149}$$

it follows that

$$\frac{da}{d\eta} + 2\eta a = 0 \tag{B.150}$$

which has the solution

$$a = \exp(-\eta^2) \tag{B.151}$$

Inserting into Eq. (B.149) and integrating over the concentration results in

$$c(\eta) = \tilde{C}_1 \int_0^\eta \exp\left(-\eta'^2\right) d\eta' + \tilde{C}_0 \tag{B.152}$$

or

$$c(z, t) = C_1 \operatorname{erf}\left(\frac{z - z_0}{\sqrt{4Dt}}\right) + C_0 \tag{B.153}$$

The solution complying to the initial conditions

$$c(z < -z_{\text{carbide}}) = c_{\text{ferrite}}$$
$$c(-z_{\text{carbide}} > z < z_{\text{Karbid}}) = c_{\text{carbide}}$$
$$c(z_{\text{carbide}} > z) = c_{\text{ferrite}}$$

can be constructed by the superposition of two solutions of the form Eq. (B.153)

$$c_1(z, t) = C_{11} \operatorname{erf}\left(\frac{z - z_{\text{Karbid}}}{\sqrt{4Dt}}\right) + C_{01} \tag{B.154}$$

$$c_2(z, t) = C_{12} \operatorname{erf}\left(\frac{z + z_{\text{Karbid}}}{\sqrt{4Dt}}\right) + C_{02} \tag{B.155}$$

with the initial conditions

$$c(z < -z_{\text{carbide}}) = c_{\text{ferrite}} - \frac{1}{2} c_{\text{Karbid}}$$
$$c(z > -z_{\text{carbide}}) = \frac{1}{2} c_{\text{carbide}}$$

and

$$c(z < z_{\text{carbide}}) = \frac{1}{2} c_{\text{carbide}}$$

$$c(z > z_{\text{carbide}}) = c_{\text{ferrite}} - \frac{1}{2} c_{\text{carbide}}$$

(see Fig. B.4).

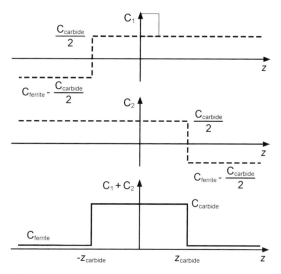

Fig. B.4 C concentration

Reference

1. E. Becker and W. Bürger, Kontinuumsmechanik, Teubner, Stuttgart, 1975

Appendix C
Laser-Induced Vaporization

C.1 Equation of Clausius–Clapeyron

The free enthalpy of a thermodynamic equilibrium system with pressure and temperature kept constant is minimal [2]:

$$G = E - TS + pV = \text{minimum} \tag{C.1}$$

This implies that if the system consists of two phases the specific free enthalpies of both phases are equal:

$$g_1 = g_2 \tag{C.2}$$
$$g = e - Ts + pv \tag{C.3}$$

with

e – specific internal energy
s – specific entropy
v – specific volume.

Because in case of phase equilibrium this applies for any temperature and the corresponding pressure it follows that

$$dg_1 = dg_2 \tag{C.4}$$

With Eq. (C.3) and the fundamental equation of thermostatics

$$de = T\,ds - p\,dv \tag{C.5}$$

it follows that

$$dg = -s\,dT + v\,dp \tag{C.6}$$

and with Eq. (C.4)

$$(s_2 - s_1) \, dT = (v_2 - v_1) \, dp \tag{C.7}$$

This yields the equation of CLAUSIUS–CLAPEYRON:

$$\frac{dp}{dT} = \frac{\Delta s}{\Delta v} \tag{C.8}$$

Setting

$$\Delta s = \frac{H_V}{T} \tag{C.9}$$

with

H_V – evaporation enthalpy.

Equation (8.4) follows.

C.2 Temperature Dependence of the Evaporation Enthalpy

The assumption that the evaporation enthalpy does not depend on the temperature is only met approximately. The total energy that is needed to evaporate melt or solid, respectively, at $T = 0\,\text{K}$ and subsequently heat the vapor up to the temperature T while keeping the pressure constant equals the energy that is needed to first heat the melt or solid up to the temperature T and then evaporate the material [1]:

$$H(0) + \int_0^T c_p^{(D)}(T) \, dT = \int_0^T c_p^{(C)}(T) \, dT + H(T) \tag{C.10}$$

with

$c_p^{(V)}$ – heat capacity of the vapor for constant pressure
$c_p^{(C)}$ – heat capacity of the condensated material (melt or solid) for constant pressure.

In case of an ideal gas $c_p^{(D)} = 5/2\,R$. With this the vaporization enthalpy is given by

$$H(T) = H(0) + \frac{5}{2} R T - \int_0^T c_p^{(C)}(T) \, dT \tag{C.11}$$

With $c_p = 5/2\,R$ it follows $H(T) = H(0)$ and the saturation pressure is given by (see Eq. (8.7))

$$p_{SV} = p_{SV,\max} \exp\left(-\frac{H_V}{RT}\right) \tag{C.12}$$

In a solid $c_p^{(C)} = 6/2\,R$ approximately holds.[1] Inserting this into Eq. (8.6) and integrating yields

$$p_{SV} \propto \frac{1}{\sqrt{T}} \exp\left(-\frac{H_V(0)}{RT}\right) \tag{C.13}$$

In case of thermal electron emission $c_p^{(C)}$ is the heat capacity of the conduction band electrons of the metal which has a very small value [2]. With $c_p^{(C)} = 0$ and the assumption that the electrons outside the metal behave like an ideal gas Eq. (8.6) yields

$$p_{SV} \propto T^{5/2} \exp\left(-\frac{H_V(0)}{RT}\right) \tag{C.14}$$

The saturation electron density is given by

$$n_{SV} \propto T^{3/2} \exp\left(-\frac{H_V(0)}{RT}\right) \tag{C.15}$$

The mean velocity of the electrons leaving the metal is proportional to $\sqrt{T}$. With the particle current density

$$j = n_{SV}\, v \tag{C.16}$$

this yields the RICHARDSON equation [2]:

$$j \propto T^2 \exp\left(-\frac{H_V(0)}{RT}\right) \tag{C.17}$$

C.3 Velocity Moments

The mean value of a velocity-dependent quantity is computed by integrating the product of this quantity and the velocity distribution function over the velocity space. The density, e.g., follows by integrating the velocity distribution function.

---

[1] Three degrees of freedom of the kinetic and three degrees of freedom of the potential energy, respectively.

The momentum density in x-direction is determined by integrating the product of $m\, v_x$ and the velocity distribution function. Velocity moments are build generally according to

$$p_{ikl}(f) = \int_{-\infty}^{\infty} \int_{-\infty}^{\infty} \int_{-\infty}^{\infty} v_x^i\, v_y^k\, v_z^l\, f(v_x, v_y, v_z)\, dv_x\, dv_y\, dv_z \tag{C.18}$$

In case of a MAXWELLian distribution, the particle density, momentum density, and the energy density are given by

$$p_{000}(f(\vec{v})) = n \tag{C.19}$$

$$p_{100}(f(\vec{v})) = m\, n\, v \tag{C.20}$$

$$\frac{m}{2}[p_{200}(f(\vec{v})) + p_{020}(f(\vec{v})) + p_{002}(f(\vec{v}))] = \frac{3}{2} m\, n\, R\, T + \frac{m\, n}{2} v \tag{C.21}$$

The conservation of density, momentum, and energy flux, respectively, while crossing the KNUDSEN layer leads to

$$p_{100}(f(\vec{v})) = n_2\, v_2 \tag{C.22}$$

$$p_{200}(f(\vec{v})) = n_2\, v_2^2 \tag{C.23}$$

$$p_{300}(f(0,\vec{v})) + p_{120}(f(0,\vec{v})) + p_{102}(f(0,\vec{v})) = n_2\, v_2\, (\frac{5}{2} R\, T_2^2 + v_2^2) \tag{C.24}$$

The left-hand sides result from determining the appropriate velocity moments[2] using the MAXWELLian distribution Eq. (8.62), whereas the moments on the right-hand side are build using the non-MAXWELLian distribution Eq. (8.63). The results can be found in [3].

References

1. G. Adam and O. Hittmair, Wärmetheorie, Vieweg, 1978
2. R. Becker, Theorie der Wärme, Springer-Verlag, 1975
3. T. Ytrehus, J. L. Potter Eds., Theory and Experiments on Gas Kinetics in Evaporisation, Int. Symp. Rarefied Gas Dynamics 1976, 10 Snowmass-at-Aspen Col., 1977

---

[2] Particle, momentum and energy density, respectively, have to be multiplied by v_x for determining the respective fluxes.

Appendix D
Plasma Physics

D.1 Some Results of Thermodynamics

The entropy is a function of the energy, the volume, and the particle numbers:

$$S = S(E, V, N_1, ... N_m) \tag{D.1}$$

The total differential of the entropy is

$$dS = \left(\frac{\partial S}{\partial E}\right)_{V,N} dE + \left(\frac{\partial S}{\partial V}\right)_{E,N} dV + \sum_{i=1}^{m} \left(\frac{\partial S}{\partial N_i}\right)_{E,V,N} dN_i \tag{D.2}$$

In case of constant particle numbers the differential of the entropy is given by the first law of thermodynamics [1]:

$$dS = \frac{\delta Q}{T} = \frac{dE + p\,dV}{T} \tag{D.3}$$

Comparison with Eq. (D.2) yields

$$\left(\frac{\partial S}{\partial E}\right)_{V,N} = \frac{1}{T} \tag{D.4}$$

$$\left(\frac{\partial S}{\partial V}\right)_{E,N} = \frac{p}{T} \tag{D.5}$$

The chemical potential is defined as

$$\mu_j = -T \left(\frac{\partial S}{\partial N_j}\right)_{E,V,N} \tag{D.6}$$

With this Eq. (D.2) becomes

$$dS = \frac{dE}{T} + \frac{p\,dV}{T} - \sum_{i=1}^{m} \frac{\mu_i}{T}\,dN_i \tag{D.7}$$

Multiplication with T and rearranging leads to

$$dE = T\,dS - p\,dV + \sum_{i=1}^{m} \mu_i\,dN_i \tag{D.8}$$

When the energy E is given as a function of S, V, and N_i Eq. (D.8) becomes

$$\left(\frac{\partial E}{\partial S}\right)_{V,N} = T \tag{D.9}$$

$$\left(\frac{\partial E}{\partial V}\right)_{S,N} = -p \tag{D.10}$$

$$\left(\frac{\partial E}{\partial N_j}\right)_{S,V,N} = \mu_j \tag{D.11}$$

Taking the derivative of Eq. (D.9) with respect to the volume V and of Eq. (D.10) with respect to the entropy S and equating the respective results yields because the order of the partial derivatives does not matter:

$$\left(\frac{\partial T}{\partial V}\right)_{S,N} = \left(\frac{\partial p}{\partial S}\right)_{V,N} \tag{D.12}$$

The free energy is defines as

$$F = E - T\,S \tag{D.13}$$

Thus

$$d(E - T\,S) = dF = -S\,dT - p\,dV + \sum_{i=1}^{m} \mu_i\,dN_i \tag{D.14}$$

Comparison of coefficients results in

$$\left(\frac{\partial F}{\partial T}\right)_{V,N} = -S \tag{D.15}$$

$$\left(\frac{\partial F}{\partial V}\right)_{T,N} = -p \tag{D.16}$$

$$\left(\frac{\partial F}{\partial N_j}\right)_{T,V,N} = \mu_j \tag{D.17}$$

Appendix D

From the first two equations it follows

$$\left(\frac{\partial S}{\partial V}\right)_{T,N} = \left(\frac{\partial p}{\partial T}\right)_{V,N} \quad (D.18)$$

The enthalpy is defined as

$$H = E + pV \quad (D.19)$$

Thus

$$dH = T\,dS + V\,dp + \sum_{i=1}^{m} \mu_i\,dN_i \quad (D.20)$$

with

$$\left(\frac{\partial H}{\partial S}\right)_{p,N} = T \quad (D.21)$$

$$\left(\frac{\partial H}{\partial p}\right)_{S,N} = V \quad (D.22)$$

$$\left(\frac{\partial H}{\partial N_j}\right)_{S,p,N} = \mu_j \quad (D.23)$$

$$\left(\frac{\partial T}{\partial p}\right)_{S,N} = -\left(\frac{\partial V}{\partial S}\right)_{p,N}. \quad (D.24)$$

The free enthalpy is given by

$$G = F + pV = E - TS + pV \quad (D.25)$$

Thus

$$d(E - TS + pV) = dG = -S\,dT + V\,dp + \sum_{i=1}^{m} \mu_i\,dN_i \quad (D.26)$$

with

$$\left(\frac{\partial G}{\partial T}\right)_{p,N} = -S \quad (D.27)$$

$$\left(\frac{\partial G}{\partial p}\right)_{S,N} = V \quad (D.28)$$

$$\left(\frac{\partial G}{\partial N_j}\right)_{T,p,N} = \mu_j \quad (D.29)$$

$$\left(\frac{\partial S}{\partial p}\right)_{T,N} = -\left(\frac{\partial V}{\partial T}\right)_{p,N}. \tag{D.30}$$

The equations (D.12), (D.18), (D.24), and (D.30) are the MAXWELL relations.

D.2 Generalization in Case of Multiply Ionized Ions

In the following some equations are generalized in the case of plasmas consisting of electrons and single as well as multiple ionized ions. The DEBYE radius is

$$r_D = \sqrt{\frac{\varepsilon_0 \, k_B \, T}{e^2 \, (n_e + \sum_i Z_i^2 \, n_i)}} \tag{D.31}$$

with

Z_i – ion charge number
n_i – ion density.

The COULOMB correction of the internal energy is

$$E_c = -\frac{1}{2} V \frac{e^2}{4 \pi \, \varepsilon_0 \, r_D} \left(n_e + \sum_i Z_i^2 \, n_i\right) \tag{D.32}$$

The COULOMB correction of the free energy is given by

$$F_c = -\frac{1}{3} V \frac{e^2}{4 \pi \, \varepsilon_0 \, r_D} \left(n_e + \sum_i Z_i^2 \, n_i\right) \tag{D.33}$$

In a state of thermodynamic equilibrium with constant temperature and pressure the free energy is minimal:

$$dF(V, T, n_e, n_0, n_i, n_{i+1}...) = 0 \tag{D.34}$$

It is assumed that the densities of all particle species are constant except the density of electrons, i-fold ionized ions, and $(i+1)$-fold ionized ions. Particle density variations are not independent of each other. When an electron and a $(i+1)$-fold ionized ion are created a i-fold ionized ion is annihilated. Thus

$$dn_e = dn_{i+1} = -dn_i \tag{D.35}$$

With this it follows

$$\frac{\partial F}{\partial n_e} - \frac{\partial F}{\partial n_i} + \frac{\partial F}{\partial n_{i+1}} = \mu_e - \mu_i + \mu_{i+1} = 0 \quad \text{(D.36)}$$

with

μ – chemical potential.

The reduction of the ionization energy of the i-fold ionized ions is

$$\Delta E_i = -\left(\frac{\partial F_c}{\partial n_e} - \frac{\partial F_c}{\partial n_i} + \frac{\partial F_c}{\partial n_{i+1}}\right) = -(\mu_{c,e} - \mu_{c,i} + \mu_{c,i+1}) = (Z_i + 1)\frac{e^2}{4\pi\varepsilon_0 r_D} \quad \text{(D.37)}$$

With Eq. (D.36) the SAHA equation in the general case of multiple ionized ions follows:

$$\frac{n_e \, n_{i+1}}{n_i} = \frac{2}{\Lambda_e^3}\frac{U'_{i+1}(T)}{U_i(T)}\exp\left(-\frac{E_i - \Delta E_i}{k_B T}\right) \quad \text{(D.38)}$$

with

U'_{i+1} – partition function of the $i+1$-fold ionized ion
U_i – partition function of the i-fold ionized ion.

The energy zero point in computing the partition function is the ground state of the respective ion. The energy E_i is the ionization energy of the i-fold ionized ion. Λ_e is the thermal DE'BROGLIE wavelength of the electrons Eq. (9.74).

Reference

1. F. Reif, Fundamentals of Statistical and Thermal Physics, McGraw Hill, 1965

Appendix E
Glossary of Symbols and Constants

The developments in the field of laser technology are highly characterized by their interdisciplinary character. In describing the basic phenomena descriptions and nomenclatures are taken from a great array of physical and engineering disciplines. Due to this a confusing diversity of notations and descriptions has evolved. The composition presented in this book follow as much as possible the norms " ... Normen über einheitliche Begriffsbestimmungen, Benennungen und Formelzeichen für physikalische Größen, über Einheiten und Einheitenzeichen sowie über mathematische Zeichen und Begriffe" that were elaborated by the *Normenausschuß Einheiten und Formelgrößen* [2].

The units used in this book are either SI units or in exceptional cases other commonly used units. The SI units ("Système International d'Unités", international system of units) can be divided into basic units (Table E.1) and derived units (Table E.2), respectively,

Table E.1 SI – basic units [2][S. 1]

Quantity	SI basic unit	
	Name	Symbol
Length	meter	m
Mass	kilogram	kg
Time	second	s
Electric current	Ampere	A
Thermodynamic temperature	Kelvin	K
Amount of substance	mol	mol
Illumination	Candela	cd

Table E.2 Derived SI units with their special names and symbols [2][S. 2]

Quantity	SI unit		Relation
	Name	Symbol	
Plane angle	radiant	rad	1 rad = 1 m/m
Solid angle	steradiant	sr	1 sr = 1 m^2/m^2
Frequency of a periodic process	Hertz	Hz	1 Hz = 1 1/s
Force	Newton	N	1 N = 1 kg · m/s^2
Pressure, mechanical stress, Pascal	Pa	Pa	1 Pa = 1 kg/m · s
Energy, work, heat	Joule	J	1 J = 1 kg · m^2/s^2
Power, heat flux	Watt	W	1 W = 1 kg · m^2/s^3
Electric charge	Coulomb	C	1 C = 1 A · s
Electric voltage	Volt	V	1 V = 1 $\frac{J}{C}$ = 1 kg · m^2/s^3 · A
Electric capacity	Farad	F	1 F = 1 $\frac{C}{V}$ = 1 s^4 · A^2/kg · m^2
Electric resistance	Ohm	Ω	1 Ω = 1 $\frac{V}{A}$ = 1 kg · m^2/s^3 · A^2
Magnetic flux	Weber	Wb	1 Wb = 1 V · s = 1 kg · m^2/s^2 · A
Magnetic induction	Tesla	T	1 T = 1 Wb/m^2 = 1 kg/s^2 · A
Inductivity	Henry	H	1 H = 1 Wb/A = 1 kg · m^2/s^2 · A^2

Table E.3 Units not belonging to SI [2][S. 3]

Quantity	SI unit		Definition
	Name	Symbol	
Pressure	Bar	bar	1 bar = 10^5 Pa
Energy in atom physics	electron volt	eV	1 electron volt is the energy that an electron gains within a potential difference of 1 V in vacuum: 1 eV = 1.6021892 · 10^{-19} J

E.1 Used Symbols

Symbol	Meaning	Einheit
a	Cylinder radius	m
a	Index: atom	–
A	Area	m^2
A	Absorptivity	%
A_T	Expansion factor	–
A_c	Phase change temperature	°C
b	Mode order	–
b	Impact parameter	m
B	Magnetic field	V s/m^2
$B(t)$	Volume of a material body	m^3
b_c	Cutting groove width	m
c	Velocity of sound	m/s
c	Specific heat capacity	J/g K
c_p	Heat capacity for constant pressure	J/g K
c_v	Heat capacity for constant volume	J/g K

Appendix E

Symbol	Meaning	Einheit
D	Diffusion coefficient	cm^2/s
D	Dielectric displacement	A s/m^2
d	Work piece thickness	m
d_m	Melt film thickness	m
e	Elementary charge	A s
e	Specific internal energy	J/m^3
e	Index: electron	
E	Electric field	V/m
E	Internal energy	J
E_{ion}	Ionization energy	J, eV
E_r	Energy of state r	J, eV
f	Focal length	m
$\vec{f}$	Volume force	N/m^3
f_1	Brennweite der ersten Teleskopoptik	m
f_2	Brennweite der zweiten Teleskopoptik	m
f_{ij}	Oszillatorenstärken	
f_{HL}	Wärmeverlustfaktor	–
F	Freie Energie	J
F_p	Druckkraft	N
$F_{r,G}$	Reibkraft des Gases	N
$F_{r,W}$	Reibkraft der Schmelze	N
G	Enthalpie	J
G	Index: Gas	
h	Rauhtiefe einer Oberfläche	m
H	Magnetische Induktion	A/m
h_M, H_m	Spezifische Schmelzwärme	J/g
h_V, H_V	Spezifische Verdampfungswärme	J/g
$\triangle_R H$	Reaktionsenthalpie	kJ/g
i	Index: Ion	
I	Intensität	W/m^2
I_0	Maximale Intensität	W/m^2
I_P	Prozeßintensität	W/m^2
I_c	Schwellintensität zur Plasmabildung	W/m^2
I_s	Schwellintensität zur Plasmaabschirmung	W/m^2
j	Stromdichte	A/m^2
k	Wellenzahl	1/m
k_0	Vakuumwellenzahl	1/m
k_i	Imaginärteil der Wellenzahl	1/m
K	Wärmeleitfähigkeit	W/m K
K_e	Elektronenwärmeleitfähigkeit	W/m K
K	Normierte Strahlqualitätszahl	–
l	LANDAU-Länge	m
$\dot{m}_{\text{ein}}$	Eintretender Materialmassenstrom	g/s
$\dot{m}_{\text{aus}}$	Austretender Materialmassenstrom	g/s
$\dot{m}_c$	Fugenmaterialmassenstrom	g/s
$\dot{m}_V$	Dampfmassenstrom	g/s

Symbol	Meaning	Einheit
M	Strahltransfermatrix	–
m	Masse	kg
m_e	Elektronenmasse	kg
M	Masse schwerer Teilchen (Atome, Moleküle)	kg
M	Dipolmoment	m^4/s
M_S	Martenstit-Start-Temperatur	°C
n	Brechungsindex	–
n	Teilchendichte	$1/m^3$
n	Realteil des komplexen Brechungsindex	–
n_e	Elektronendichte	$1/m^3$
n_i	Ionendichte	$1/m^3$
n_a	Atomdichte	$1/m^3$
n_c	Komplexer Brechungsindex	–
n_{SD}	Sättigungsdampfdichte	$1/m^3$
$\vec{n}$	Normaleneinheitsvektor	–
N	Teilchenzahl	
N_D	Anzahl der Elektronen in einer DEBYE-Kugel	
p	Dipolmoment	A s m
p	Statischer Druck	N/m^2
p_K	Kapillardruck	N/m^2
p_{SD}	Sättigungsdampfdruck	N/m^2
P	Polarisation	$A\,s/m^2$
P_r	Wahrscheinlichkeit eines Zustandes	–
p_{Fl}	Gemittelter statischer Druck der Schmelze	N/m^2
p_u	Umgebungsdruck	N/m^2
p_0	Schneidgasdruck	N/m^2
$\widehat{p}$	Ruhedruck, Staudruck	N/m^2
$\overline{p}$	Gemittelter statischer Druck in der Fuge	N/m^2
P_c	Schneidleistung	W
P_{HL}	Wärmeverlustleistung durch Wärmeleitung	W
P_c	Kapillarleistung	W
P_L	Laserleistung	W
$P_{L,\text{ein}}$	Eingestrahlte Laserleistung	W
P_R	Reaktionsleistung	W
P_V	Verlustleistung	W
P_c	Schweißleistung	W
q	Komplexer Strahlparameter	–
Q	Quellstärke einer Quellströmung	m^3/s
$\dot{q}$	Flächenbezogener Wärmestrom	W/m^2
r	Amplitudenreflexionsfaktor	–
r	Radius	m
r_D	DEBYE-Radius	m
r_{DR}	Bohrlochradius	m
r_s	Effektive Abschirmlänge	m
R	Reflexionsgrad	%
s	Bearbeitungstiefe	m
s	Spezifische Entropie	$\frac{J}{g}$
s_c	Schnittfugentiefe	m

Appendix E

Symbol	Meaning	Einheit
s_{DR}	Bohrlochtiefe	m
s_w	Schweißnahttiefe	m
$\vec{S}$	POYNTING-Vektor	W/m²
S	Entropie	J
S	Index: Schmelze	–
t	Zeit	s
t	Amplitudentransmissionsfaktor	–
$\vec{t}$	Tangentialeinheitsvektor	–
$\vec{t}$	Oberflächenkräfte	N/m²
t_h	Haltezeit	s
t_k	Abkühldauer	s
t_L	Pulsdauer	s
$\hat{T}$	Spannungstensor	N/m²
T	Temperatur	K
T_H	Härttemperatur	°C
T_M	Schmelztemperatur	K
T_P	Prozeßtemperatur	K
T_V	Verdampfungstemperatur	K
T_0	Ruhetemperatur des Gases	K
T_∞	Umgebungstemperatur	K
T_z	Zündtemperatur	K
u	Wechselwirkungsenergie	J, eV
U	Zustandssumme der inneren Freiheitsgrade	–
V	Index: vapor	–
v	Geschwindigkeit	m/s
v	Spezifisches Volumen	m³/kg
v_c	Schneidgeschwindigkeit	m/s
v_{DR}	Bohrgeschwindigkeit	m/s
v_F	FERMI-Geschwindigkeit	m/s
v_{TH}	Mittlere thermische Geschwindigkeit	m/s
v_G	Gasgeschwindigkeit	m/s
v_H	Vorschubgeschwindigkeit beim Härten	m/s
v_{krit}	Kritische Abkühlgeschwindigkeit	°C/h
v_w	Schweißgeschwindigkeit	m/s
V_W	Volumen der wärmebeeinflußten Zone	m³
$\overline{v}_M$	Mittlere Schmelzaustrittsgeschwindigkeit	m/s
w	Radius des Laserstrahls	m
w	Leistungsdichte	W/m³
w_0	Strahltaillenradius des Laserstrahls	m
w_F	Fokusradius des Laserstrahls	m
w_L	Radius des Laserstrahls an der Bearb.optik	m
x	Verhältnis Dampfdichte/Sättigungsdampfdichte	–
x_v	Laminare Vorstrecke	m
y_H	Breite der Härtspur	mm
z_D	Diffusionslänge	μm
z_H	Härttiefe	mm
z_p	Verhältnis Dampfdruck/Sättigungsdampfdruck	–
z_R	RAYLEIGH-Länge des Laserstrahls	m

Symbol	Meaning	Einheit
Z	Zustandssumme	–
α	Wärmeübergangskoeffizient	$W/m^2 \cdot K$
α	Absorptionskoeffizient	$1/m^1$
α	Atomare Polarisierbarkeit	m^3
α	Energiebeiwert	–
α_B	BREWSTER-Winkel	–
β	Impulsbeiwert	–
β	Absorptionsindex	–
δ	Dämpfungskonstante	$1/m$
δ_M	Schmelzfilmdicke	m
δ_{opt}	Optische Eindringtiefe der Laserstrahlung	nm
δ_w	Wärmeeindringtiefe	mm
ε	Dielektrizitätskonstante	–
ε	Mittlere Elektronenenergie	J, eV
ε	Spezifische Gesamtenergie	J/m^3
ε_F	FERMI-Energie	J, eV
η	Dynamische Viskosität der Schmelze	g/m s
Γ	Plasmaparameter	–
κ	Temperaturleitfähigkeit	m^2/s
κ	Imaginärteil des komplexen Brechungsindex	–
λ	Wellenlänge	m
λ	Mittlere freie Weglänge	m
λ	Volumenviskosität	$N s/m^2$
Λ	DE'BROGLIE-Wellenlänge	m
Λ_c	COULOMB-Logarithmus	–
μ	Magnetische Permeabilität	–
μ	Chemisches Potential	J
μ	Plasmaparameter	–
ν	Frequenz	$1/s$
ν_m	Impulsübertragungsfrequenz	$1/s$
ρ	Massendichte	g/m^3
ρ	Raumladungsdichte	$A s/m^3$
σ	Oberflächenspannung	N/m
σ	Elektrische Leitfähigkeit	A/V m
σ_e	Elektronenleitfähigkeit	A/V m
σ_m	Impulsübertragungsquerschnitt	m^2
σ_{sp}	SPITZER-Leitfähigkeit	A/V m
τ	Scherspannung	N/m^2
τ_W	Schubspannung der Schmelze	N/m^2
τ_G	Schubspannung des Schneidgases	N/m^2
ϕ	Elektrisches Potential	V
ϕ	Potential einer Potentialströmung	m^2/s
φ	Einfallswinkel	rad
φ_B	BREWSTER-Winkel	rad
θ	Divergenzwinkel, Einfallswinkel der Strahlung	rad
ω	Kreisfrequenz	$1/s$
ω_p	Plasmafrequenz	$1/s$
ξ	Zustandssumme der Translationsfreiheitsgrade	–

Appendix E

E.2 Physical Constants

Symbol	Meaning	Value	
c	Vacuum velocity of light	$2.998 \cdot 10^8$	m/s
e	Elementary charge	$1.602 \cdot 10^{-19}$	C
g	Local acceleration of gravity	9.81	m/s^2
h	PLANCK's constant	$6.626 \cdot 10^{-34}$	J$\cdot$s
$\hbar$	$\frac{h}{2\pi}$	$1.05456 \cdot 10^{-34}$	J$\cdot$s
k	BOLTZMANN's constant	$1.381 \cdot 10^{-23}$	J/K
L	LORENZ number		
$L(\text{Al})$	LORENZ number of Al	$2.4 \cdot 10^{-8}$	V^2/K^2
$L(\text{Fe})$	LORENZ number of Fe	$2.8 \cdot 10^{-8}$	V^2/K^2
m_e	Electron mass	$9.10956 \cdot 10^{-31}$	kg
m_p	Proton mass	$1.67261 \cdot 10^{-27}$	kg
R	Molar gas constant	8.3143	kJ/kmol K
Z_0	Vacuum wave impedance	$3.767 \cdot 10^2$	$\frac{V}{A}$
ϵ_0	Vacuum dielectric constant	$8.854 \cdot 10^{-12}$	A$\cdot$s/V$\cdot$m
μ_0	Magnetic permeability	$4\pi \cdot 10^7$	V$\cdot$s/A$\cdot$m
σ	STEFAN–BOLTZMANN constant	$5.670 \cdot 10^{-8}$	W/m$^2 \cdot$K

E.3 Characteristic Numbers

Number	Name	Formula
Nu	NUSSELT number	$\alpha l / \lambda_{fl}$
Pe	PÉCLET number	$vl/a = \text{Re} \cdot \text{Pr}$
Pr	PRANDTL number	ν/a
Re	REYNOLDS number	vl/ν

with

λ_{liquid} heat conductivity of the liquid
l characteristic length.

E.4 Reference State

In practice there are quite a lot of reference states being used that accordingly have to be indicated or cited appropriately. Such a state of a solid, liquid or gaseous substance is characterized by a reference temperature T_{ref} and a reference pressure p_{ref}. As an example the norm state and the norm volume according to DIN 1343 [2][S. 142] are given here: In the norm state a substance has the norm temperature of $T_n = 273.15$ K ($t_n = 0\,°\text{C}$) and the norm pressure $p_n = 1.01325$ bar. One mol of an ideal gas has the norm volume:

$$V_{m,0} = (22.41410 \pm 0.00019) \cdot 10^{-3} \frac{m^3}{mol} \quad (E.1)$$

with

$V_{m,0}$ molar norm volume.

E.5 Material Constants

Temperature values given in Celsius or Kelvin, respectively, are related by

$$273.2\,K = 0°C \quad (E.2)$$

The temperature dependence of the dynamical viscosity can be determined by using

$$\eta(T) = \eta_0 \exp(E/T) \quad (E.3)$$

The material parameter values listed below for Al and Cu are taken from [1], [8], [12]. If not stated explicitly the values for Fe, steel, and stainless steel, respectively, are taken from [8].

If not otherwise stated the material parameter values used in the text are those compiled below.

Al	Temperature	Value	Unit
Melting temperature T_M		933	K
Evaporation temperature T_V		2793	K
Melting enthalpy h_M		$3.77 \cdot 10^5$	J kg^{-1}
Evaporation enthalpy h_V		$1.172 \cdot 10^7$	J kg^{-1}
Mass density ρ	293 K	$2.7 \cdot 10^3$	kg m^{-3}
	933 K, liquid	$2.385 \cdot 10^3$	kg m^{-3}
Specific heat capacity c	293 K	900	J kg^{-1} K^{-1}
	673 K	1076	J kg^{-1} K^{-1}
Heat conductivity K	293 K	238	W m^{-1} K^{-1}
	673 K	238	W m^{-1} K^{-1}
	933 K, liquid	94.03	W m^{-1} K^{-1}
	1273 K, liquid	105.35	W m^{-1} K^{-1}
Electrical conductivity σ	293 K	$3.74 \cdot 10^7$	Ω^{-1} m^{-1}
	673 K	$1.46 \cdot 10^7$	Ω^{-1} m^{-1}
	1273 K, liquid	$3.45 \cdot 10^6$	Ω^{-1} m^{-1}
Surface tension σ	933 K, liquid	0.914	N m^{-1}
$\frac{d\sigma}{dT}$		$-0.35 \cdot 10^{-3}$	N m^{-1} K^{-1}
Dynamic viscosity η	933 K, liquid	$1.3 \cdot 10^{-3}$	N s m^{-2}
η_0		$0.453 \cdot 10^{-3}$	N s m^{-2}
YOUNG's modulus E		$2.67 \cdot 10^3$	Nm^{-2}
POISSON's ratio ν		0.3	

Appendix E

Cu	Temperature	Value	Unit
Melting temperature T_M		1357	K
Evaporation temperature T_V		2833	K
Melting enthalpy h_M		$2.07 \cdot 10^5$	$J\,kg^{-1}$
Evaporation enthalpy h_V		$4.65 \cdot 10^6$	$J\,kg^{-1}$
Mass density ρ	293 K	$8.96 \cdot 10^3$	$kg\,m^{-3}$
	1356 K, liquid	$8.0 \cdot 10^3$	$kg\,m^{-3}$
Specific heat capacity c	(293 K	385	$J\,kg^{-1}\,K^{-1}$
	1273 K	473	$J\,kg^{-1}\,K^{-1}$
Heat conductivity K	(293 K	394	$W\,m^{-1}\,K^{-1}$
	1310 K)	244	$W\,m^{-1}\,K^{-1}$
	1356 K, liquid	165.6	$W\,m^{-1}\,K^{-1}$
	1873 K, liquid	180.4	$W\,m^{-1}\,K^{-1}$
Electrical conductivity σ	(293 K	$5.9 \cdot 10^7$	$\Omega^{-1}\,m^{-1}$
	1250 K	$1.2 \cdot 10^7$	$\Omega^{-1}\,m^{-1}$
	1356 K, liquid	$5.0 \cdot 10^6$	$\Omega^{-1}\,m^{-1}$
	1873 K, liquid	$3.95 \cdot 10^6$	$\Omega^{-1}\,m^{-1}$
Surface tension σ	1356 K, liquid	1.285	$N\,m^{-1}$
$\frac{d\sigma}{dT}$		$-0.13 \cdot 10^{-3}$	$N\,m^{-1}\,K^{-1}$
Dynamic viscosity η	933 K, liquid	$4.0 \cdot 10^{-3}$	$N\,s\,m^{-2}$
η_0		$0.3009 \cdot 10^{-3}$	$N\,s\,m^{-2}$
YOUNG's modulus E		3666.8	$N\,m^{-2}$
POISSON's ratio ν		0.3	

Fe	Temperature	Value	Unit
Melting temperature T_M		1673 (1803 [4])	K
Evaporation temperature T_V		3008 (3003 [7])	K
Melting enthalpy h_M		$2.7 \cdot 10^5$ ($3 \cdot 10^5$ [5])	$J\,kg^{-1}$
Evaporation enthalpy h_V		$6.370 \cdot 10^6$	$J\,kg^{-1}$
Mass density ρ	293 K	$7.87 \cdot 10^3$	$kg\,m^{-3}$
Specific heat capacity c	293 K	456 [4]	$J\,kg^{-1}\,K^{-1}$
	1073 K	791	$J\,kg^{-1}\,K^{-1}$
	T_M	754 – 838 [5]	$J\,kg^{-1}\,K^{-1}$
Heat conductivity K	293 K	74 [7]	$W\,m^{-1}\,K^{-1}$
	T_M	40	$W\,m^{-1}\,K^{-1}$
	1073 K	29.7	$W\,m^{-1}\,K^{-1}$
Electrical conductivity σ	293 K	$9.9 \cdot 10^6$	$\Omega^{-1}\,m^{-1}$
	1073 K	$9.45 \cdot 10^6$	$\Omega^{-1}\,m^{-1}$
	1809 K, liquid	$7.22 \cdot 10^5$	$\Omega^{-1}\,m^{-1}$
Surface tension σ	T_M	1.872	$N\,m^{-1}$
$\frac{d\sigma}{dT}$		$-0.49 \cdot 10^{-3}$	$N\,m^{-1}\,K^{-1}$
Dynamic viscosity η	1809, liquid	$5.5 \cdot 10^{-3}$	$N\,s\,m^{-2}$
η_0		$0.453 \cdot 10^{-3}$	$N\,s\,m^{-2}$
YOUNG's modulus E		$4.979 \cdot 10^3$	$N\,m^{-2}$
POISSON's ratio ν		0.3	

Steel	Temperature	Value	Unit
Melting temperature T_M		1744 [4]	K
		1673 [11]	K
Evaporation temperature T_V		2773 [4]	K
		2945 [11]	K
Melting enthalpy h_M		$2.05 \cdot 10^5$ [4]	$J\,kg^{-1}$
		$2.81 \cdot 10^5$ [11]	$J\,kg^{-1}$
Evaporation enthalpy h_V		$6.246 \cdot 10^6$ [11]	$J\,kg^{-1}$
Mass density ρ	293 K	$7.900 \cdot 10^3$ [4]	$kg\,m^{-3}$
Specific heat capacity c	293 K	510 [7, 11]	$J\,kg^{-1}\,K^{-1}$
	293 K	490 [4]	$J\,kg^{-1}\,K^{-1}$
	T_M	754 − 838 [5]	$J\,kg^{-1}\,K^{-1}$
Heat conductivity K	293 K	45 [7]	$W\,m^{-1}\,K^{-1}$
Surface tension σ	T_M	1.7	$N\,m^{-1}$
η_0		$0.453 \cdot 10^{-3}$	$N\,s\,m^{-2}$

Stainless steel	Temperature	Value	Unit
Melting temperature T_M		1723 [4]	K
		1801 [10]	K
Evaporation temperature T_V		2673 [10]	K
Melting enthalpy h_M		$2.32 \cdot 10^5$ [10]	$J\,kg^{-1}$
		$2.81 \cdot 10^5$ [11]	$J\,kg^{-1}$
Evaporation enthalpy h_V		$6.229 \cdot 10^6$ [10]	$J\,kg^{-1}$
Mass density ρ	293 K	$7.9 \cdot 10^3$ [4]	$kg\,m^{-3}$
	293 K	$7.8 \cdot 10^3$ [10]	$kg\,m^{-3}$
Specific heat capacity c	293 K	504 [3]	$J\,kg^{-1}\,K^{-1}$
	T_M	570–680 [6]	$J\,kg^{-1}\,K^{-1}$
Heat conductivity K	293 K	14 − 16 [6, 9, 4]	$W\,m^{-1}\,K^{-1}$
	900 K	18 [6]	$W\,m^{-1}\,K^{-1}$
	800 K	29.7	$W\,m^{-1}\,K^{-1}$
	800 K	15 [3]	$W\,m^{-1}\,K^{-1}$

References

1. E.A. Brandes (ed.): Smithells Metals Reference Book, Butterworth, London, 1983
2. DIN, Einheiten für Begriffe und physikalische Größen: Normen, Edition 7, Beuth Verlag, 1990
3. DVS, Handbuch der Kennwerte von metallischen Werkstoffen zur FEZEN–Werkstoff–Datenbank; Bd.: 2: Hochlegierte Stähle und Nichteisenmetalle, DVS-Verlag, Oberhausen, 1990
4. K. Gieck, Technische Formelsammlung, Gieck-Verlag, Heilbronn, 1984
5. Gmelin-Durrer, Metallurgie des Eisens, Springer-Verlag, Berlin, 1978
6. P. D. Hervey, Engineering Properties of Steel, American Society For Metals, 1982
7. H. Kuchling, Taschenbuch der Physik, Hari Deutsch, Thun, 1985
8. Mende/Simon: Physik. Gleichungen und Tabellen, Wilhelm Heine, München, 1976
9. Thyssen-Edelstahlwerke A.G., Remanit-Werkstoffblätter, Thyssen-Edelstahlwerke A.G., 1987
10. J. Ruge, Handbuch der Schweißtechnik, Springer-Verlag, Heidelberg, 1980
11. H. G. Treusch, Geometrie und Reproduzierbarkeit einer plasmaunterstützten Materialabtragung durch Laserstrahlung, 1985
12. R. C. Weast, Handbook of Chemistry and Physics, CRC Press, 1990

Appendix F
Übersetzung der Bildbeschriftungen

deustch; englisch

Figure.1.1:

1. Laserstrahl: laser beam
2. Wellenlänge: wavelength
3. Leistung: power
4. Pulsdauer: pulse duration
5. Strahlqualität: beam quality
6. Polarisation: polarization
7. Modenordnung: mode order
8. Räumliche-zeitliche Fluktuationen: spatio-temporal fluctuations
9. Strahlformung: beam shaping
10. Brennweite: focus length
11. Apertur: aperture
12. Abbildunsgfehler: aberrations
13. Materialeigenschaften: material properties
14. Absorption: absorption
15. Wärmeleitfähigkeit: heat conductivity
16. Dichte: density
17. Wärmekapazität: heat capacity
18. Schmelenthalpie: melt enthalpy
19. Verdampgungsenthalpie: evaporation enthalpy
20. Werkstückgeometrie: workpiece geometry
21. Dynamische Prozess: dynamical processes
22. Schmelze: melt
23. Oberflächenspannung: surface tension
24. Viskosität: viscosity
25. Dampf: vapor
26. Dampfdichte: vapor density
27. Elektronendichte: electron density
28. Temperatur (-gradient): temperature (and gradients)
29. Plasmaansorption: plasma absorption

Figure 2.1:

1 Einfallsebene: plane of incidence

Figure 2.4:

1 Reflexion R: reflection R
2 Einfallswinkel α: angle of incidence α
3 s-Polarisation: $\perp$ polarization
4 p-Polarisation: $\parallel$ polarization
5 unpolarisiertes Licht: unpolarized light

Figure 2.5:

1 Reflexion R: reflection R
2 Einfallswinkel α: angle of incidence α
3 s-Polarisation: $\perp$ polarization
4 p-Polarisation: $\parallel$ polarization
5 unpolarisiertes Licht: unpolarized light

Figure 2.6:

1 Einfallenden Welle: incident wave
2 Reflektierte Welle: reflected wave

Figure 2.7:

1 Intensitätsverteilung der einfallenden Strahlung: intensity distribution of the incident laser beam
2 Modell der Schneidfront: cutting front
3 Absorptionsverteilung auf der Schneidfront: distribution of the absorbed intensity

Figure 3.1:

1 Kreisfrequenz $\omega(s^{-1})$: circular frequency $\omega(s^{-1})$

Figure 3.2:

1 Kreisfrequenz $\omega(s^{-1})$: circular frequency $\omega(s^{-1})$

Figure 3.3:

1 Kreisfrequenz $\omega(s^{-1})$: circular frequency $\omega(s^{-1})$
2 Reflexion R: reflection R

Figure 3.4:

1 ω_p/ω: normalized plasma frequency ω_p/ω
2 Reflexion R: reflection R

Figure 3.5:

1 Reflexionskoeffizient R: reflectivity R
2 Wellenlänge $\lambda/\mu m$: wavelength $\lambda[\mu m]$

Appendix F

3 Aluminium: Al
4 Experiment: experiment
5 Drude Theorie: Drude theory

Figure 3.6:

1 Absorption (%): absorption (%)
2 rostfreier Stahl: stainless steel
3 Eisen: Fe
4 Temperature (°C): temperature (°C)
5 Elektrischer Widerstand $\rho(\mu\Omega\ cm)$: specific electric resistivity $\rho\ (\mu\Omega\ cm)$

Figure 3.7:

1 Material: Fe: material: Fe
2 Temperatur (°C): temperature (°C)

Figure 3.8:

1 Material: Fe: material: Fe
2 Stossfrequenz $\nu(s^{-1})$: collision frequency $\nu(s^{-1})$

Figure 3.9:

1 Absorption (%): absorption (%)
2 CO Laser: CO laser
3 CO_2 Lser: CO_2 laser
4 Experiment: experiment
5 Theorie: theory
6 Temperatur (°C): temperature (°C)
7 rostfreier Stahl: stainless steel
8 Eisen: Fe

Figure 3.10:

1 Absorption (%): absorption (%)
2 $Nd:YAG$ Laser: $Nd:YAG$ laser
3 CO_2 Laser: CO_2 laser
4 Stahl 35CD4: steel 35CD4
5 polierte Oberfläche: polished surface
6 CO_2 Laser: CO_2 laser
7 Temperature: temperature

Figure 3.11:

1 Absorption: absorption
2 Oberfläche geschliffen: grinded surface
3 Oberfläche poliert: polished surface
4 CO_2 Laser: CO_2 laser
5 Temperatur (°C): temperature (°C)

Figure 3.12:

1 einfallende Intensität: incident intensity
2 direkt reflektieret Intensität: directly reflected intensity
3 Linse: lens
4 diffus reflektierte Intensität: diffusely reflected intensity

Figure 4.1:

1 Rand: ∂B

Figure 4.2:

1 Oberflächenquelle bei $z = 0 + \epsilon$: surface heat source at $z = 0 + \epsilon$
1 Oberflächenquelle bei $z = 0 - \epsilon$: surface heat source at $z = 0 - \epsilon$

Figure 4.7:

1 Temperatur T (K): temperature T (K)
2 Parameter: Zeit t: parameter: time t

Figure 4.9:

1 Temperatur T (K): temperature T (K)
2 Parameter: Geschwindigkeit v (m/s): parameter: velocity v (m/s)

Figure 4.11:

1 Al fest: Al solid
2 Al flüssig: Al liquid

Figure 5.1:

1 Spannung: stress
2 Dehnung: strain
3 Plastische Dehnung: plastic strain
4 Fließgrenze: yield limit

Figure 6.3:

1 Volumen: volume
2 Ferrit: ferrite
3 α-Eisen: α-Fe
4 γ-Eisen: γ-Fe
5 δ-Eisen: δ-Fe
6 δ-Ferrit: α-ferrite
7 krz: cbc
8 kfc: cfc
9 Austenite: austenite
11 flüsig: liquid
12 Temperatur [$°C$]: temperature [$°C$]

Appendix F 599

Figure 6.6:

1 Normierte Konzentartion c': normalized concentration c'
2 Normierte Ortskordinate z': normalized coordinate z'
3 parameter: t': parameter: t'

Figure 7.1:

1 G: V
2 S: L

Figure 7.2:

1 G: V
2 S: L

Figure 7.7:

1 Gas: gas
2 Schmelzgrenze: melt boundary

Figure 8.1:

1 Wärmebad: heat bath
2 $j_{D \to S}$: $j_{V \to L}$
3 $j_{S \to D}$: $j_{L \to V}$
4 Dampf: vapor
5 Schmelze: melt
6 p_D: p_V
7 T_D: T_V
8 ρ_D: ρ_V
9 g_D: g_V
10 p_S: p_L
11 T_S: T_L
12 ρ_S: ρ_L
13 g_S: g_L

Figure 8.2:

1 Nichtgleichgewicht: non-equilibrium
2 Gleichgewicht: equilibrium
3 $g(V_x)/n$: velocity distribution function $\dfrac{g(v_x)}{n}$
4 $V_x[a.u.]$: x-component of velocity $v_x[a.u.]$

Figure 8.3:

1 Dampf: vapor
2 Schmelze: melt
3 Fest: solid
4 Laser: laser

Figure 8.4:

1 Temperature (K): temperature (K)
2 Intensität (W cm^{-2}): intensity (W cm^{-2})

Figure 8.5:

1 Druck (bar): pressure (bar)
2 Intensität (W cm^{-2}): intensity (W cm^{-2})

Figure 8.6:

1 Metalldampfdichte (cm^{-3}): metal vapor density (cm^{-3})
2 Intensität (W cm^{-2}): intensity (W cm^{-2})

Figure 8.7:

1 Dampfgeschwindigkeit (cm s^{-1}): vapor velocity (cm s^{-1});
2 Intensität (W cm^{-2}): intensity (W cm^{-2})

Figure 8.8:

1 Abtraggeschwindigkeit (cm s^{-1}): ablation velocity (cm s^{-1})
2 Intensität (W cm^{-2}): intensity (W cm^{-2})

Figure 8.9:

1 Abdampfrate (cm^{-2}s^{-1}): evaporation rate (cm^{-2}s^{-1})
2 Intensität (W cm^{-2}): intensity (W cm^{-2})

Figure 8.12:

1 Schmelze: melt
2 Knudsenschicht: Knudsen-layer
3 Dampf: vapor
5 $j_{S \to D}$: $j_{L \to V}$
6 $j_{D \to S}$: $j_{V \to L}$

Figure 9.1:

1 Laserintensität I (10^7W cm^{-2}): laser intensity I (10^7 W cm^{-2})
2 Reflexion R: reflection R

Figure 9.2:

1 Plasma-Abschirmbereich: plasma shielding regime
2 Bearbeitungsbereich: processing regime
3 Laser: laser
4 Werkstück: work piece

Figure 9.3:

1 Debye-Radius $r_D(m)$: Debye-radius $r_D(m)$
2 n_e (m^{-3}): electron density n_e (m^{-3})

Appendix F

Figure 9.4:

1 Zustandssumme U: partition function U
2 T (K): temperature T (K)

Figure 9.5:

1-6 Parameter: Druck (Mpa): parameter: pressure (MPa)

Figure 9.6:

1 Parameter: Druck (Mpa): parameter: pressure (MPa)
2 $v_m(s^{-1})$: momentum transfer frequency $v_m(s^{-1})$

Figure 9.7:

1 Parameter: Druck (Mpa): parameter: pressure (MPa)

Figure 9.8:

1 parameter: pressure
2 $\omega_p(s^{-1})$: plasma frequency $\omega_p(s^{-1})$

Figure B.4:

1 Karbid: carbide
2 Ferrit: ferrite

Index

A
Ablation, 2
Absorbed radiation energy, 15
Absorption, 5, 8, 43
    index, 16
    length, 58
Alloying additives, 81

B
Beam
    cross section, 1
    quality, 1
Bernoulli's law, 85
Beschichten, 471
Bessel-function, 53
Boltzmann-equation, 98
Boundary layer
    approximation, 87
    flow, 77, 86
Brewster angle, 9
Brewster effect, 10
Brillouin-scattering, 140

C
Chapman-jouguet-condition, 105
Clausius-Clapeyron, 93
Clausius-Mosotti-equation, 22
Collision cross section, 132
Collision frequency, 35
    electron lattice collision, 35, 39
Collisional radiative (CR)-model, 146
Conductivity
    DC, 34
    electric, 16, 33
    electron heat conductivity, 135
    heat conductivity, 16
Convection of melt, 43, 77
Coulomb-corrections, 124, 126
Coulomb-interaction, 26, 124
Coulomb-logarithm, 118
Coulomb-potential, 121
Cutting front, 12

D
De'broglie-wavelength, 121, 128
Debye-approximation, 126
Debye-radius, 116, 126, 129
Debye-screening, 133
Debye-shielding, 115
Debye-sphere, 134
Degeneracy, 122
Dielectric
    constant, 27, 29
    permeability, 22
Differential equation
    linear, 45
    parabolic, 44
Dirichlet-boundary condition, 44
Drude-model, 30, 32, 34, 35

E
Effective electron mass, 34
Electromagnetic wave, 27
Electron conductivity, 135
Electron plasma frequency, 26
Electronic polarizability, 21
Elementary gas theory
    kinetic, 45
Energy
    absorbed, 1
    band, 31
    conservation, 100
    excitation, 124
    free, 119, 124, 126
    internal, 119
    ionization, 121
    transport, 43
        equation, 61

Enthalpy, 46
Entropy, 124, 126
Error function, 50
Evaporation enthalpy, 94
Exponential integral, 52

F
Fermi
　energy, 46
　statistic, 30, 32
　velocity, 46
Flow dynamics, 3
Focusing optics, 1
Fresnel formula(e), 5, 39
Friction, 43

G
Green's function, 55, 59

H
Hagen-Rubens-relation, 29
Hartree-Fock-approximation, 115
Heat
　conduction, 1, 43
　conductivity, 16
　energy, 15
Heaviside step function, 49
Helmholtz vortex theorem, 81
Hybridschweiβen, 471

I
Impurities, 81
Incompressible fluid, 44
Index
　absorption, 34
　refraction, 16, 26, 28, 29, 34
Initial boundary value problem, 44
Initial conditions, 44
Intensity distribution, 59
Interaction potential, 131
Inverse bremsstrahlung, 140
Isotherm, 50

K
Knudsen-layer, 98, 104, 108, 109

L
Landau-length, 118
Laplace-equation, 82
Laser, 1
　alloying, 2
　material processing, 1
　surface alloying, 77
　technology, 1
　induced vaporization, 93, 99
Lattice vibrations, 45
Law of cosines, 83
Law of mass action, 127
Line source, 51
Lorentz-Lorenz-law, 22
Lorentz-model, 26
Lorentz-plasma, 135
Lorenz-number, 46
LSD-wave, 113
LTE-model, 145
Lyddane-Sachs-Teller-relation, 24

M
Magnetic field, 30
Maiman, 1
Mass
　effective, 32
　transport equation, 44
Material parameters, 16
Material processing, 1
Maxwell-distribution, 95, 142
Maxwell equations, 5, 17
Maxwellian velocity distribution, 139
Maxwell relation, 26
Mean field approximation, 115
Mean thermal speed, 96
Mechanisms of heat transport, 45
Melt, 2
Melting enthalpy, 57
Momentum
　transfer cross section, 139
　transfer frequency, 26
Monte Carlo-simulation, 98
Multiphoton ionization, 140

N
Navier-Stokes-equation, 44, 87
Newtonian fluid, 79
None equilibrium state, 15
Non thermal equilibrium evaporation, 3
Normal dispersion, 23

O
Ohm's law, 32
Optical penetration depth, 24
Oscillator strength, 21

P
Partition function, 118
Permanent dipole moment, 20
Phase equilibrium, 93
Phonon, 45

Photon, 16
Planck's constant, 121
Planck's law, 145
Plane potential flow, 77
Plasma, 3, 25
    absorption, 3
    enhanced coupling, 114
    ideal, 120
    parameter, 118
    waves, 30
Point source, 49
Poisson-equation, 116
Polarization, 6
    electronic, 20
    ionic, 20
Potential
    chemical, 119, 124, 126
    scalar, 82
Poynting-vector, 9, 28
Pressure, 126
    ionization, 122

Q
Quantum electrodynamics, 16
Quasi free electron gas, 33

R
Raman-scattering, 140
Rankine-Hugoniot-jump conditions, 104
Rayleigh-scattering, 140
Reynolds-number, 89
Roughness, 38

S
Saha-equation, 3, 115, 127
Saturation
    vapor density, 95
    vapor pressure, 94

Shear stress, 89
Shock-front, 3
Skin depth, 30
Snell's law, 6
Source, 82
Specific heat, 16
Spitzer-equation, 135
Stagnation point, 85
Statistic, 118
Stefan-Boltzmann law, 145
Stirling-formula, 123
Stress tensor, 80
Superposition, principle of, 47
Surface
    force, 79
    tension, 2

T
Thermal equilibrium plasmas, 3
Thermodynamic, 118
Thomson-scattering, 140
Transmission, 5

V
Velocity of sound, local, 104
v. Neumann-boundary condition, 44

W
Wavelength, 1
Wave number, 27
Wave vector, 5
Werkzeugmaschinen, 471
Wiedemann-Franz-law, 35

Printing: Ten Brink, Meppel, The Netherlands
Binding: Stürtz, Würzburg, Germany